Meteorology

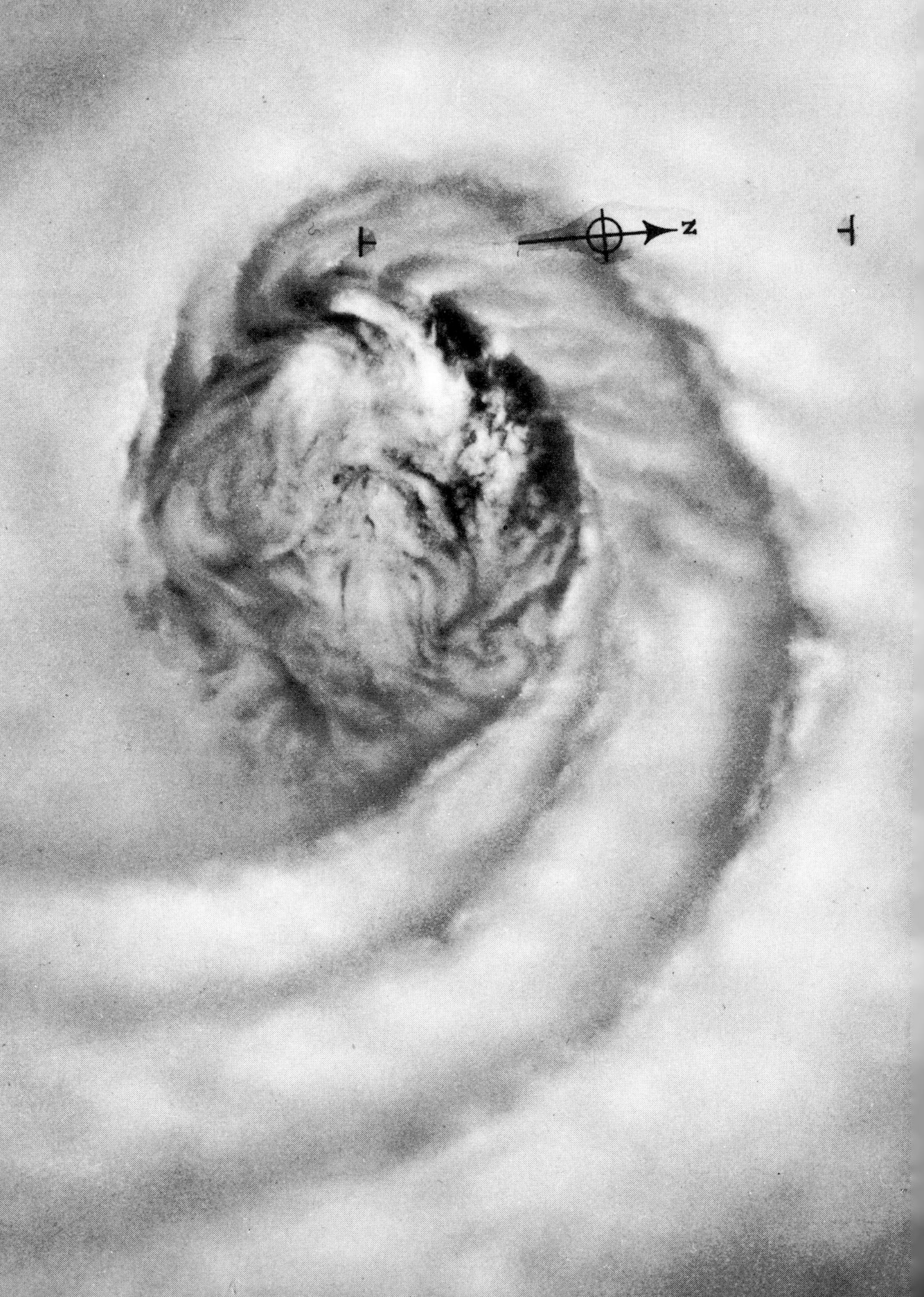

Marvelous are the offices and wonderful is the constitution of the atmosphere. Indeed, I know of no subject more fit for profitable thought on the part of the truth-loving, knowledge-seeking student, be he seaman or landsman, than that afforded by the atmosphere and its offices. Of all parts of the physical machinery, of all the contrivances in the mechanism of the universe, the atmosphere, with its offices and its adaptations, appears to me to be the most wonderful, sublime, and beautiful. (MAURY, THE PHYSICAL GEOGRAPHY OF THE SEA.)

U-2 photograph of Typhoon Ida, 1958.
(Directorate of Scientific Services, Air Weather Services)

Meteorology

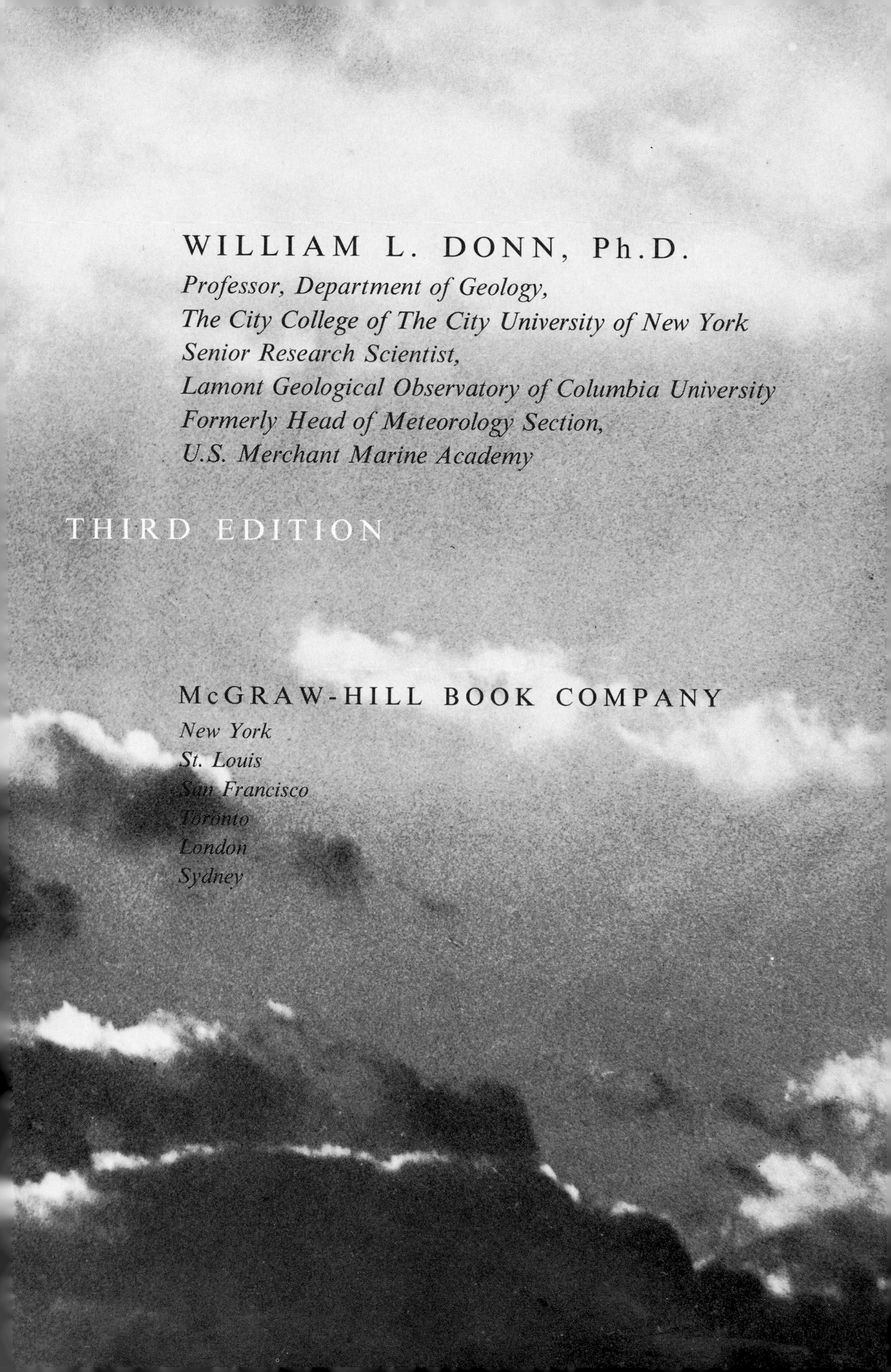

WILLIAM L. DONN, Ph.D.
Professor, Department of Geology,
The City College of The City University of New York
Senior Research Scientist,
Lamont Geological Observatory of Columbia University
Formerly Head of Meteorology Section,
U.S. Merchant Marine Academy

THIRD EDITION

McGRAW-HILL BOOK COMPANY
New York
St. Louis
San Francisco
Toronto
London
Sydney

METEOROLOGY

Library of Congress Catalog Card Number: 65-17492

ISBN 07-017598-5

789101112 HDMM 75432

To Tina

MY MOTHER

and to

RENÉE, MATTHEW, AND TARA

Eye of Hurricane Betsy, September 2, 1965, as photographed from more than eleven miles high. (U.S. Air Force)

PREFACE

This book is primarily a general introductory text on meteorology. It also presents all important marine applications of the subject. Because the oceans cover more than seven-tenths of the earth's surface, they exert an important control over the atmosphere. They also provide a main "highway" for amateur and professional mariners, who play a large part in the gathering of weather observations. Hence, two specific chapters on marine weather problems and the oceans are included, in addition to the appropriate maritime aspects given in many other chapters.

To bring this edition up to date and into clear focus, more than half of the original text has been completely rewritten. New chapters include the "Heat Energy of the Atmosphere," "The General Circulation of the Atmosphere," and "Weather at Sea," the latter being a special application of the subject for the seaman, whether amateur or professional. Other than retention of the original name, the chapter on the oceans has been so enlarged and rewritten that it may be considered new. The original chapter on weather coding has been deleted because all the information is so readily available in card or manual form from the U.S. Weather Bureau and also because frequent code changes tend to make such a section obsolete rather quickly. Information of basic importance to understanding or preparing weather maps, however, is included in the chapter on weather analysis and interpretation.

Questions and exercises have been added to the end of each chapter. Even if not used directly, these may serve as a guide to the instructor in the preparation of different or additional study material. They have also been prepared so as to give the informal reader a guide to most of the basic ideas and information in each chapter.

As in past editions, the presentation of material is so arranged that a less technical reading or use of the book can be made without affecting continuity. Although Chap. 5, "Vertical Equilibrium of the Atmosphere," and the sections on wind theory as well as Chap. 12 "The General Circulation of the Atmosphere," are basic to a good physical understanding of the subject, they may be omitted without seriously affecting the continued development.

William L. Donn

CONTENTS

A Tiros VI photograph taken over the North Atlantic Ocean. (U.S. Weather Bureau)

Meteorology

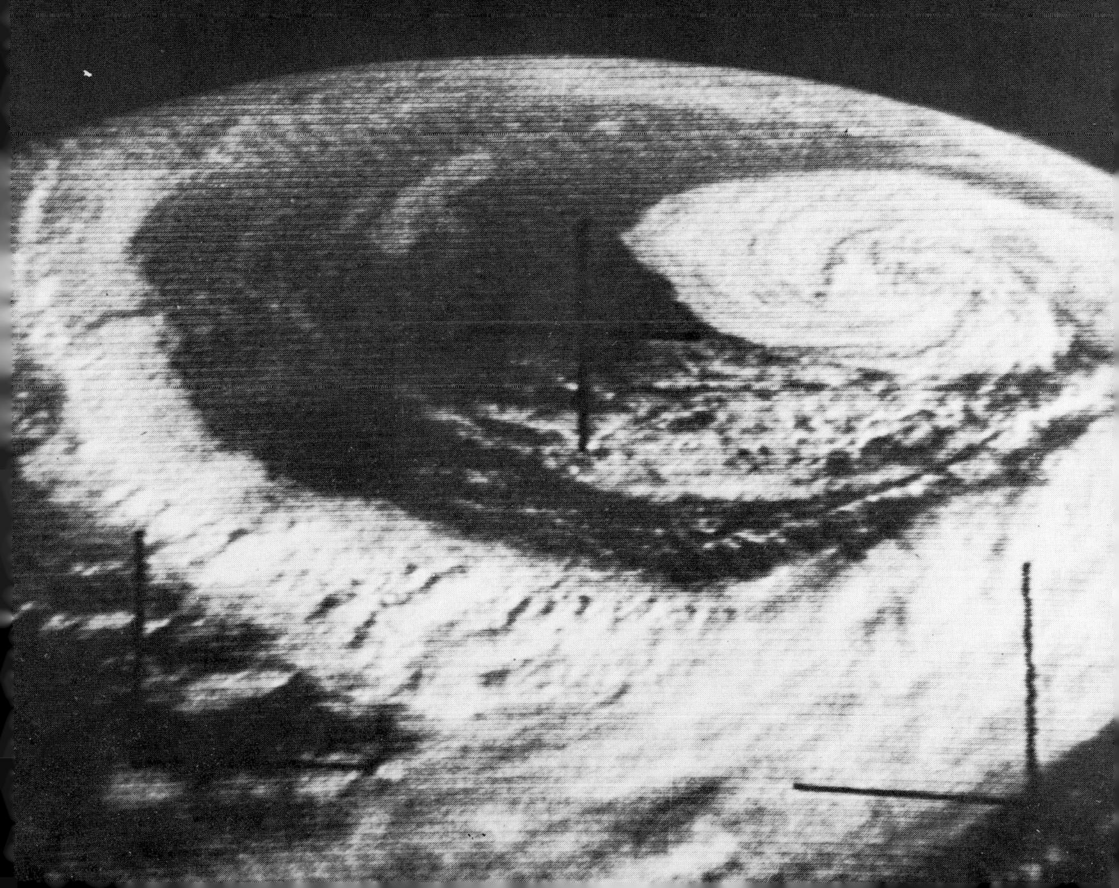

CHAPTER 1

Nature of the Atmosphere

Introduction

In recent years the study of the air has become so enlarged that its every aspect, from the ground surface to outer space, has come under detailed investigation. This grand body of knowledge is now called *atmospheric science,* and the older term *meteorology* is more commonly applied to the lower atmosphere and its continuously occurring changes. The daily variations of the different conditions of the lower air are known as *weather.*

Although any general study should include at least a brief look at the entire subject, we will emphasize in this book the meteorology of the lower atmosphere, for this is the region of most direct concern to all of us. Despite the emphasis on marine applications and the frequent use of marine examples of the subject, it must be realized that the atmosphere is global, so that its study cannot be artificially divided into marine and continental meteorology. And the applications that were mainly professional not long ago are now possibly as important in amateur seamanship and navigation in view of the tremendous growth in small boat use wherever navigable water of any kind can be found.

The boat owner or ship's officer who can figure the wind, waves, and fog, among other things, can complete his voyage in the shortest time and with the greatest assurance of safety to himself, as well as to his vessel and passengers. The proper use of meteorology can also be of great value to the merchant marine officer in determining the proper care and ventilation of cargo.

In addition to using weather knowledge, the merchant marine and navies of the world help create the knowledge of weather for further use on both land and sea. Weather is global in its origin and movement. Since the earth is about 70 percent covered by the seas, reports by skilled marine observers provide the only practical way at present of obtaining the necessary global summary of the weather elements. These observations are necessary for the preparation of daily weather maps for immediate forecasts on land and sea as well as for the preparation of the marine pilot charts that show average or expected conditions, often so valuable in the planning of a route. The critical weather forecasts for the marine airways of the world also depend to a considerable extent on the observations made on surface vessels.

The reader will not become an expert weather forecaster from this general study, for that requires years of specialized training. But he will gain sufficient knowledge concerning his atmospheric environment to be of invaluable aid, whether navigating small craft or large, or simply planning a picnic in the open.

This book is therefore intended to include such material as will enable the reader:

1. To obtain a good understanding of the weather changes and their causes.
2. To take weather observations properly and accurately.
3. To make local short-period weather predictions based on his own and other observations received by radio.
4. To relate weather information to the problems of seamanship and navigation.

Functions and Importance of the Atmosphere

We take the atmosphere very much for granted. In so doing, many of its important functions are usually overlooked, just as the forest is often invisible because of the trees. From the biological viewpoint alone, the atmosphere is of tremendous significance. The air—particularly the oxygen component—supports life as we know it. Most living organisms, whether advanced or primitive in development, require oxygen for survival. Although the manner of obtaining this gas may vary, the purpose for which it is absorbed is the same. Not only does the atmosphere supply the vital oxygen, but it also provides living things with the equally important *fresh* water. The importance of our atmosphere to biologic existence cannot be doubted.

Geographically, the contours of the earth's surface, with its timeless changes, are the result of that greatest of sculptors, the atmosphere. From time immemorial the effect of the wind, rain, and running water resulting from rain has been to modify continuously the relief features of the earth. Flat areas are made more rugged; rugged areas are rounded and flattened through the ceaseless operations of these atmospheric agencies. We have only to regard the moon to be convinced of this effect. That desolate world no longer maintains an atmosphere. As a consequence, the features it exhibits are static and changeless. No molding of the lunar landscape occurs. The shape of our coastlines is under continuous modification through the work of the air. The winds that prevail therein cause waves and currents whose action is to smooth irregular, rugged coastlines and to make more irregular the appearance of those originally straight and uniform.

Much of our great natural wealth is dependent on the conditions of the atmosphere. The accumulation of huge mineral and ore reserves has frequently been the direct result of the amount of water in the air. Where abundant supplies of moisture have existed, great accumulations of economically valuable minerals have been deposited through the chemical actions of rainwater soaking into the rock. And conversely, in arid regions, the increased evaporation of ground water leaves behind valuable mineral deposits. Again, from an economic standpoint, international trade could not have flourished nor existed prior to steam power, were it not for the wind that filled the sails for the ancient mariners.

Many of the natural wonders of the world are products of atmospheric agencies. The brilliant colors and weird carvings of the rock mantle, the huge natural bridges, and awe-inspiring canyons and hosts of other features are the results of the same forces already mentioned. The beauty of the sky itself and the phenomena displayed above us are caused directly by the optical properties of the air. Refraction, diffusion, and dispersion of light within the atmosphere give rise to the colorful sunsets and sunrises, mirages, rainbows, etc. The phenomenon of daylight is merely the diffusion of sunlight by the minute particles that constitute the atmosphere. Were there no air, constant darkness of the sky would prevail, despite the brilliance of the sun.

Lastly, we cannot overlook that very important feature, the continuous changing conditions within the air—the weather—which exerts such a powerful influence on all things, animate and inanimate. Our atmosphere is certainly deserving of a large amount of attention, interest, and study.

The Exploration of the Atmosphere

Both the surface and the lower portion of the atmosphere are observed by means of fairly conventional instruments, the description and use of which constitute one important phase of this book. But the air high over our heads (and until recently beyond the reach of man) must be investigated by much more difficult methods. Until the past couple of decades much of our knowledge of the upper atmosphere was derived from observation of special atmospheric effects. These include auroras, meteorites, and atmospheric spectra that are all optical phenomena of the atmosphere and are described in more detail in Chap. 18, Optical Features of the Atmosphere.

Modern experimental procedures have provided more direct methods for studying the chemical composition and physical properties of the upper air. Sounding balloons regularly carry aloft instruments that are capable of transmitting information regarding temperature, pressure, and humidity to radio receivers in the laboratory. Such operations are effective to above 100,000 feet. The radiosonde, which telemeters the most complete information on the vertical distribution of temperature, pressure, humidity, and indirectly, wind, is illustrated in Fig. 1 · 1. Information on

Fig. 1 · 1 Launching a radiosonde on an ocean weather ship. The balloon in the foreground can carry the instrument, held by the man in the middle ground, to about 100,000 feet. (C. A. Woolum, U.S. Weather Bureau)

Fig. 1·2 The launching of a Tiros *weather satellite—contained in the nose cone in the top section of the rocket.* (U.S. Weather Bureau)

wind motion up to about the same elevation is obtained through the use of smoke shells that are programmed to explode at preset levels. The wartime V-2 rockets have become some of the most important of atmospheric probes carrying instrument payloads to elevations of 75 miles. The most important present method of determining temperatures and often wind speeds in the upper atmosphere is by means of grenades fired to explode at particular elevations. Because the speed of sound in air depends on the temperature, the arrival time of the explosive sounds at ground recording stations can be interpreted in terms of temperature.

The use of radio signals is one of the oldest experimental means of atmospheric study. Long-wave radio communication is made possible by the reflection of radio waves from the high, ionized layers. The height and number of such layers is worked out from the geometry of the wave paths when the locations of the source and the receiver are known. And most recently, the dramatic innovation of deep space probes and artificial earth satellites has provided a potential for the study of the upper and also the lower air that is not yet fully realized. Strangely, the high-elevation *Tiros* weather satellites are very useful in the meteorology of the lower air because of the vast areal cloud coverage provided by the composite television pictures transmitted from these complicated space messengers. This is particularly important information over the large ocean expanses not frequented regularly by either commercial or naval vessels. Tremendous meteorological dividends have already been obtained both in the outlining of normal weather systems and in the discovery of unsuspected violent tropical storms.

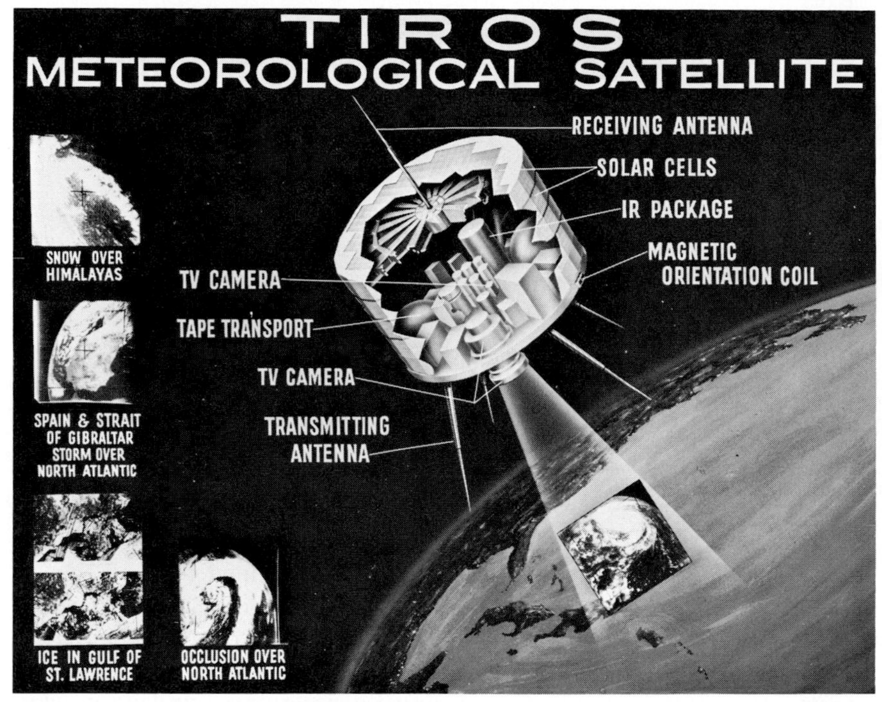

Fig. 1 · 3 Schematic view of the operation of a Tiros *weather satellite together with some examples of resulting photographs (insets).* (U.S. Weather Bureau)

Figure 1 · 2 shows the launching of a *Tiros* weather satellite which is in the forward or nose section of the carrier rocket. A schematic view of the operational procedure is shown in Fig. 1 · 3, which includes some actual photographic results in the insets.

The Composition of the Atmosphere

It is interesting to note in passing that the composition of the atmosphere has not always been the same during the long 5-billion-year history of the earth. According to most recent theory, the primordial earth had no gaseous cover because it formed by the accumulation of relatively small, cold, solid objects. As the earth's interior became heated from contraction, radioactivity, and chemical reactions, the volatile components were expelled to form the atmosphere and also the oceans. The composition of the early atmosphere was quite different, particularly as regards oxygen, which has been added slowly during the earth's history through the process of photosynthesis, in which plants produce oxygen.

At present, the air is a mixture of gases, and with the exception of water vapor, the relative proportions of the gases are quite constant to a height of more than 40 miles. Because the temperature of the lower atmosphere varies about the critical point at which the state of water reverses between vapor and liquid, this component is very variable in atmospheric content.

Although hardly apparent to us, owing to its gaseous nature, the atmosphere has a mass of 5.6 by 10^{14} tons. About 99 percent of this is nitrogen and oxygen, and most of this percentage is nitrogen, which really makes up the bulk of the atmosphere. The composition of the atmosphere according to both volume and mass, the former measure being the more commonly used, is expressed in Table 1 · 1. The proportions given of course

*Table 1 · 1 Nonvariable Composition of the Atmosphere**

COMPONENT	VOLUME, %	MASS, %
Nitrogen	78.084	75.51
Oxygen	20.946	23.15
Argon	0.934	1.28
Carbon dioxide	0.033	0.046

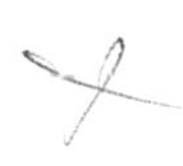

* Very minor amounts of neon, helium, methane, krypton, nitrous oxide, hydrogen, ozone, xenon, nitric oxide, and radon compose the remaining fractional percent.

change if air is considered with its water vapor composition. This varies considerably with location and with time at the same location. Air can be practically barren of water vapor at some places and can contain up to 4 percent (moist air volume) at others. A general global average of 1 percent of moist air volume is often used for certain estimates. Although the molecular weight of water vapor is less than that of the other gases in the air, and it is thus lighter in weight, it is nevertheless concentrated near the ground. Ninety percent of this vapor lies in the lowest few miles of the atmosphere and decreases rapidly from the surface upward. This is less strange when it is realized that the direct source of atmospheric water is the world ocean and that air temperatures aloft are too low to maintain water in its gaseous state. We will return to this subject in the chapter on humidity of the air.

It is also interesting to note that the carbon dioxide content has increased during the past century. Measurements from 1860 to 1950 indicate a continuous increase of about 9 percent beginning about 1900. At least a good part of this increase has been explained by the tremendous increase in the burning of carbon and hydrocarbon fuels, the by-product of which is the release of carbon dioxide.

In addition to these regular constituents, the air supports several types of material which, although in the air, are nevertheless foreign to it. Some of these are salt crystals, dust particles, and water droplets. The air has been defined as a mixture of vapors or gases. Obviously, then, these particles in the air cannot be considered as part of the gaseous atmosphere. A speck of dust in one's eye is not part of the eye merely because of its presence; just as the speck of dust may have an important effect on one's eye, so do the foreign particles in the atmosphere have a very basic influence on it and the changes therein. This importance will be brought out later.

The effect of water droplets in the air is somewhat more obvious. When the water vapor, which is part of the air, changes its state from a gas to a liquid or solid, the resulting water or ice particles become foreign to the air and provide our clouds, fog, rain, snow, dew, etc. Hence, water has a somewhat dual characteristic in the air, depending on its state, and is also of very fundamental importance in determining weather conditions.

Height and Structure of the Atmosphere

The atmosphere can be considered as a huge envelope of gases that surrounds and permeates the earth and that extends upward with continuously decreasing density. There is no sharply defined upper limit to the air; it simply "peters out" ever more slowly, until it merges imperceptibly with the extensive "atmosphere" of the sun. The decrease of density is so rapid at first that one-half of the entire atmosphere lies beneath a height of 3½ miles (or about 18,000 feet); half of what is left lies within the next 3½ miles, or three-fourths of the total lies beneath 7 miles. The increase in density toward the ground is accounted for by the gravitational compression of the air from the overlying portions. At the temperature and pressure of 0°C (32°F) and 76 centimeters (29.91 inches) of mercury the density of air at the earth's surface, resulting from this compression, is 1.3 kilograms per cubic meter (0.08 pound per cubic foot).

Although air at great elevations above the ground has less matter in a given volume than can be found in the best evacuated laboratory containers, the density is nevertheless great enough to be made known by observation and experiment. Thus, although orbiting at many hundreds of miles from the earth's surface, artificial satellites are soon slowed by friction with the rarefied air and fall back toward the ground.

Despite the fact that the atmosphere consists of a uniform mixture of gases to at least 40 miles and is so tenuous above only a few miles, it is not physically homogeneous but has a concentrically layered structure. Superimposed on the physical structure is a compositional layering which is also recognized out to hundreds of kilometers.

The structure of the atmosphere can be described according to a number of different reference conditions. For example, the thermal stratification, which is most important from the standpoint of weather changes, results from the interplay of the varied sources of direct heat in the atmosphere. Superimposed on this basic physical structure are the effects related to ionization, which produces atmospheric shells not directly related to the temperature structure. Then, in the upper atmosphere, there are spheres identified by compositional rather than by physical differences.

Because knowledge of the details of the upper air is very recent and because this region is still under very active investigation, no universally accepted structural standards have been set. However, Fig. 1 · 4 illustrates some of the basic physical aspects of the atmosphere. The lowest layer of importance that contains the bulk of the air is the *troposphere,* characterized by a nearly uniform decrease in temperature (shown by the sloping temperature curve in Fig. 1 · 4). Most of the weather changes in the air are limited to this lowest layer, in great part because of the decrease of temperature with elevation. Although the troposphere has an average thickness of about 7 miles, it is thickest at the equator and thinnest at the poles.

Above the troposphere lies the *stratosphere,* characterized by a vertically isothermal (equal temperature) structure in the lower portion followed by increasing temperatures in the upper portion. The *mesosphere,* lying above the stratosphere, is identified by a strong temperature decrease from the maximum temperature zone in its lowest portion just above the stratosphere. The outermost shell is the *thermosphere* in which the temperature rises rapidly from an initial constant low value at about 50 miles to a constant high value above about 120 miles. The upper boundaries of each of the three lower layers are known as the *tropopause, stratopause,* and *mesopause,* respectively, the latter marking the region of lowest temperature in the atmosphere. We will return to a further consideration of the temperature distribution in the next chapter.

The thermosphere, which extends upward until merging with the solar atmosphere many thousands of miles above the earth's surface, is partly ionized throughout much of its extent. In the lower part of the thermosphere, enriched ion zones exist in the form of distinct ionized layers whose positions and designations are included in Fig. 1 · 4. Long-distance radio communication is made possible by one or multiple reflections of shortwave radio beams from these ionized shells. The term *ionosphere* has been used to describe that portion of the upper atmosphere which includes the ion-rich layers.

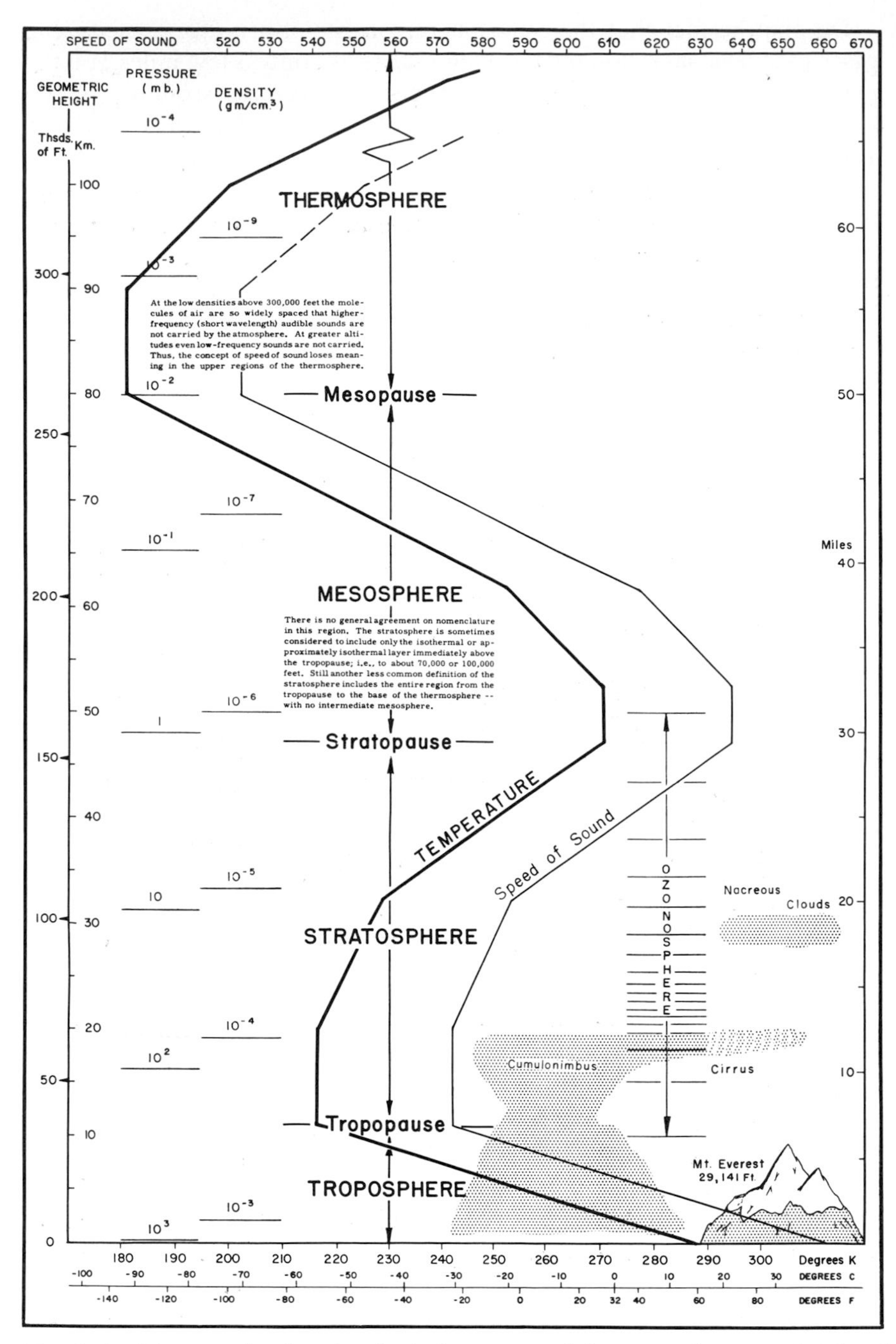

Fig. 1·4 Diagram showing the physical structure of the atmosphere to a height of 350,000 feet (66 miles). (U.S. Navy Weather Research Facility)

Although atmospheric composition has been noted as being quite uniform up to an elevation of 40 to 50 miles, the atmosphere above this, despite its very rarefied nature, also exhibits a concentric layering of a compositional character in addition to the physical stratification described above. The gases that predominate in the outer layers are indicated in Fig. 1 · 5, which shows the currently held picture of the compositional arrangement obtained as experimental data in the United States *Explorer* satellite program. The way in which the constituents of the atmosphere vary with elevation in the lower 150 miles is shown in Fig. 1 · 6.

An important supplement must be added to our picture of the uniform, lower 50 miles of atmosphere. The region surrounding the stratopause

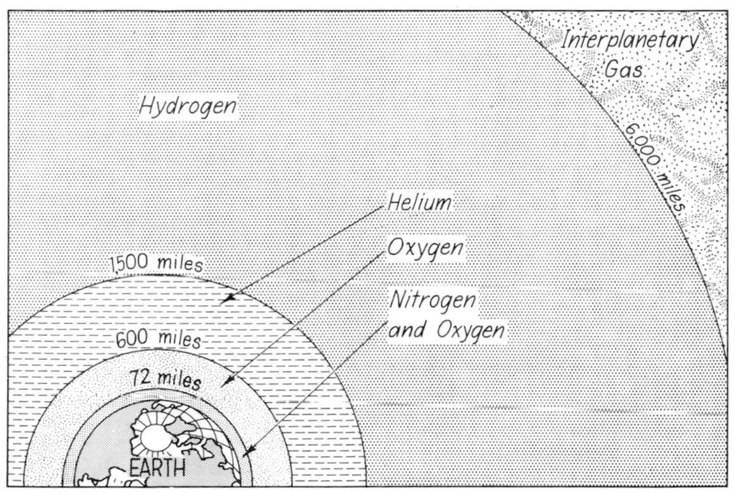

Fig. 1 · 5 Diagram showing the compositional layering of the atmosphere based on information of the Explorer *satellite program of the National Aeronautics and Space Administration.*

contains a relatively high proportion of oxygen in the form of ozone, whose molecular composition is given as O_3 compared to O_2 for oxygen. In thickness, the ozone-rich layer is about 10 to 20 miles, the value being approximate because no sharp boundaries are present. It is interesting that the atmosphere is so rarefied at this height that if all of the ozone present in this region were compressed under sea-level pressure conditions, a layer only a fraction of an inch thick would form. The great importance of this layer, despite the low mass, lies in its very high absorption of ultraviolet rays from the sun. Ozone, which is created from oxygen by the effect of this short-wave radiation, then becomes quite opaque to the further transmission of these rays. In view of the lethal effects of concentrated ultraviolet radiation, the presence of ozone is very fortunate.

Modern space-age science requires that we also examine the earth's outer atmosphere or the region in which factors other than temperature become important in determining the physical conditions. One classification scheme is shown in Fig. 1 · 6, in which most of the thermosphere (the uppermost layer in Fig. 1 · 4) is also called the exosphere because the high temperatures and very low particle density permit the escape of some atoms and molecules from the earth's gravitational field. The upper part of the thermosphere is also called the magnetosphere or protosphere because the earth's magnetic field is here more important than the gravitational field in controlling the behavior of the chief constituent—protons.

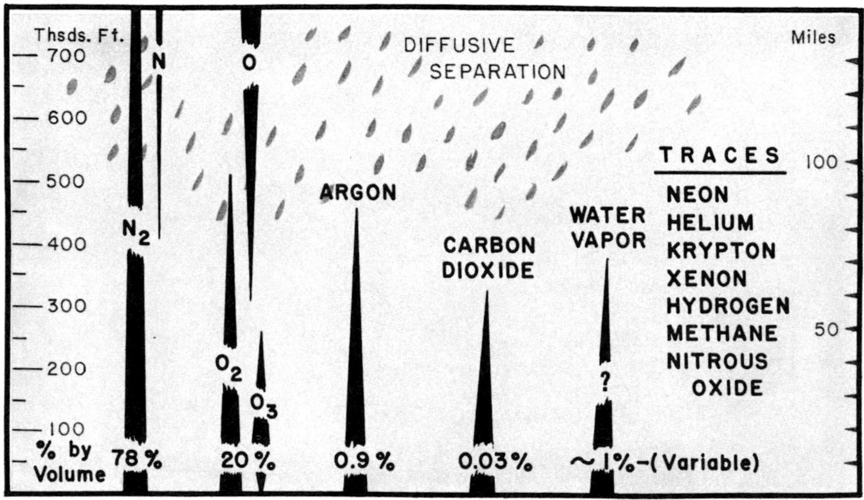

Fig. 1 · 6 The variation of atmospheric constituents with elevation. (U.S. Navy Weather Research Facility)

Weather Elements

The word *weather* refers to the short-period variations in the atmosphere. But just what varies? There are certain properties of the atmosphere on which observations are made that are subject to constant change, their state at any time determining the state of the weather. These variables are:

1. Temperature of the air
2. Humidity of the air
3. Horizontal visibility (fog, etc.)
4. Clouds and state of the sky
5. Kind and amount of precipitation
6. Atmospheric pressure
7. Winds

Although other strictly observational features are frequently added to this list, these are the seven basic weather elements that will be studied, together with the factors affecting and related to them. It is principally on these elements that observations and reports are made. The actual observations and reports may be brief or complex depending on the instruments available and the purpose for which they are made. Of the seven elements listed, two are of particular importance in making observations for the purpose of local short-period predictions, when no reports are obtainable. These are the clouds and the wind direction. One of the purposes of this book is to teach the marine observer enough about the

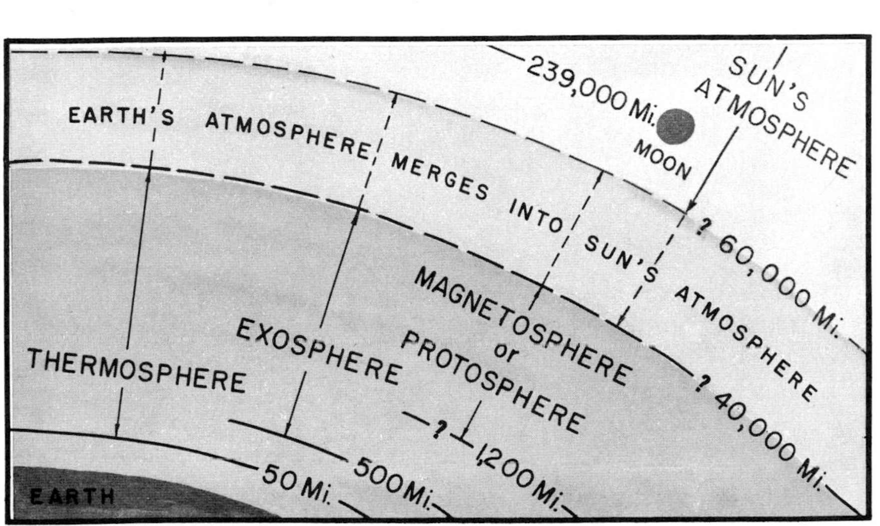

Fig. 1 · 7 Nomenclature scheme for the earth's outer atmosphere. (U.S. Navy Weather Research Facility)

processes underlying weather changes to enable him to use his observations to advantage in foretelling weather conditions.

These weather elements are not to be considered as separate entities. On the contrary, they are closely interrelated. Of the seven mentioned, temperature is the most basic and fundamental, and its variations cause changes in the other elements, the results of which we know as weather.

It is well known that warm air can "hold" more water vapor than cold air. In many drying processes the blowing of hot air on the object greatly expedites evaporation of moisture. If warm air is cooled sufficiently, the excess water vapor that was present must settle out as liquid water droplets. Thus, temperature changes in the air are the direct cause of humidity variations, which in turn yield clouds, fog, and precipitation.

The temperature variations are also responsible for pressure differences. If a part of the air becomes relatively warmer than the surrounding air, it will expand, become lighter, and tend to rise from the earth. Cold air surrounding the warmer air will be relatively heavy. Consequently, the warm air will have less weight and show less downward pressure than the surrounding cold air and will therefore (other things being equal) show lower barometer readings.

Now, whenever such a condition arises, there will clearly be a heaping up of air in one locality and a deficiency in a neighboring area. For example, consider an inverted bowl of water: when the bowl is removed, the water will flow outward from the center. Similarly, a flow of air tends to start along the earth's surface from the region of higher pressure (heavier air) to the region of lower pressure (lighter air). This horizontal movement of air is called *wind.* These rather simple descriptions will be expanded considerably in the following chapters.

EXERCISES

1·1 Explain what is meant by the secondary nature of our atmosphere.

1·2 Construct a table giving in the first column the method of exploration of the atmosphere; in the second column, the approximate maximum elevation reached; and in the third column, the information obtained by each method.

1·3 (*a*) What is the atmospheric composition by mass (in tons) of nitrogen and oxygen? (*b*) If combustion of some inflammable material occurs in a 1-cubic-foot container sealed except for a connection to a vessel of water, how much water will rise into the container?

1·4 (*a*) Compare the vertical distribution of water vapor in the atmosphere with that of all the other components. (*b*) Why do statements of atmospheric composition usually refer to dry air?

1·5 Why is water vapor concentrated near the earth's surface in view of its relatively low molecular weight?

1·6 Why does water vapor occur in all three states of matter in the atmosphere?

1·7 (*a*) Why does the density of the atmosphere decrease from the earth's surface upward, instead of being uniform? (*b*) What is the approximate density, in both metric and English units, at 3½ and 7 miles above the ground? (*c*) In a ground-floor room with dimensions 20 by 20 by 10 feet, what is the weight of the air in pounds (at 32°F and 29.91 inches)?

1·8 (*a*) What is the basic cause of the physical structural layering of the atmosphere? (*b*) What is the significant physical difference between the troposphere and stratosphere?

1·9 List the structural divisions of the atmosphere together with their elevation ranges and main thermal features, indicating which division is characterized by strong weather changes.

1·10 How do the forces affecting the outermost part of the atmosphere differ from those controlling the lower portion?

CHAPTER 2

Heat Energy of the Atmosphere

The continuous changes in atmospheric conditions, known as weather, may involve only the most gentle of breezes or the most violent of storms. The energy which drives this atmospheric engine is heat energy. In order to understand the causes of the never-ending weather effects, we must have some insight into the nature of heat and the source of atmospheric heat.

Heat and Temperature

Heat and temperature are closely related properties that often become confused with each other. Thus, a *hot* object is one with high temperature rather than high heat content. Temperature can be both measured and defined rather easily. Heat measurement is more involved and tends to evade simple definition.

One way of distinguishing between them is as follows: Suppose identical quantities of copper and of water at the same temperature are heated by two identical burners. Although the same amount of heat is being transmitted to the two materials, the temperature of the copper will rise much faster than that of the water. If the burners are then removed and the copper and water placed in contact, heat will flow from the higher-temperature copper to the water despite their both having received the same amount of heat from the burners.

As a result of heat absorption, the molecules of a substance receive increased energy which increases their speed of motion. The experiment referred to shows that different materials require different amounts of heat in order to reach the same degree of internal energy or molecular motion. Technically, temperature can be defined as the measure of the average energy of molecular motion (not including molecular vibrations). We are usually concerned with effects resulting from differences in temperature because heat is transferred only from higher to lower temperatures.

PROCESSES OF HEAT TRANSMISSION. Heat energy is transmitted from place to place by the processes of *radiation, conduction,* and *convection.* Each of these three processes has an important function in determining processes affecting weather changes.

Radiation. This is the process of heat transfer in wave form, without the use or necessity of a transmitting medium. For example, the insolation (radiant energy) received by the earth comes from the sun by radiation in wave form through the emptiness of space. When we stand near a very hot object, the intense heat felt is mostly the result of heat rays radiated by the hot object. By the use of infrared (or heat-sensitive) film in a camera in a dark room, it is possible to obtain an excellent photograph of an ordinary hot pressing iron merely by the heat rays radiated from the iron.

Conduction. This is the process of heat transmission through a medium by contact of the minute particles of which the medium is composed. For instance, if one end of a metal rod is heated, the other end will soon

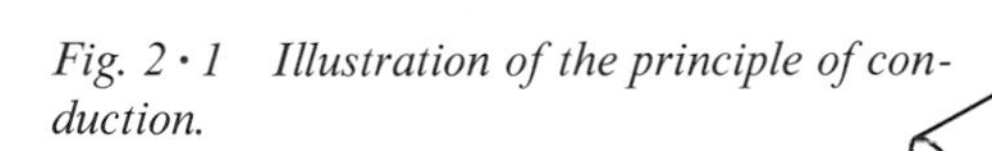

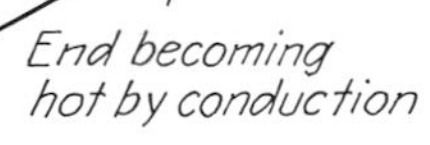

Fig. 2 · 1 Illustration of the principle of conduction.

become hot. This is accomplished by conduction, in which the heat energy is passed (or conducted) along the rod from the heated to the cold end by the molecules within the rod. See Fig. 2 · 1.

Convection. This is the process of heat transmission by the actual motion of the heated material. When the air over a radiator is heated, it expands and rises to the ceiling. This motion is evident by the distortion of the light coming through the window and passing through the rising air. Similarly, images seen across an airport runway often waver owing to the disturbance caused by the warmed air rising from the heated runway. In the same way, when water or any liquid is heated, the warm liquid at the bottom rises to the top and is replaced by cooler descending water. Gases exhibit convection as a means of heat transfer more so than liquids, and liquids more so than solids. In fact, convection, or the rising and falling of warm and cold air masses, respectively, is one of the most important and fundamental processes in the atmosphere.

The typical convection system is illustrated in Fig. 2 · 2, where the fluid

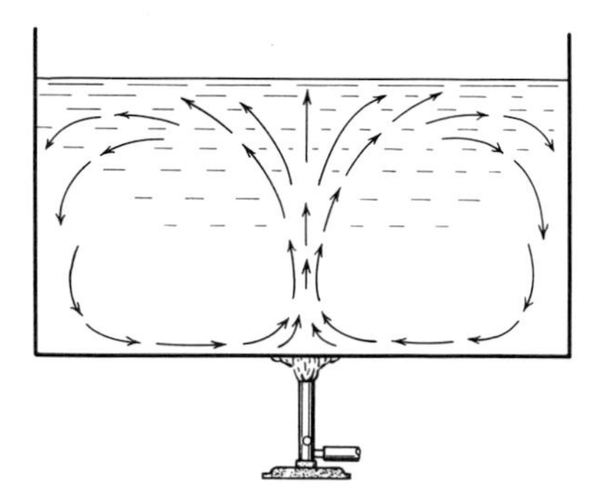

Fig. 2 · 2 Illustration of the principle of convection.

over the flame becomes overheated and rises as it becomes lighter through expansion. The surrounding colder liquid falls and moves in to replace the rising material.

We can summarize these processes by a crude analogy. Let us consider that an object is given to the first person in a line of people, with orders to transmit it to the last man in the line. The first man may simply throw the object to the last one, with no one else involved; that would be radiation. Or he may pass it along from one person to the next, which would be conduction; or he may himself walk over and hand it to the last man, which would be similar to convection.

The Heating of the Atmosphere

Although solar radiation is the ultimate source of atmospheric energy, surprisingly most of the direct heat comes not from the sun but from the earth. To understand this paradox, we must examine first the nature of the solar radiation and then the transformations and disposal of this radiation after it reaches the earth.

INSOLATION. The radiant energy received from the sun is called insolation. This energy is spread over a very broad band of wavelengths known as the *solar spectrum,* which from the long- to the short-wave end consists of radio waves, microwaves, infrared waves, visible light (red to violet), ultraviolet rays, X rays, and gamma rays. In addition, there are the bursts of corpuscular material in the form of electrons and other ions which occasionally stream out from the sun. About 90 percent of the total energy is concentrated in the very narrow visible portion of the spectrum, so that it is strangely the visible light which indirectly provides most of the heat energy to the atmosphere (see Fig. 2 · 3).

After careful measurements of solar radiation over many years, and after allowance for the loss in passing through the atmosphere, a precise determination of the absolute amount of solar energy received at the outer

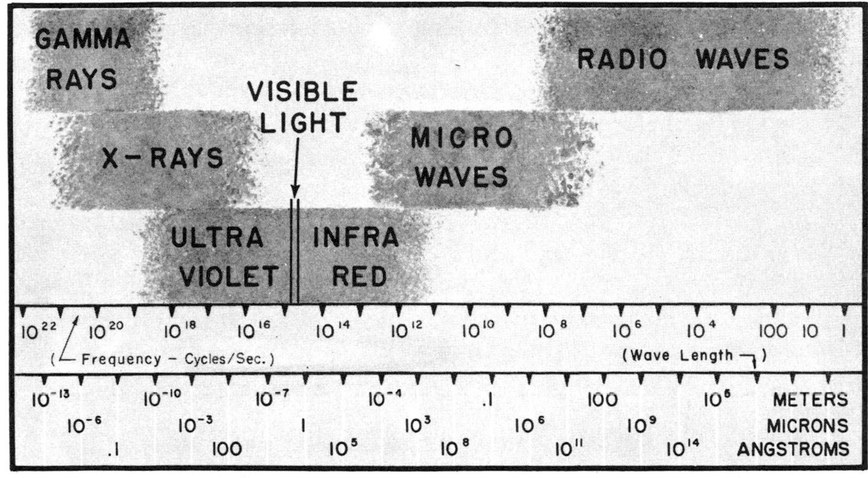

Fig. 2 · 3 Spectrum of solar radiation. The wavelengths are given in several of the common units of science. (A meter is 3.39 feet; a micron is 1/1,000 *of a millimeter or* 1/25,000 *of an inch; an angstrom is one-billionth of a millimeter.)*

edge of the earth's atmosphere has been made. This value, known as the *solar constant,* which is defined as the energy falling in one minute on a surface one centimeter square held normal to the sunlight at the mean distance of the earth from the sun, is given as 1.94 calories per square centimeter per minute. When the invisible portions of the spectrum are considered more carefully, the value is increased to 2.0 calories per square centimeter per minute.

DISPOSAL OF INSOLATION. As the result of reflection by clouds and other particles in the atmosphere, and the irregular reflection by water, ice, and variable ground surfaces, 35 to 40 percent of insolation is returned to space with no appreciable effect on the atmosphere. This is known as the reflectivity or *albedo* of the earth. If we consider the remaining portion (57 percent) of the solar constant which is not reflected, or the usable insolation, as 100 percent, it is found that one-quarter is absorbed in the atmosphere and three-quarters penetrates to the earth's surface and is there absorbed.

The absorption in the atmosphere is primarily accomplished by ozone, which removes most of the ultraviolet radiation, and water vapor, which is very absorbent in the infrared region of the spectrum. Nitrogen and oxygen, composing the bulk of the atmosphere, are practically transparent to solar radiation. Also, visible light, which is the major portion of solar radiation and which is the chief source of atmospheric energy, passes through the atmosphere without sensible absorption.

The atmosphere, the lower air in particular, thus derives most of its heat energy directly from below, following terrestrial heating primarily from absorption of visible radiation. This is accomplished in several ways, described below in decreasing importance.

Radiation. In the course of absorption and reradiation, energy is downgraded to longer wavelengths having lower intensities. Thus the earth absorbs solar short wavelength energy (primarily the visible portion of the spectrum) and reradiates long wavelength energy in the infrared portion of the spectrum. Water vapor and carbon dioxide are the only important absorbers of this long-wave radiation; and of the two, water vapor is much more influential because it absorbs energy over most of the infrared spectrum radiated by the earth. Carbon dioxide is more selective and absorbs in a narrower band. The rapid nocturnal fall of temperature over deserts and mountains is thus explainable by the dryness of the air in these regions.

It is noteworthy that after the atmosphere becomes warmed through infrared absorption, it also radiates heat, with much radiation directed back to the earth's surface. This secondary warming of the earth is known as the *greenhouse effect* and is particularly strong on cloudy nights, when the surface air is always warmer than on clear nights when part of the radiation is lost.

Latent Heat of Condensation. It is well known that evaporation is a cooling process because energy is required to transform water from a liquid to a gaseous state. Specifically 540 calories of heat is required to vaporize 1 gram of water, without any temperature change occurring. That this is a substantial amount of heat is clear when it is remembered that only 1 calorie is required to raise 1 gram of water by 1° Celsius (formerly called centigrade).

Since the earth's surface is close to 71 percent water-covered, evaporation is a most important process. The energy for this comes from the shortwave insolation absorbed by the water. When condensation of water occurs in the atmosphere (clouds, fog, etc.), the 540 calories per gram are then released in the atmosphere and provide an important warming process.

Conduction and Convection. A smaller, but significant amount of heat transfer in the lowest layer of air takes place by conduction. This can operate in either direction; that is, air may transfer heat to the earth if the latter is colder, or may gain heat if the earth is warmer. Because air, or any gas, is a poor heat conductor, this process is restricted to the air essentially in direct contact with the land or water surface beneath. This heat is then transferred aloft by convection, or *turbulent transfer.*

THE HEAT BALANCE OF THE ATMOSPHERE. It is interesting and important to note that despite local and relatively short-period heat and temperature changes in the atmosphere, the air as a whole is becoming neither warmer nor colder. A nice balance exists between the amount of heat gained from the absorption of solar and terrestrial energy and the amount of heat lost by the atmosphere through radiation and other methods. In fact, a numerical budget can be set up to show that the total heat energy absorbed from solar radiation by the terrestrial system is just balanced by the radiation to space primarily by the atmosphere and to some extent by the earth's surface.

THE IRREGULAR HEATING OF THE ATMOSPHERE. Despite the uniformity in intensity of insolation reaching the earth, the atmosphere is heated very irregularly. In part, the nonuniformity in heating is systematic, being related to the earth's curvature and motions, and in part the effect is a consequence of the nature of the earth's surface. Because the heat of the air is so important in providing the energy for all atmospheric processes and conditions—"weather" being the short-term and "climate" the long-term condition—any irregularity in the heating process must be of great importance in both weather and climate. Since the lower air is primarily heated from below, its uneven heating must follow the uneven heating of the earth's surface, the causes for which can be classified as follows: (1) systematic causes affecting the amount of insolation reaching the earth's surface and (2) variations in the nature of the surface. Owing to their importance in understanding atmospheric conditions, these factors merit additional consideration.

1. Several factors contribute to the varying amounts of insolation reaching the earth's surface:
   a. The most important is the variation of the angle of incidence of the sun's rays, or the varying angle at which the insolation strikes the earth's surface. Since the sun's position in the heavens shifts every day as a result of its daily and yearly paths, this angle must differ considerably from place to place; and at the same place it must change depending on the time. When the sun's rays strike the earth at a high angle, their intensity is much greater than when

impinging at a lesser angle, since the same beam must cover a much larger area. Figure 2 · 4 illustrates this variation at a given place at different times of the year. Figure 2 · 5 shows this variation at different parts of the earth at the same time.

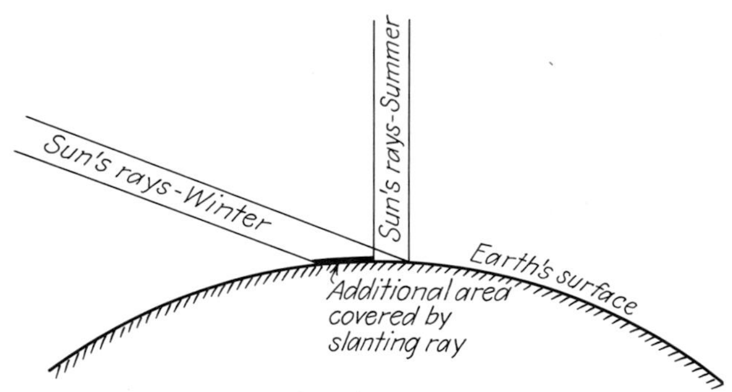

Fig. 2 · 4 The seasonal variation of insolation received at a given place as the result of the seasonal variation in the latitude of the sun.

b. The thickness of the atmosphere traversed by the sun's rays varies widely, also as a consequence of the variation in the angle of incidence. Solar energy coming in at a low altitude is filtered out to a much greater extent by the air and the foreign particles present than when approaching from near the zenith. Consider the weakness of the sun at sunset compared to the blazing noon-

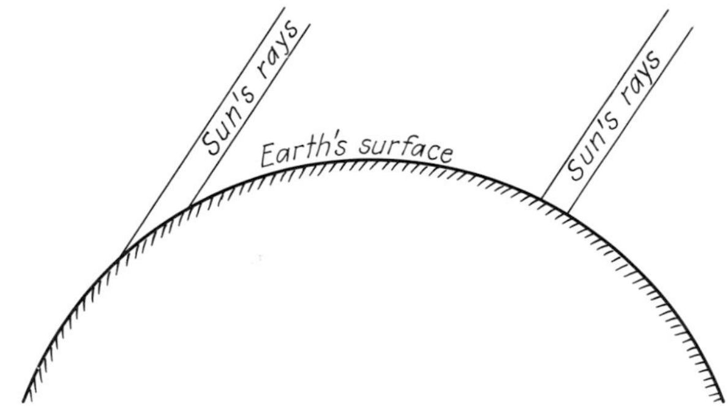

Fig. 2 · 5 Variation in insolation over the earth's surface at the same time owing to the curvature of the earth.

time sun. Figure 2 · 6 indicates this effect. Actually, about thirty-eight times more air is traversed by the tangential rays.

c. Foreign particles such as dust and clouds show pronounced variation in the atmosphere with locality and time. The dust in the air over mid-ocean is at a minimum, whereas the industrial city atmosphere exhibits a maximum dust count. The dust and water particles are of great importance in filtering out much of the sun's energy by absorption, reflection, and scattering.

d. The period of insolation is rarely constant at any one place. Aside from protracted periods of cloudiness, which of course obscure the sun, the length of the daylight period shows great seasonal change. Summer days are twice (or more, depending on the latitude) the duration of winter days and greatly affect the amount of energy reaching any one area.

2. The second basic factor concerning irregular absorption of insolation by the earth is the result of the differing composition of the earth's surface. Even if the sun's energy were equally received at the earth's surface, the great variation in composition of the earth would cause considerable differences in the amount absorbed (which later heats

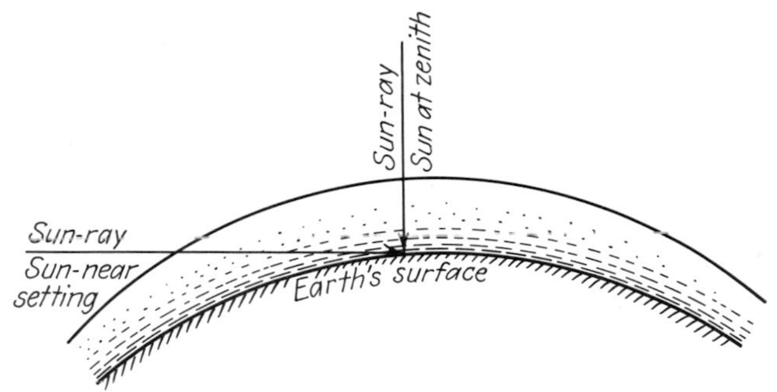

Fig. 2 · 6 Schematic diagram showing the variation in thickness of the atmosphere traversed by the tangential and vertical rays of the sun.

the air). This difference of the earth's composition is of extreme importance in determining weather and climate.

The most notable difference is that between land and water. Of the two, water is very conservative, thermally. It always lags behind temperature changes on land, therefore being cooler in summer and warmer in winter. There are several reasons for this.

a. Water is relatively transparent to sunlight, being penetrable to about 100 feet. Hence, while the opaque land surface concentrates all of the absorbed energy in a very shallow layer, producing a relatively large temperature increase, the transparent water spreads the warming effect over a much thicker layer. In the reverse process, the shallow land layer can radiate and conduct its heat more rapidly than the deeper water.

b. Water has a much higher *specific heat* than rock or soil, this being the heat necessary to warm one gram of a substance one degree

Celsius. For water the specific heat is 1, which is among the highest of all such values. Thus, compared with land temperatures, water temperatures rise slowly as water is heated and then fall slowly because much more heat must be lost before a decrease in temperature occurs.

c. Water is very mobile, being a fluid. Convection, plus the effects of waves and currents, distributes temperature differences to great depths so that surface temperatures tend to fluctuate slowly compared to those on the static land surface.

d. In the evaporation process described earlier, a large amount of energy is utilized in the conversion of liquid water to water vapor without any temperature increase occurring. This process, which is more important under the warm summer sun, tends to slow the rate of increase of temperature in all large bodies of water.

e. Ocean currents tend to equalize horizontal differences in temperature, thereby minimizing the extremes that might otherwise occur. For example, currents like the *Gulf Stream* off eastern North America and the *Kuroshio* off eastern Asia transport huge quantities of warm water to higher latitudes. As part of this ocean circulation process, cooler water from high latitudes flows toward lower latitudes. This redistribution of heat by ocean currents is of great significance in the heat budget of the entire earth.

The combination of all the above effects produces the striking difference between water and land temperatures—and hence between marine and continental air temperatures. This effect is considered further in the following chapter.

EXERCISES

2·1 Define temperature.

2·2 Briefly describe the three methods of heat transmission.

2·3 Define the solar constant and give currently accepted values.

2·4 Define and give the value of the earth's albedo.

2·5 In view of the earth's albedo and the absorption in the atmosphere, what percent of the energy indicated by the solar constant reaches the earth's surface where it is perpendicular to the sun's rays? Give the amount in calories per square centimeter per minute, also.

2·6 (*a*) Which three gases in the atmosphere are the chief absorbers of radiant energy and indicate the spectral band of greatest absorption? (*b*) Of these, which is most important to the temperature of the troposphere?

2·7 Explain the greenhouse effect.

2·8 What are the principal processes by which the troposphere is heated.

2·9 Explain the heat balance of the atmosphere, including the nature of the observation that indicates that the balance exists.

2·10 Why are the annual temperature variations of large bodies of water less than those of land areas?

2·11 Explain the basic reason for the decrease of temperature from equator to the poles.

CHAPTER 3

Temperature

Temperature is a fundamental weather element. In response to the irregular disposal of the sun's energy (insolation) the air temperatures show variations between wide extremes. These variations in turn cause other significant weather changes. We shall consider first the instruments and methods of measuring temperature and then proceed to the nature of temperature variations, examining the periodic, the horizontal, and the vertical air temperature changes. Then the extremely important temperature variations that result whenever a mass or masses of air engage in vertical motion will be studied.

Temperature Instruments

THERMOMETERS. The temperature of an object is measured by a thermometer, which is a sealed glass tube having a very small opening —the bore—running through the center of it from top to bottom. The bore of the tube is greatly enlarged into a bulb-shaped opening within the bottom of the tube, as indicated in Fig. 3 · 1. The bulb end of the tube is filled with a liquid, usually mercury or alcohol, which rises into the narrow bore. The space above the liquid is a vacuum. When the temperature changes, the liquid within the thermometer expands and rises or contracts and falls, depending on whether the temperature increases or decreases.

The outer glass surface of the thermometer is etched in the form of a graduated temperature scale. Hence, by reading the height of the upper surface of the liquid in the bore of the thermometer, the existing temperature is obtained.

MAXIMUM THERMOMETERS. It is often necessary to determine not only the current temperature, but also the highest temperature reached during a given period. For this purpose a type of registering thermometer known as a maximum thermometer is used. This is almost identical with the standard thermometer described above, with one exception. Just above the bulb of the thermometer, the bore narrows very abruptly for a short space. This leaves a constriction of the bore to a very thin channel, much thinner than that in the rest of the tube, as shown in Fig. 3 · 3*b*. When the temperature rises, the mercury in the bulb expands. The force of the expansion is sufficient to force the mercury through the constriction in the tube, causing it to rise higher in the bore. When the temperature reaches its maximum and then starts decreasing, the mercury below the constriction contracts into the bulb. But the narrowness of the constriction prevents the mercury above from falling through from its weight alone. Consequently the mercury above the constriction remains in the position it took at the highest temperature, and the top of the column of liquid indicates the maximum temperature reached.

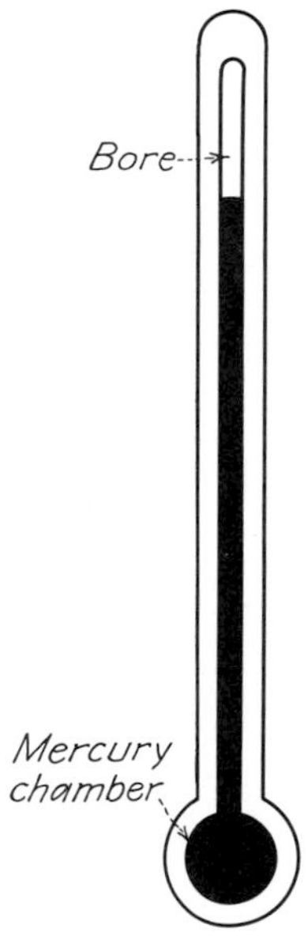

Bore
Mercury chamber

Fig. 3 · 1 Principle of construction of thermometers.

Although the mercury thread in the bore also contracts with a fall in temperature, this contraction is so minute that it may be considered negligible.

The clinical thermometer is a common example of the maximum-registering type. To reset the thermometer after it is read, it is generally whirled around rapidly, and the effect of whirling or shaking forces the mercury back through the constriction and into the bulb of the thermometer.

MINIMUM THERMOMETERS. Minimum thermometers are used to register the lowest temperature reached during a given period. The minimum thermometer resembles the regular thermometer in appearance, except that it always contains a liquid of low density such as alcohol, instead of mercury. In addition, within the liquid in the bore of the tube, there is a thin glass rod, shaped somewhat like a dumbbell, called the *index*. This is indicated in Fig. 3 · 3*a*. When the temperature decreases, the liquid contracts so that its upper surface pulls the index down from the effect of *surface tension.*

When the temperature rises again, the alcohol flows around the index

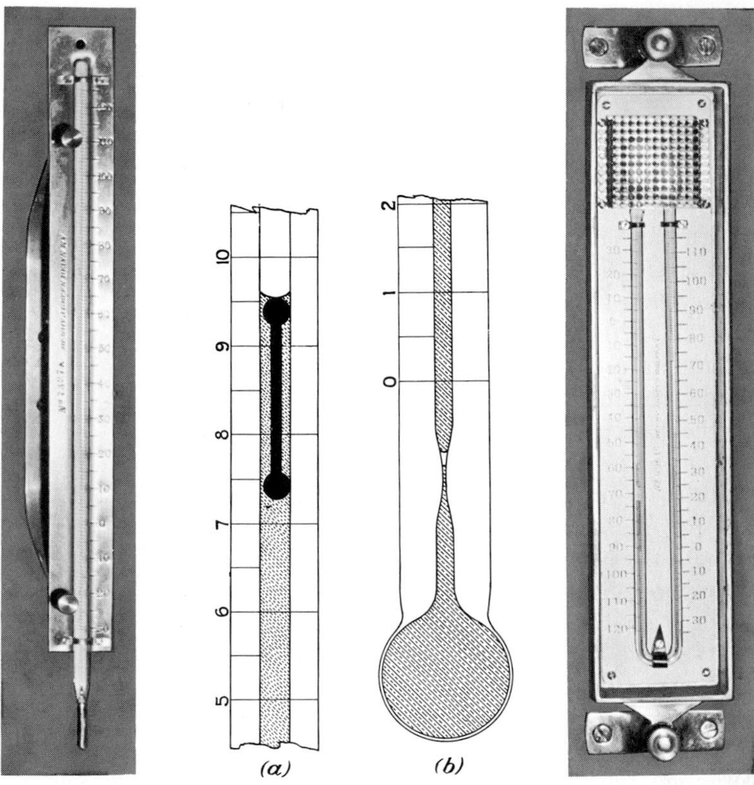

Fig. 3 · 2 Standard exposed thermometer and common type of support.

Fig. 3 · 3 (a) Construction of minimum thermometer showing index; (b) Construction of maximum thermometer showing constriction in bore.

Fig. 3 · 4 Six's thermometer showing indexes above the dark mercury thread in both arms.

and rises in the tube, leaving the index at the lowest point to which the liquid surface descended. Hence, the upper surface of the glass index marks the lowest or the minimum temperature reached, at the same time that the alcohol surface itself always indicates the current temperature. Obviously, the minimum thermometer should always be kept in a horizontal position, or the glass index will fall through the liquid to the bottom of the tube.

To reset the thermometer after a reading is taken, the instrument is inverted with the bulb end uppermost, until the index falls to the surface of the alcohol under its own weight; it is then restored to a horizontal position.

SIX'S THERMOMETER. Since Six's thermometer is so common aboard ship, it will be explained in detail. It is a combination maximum-minimum thermometer commonly used because of its convenience. As shown in Fig. 3 · 4, the instrument is a U-shaped glass tube with sealed, widened ends masked here by the protective screen. The lower portion of the U is filled with mercury. The remainder of the left-hand horn of the U is filled with alcohol, including the widened upper portion. The right-hand side of the U above the mercury is also filled with alcohol except that part of the expanded chamber contains a gas above the alcohol.

Above the surface of the mercury, within the alcohol in both arms of the thermometer, is a colored glass index containing a thin iron pin in the center. To prevent the index in either arm from falling through the alcohol to the surface of the much denser mercury, a thin spring protrudes from the index and presses against the inner glass wall. The pressure of this wire keeps the pin suspended in whatever position it is pushed by the mercury surface.

When the temperature increases, the alcohol in the left side, in expanding, forces the mercury and the alcohol column to the right of it into the gas chamber in the upper right bulb of the tube. The gas thereby becomes compressed. As the mercury rises in the right-hand column, its high density causes the glass index above it to be pushed higher in the tube. When the temperature decreases, the alcohol in the left side contracts and the gas therefore expands, forcing all the liquids back toward the left side of the tube. As explained previously, the glass index will remain suspended in the alcohol as the mercury recedes, and its *lower surface* indicates the *maximum* temperature reached.

As the temperature continues to fall, the alcohol in the left-hand arm of the U contracts further, allowing further expansion of the gas in the upper right chamber. The expansion of the gas now forces the mercury into the left-hand arm of the U. The overlying index in this arm is now raised by the rising surface of the mercury.

When the temperature has reached its lowest point and begins to increase again, the mercury is once more forced to the right, leaving the glass index up in the left side of the tube, with its *lower surface* marking the *minimum* temperature reached. It should be seen from this that the side indicating minimum temperatures will have the highest temperature readings on the bottom of the arm of the U and the lowest on top, in contrast to the maximum side of the thermometer which reads from the bottom up, in the normal fashion. The surface of the mercury in both arms of the thermometer will always show the same readings and will indicate the current temperature. The fact that the glass index arms are reset to the mercury surface by means of a magnet explains the reason for the iron core within the glass indexes.

THE THERMOGRAPH. The last instrument to be considered is the thermograph. This is a purely mechanical device consisting essentially of a

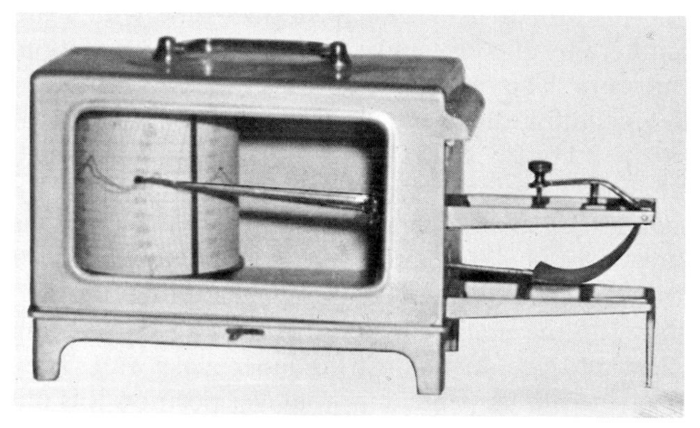

Fig. 3 · 5 Thermograph.

metallic element whose curvature varies with the temperature. One end of the curved sensitive element is connected to a long movable lever arm which contains an inked pen at its end. The arm, in turn, rests on a cylindrical drum which rotates by means of an inner clockwork. A sheet of graph paper is wrapped around the drum and is divided into days and hours, horizontally, and temperature in degrees, vertically.

Changes in temperature cause variations in curvature of the sensitive

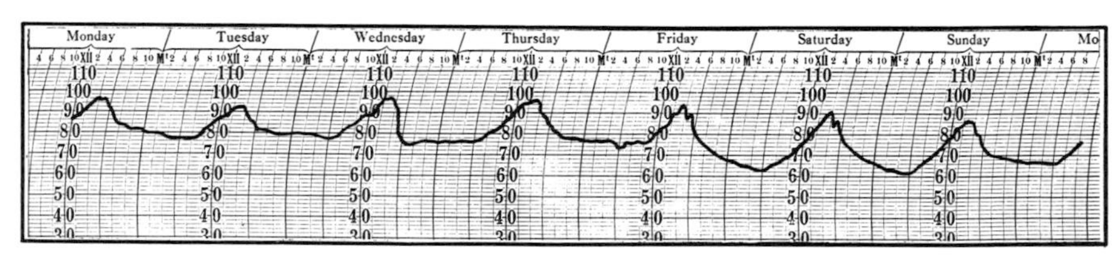

Fig. 3 · 6 Weekly thermograph record. (Note hours of maximum and minimum temperatures.)

metallic element which cause deflection of the long pen arm upward or downward, depending on the nature of the change. The pen inscribes an inked line on the sheet surrounding the drum. As it does so, the drum is slowly rotating; hence a line is traced which indicates the temperature at any time during the recording interval, as well as the current temperature.

Thus the thermograph gives a continuous record of the temperature, combining the results of the other instruments. Being mechanical, however, the thermograph may vary from correct readings at intervals and have to be reset by a thumbscrew provided for that purpose, after being compared to an accurate thermometer.

LOCATION OF TEMPERATURE INSTRUMENTS. In placing temperature instruments properly to record or register air temperatures accurately, several factors should be taken into consideration:

1. The instruments should be shaded from direct sunlight. The air temperature is desired and not the temperature of the sun's rays falling on the instruments. Thermometers should also be sheltered from radiation from walls, bulkheads, or any other source of heat.
2. Good ventilation is required. If the air is not in motion in the vicinity

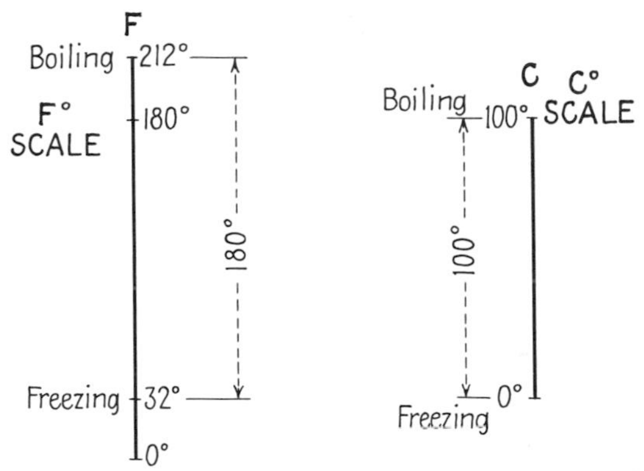

Fig. 3 · 7 Comparison of Fahrenheit with Celsius scales.

of the instruments, they will not indicate the true air temperature. Hence, any shelter used for thermometers should allow the air to pass through it without obstruction. For this purpose, such shelters are constructed with louvered sides.

Thermometer Scales

There are two common systems of scales or units used in measuring temperatures: Fahrenheit and Celsius. To compare the two systems we note that the boiling point of water is arbitrarily put at 212° on the Fahrenheit (F) scale and at 100° on the Celsius (C) scale. The freezing point of water is put at 32° on the Fahrenheit scale and 0° on the Celsius scale (Fig. 3 · 7). (The term *Celsius* replaces *centigrade* as the designation for the metric thermal scale in accordance with an international convention on weights and measures.)

Hence, between freezing and boiling points on the Fahrenheit scale there is a range (from 32° to 212°) of 180°. However, from freezing to boiling (0° to 100°) on the Celsius scale, there is a range of 100°. Thus the relationship between the two scales is 180 to 100, or 9 to 5. For every 9°F, we have 5°C.

To convert from Fahrenheit to Celsius, it is clear that we must take 5/9 of the Fahrenheit (the larger figure) to obtain the Celsius value. But before we take 5/9 of the Fahrenheit figure, we must reduce the freezing points on both to the same level. Thus, if we subtract 32° from the Fahrenheit figure, we obtain the freezing point on both scales starting at zero. Then, 5/9 of the Fahrenheit figure after 32° has been subtracted will give the Celsius equivalent. Hence,

$$C = (F - 32)5/9$$

Conversely, to change from Celsius to Fahrenheit, we first multiply the Celsius figure by 9/5 to change the units to Fahrenheit and then add 32°, since the Fahrenheit freezing point is 32° above zero.
Thus,

$$F = 9/5C + 32$$

A temperature scale based on *absolute zero,* at which no random molecular motion exists, is necessary in studies in the physics of the atmosphere. Celsius units are used in this system, but zero is at −273°C. Temperature values on the absolute scale are given in *degrees Kelvin,* or *degrees absolute;* hence 0°C = 273°K.

Table 3·1 Equivalent Temperature (Celsius and Fahrenheit)

°C	°F	°C	°F	°C	°F	°C	°F	°C	°F
−10	14.0	0	32.0	10	50.0	20	68.0	30	86.0
−9	15.8	1	33.8	11	51.8	21	69.8	31	87.8
−8	17.6	2	35.6	12	53.6	22	71.6	32	89.6
−7	19.4	3	37.4	13	55.4	23	73.4	33	91.4
−6	21.2	4	39.2	14	57.2	24	75.2	34	93.2
−5	23.0	5	41.0	15	59.0	25	77.0	35	95.0
−4	24.8	6	42.8	16	60.8	26	78.8	36	96.8
−3	26.6	7	44.6	17	62.6	27	80.6	37	98.6
−2	28.4	8	46.4	18	64.4	28	82.4	38	100.4
−1	30.2	9	48.2	19	66.2	29	84.2	39	102.2

Periodic Temperature Variations

It is a matter of common observation and knowledge that the air temperature shows periodic variations from high to low peaks. These variations can be differentiated into daily and annual changes. The daily maximum temperatures occur in midafternoon, between 1 and 4 P.M. on the average, after the earth transmits to the air the excess noon time insolation. Then, just before sunrise, on the average of 4 to 6 A.M., after the earth and air have been cooling all night, the lowest or minimum temperatures are experienced (see Fig. 3 • 6).

On a larger scale, the yearly variations follow the daily ones. The end of July, a full month after the summer solstice, shows the highest temperatures, and the end of January shows the lowest. These are mean conditions. During December, when the greatest deficiency in insolation occurs, the earth is transmitting to the air much of the heat received a month earlier. It should be remembered that these annual variations are for the Northern Hemisphere only, the seasons being the reverse in the Southern.

Horizontal Temperature Variation

The most fundamental horizontal temperature variation is the slow decrease in air temperatures poleward (north or south) from the equator. This is the normal effect of latitude on temperature, as a result of the increasing slant of the sun's rays resulting from the curvature of the earth.

However, this is a uniform change and, should the earth's surface have been perfectly homogeneous in composition, i.e., all water, then the air temperatures at a given latitude would be much the same, regardless of longitude. But it is quite evident that temperatures along a given parallel of latitude vary widely. This recalls the effect of the irregular composition of the earth's surface on temperature. In the summer season, the land will be definitely warmer than the ocean at the same latitude. Then in the winter, the land, cooling off much faster than the water, will show much lower temperatures.

Further, in response to the transportation of huge quantities of warm and cold water by ocean currents, the temperature over the oceans themselves is rarely uniform. Only over large ocean areas that have no northward or southward currents will the air temperature be uniform.

To show the horizontal distribution of air temperature most conveniently, *isotherms* are used. Isotherms are lines connecting points of equal temperature. Consequently, if the earth were uniform in composition, isotherms would be straight east-west lines, similar to parallels of latitude. The isotherms representing the highest temperatures would be near the equator. The temperature change in a direction normal to the isotherms is called the horizontal *temperature gradient.*

Isotherms are rarely straight east-west lines. Their distribution on a world map is quite irregular and varies greatly from winter to summer and from hemisphere to hemisphere. For an isotherm to remain on equal temperature points, it must be deflected away from the equator in the wintertime when passing from land to ocean, and toward the equator in the summer. The reason for such a bending or deflecting of isotherms can be clearly seen from the discussion above. If we assume that the winter temperature along an east-west line over the land is 30°, then the adjacent ocean will be warmer (in winter). Hence, the isotherm must bend poleward to remain on points having a 30° temperature.

Similarly, in the summer, with the oceans colder, the isotherms must deflect equatorward to discover equivalent warm temperatures. Ocean currents, depending on whether they are cold or warm, may add to or detract from this isotherm deflection. These effects are shown clearly on the world isotherm charts for January and July (Figs. 3 · 8 and 3 · 9). A line known as the *heat equator* is sometimes drawn connecting places having the highest average temperatures. Owing to the greater land areas in the tropics of the Northern Hemisphere compared to those of the Southern, the annual heat equator, although somewhat irregular, lies to the north of the geographic equator.

The influence of ocean currents in regulating air temperatures is well shown by the appearance of isotherms on opposite sides of North America. The effect of the warm Atlantic Gulf Stream is very striking in Fig. 3 · 8, which depicts Northern Hemisphere winter conditions. North of about 35° the isotherms bend northeastward very strongly. Note that the same isotherm that crosses New York City (about 40°) also crosses Iceland and northern Norway. The Gulf Stream thus amplifies the normal poleward winter deflection over water.

On the other hand, the isotherms extending westward from California experience very little northward deflection because of the cold California Current which tends to equalize ocean and land air temperatures during the winter. Note that in the summer in these two regions (Fig. 3 · 9) the reverse effect occurs. East Coast isotherms are but little deflected because of the tendency toward equalization of land and water air temperatures by the warm Gulf Stream, whereas the cold California Current causes a strong southward isothermal deflection.

Obviously, the greatest horizontal temperature variation from summer to winter will occur over the largest land mass, which mass will, by virtue of its intense heating, become extremely warm in summer. Then, since land masses rapidly lose their heat in winter, such a large mass would show a very pronounced temperature decrease. For example Asia becomes very warm in summer and very cold in winter. Note Fig. 3 · 10.

Just the opposite condition exists over the southern part of the South Pacific and Atlantic. Here, the large expanse of ocean is undisturbed by land, being nearer to the ideal case than any other part of the earth. In this region the isotherms are straight east-west lines, indicating a gradual and uniform temperature decrease from north to south. The temperature con-

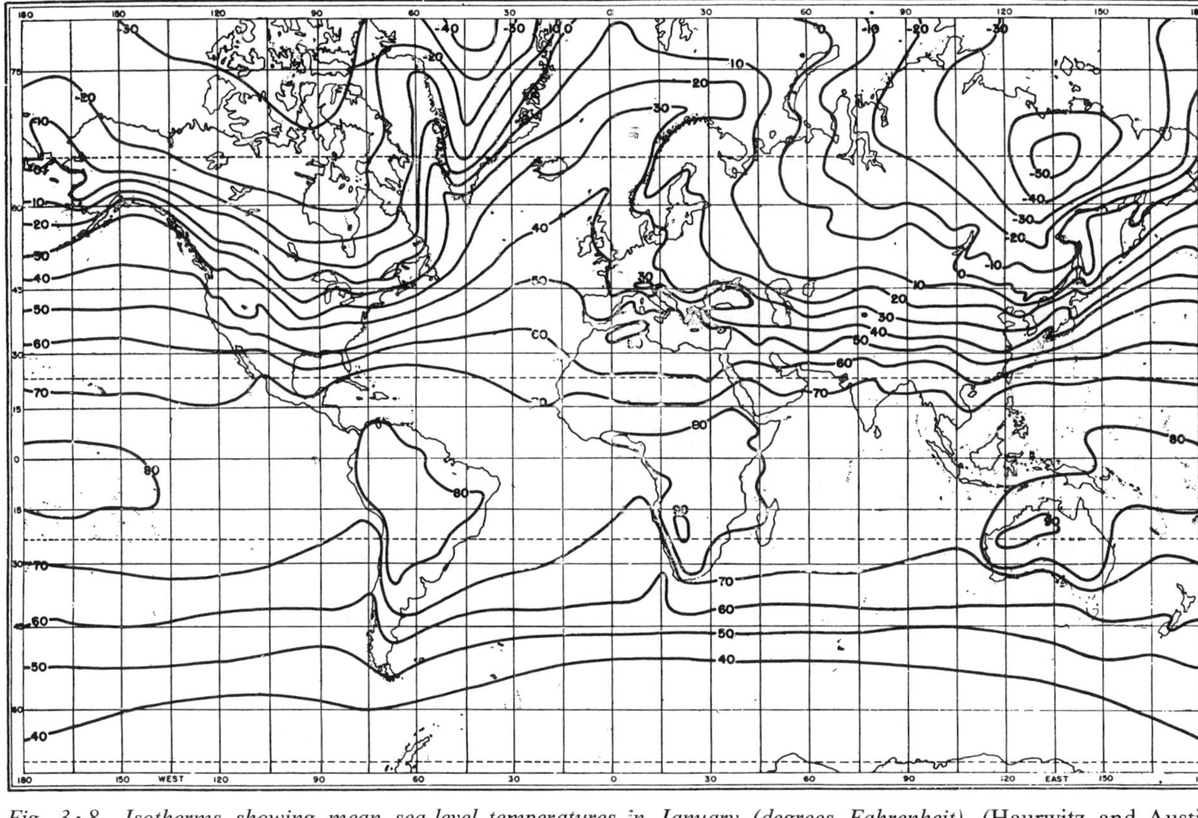

Fig. 3·8 Isotherms showing mean sea-level temperatures in January (degrees Fahrenheit). (Haurwitz and Austin, Climatology)

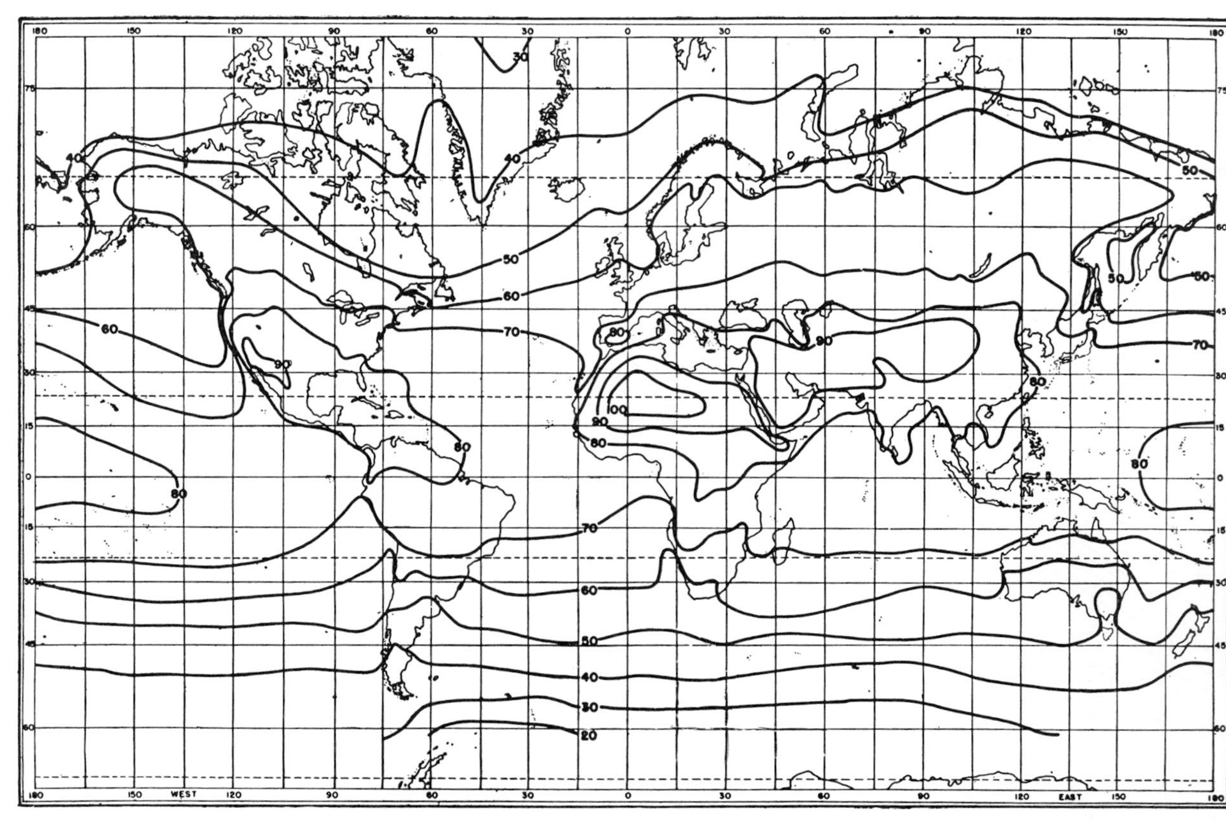

Fig. 3·9 Isotherms showing mean sea-level temperatures in July (degrees Fahrenheit). (Haurwitz and Austin, Climatology)

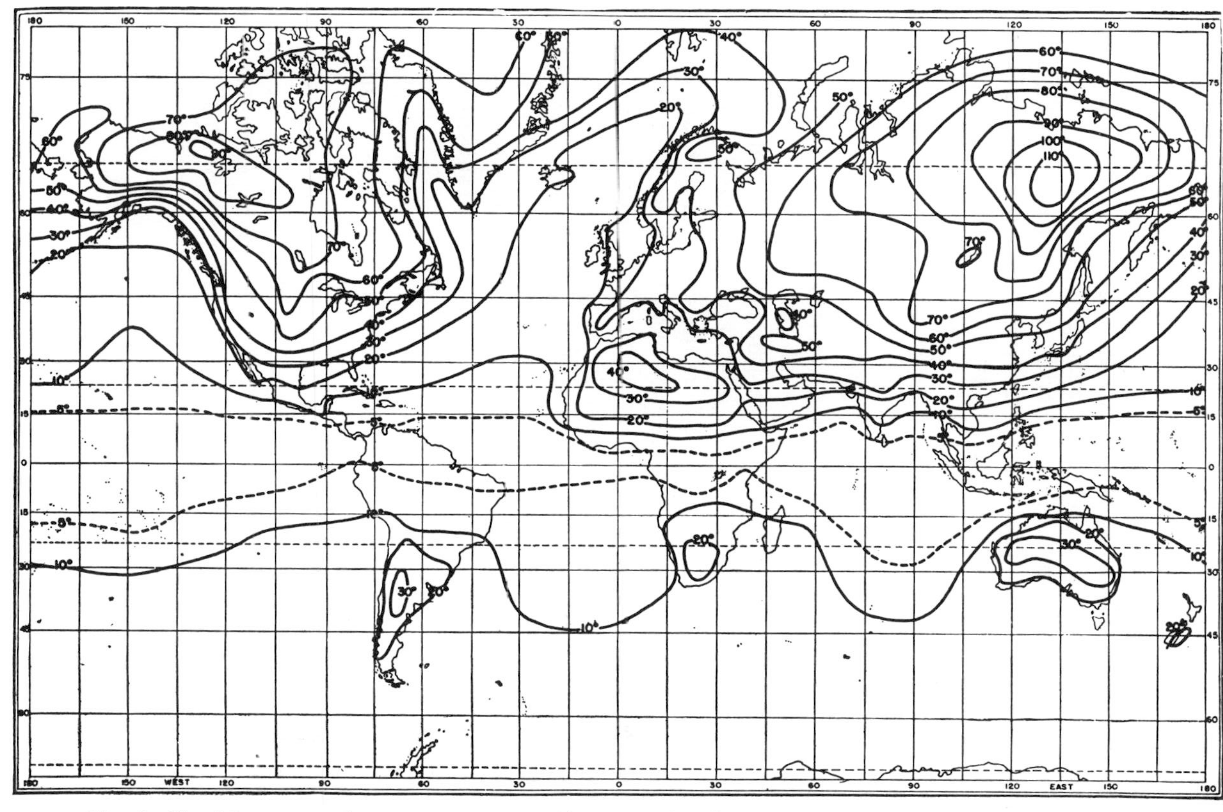

Fig. 3·10 Mean annual temperature range (degrees Fahrenheit). (Haurwitz and Austin, Climatology)

ditions in each of the above regions have a very pronounced effect on the weather and climate of the parts of the earth affected by them. That effect will be considered in a later chapter.

It is always interesting to note some of the global extremes in temperature. The lowest valid temperature ever registered was −127°F at Vostok Station, Antarctica. One of the lowest average temperatures for one month was at Verkhoyansk, Siberia, being −64°F. Yellowstone Park holds the low record in the United States with a reading of −66°F.

The high temperature mark for the world was recorded in North Africa in Tripoli, where the mercury reached 136°F. Death Valley in the United States follows closely with a record of 134°F.

Vertical Temperature Variation

THE VERTICAL TEMPERATURE GRADIENT—LAPSE RATE. As one ascends in the atmosphere, steadily decreasing temperatures are encountered. This decrease of temperature with higher altitudes in the air is known as *vertical temperature gradient*. There are three essential causes for this decline in temperature. (1) The major source of heat for the air is the earth. Clearly, then, with increasing distance from the source of its heat, the air's temperatures must decline. (2) The density of water vapor decreases with elevation so that less heat can be held in the air. (3) Temperature decreases result from expansion of air rising from the earth's surface (see page 43).

However, although air temperatures decrease vertically, there is nothing constant about the rate at which this temperature drop occurs. In fact, the only constant thing about the vertical temperature gradient is that it varies. As stressed previously, the heating of the earth, and subsequently of the air, varies so widely from time to time and from place to place that there can be no uniform rate at which the temperature decreases. The only way to determine this rate is to obtain air temperature readings at different elevations by means of airborne instruments.

LAPSE RATE. The actual figures obtained by observations of the vertical temperature decrease or gradient are known as the *lapse rate.* As stated, this lapse rate varies through a wide range. But at a given place the different lapse-rate figures that have been observed can be averaged, giving the *normal lapse rate.*

Although the lapse rate at a given time and place can be obtained only by observation, the average or normal rate is well known:

3.5°F per 1,000 feet
6.5°C per kilometer

Notice that the lapse rate refers to temperature conditions existing in a stationary column of air at a given place and time. This air should not be considered as having any vertical motion, either upward or downward.

INVERSIONS. Occasionally at some altitude the temperature abruptly increases instead of decreasing. This can occur only if a warm layer of air overlies a colder layer. The temperature will fall with increasing altitude in the cold layer and will rise suddenly as the warm layer is encountered. After a short vertical distance, the temperature in the warmer layer will also continue to fall. The condition in which this abrupt rise instead of fall in temperature occurs in the air is an *inversion.* An inversion may result (1) when the air near the ground cools off faster than the overlying air, because of heat loss to the earth which might be cold, (2) from an actual warm layer passing over a lower cold one, (3) from warming by subsidence or falling, or (4) from turbulence. The tropopause, described earlier and illustrated in Fig. 1 · 4, is a level of major temperature inversion between the troposphere, characterized by declining temperatures, and the overlying isothermal stratosphere.

Effect of Vertical Air Motion on Temperature

CAUSES OF VERTICAL AIR MOTION. In addition to horizontal movement of the air, there is often pronounced vertical movement. The amount of air exhibiting this rising or falling motion depends on the force that initiates the movement. The causes of vertical air motion can be separated into four distinct influences: (1) heating and cooling of isolated parts of the air, (2) topographic uplifting, (3) the effects of fronts or cold-air wedges, and (4) horizontal convergence and divergence.

1. If a portion of the earth's surface becomes highly heated, the overlying air also becomes heated. Upon being heated, air expands, becomes lighter than adjacent air, and tends to rise. Air over a cold surface will become cold, hence heavier than the surrounding warmer air, and will tend to sink. This is, on a much larger scale, the same process that occurs in a room heated by a radiator where the hot air rises in a vertical stream toward the ceiling. Although it is difficult to see these rising and falling air columns, their presence is well known to air travelers, who constantly encounter so-called "bumpy air," or "air pockets," which are nothing more than air columns in vertical motion.
2. The topographic effect is easily pictured. Air in motion, approaching a mountain or a ridge, will ascend the windward slopes and descend on the leeward sides of the elevation.
3. *Fronts* are the bounding surfaces between different air masses. For example, a cold air mass moves southward until it meets a warmer moving mass of air; the line of separation between them is a front. However, it is not a vertical boundary between the two, but a sloping

one. The cold air, being heavier, will flow in the form of a wedge beneath the warmer air and force the latter up over the cold air, as indicated in Fig. 3 · 11. Consequently, there will be a continuous mass

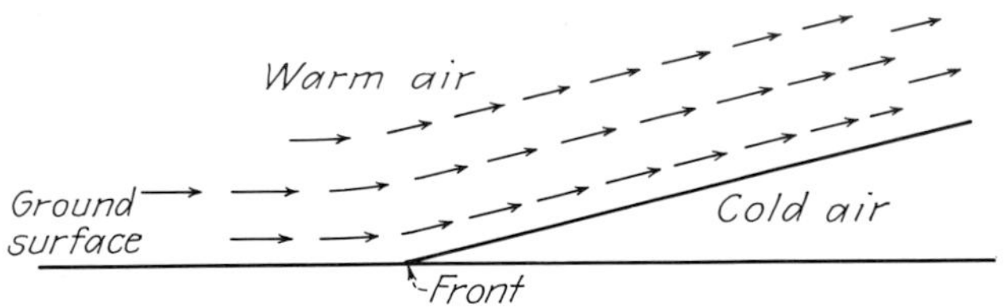

Fig. 3 · 11 The flow of warm air over an intruding wedge of cold air.

of warm air rising over the cold beyond the meeting point of the two air masses. The extreme importance of this process will be explained in greater detail later.

4. For various reasons to be touched on in this book, air in horizontal

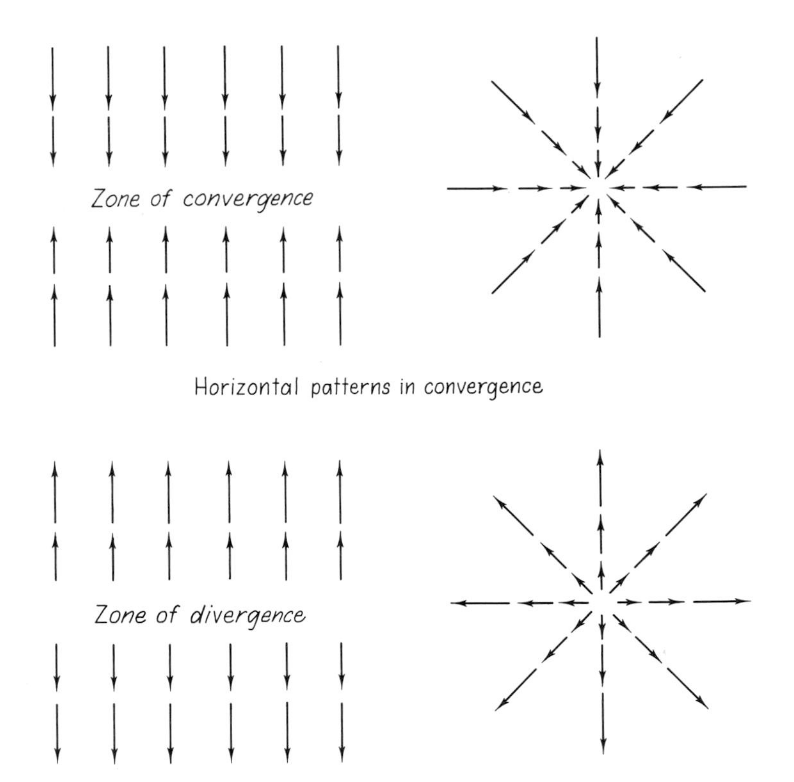

Fig. 3 · 12 Horizontal air-flow patterns in convergence and divergence.

motion may be forced to *converge* or *diverge* as shown by the net airflow patterns in Fig. 3 · 12. Note that the diagrams do not necessarily show true air motion but may only indicate the component of motion producing either convergence or divergence. Because air cannot accumulate from convergence or form a vacuum from divergence, it follows that ascending and descending air must occur, respectively. Figure 3 · 13 illustrates vertical schematic views of convergence and divergence with related vertical motions.

Regardless of the influences causing air to rise or fall, once air does so, there occur very profound effects on temperature and consequent weather. *The temperature change in vertically moving air is one of the most fundamental and important processes in the atmosphere.*

RESULTING TEMPERATURE EFFECT—ADIABATIC CHANGES. We have seen that relatively local masses of air can be forced to rise and fall in the atmosphere as a result of one of the methods just considered. During this process,

5.5°F

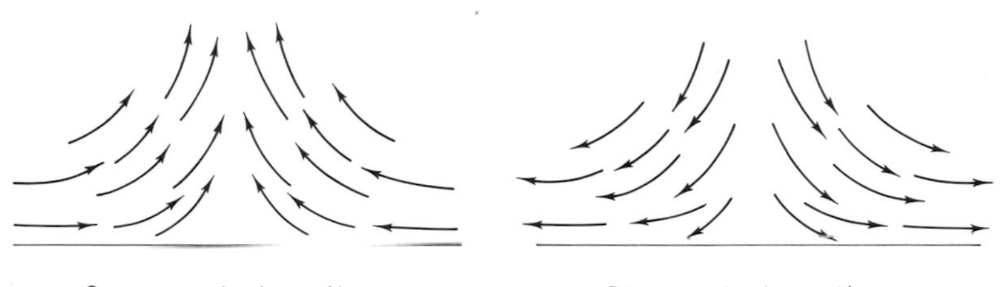

Fig. 3 · 13 Vertical cross sections showing air motion in convergence and divergence.

the greater part of the atmosphere can be considered more or less calm and motionless as regards vertical movement.

Let us consider a mass of air on the earth's surface. Assume that this air is forced upward from any one of the four causes of uplift described in the preceding section. As this air rises, the pressure on it grows less, causing it to expand. This principle is very familiar in the case of ordinary gas-filled balloons, which, when released, rise in the air and expand continuously until they burst. The process is shown in Fig 3 · 14, where the horizontal arrows indicate the decreasing pressure on a rising air mass. The arrow lengths are proportional to the air pressure and indicate the decrease of pressure with altitude.

Remember that the atmosphere is an ordinary mixture of gases. As a gas expands, it uses up heat energy from within itself to supply the energy necessary for expansion. If no external source of heat is available, the expanding gas or air must cool. We have previously noted that a rising or falling air mass can be considered as isolated from the rest of the air. Practically no heat can be transmitted to the rising or falling air mass either by the earth or by the surrounding air, for air is well known as a good insulator (or poor conductor) of heat.

An excellent and common example of this process occurs in the ordinary bicycle or automobile tire. When the valve is opened, the compressed air in the tire will escape very rapidly. As it does so, the air expands very rapidly with a consequent rapid decrease in temperature. This is a rather familiar cooling process. A very important practical application of adiabatic cooling is the common refrigeration mechanism consisting of a compressor unit and cooling chamber. After compression, the gas escapes into a tubular coil wherein rapid expansion occurs. The cold pipes then absorb heat from the larger volume containing material to be refrigerated.

In the same manner, then, rising air grows continuously cooler, and

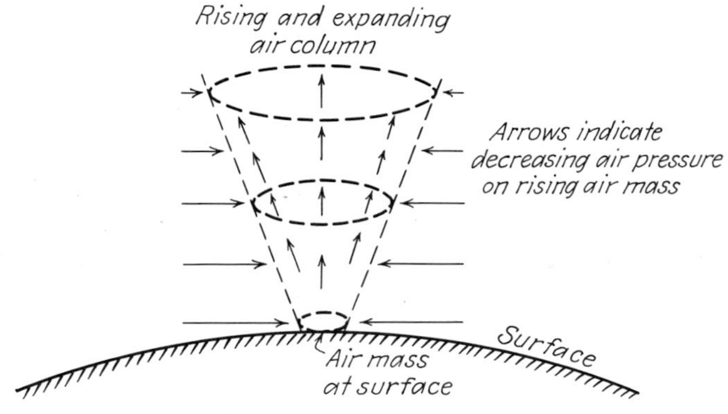

Fig. 3 · 14 The expansion of a column of air rising from the earth's surface. (This is schematic since the lengths of the horizontal arrows are not exactly proportional to actual pressure.)

since the surrounding air pressure falls off uniformly, expands uniformly. If the air expands at a uniform rate, it will cool at a uniform and constant rate.

Oppositely, air that is descending will encounter increasing pressure from the surrounding air and will be compressed. Whenever a gas is compressed, the work done on it will cause its temperature to rise. When air is pumped into a tire, the temperature rises noticeably as a result of the compression. When air subsides or sinks in the atmosphere, it will therefore always grow warmer and will do so at the same rate at which rising air cools when rising an equivalent distance.

This change in the temperature of a gas or the air, due only to the change in pressure on the air, is called an *adiabatic change.* The rate at which the temperature changes as air rises or falls is called the *adiabatic lapse rate.*

This rate is constant for dry air. If the original temperature of the air is known and the distance through which it rises or falls is known, then the resulting temperature can be calculated by means of the adiabatic rate. The rate for dry air, known as the *dry adiabatic lapse rate,* is

5.5°F per 1,000 feet
10°C per kilometer

If a mass of air rises 1,000 feet, whether vertically or at an angle through the effects of ground slopes or cold-air wedges, the mass will cool by 5.5°F. The air rising at an acute angle will not of course cool so rapidly as that rising vertically.

To summarize: As a result of thermal effects, of mountains, of convergence or divergence and of cold-air wedges, relatively small parts of the atmosphere are often forced into vertical motion. *Whenever air rises, it cools* at the adiabatic rate; whenever air falls, it warms at the same rate. This change in temperature has nothing whatsoever to do with the surrounding air conditions or with the earth. This adiabatic rate should not be confused with the lapse rate, which refers to the general decrease of temperature with altitude in the large quiet atmosphere. If air is in the process of rising or falling, its temperature is determined by the adiabatic rate. If air has no vertical motion, its temperature at any elevation is determined by the lapse rate. Thus, the adiabatic lapse rate can really be considered a special and constant case of the lapse rate, applied only to air having vertical motion.

EXERCISES

3·1 Distinguish among the temperature units Fahrenheit, Celsius, and Kelvin, giving the values of the freezing and boiling points of water in each system.

3·2 (*a*) Convert −40°C to °F and °K. (*b*) Convert 98.6°F to °C and °K.

3·3 Since the thermograph produces a continuous record of temperature, why is it not used as a standard air-temperature instrument?

3·4 Explain the time lag in the annual and diurnal maximum and minimum temperatures.

3·5 (*a*) Using Fig. 3 · 8, construct a graphical temperature profile from 75°N to 75°S along the meridians 90°W and 120°E. (*b*) Explain the difference between the temperature gradients of Northern and Southern Hemispheres that show up in your profiles.

3·6 (*a*) During the Northern Hemisphere winter, in which direction should isotherms be deflected in passing seaward from the land? Explain. (*b*) Explain the difference in the behavior of isotherms off the eastern and western coasts of the United States in winter. (Refer to Fig. 3 · 8.)

3 · 7 (*a*) Construct a graph of temperature versus elevation from the following data:

ELEVATION, ft	TEMPERATURE, °F
1,400	62
1,200	63
1,000	64
800	62
600	64
400	66
200	68
0	70

(*b*) In the first 800 feet what is the lapse rate in degrees F per 1,000 feet? (*c*) What is the lapse rate in the same units from 1,200 to 1,400 feet? (*d*) Explain the negative lapse rate between 800 and 1,000 feet.

3 · 8 Why does air temperature in the troposphere decrease with elevation?

3 · 9 Distinguish between the lapse rate and the adiabatic lapse rate.

3 · 10 Imagine that a unit of air at a temperature of 75°F rises up a mountain range that is 8,000 feet high on the windward side and which decends to 4,000 feet on the lee side. If the air remains dry, what will its temperature be when it crosses the top of the range and what will it be when it descends to the base on the lee side?

CHAPTER 4

Humidity and Water Vapor

Nature of Atmospheric Water Vapor

We noted in Chaps. 1 and 3 that the average temperatures over the earth's surface and therefore in the lower atmosphere cover the range in which water can appear in each of three states: solid, liquid, or gas. It is only in the gaseous state that water is a true component of the air. However, as noted, this composition is extremely variable, ranging from perfectly "dry" air to air that contains water vapor up to 4 percent by volume. Although the variation is primarily due to temperature differences, an important secondary cause is the availability of water for evaporation into the air.

It should be noted that water vapor is colorless, odorless, and tasteless. The white mass of water droplets seen escaping from whistles or stacks, etc., is not steam or water vapor. It is rather a mass of minute water droplets, so small in size that they rapidly evaporate into true vapor. It will be observed, however, that between the exhaust stack and the white cloud of escaping droplets there is always a small clear area just above the exhaust opening. This clear space actually contains the true water vapor. Upon entering the cooler air, this vapor is suddenly chilled and turns to minute liquid droplets. The same effect is observed when one breathes rapidly in cold weather, as the moisture in the breath condenses on meeting the cold air.

The term *humidity* is usually used to describe some particular water vapor property of the atmosphere, as will be described later in this chapter. In formal meteorology there are many types of humidity, although in common lay usage humidity generally refers to *relative humidity,* also described below.

For the most part, the water vapor in the atmosphere is derived by evaporation from the oceans, with the evaporation being greatest in low latitudes. The weather processes to be considered in the remainder of the book provide the mechanism for maintaining a continuous return of water to the ocean. The entire process, known as the *hydrologic cycle,* maintains a definite water balance in the atmosphere. Despite the seemingly large amount of water vapor in the atmosphere as a whole, it is interesting to note that if all this water were to be precipitated out, it would add a layer only about 1 inch thick to the oceans.

Relationship between Temperature and Humidity

We have already noted that the atmospheric water content is closely related to temperature. Warm air can, and usually does, contain more water vapor than cool air.

CAPACITY. The amount of water vapor that can be contained in the air depends directly and only upon the temperature of the surrounding air. The maximum quantity of vapor that can be supported at a given temperature is called the capacity. Usually, an insufficient supply of water vapor exists for the capacity to be reached. When the capacity is achieved, the air is said to be *saturated.*

SATURATION. From the discussion above it appears that there are two methods of accomplishing saturation in the atmosphere: (1) If the temperature, and hence the capacity, remains the same, saturation may be brought about by increasing the amount of water vapor in the air through evaporation from some source. (2) If the temperature should decrease, then the capacity decreases accordingly until the capacity will just equal the actual amount of water vapor in the air, thereby achieving saturation. Of these the latter, or the decrease of temperature, is the more important natural process of saturating the air. At any time, the difference between the capacity of the air (or saturation value) and the actual water vapor content is called the *saturation deficit.*

DEW POINT. If the temperature should continue to fall below the point at which the air became saturated, there will be an excess amount of water vapor present compared to the capacity of the air at the new lower temperature. Consequently, this excess (represented by the difference between the amount in the air and the new capacity) will change its state to water droplets or ice particles. The temperature at which saturation occurs, or the temperature at which a change from water vapor to liquid water will occur on further cooling, is called the *dew point.* The process by which the water vapor changes to liquid water is called *condensation.*

The dew-point temperature of the air at any time can be obtained by a simple experiment. If a metallic or thin glass container is filled with water containing lumps of ice, the water will become continuously colder as the ice melts. To ensure uniform water temperatures, the liquid should be stirred. When the water and hence the air in contact with the container are cooled to the dew point, beads of water (sweat) will form on the outer surface of the glass. Thus, if the temperature of the ice water is taken the moment the first droplets of water appear, the thermometer reading will be the dew point of the air.

The dew point of rising air decreases with altitude at the rate of about 1°F for every 1,000 feet (2°C per kilometer), since the water vapor concentration per unit volume decreases as the air expands.

A great value of this dew-point property of the air is its constancy. As long as the water content of a given quantity of the air remains constant, the dew point of that air will remain practically constant. In a later consideration of weather analysis, the importance of this conservative property will be referred to again.

Since the cooling of the air is all-important in determining humidity changes, we may note briefly the causes of this temperature decrease. Any one or a combination of the following is effective in reducing the temperature of the air:

1. The air may be cooled adiabatically by rising and expanding.
2. It may be cooled by contact with a cold surface beneath.
3. The mixing of warm and cold air masses results in a lowering of the temperature of the warmer mass.
4. Radiation by the air itself results in cooling of the air.

The actual results dependent on the above processes will be considered in a later chapter.

RECIPROCAL TEMPERATURE-HUMIDITY RELATIONSHIP. Although at any time and place the humidity of the air varies with local temperature changes, it is important to remember the extremely important control that water vapor exerts on the total heat balance of the atmosphere. This importance is twofold, involving both the tremendous greenhouse effect of water vapor in absorbing long-wave reradiation of the earth and the transfer of heat through the release of latent heat of condensation gained at the time of evaporation. In the latter process, heat gained in tropical regions can be released in the condensation process to warm the atmosphere in the higher latitudes. This is particularly important in the winter season when little direct insolation is available in the higher latitudes.

Distribution of Water Vapor in the Atmosphere

The distribution of water in the atmosphere is directly related to the distribution of temperature. Atmospheric water vapor decreases in a nonuniform way from the equator to the poles in both hemispheres as a consequence of the latitudinal temperature gradient described in Chap. 3.

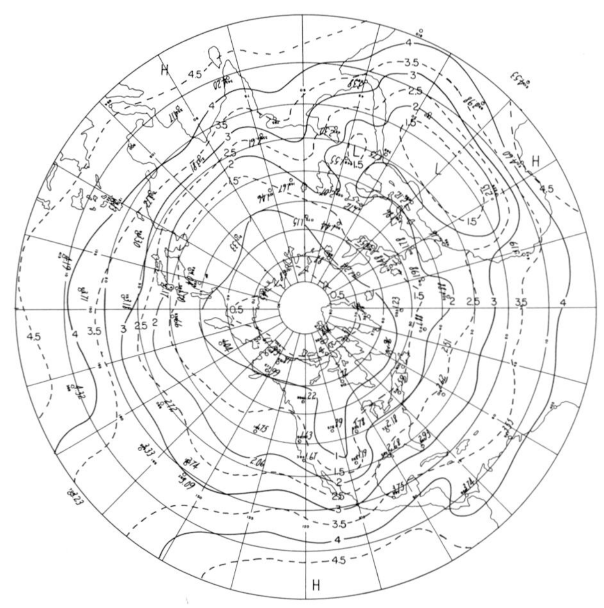

Fig. 4 • 1 Distribution of precipitable water vapor in the Northern Hemisphere shown by lines of equal water vapor content in units of grams per square centimeter. (J. Peixoto, Scientific Report 3, Massachusetts Institute of Technology General Circulation Project)

This water vapor gradient is clearly shown in the latitudinal precipitation statistics which show for example that precipitation in the equatorial belt of the Northern Hemisphere averages about 73 inches a year whereas in the polar zone it is only about 4.5 inches a year.

The horizontal moisture distribution also varies according to the nature of the underlying surface. Thus, marine air is commonly at about 80 percent of saturation while continental air, as over deserts, may be only 20 percent saturated. Again, just as temperature varies with seasons as well as with location, so the water vapor density also changes within relatively wide limits. One need only recall the difference between a warm, humid summer day and the cold, dry days of winter to appreciate this.

Data for the construction of charts of water vapor distribution are not nearly as easy to obtain as those for surface temperatures shown in Figs. 3 · 8 to 3 · 10. Using the best available data, scientists at the Massachusetts Institute of Technology have made a very complete study for the year 1950 for the Northern Hemisphere. The distribution and amount of "precipitable" water vapor (Fig. 4 · 1) clearly show the effect of land and water on both the values and shapes of the lines of equal water vapor content.

Although water vapor is much lighter than air, the combination of higher temperatures and the ocean source tends to restrict it to the lower atmosphere, as noted earlier. The upper part of the troposphere is nearly devoid of water vapor. The decrease of water vapor with elevation is quite important in many meteorological processes, and the way in which this decrease occurs can be seen in Table 4 · 1.

Table 4 · 1 Decrease of Water Vapor with Elevation

HEIGHT, km	(ft)	WATER VAPOR CONTENT, %
0	0	1.3
1	3,281	1.0
2	6,562	0.69
3	9,843	0.49
4	13,124	0.37
5	16,405	0.27
6	19,686	0.15
7	22,967	0.09
8	26,248	0.05

Evaporation

No discussion of humidity would be complete without a consideration of the process by which the air acquires its water content. *Evaporation* is the process by which water in a liquid state is changed to vapor. The water vapor in the air is obtained by evaporation from the surface waters of the earth. Subsequent condensation and precipitation return this water to the earth, completing a continuous cycle.

However, evaporation does not take place at a constant rate, regardless of the supply of surface water available. There are many factors that retard or promote the rate and amount of evaporation.

FACTORS AFFECTING EVAPORATION. 1. *Temperature.* The rate of evaporation varies directly with the temperature of the water. As the water temperature increases, the vapor pressure of the water, or the ability of the water particles to "fly off" into the air, increases rapidly. It is common knowledge that hot water will evaporate faster than cold water.

2. *Relative Humidity.* When the air above the water is dry or has a low relative humidity (see p. 53), evaporation will clearly be greater than when air with a high relative humidity overlies the water surface.
3. *Wind.* Wind is an important aid in evaporation in that it replaces the moist air near the water with dry air. A minimum wind velocity is required to remove the moist air completely. Any further increase in velocity is of no greater value. Further, over ocean surfaces the more the vertical gustiness of the wind, the greater will be the evaporation. Obviously, simple horizontal air motion will result only in the transposition of moist air above the ocean surface, without the introduction of drier air from aloft.
4. *Composition of Water.* Evaporation varies inversely with the salinity of the water, proceeding at a greater rate from fresh water than from salt water. Under equivalent conditions, ocean water will evaporate about 5 percent more slowly than fresh water.
5. *Area of Evaporation.* If two volumes of water are equal, evaporation will be greater for the one having the larger exposed surface.

LATENT HEAT OF VAPORIZATION. Whenever water evaporates, a large amount of heat is absorbed. This heat energy is required for the change of state from liquid to gas. Since this heat is used only to effect the transition to a vapor state and has no effect on the temperature of either the liquid or the vapor, it is known as *latent heat.* When water reaches the boiling point, it remains at a temperature of 100°C (or 212°F), until all the water has boiled off. The heat absorbed after the liquid reached 100°C is employed in the change of state, and the resulting steam or water vapor is also at a temperature of 100°C.

We are all familiar with the cooling effect produced by evaporation. Why are we afforded relief in hot humid weather by fanning, if more hot air is brought in contact with the skin? Fanning (wind motion) stimulates evaporation. The latent heat absorbed in this process cools the skin. In the same way a porous earthenware water jug remains cool in hot weather owing to the loss of heat from the evaporation at its moist surface. Specifically, the latent heat of vaporization is the heat necessary to change one gram of water under normal conditions at 100°C to one gram of vapor at 100°C, and equals 540 calories per gram of water. At 0°C, this value is close to 600.

It is extremely important to note that when water vapor condenses, this latent heat is liberated. Consequently, during condensation in the atmosphere, heat is released. This process, as noted earlier, is very important in the heat balance of the atmosphere.

COMPARISON OF EVAPORATION WITH PRECIPITATION. We have already referred to the hydrologic cycle as the mechanism whereby the atmosphere maintains an essentially fixed water vapor budget. We have also noted that water transferred from the oceans to the atmosphere in one region is not necessarily returned in the same amount in that region, but is often returned elsewhere. This is important in understanding the heat as well as the water balances of the atmosphere.

A most instructive way to understand this effect is to consider the difference between evaporation and precipitation ($E - P$) as has been

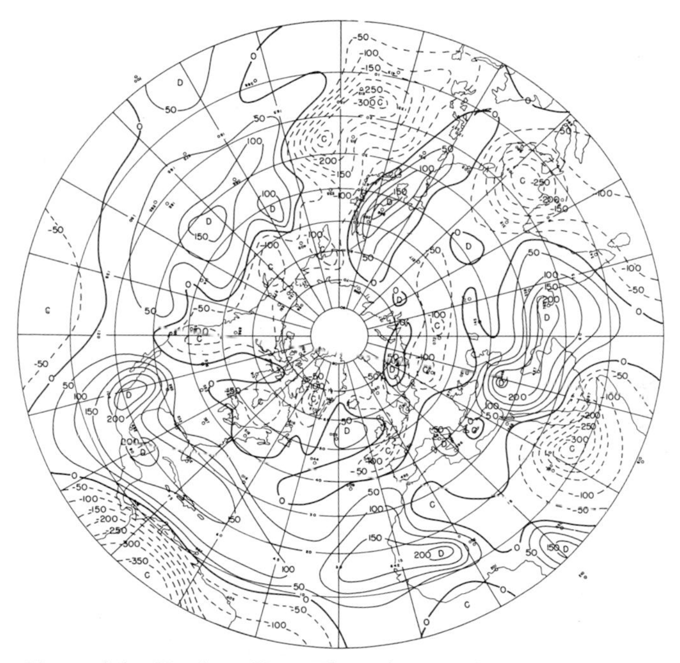

Fig. 4 • 2 Chart of the Northern Hemisphere showing the water budget of the atmosphere in terms of the difference: evaporation − precipitation (E − P). Solid lines indicate evaporation exceeds precipitation by the amounts shown; broken lines indicate precipitation exceeds evaporation by the amounts shown. (H. Lufkin, Scientific Report 4, Massachusetts Institute of Technology General Circulation Project)

done for the year 1950 for the Northern Hemisphere. In Fig. 4 • 2, the solid lines show positive values—where evaporation exceeds precipitation by the amounts indicated—and the broken lines give negative values—where evaporation is less than precipitation by the amounts indicated. The units are centimeters of water per year. Note that the desert regions are positive; hence, moisture is actually transferred from ground to air in such places owing to the low rainfall. The oceans are variable in this effect for reasons that will become clear in later chapters.

Humidity Measurements

ABSOLUTE HUMIDITY. So far, we have considered the quantity of water vapor in the air, but only as it is governed by the temperature. One of the methods of expressing the actual water content of the air is to give the absolute humidity, which is the weight of water vapor in a particular volume of air. More specifically, it is the weight of water vapor per unit *volume* of air. The unit volume generally used is either the cubic foot or the cubic meter. Hence if we extract the amount of water vapor in 1 cubic foot of ordinary air and weigh this water, the result expressed in grains per cubic foot would be the absolute humidity. If cubic meters are used, the weight is given in grams.

We can see that the absolute humidity will vary if the air expands or contracts, even though the water vapor itself is constant in amount. Suppose, for example, that 10 grains of water vapor exist in 1 cubic foot of air. The absolute humidity is then 10 grains per cubic foot. If this air should for any reason (an increase in temperature) expand and occupy 2 cubic feet, the amount of water vapor left in 1 cubic foot would now be only 5 grains, the absolute humidity equaling the 5 grains measurement, although the water content has actually remained the same.

Table 4 · 2 shows the absolute humidity (grains of water vapor per cubic

Table 4 · 2 Absolute Humidity Values for Saturated Air

TEMPERATURE, °F	WATER VAPOR, grains per cu ft	TEMPERATURE	WATER VAPOR, grains per cu ft
0	0.479	50	4.108
5	0.613	55	4.891
10	0.780	60	5.800
15	0.988	65	6.852
20	1.244	70	8.066
25	1.558	75	9.460
30	1.942	80	11.056
35	2.375	85	12.878
40	2.863	90	14.951
45	3.436	95	17.305

NOTE: 1 grain = 0.002 oz or 0.0648 g.

foot of air) for saturated air at different temperatures and indicates strongly the control exerted by air temperatures over humidity.

SPECIFIC HUMIDITY. This is a more constant property of the air and has come into use in meteorology as a result of the variability of absolute humidity. The specific humidity is the weight of water vapor per unit *weight* of *moist* air, expressed as grams of water vapor per kilogram of air, or as grains of water vapor per pound of air. Obviously, if we are to extract and weigh the water vapor in a given weight such as a pound of air, regardless of what happens to the temperature and volume of the air, a pound is still a pound and will contain the same mass of air regardless of any volume changes. We simply include a larger volume of air to obtain a pound, should the air expand between measurements. Only by an actual variation of the water content of the air will the specific humidity change. This too is an extremely important property as regards weather analysis. See page 56 for the method of determining specific humidity.

MIXING RATIO. This is another fairly constant or conservative property of the air that has come into widespread use. It is the weight of water vapor per unit weight of *dry* air and is expressed in units similar to those for specific humidity. The numerical difference between these properties is very small. See page 56 for the method of determining the mixing ratio.

RELATIVE HUMIDITY. Generally, when humidity is mentioned, it is relative humidity to which reference is made. Relative humidity is the ratio of the amount of water vapor in the air to the amount the air can hold at that temperature (or the capacity). This ratio is always expressed as a percentage.

If the air has 40 grains of water vapor per pound and can hold at that temperature 50 grains per pound (the capacity), then the relative humidity is $^{40}/_{50}$, or 80 percent. The relative humidity is therefore a very descriptive atmospheric property. Sticky, muggy days are those with a high relative humidity. Evaporation is very slow since the air is nearly saturated.

It is clear that the relative humidity must change whenever the amount of water vapor in the air changes and whenever the capacity of the air changes. Thus, the relative humidity varies inversely with the temperature. A decrease in temperature causes corresponding capacity decrease. If the capacity decreases, the relative humidity increases as the air is brought nearer the saturation point. When the temperature, hence the capacity, decreases such that the relative humidity is 100 percent, the air will be saturated, and the temperature at which this humidity is reached is the dew point. Further cooling causes condensation.

VAPOR PRESSURE. Although the subject of atmospheric pressure will be examined in a later chapter, we must note here that a small part of the total air pressure at any time and place is composed of the contribution made by the water vapor present. This value, known as the *partial pressure* of water vapor, is a variable depending on the existing amount of this gas. At any temperature there is a maximum vapor pressure or *saturation vapor pressure* corresponding to the saturation values given, for example, in Table 4·2. Another way to consider the saturation deficit, described earlier, is to express it as the difference between the saturation vapor pressure and the partial pressure at a given time and temperature.

The second column in Table 4·4 gives the saturation vapor pressure of

Table 4 · 3 Relative Humidity in Per Cent

Air temp., °F	Depression of wet-bulb thermometer, °F																																		
	1	2	3	4	5	6	7	8	9	10	11	12	13	14	15	16	17	18	19	20	21	22	23	24	25	26	27	28	29	30	31	32	33	34	35
0	67	33	1																																
5	73	46	20																																
10	78	56	34	13																															
15	82	64	46	29	11																														
20	85	70	55	40	26	12																													
25	87	74	62	49	37	25	13	1																											
30	89	78	67	56	46	36	26	16	6																										
35	91	81	72	63	54	45	36	27	19	10	2																								
40	92	83	75	68	60	52	45	37	29	22	15	7																							
45	93	86	78	71	64	57	51	44	38	31	25	18	12	6																					
50	93	87	80	74	67	61	55	49	43	38	32	27	21	16	10	5																			
55	94	88	82	76	70	65	59	54	49	43	38	33	28	23	19	11	9	5																	
60	94	89	83	78	73	68	63	58	53	48	43	39	34	30	26	21	17	13	9	5	1														
65	95	90	85	80	75	70	66	61	56	52	48	44	39	35	31	27	24	20	16	12	9	5	2												
70	95	90	86	81	77	72	68	64	59	55	51	48	44	40	36	33	29	25	22	19	15	12	9	6	3										
75	96	91	86	82	78	74	70	66	62	58	54	51	47	44	40	37	34	30	27	24	21	18	15	12	9	7	4	1							
80	96	91	87	83	79	75	72	68	64	61	57	54	50	47	44	41	38	35	32	29	26	23	20	18	15	12	10	7	5	3					
85	96	92	88	84	81	77	73	70	66	63	59	57	53	50	47	44	41	38	36	33	30	27	25	22	20	17	15	13	10	8	6	4	2		
90	96	92	89	85	81	78	74	71	68	65	61	58	55	52	49	47	44	41	39	36	34	31	29	26	24	22	19	17	15	13	11	9	7	5	3
95	96	93	89	86	82	79	76	73	69	66	63	61	58	55	52	50	47	44	42	39	37	34	32	30	28	25	23	21	19	17	15	13	11	10	8
100	96	93	89	86	83	80	77	73	70	68	65	62	59	56	54	51	49	46	44	41	39	37	35	33	30	28	26	24	22	21	19	17	15	13	12
105	97	93	90	87	84	81	78	75	72	69	66	64	61	58	56	53	51	49	46	44	42	40	38	36	34	32	30	28	26	24	22	21	19	17	15
110	97	93	90	87	84	81	78	75	73	70	67	65	62	60	57	55	52	50	48	46	44	42	40	38	36	34	32	30	28	26	25	23	21	20	18
115	97	94	91	88	85	82	79	76	74	71	69	66	64	61	59	57	54	52	50	48	46	44	42	40	38	36	34	33	31	29	28	26	25	23	21
120	97	94	91	88	85	82	80	77	74	72	69	67	65	62	60	58	55	53	51	49	47	45	43	41	40	38	36	34	33	31	29	28	26	25	23
125	97	94	91	88	86	83	80	78	75	73	70	68	66	64	61	59	57	55	53	51	49	47	45	44	42	40	38	37	35	33	32	30	29	27	26
130	97	94	91	89	86	83	81	78	76	73	71	69	67	64	62	60	58	56	54	52	50	48	47	45	43	41	40	38	37	35	33	32	30	29	28

Table 4·4 Temperature of Dew Point

Air temp., °F	Vapor pressure, in.	Depression of wet-bulb thermometer, °F																																	
		1	2	3	4	5	6	7	8	9	10	11	12	13	14	15	16	17	18	19	20	21	22	23	24	25	26	27	28	29	30	31	32	33	34	35
0	0.0383	−7	−20																																	
5	0.0491	−1	−9	−24																																
10	0.0631	5	−2	−10	−27																															
15	0.0810	11	6	0	−9	−26																														
20	0.103	16	12	8	2	−7	−21																													
25	0.130	22	19	15	10	5	−3	−15	−51																											
30	0.164	27	25	21	18	14	8	2	−7	−25																										
35	0.203	33	30	28	25	21	17	13	7	0	−11	−41																								
40	0.247	38	35	33	30	28	25	21	18	13	7	−1	−14																							
45	0.298	43	41	38	36	34	31	28	25	22	18	13	7	−1	−14																					
50	0.360	48	46	44	42	40	37	34	32	29	26	22	18	13	8	0	−13																			
55	0.432	53	51	50	48	45	43	41	38	36	33	30	27	24	20	15	9	1	−12	−59																
60	0.517	58	57	55	53	51	49	47	45	43	40	38	35	32	29	25	21	17	11	4	−8	−36														
65	0.616	63	62	60	59	57	55	53	51	49	47	45	42	40	37	34	31	27	24	19	14	7	−3	−22												
70	0.732	69	67	65	64	62	61	59	57	55	53	51	49	47	44	42	39	36	33	30	26	22	17	11	2	−11										
75	0.866	74	72	71	69	68	66	64	63	61	59	57	55	54	51	49	47	44	42	39	36	32	29	25	21	15	8	−2	−23							
80	1.022	79	77	76	74	73	72	70	68	67	65	63	62	60	58	56	54	52	50	47	44	42	39	36	32	28	24	20	13	6	−7	−53				
85	1.201	84	82	81	80	78	77	75	74	72	71	69	68	66	64	62	61	59	57	54	52	50	48	45	42	39	36	32	28	24	19	12	3	−12		
90	1.408	89	87	86	85	83	82	81	79	78	76	75	73	72	70	69	67	65	63	61	59	57	55	53	51	48	45	43	39	36	32	28	24	19	11	1
95	1.645	94	93	91	90	89	87	86	85	83	82	80	79	78	76	74	73	71	70	68	66	64	62	60	58	56	54	52	49	46	43	40	37	33	29	24
100	1.916	99	98	96	95	94	93	91	90	89	87	86	85	83	82	80	79	77	76	74	72	71	69	67	65	63	61	59	57	55	52	50	47	44	41	37
105	2.225	104	103	101	100	99	98	96	95	94	93	91	90	89	87	86	84	83	82	80	78	77	75	74	72	70	68	67	65	63	61	58	56	54	51	48
110	2.576	109	108	106	105	104	103	102	100	99	98	97	95	94	93	91	90	89	87	86	84	83	81	80	78	77	75	73	72	70	68	66	64	62	60	57
115	2.975	114	113	112	110	109	108	107	106	104	103	102	101	99	98	97	96	94	93	92	90	89	87	86	84	83	81	80	78	76	75	73	71	69	67	65
120	3.425	119	118	117	115	114	113	112	111	110	108	107	106	105	104	102	101	100	98	97	96	94	93	92	90	89	87	86	84	83	81	80	78	76	75	73
125	3.933	124	123	122	121	119	118	117	116	115	114	112	111	110	109	108	106	105	104	103	101	100	99	97	96	95	93	92	90	89	88	86	84	83	81	80
130	4.504	129	128	127	126	124	123	122	121	120	119	118	116	115	114	113	112	110	109	108	107	106	104	103	102	100	99	98	96	95	94	92	91	89	88	86

water over a broad temperature range and once again emphasizes the dependence of water vapor composition on air temperature. The units of vapor pressure given here are in inches of mercury, which will be explained further in the chapter on atmospheric pressure. Another way of expressing relative humidity is to consider it as the ratio of partial pressure to saturation vapor pressure.

INTERRELATIONSHIP OF HUMIDITY MEASUREMENTS. Direct determinations of many of the humidity measures just described involve laborious laboratory procedures which would not be feasible in meteorological analysis. However, meteorological theory provides a convenient numerical relationship of several of these measures so that the more difficult ones can be obtained from those easier to determine. For example, the relative humidity can be evaluated very quickly by means of the psychrometer (described in the following section) and Table 4·3. From the relative humidity, the partial pressure e is determined, because as noted in the last paragraph, relative humidity is equal to the ratio of the partial pressure e to the saturation vapor pressure E. The latter is obtained from a table like Table 4·4 after the temperature is observed. Thus,

$$RH = \frac{e}{E} \times 100$$

and

$$e = \frac{RH \times E}{100}$$

(The multiplication and division by 100 is required in these formulas because relative humidity is expressed in percent.)

Once the partial pressure of water vapor is obtained from the above simple relationship, other quantities such as the absolute humidity, the specific humidity, and the mixing ratio can also be obtained quickly from the following formulas:

$$\text{Absolute humidity} = 217e/T \text{ in grams per cubic meter}$$

where T is absolute temperature

$$\text{Specific humidity} = 623e/p \text{ (more exactly, } 623e/p - 0.377e) \text{ in grams per kilogram}$$

where p is atmospheric pressure, and

$$\text{Mixing ratio} = 623e/p - e \text{ in grams per kilogram}$$

Humidity Instruments

The measurement of humidity in the air is known as *hygrometry,* and the instruments used fall under the general name of *hygrometers.* Many types of hygrometers have been developed for the purpose, but only the common types will be considered.

PSYCHROMETER OR WET- AND DRY-BULB HYGROMETER. The most common of the humidity instruments is a hygrometer consisting of a support to which are attached two ordinary, accurate, mercury thermometers. One of the thermometers has a thin layer of muslin wrapped around the bulb, which is kept wet when the instrument is in use. This is therefore called the *wet-bulb* thermometer and the other the *dry-bulb* thermometer. The muslin should never be wet with salt water.

The dry-bulb thermometer will show the current air temperature. However, as the moisture in the muslin of the wet bulb evaporates, latent heat is absorbed by the evaporating moisture, causing the temperature of the wet-bulb thermometer to fall. The faster the rate of evaporation, the more heat is absorbed from the covered wet bulb and the lower the wet-bulb reading falls below that of the dry. Now the rate of evaporation always depends on the degree of saturation of the air, which is relative humidity. Hence the difference in temperature between the wet- and dry-bulb thermometers is a measure of relative humidity.

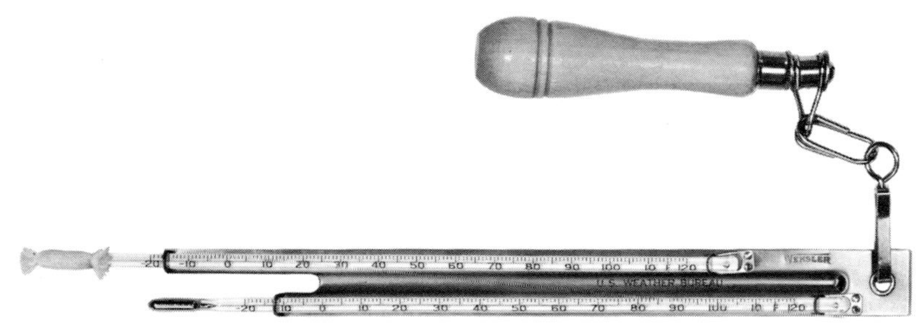

Fig. 4 · 3 A type of sling psychrometer. (Belfort Instrument Company)

To ensure proper evaporation from the wet-bulb thermometer, the air surrounding it should be continuously replaced. Otherwise evaporation would slow down as the stagnant air surrounding the wet bulb becomes saturated. Therefore the psychrometer should be fanned to replace this air. Readings of the wet bulb should be taken at intervals of 10 to 20 seconds until two successive readings of the wet bulb show the same temperature.

To avoid fanning, an instrument called the *sling psychrometer* is used. Here, the whole instrument is whirled rapidly until the lowest wet-bulb reading is obtained. Another convenient psychrometer allows air to be drawn over the thermometer bulbs by means of hand pressure on a rubber bulb. This avoids the whirling, which is often difficult to do in confined quarters, such as cargo holds (see Fig. 4 · 4). *Psychrometry* is another name commonly applied to the process of humidity measurement.

From these facts, tables have been developed to obtain both the relative humidity and the dew point of the air (see Tables 4 · 3 and 4 · 4). It is simply necessary to note the dry-bulb reading and the difference between the wet and the dry, known as the *depression* of the wet bulb, to obtain these properties of the air from the tables.

An example indicates how the tables are to be used: Assume the dry bulb to read 80°F. Assume the wet bulb after whirling or fanning to read 70°F. Then the depression of the wet bulb is 10°F.

In the left-hand column of Table 4 · 3, we find 80°F for the dry-bulb reading. Across the top of the table we find the difference of 10°F. This yields a reading of 61 percent for the relative humidity.

Table 4 · 4 gives dew-point figures using the same dry bulb and depression of the wet-bulb readings. For the figures given above, the dew point is 65°F. Relative humidity and dew point for intervening values in Tables 4 · 3 and 4 · 4 can be found by simple interpolation.

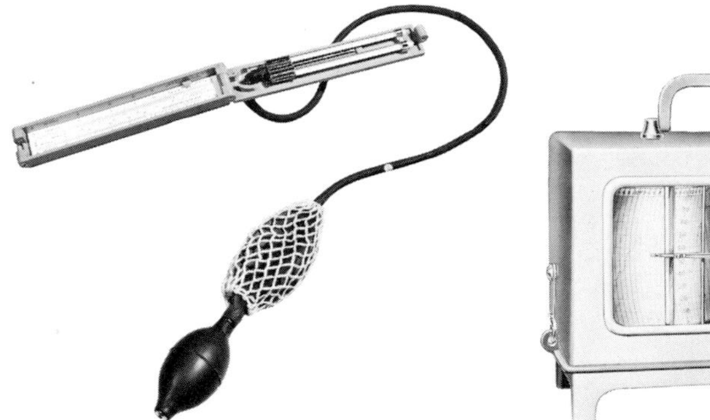

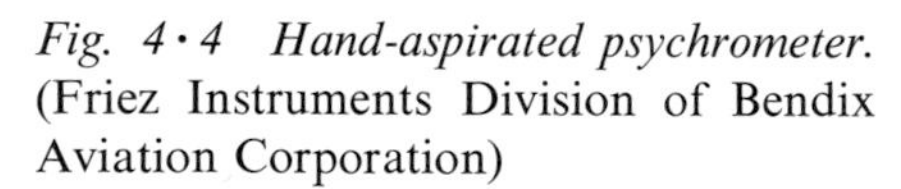

Fig. 4 · 4 Hand-aspirated psychrometer. (Friez Instruments Division of Bendix Aviation Corporation)

Fig. 4 · 5 Hygrograph. (Belfort Instrument Company)

HYGROGRAPH. This instrument makes a continuous record of humidity. The sensitive element is a bundle of ordinary blond human hair which expands or contracts in length as the humidity increases or decreases. By means of delicate springs and levers, this change in the length of the hair is communicated to a long pen arm which inscribes a trace on a rotating drum. In appearance the instrument resembles the thermograph. The graph chart is calibrated vertically in relative humidity from 0 to 100 percent. In Fig. 4 · 6, records from a thermograph and hygrograph are compared for the same location and time. The inverse relationship between temperature and relative humidity is very striking in this illustration.

HYTHEROGRAPH. The hytherograph is commonly observed on merchant ships in connection with Cargocaire units. This instrument is a combination of the thermograph and the hygrograph, the two units being placed in the same case. The chart encircling the revolving drum is divided into two

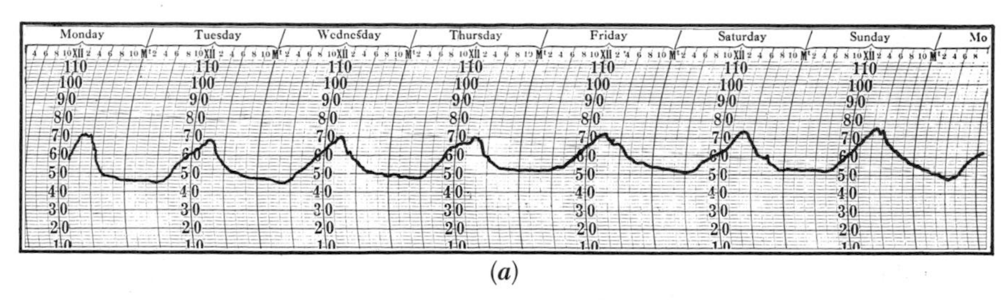

(*a*)

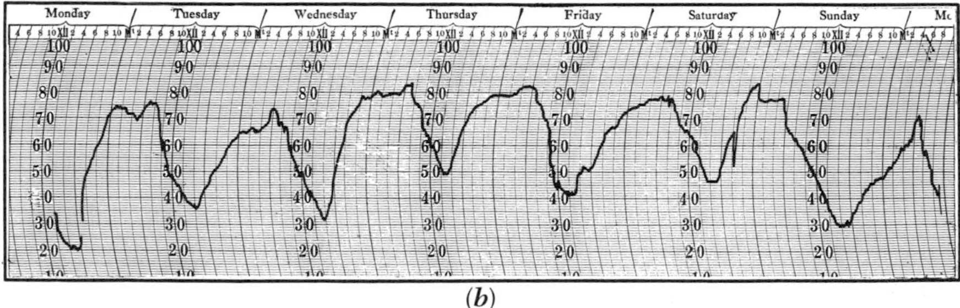

(*b*)

Fig. 4 · 6 Comparison of thermogram (a) and hygrogram (b) for same week. Note the inverse relationship between times of temperature and relative humidity extremes.

horizontal sections, the upper indicating temperature and the lower showing the relative humidity. The pens attached to the temperature and humidity elements inscribe their traces on the upper and lower sections of the chart, respectively (Fig. 4 · 7).

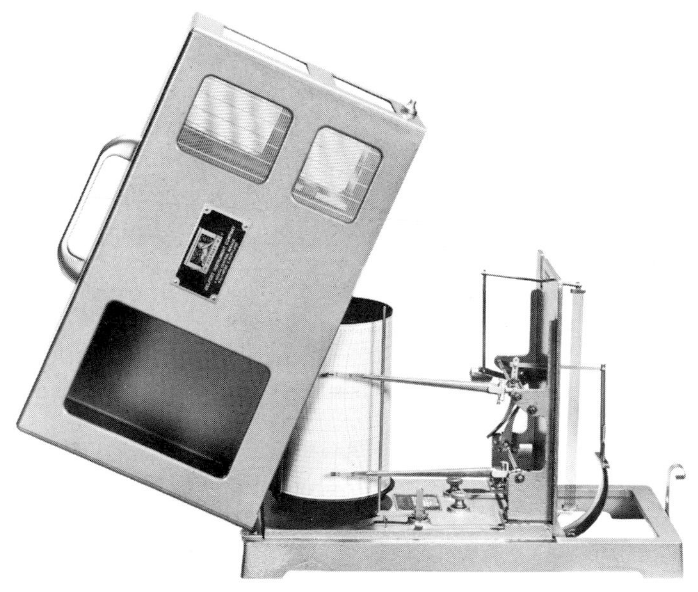

Fig. 4 · 7 Hytherograph. (Friez Instruments Division of Bendix Aviation Corporation)

Humidity and Cargo Ventilation

To cope with the problem of ventilating cargo holds, a knowledge of humidity and its measurement is of utmost value. Although a full treatment of the subject is given in texts dealing with the handling and stowage of cargo, some of the more essential points involving humidity can be considered here.

Varied climates are encountered by vessels in following different trade routes. Marine air is naturally more humid than continental air in the same latitude owing to the large amounts of water available for evaporation. Consequently, vessels in the tropics encounter air with a very high absolute or specific humidity as a result of the high capacity of the hot tropical air. The air in the higher middle latitudes will have a much lower humidity.

In proceeding from one climate to another of contrasting type, very significant temperature and humidity changes are encountered, which, if not properly anticipated and understood, may seriously affect the cargo. Sweating or condensation of moisture from saturated air is one of the worst evils involved. From our discussion of humidity, it is clear that, if tropical air is admitted to a hold still chilled from a previously visited colder region, the entering air may be cooled rapidly to and below the dew point. The resulting moisture will be deposited directly on the cargo, as well as on the interior surfaces of the hold, and will later drip down onto the cargo.

A vessel loaded in the tropics and steaming to higher latitudes should have the holds ventilated frequently. Otherwise the air there will gradually become chilled and saturated. Since such air contains a naturally large amount of water vapor, it will release large amounts of "sweat" on the cargo and on the cooling bulkheads. If the air is changed very suddenly, a similar effect will result from the rapid mixture of warm and cold air.

The passage of a vessel across warm and cold ocean currents may produce humidity changes similar to those experienced in sailing through marked differences in latitude.

EXERCISES

4 · 1 Explain the relationship between temperature, capacity, and relative humidity.

4 · 2 Why is the dew point a conservative property of the atmosphere?

4 · 3 (*a*) How is the distribution of atmospheric moisture related to the air temperature? (*b*) Construct a graphical profile showing the change of water vapor from the equator to the poles along the 120°W meridian. Use Fig. 4 · 1 for the data source, and compare the result with that in answer to question 3 · 5.

4·4 (*a*) From the data in Table 4·1, plot a curve showing the variation of water vapor with elevation and discuss the rate at which this variation takes place. (*b*) Why does this change occur when the other components of the atmosphere show little percentage variation in the lower atmosphere?

4·5 In general, in what region does evaporation exceed precipitation, and the reverse?

4·6 (*a*) What is the saturation water vapor pressure in inches of mercury at 50°F? (*b*) If the partial water vapor pressure is 0.180 inch at 50°F, what is the relative humidity? (*c*) If the relative humidity is 65 percent at a temperature of 70°F, what is the partial pressure? What is the absolute humidity in grains per cubic foot (Tables 4·2 and 4·4)?

4·7 (*a*) Explain the principle of the sling psychrometer. (*b*) What values can be determined directly, and what can be determined by simple computation using this instrument?

4·8 What are the chief factors that control the rate of evaporation?

CHAPTER 5

Vertical Equilibrium of the Atmosphere

The relationship between heat and mechanical energy involves some of the most fundamental laws of nature (the laws of thermodynamics). The atmosphere, which is often referred to as a great heat engine, has its behavior strongly controlled by these laws as reflected in changes of temperature, pressure, density, and water vapor. In this chapter we shall be mostly concerned with conditions related to the variation of these properties along the vertical, and we shall examine the important equilibrium aspects of the atmosphere. The subjects to be studied are basic to a real understanding of the behavior of the atmosphere.

Lapse Rate and Adiabatic Lapse Rates

The relationship between the lapse rate and the adiabatic lapse rates is extremely critical in the determination of the vertical equilibrium conditions to be described in this chapter. We must therefore reexamine these temperature properties in somewhat greater depth.

VARIATIONS OF THE LAPSE RATE. We noted in Chap. 3 that the lapse rate varies with both time and location. It is also important to note that the lapse rate may vary along the vertical at a given time and place. The variations may be abrupt, as in the case of temperature inversions, or they may be gradual. Figure 5 · 1, which is a graph of temperature against altitude, illustrates a number of possible lapse rates. Curve *A* represents a constant lapse rate. Curve *B* illustrates a lapse rate that grows smaller with increase in altitude, while curve *C* indicates a lapse rate that increases with altitude. Curve *D* depicts a lapse rate that is constant and the same above and below an inversion level, and *D′* shows a steeper lapse rate above the inversion level. Curve *E* indicates a lapse rate that begins with a ground inversion, a common occurrence when the ground cools rapidly at night, thereby cooling the air near the ground to a temperature below that at a slightly greater altitude.

DRY AND MOIST ADIABATIC LAPSE RATES. The adiabatic lapse rate of

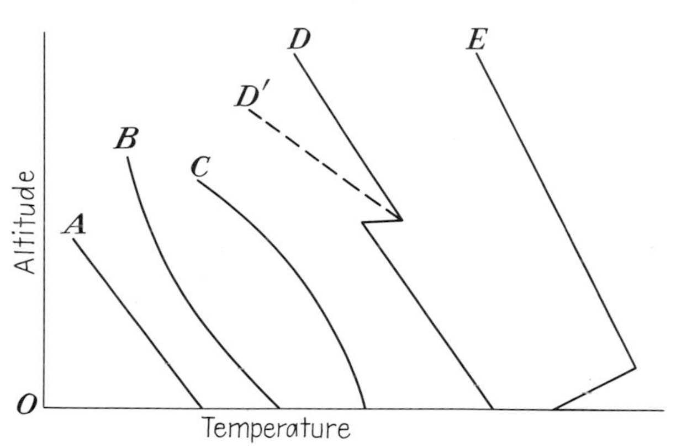

Fig. 5 · 1 Examples of commonly occurring lapse rates.

5.5°F per 1,000 feet described in Chap. 3 applies to the temperature change in dry air having a vertical component of motion. Dry air in this case means air in which the relative humidity is less than 100 percent. When air is saturated, having a relative humidity of 100 percent, adiabatic temperature changes (resulting from the expansion or the contraction of air) occur at a different and somewhat variable rate depending on temperature, pressure, elevation, and moisture content. Saturated air is also termed *wet* or *moist* air.

Assume that a mass of moist air is set in vertical motion by any of the conditions explained earlier. As the moist air rises, expansion results in cooling, and cooling of saturated air causes condensation. But in the process of condensation, latent heat is always liberated by the condensing water vapor. Consequently, as saturated air rises and cools adiabatically, latent heat is returned to the air as long as condensation continues. The amount of heat liberated depends on the amount of condensation, which may vary. On the average, near the ground rising moist air liberates heat at a rate which warms the air by 2.3°F per 1,000 feet. However, rising air cools normally at a rate of 5.5°F per 1,000 feet. If saturated, however, the moist air regains 2.3° from condensation and leaves a net cooling effect of 3.2°F per 1,000 feet. The average rate of cooling for saturated air in the low atmosphere is thus 3.2°F per 1,000 feet.

The saturated adiabatic rate is sometimes referred to as the *moist,* or *pseudo,* adiabatic rate. It is emphasized that this rate may show considerable variation. In the upper portion of the troposphere the water vapor content is so low, as is evident in Table 4 · 2, that the latent heat released in condensation becomes almost negligible. Also, at very low tempera-

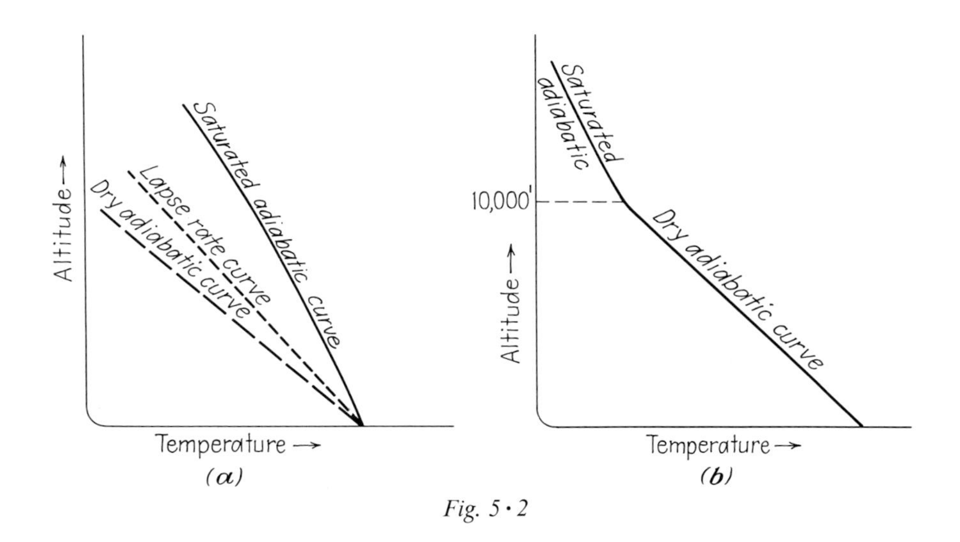

Fig. 5 · 2

tures, such as in high latitudes, the release of latent heat during condensation is very small. Thus, at high elevations and low temperatures, the saturated and dry adiabatic lapse rates are nearly equivalent.

When moist air subsides, the resulting adiabatic warming may follow either dry or moist conditions. If the water droplets are carried downward with the air, without evaporating, the air warms according to the dry adiabatic lapse rate. But if evaporation occurs, latent heat is absorbed from the air and the rate of warming decreases by an amount that depends on the amount of evaporation.

From a consideration of the two rates, it is clear that a mass of air can and does cool according to both the dry and moist adiabatic rates along different portions of its path. A parcel of air may be originally dry before uplift begins, regardless of the cause of the uplift. The rising air then cools according to the dry adiabatic rate. If the air rises sufficiently, it will ultimately cool to the dew point and become saturated. Further rising will result in cooling according to the saturated or moist adiabatic rate. Hence, rising moist air within clouds usually exhibits pseudo, or moist, adiabatic conditions.

ADIABATIC AND LAPSE RATES COMPARED. From the values given, we note that the dry adiabatic rate is greater than the saturated adiabatic or the normal lapse rates. The normal lapse rate is only slightly more than the low-level moist adiabatic. Remember, however, that the lapse rate varies widely with time and location, being sometimes less than the saturated adiabatic and at other times more than the dry adiabatic. We shall see later in this chapter just how the erratic behavior of the lapse rate affects atmospheric equilibrium.

The graphic relationship of these three temperature rates often aids in understanding them. In Fig. 5 · 2, altitude is shown on the vertical axis, increasing from sea level upward; temperature is shown on the horizontal axis, increasing to the right. In (*a*), the dry adiabatic has the smallest slope, showing the greatest temperature change with altitude. The adiabatic curve in (*b*) shows the case in which dry air, after rising 10,000 feet, becomes saturated and then cools as shown by the saturated adiabatic curve. Note that the saturated adiabatic curves in Fig. 5 · 2 show a change in slope because of the decrease in water vapor content with increasing clevation.

Stable and Unstable Equilibrium

Air which is at rest as regards motion along the vertical (or along a vertical component) is considered to be in vertical equilibrium or balance. As noted in Chap. 3, air may be given vertical motion as the result of (1) ground temperature differences, (2) frontal effects, (3) topographic irregularities, and (4) convergence and divergence. The rate and amount of vertical motion that develop depend to a great extent on the type of equilibrium prevailing in the air. Although equilibrium here is a mechanical balance between physical forces, the nature of the balance is determined by thermal conditions in the atmosphere, in particular the relation between the lapse rate and the adiabatic lapse rate. An analogy with more familiar objects may be considered first to help clarify the meaning of equilibrium as used here.

Consider two unbroken pieces of blackboard chalk at rest on a table. One piece may be lying flat on the table, and the other may be standing vertically on its small circular cross section. In both cases the pieces of chalk are in equilibrium with the upward and downward forces acting on them. If one end of the flat-lying chalk is raised slightly and released, it promptly returns to its original position. This relationship between the chalk and related forces is said to be a stable one, or the chalk is in stable equilibrium. However, if the chalk standing vertically is tilted and released, rather than returning to the vertical it falls at an accelerating rate until it lies flat on the table. The vertical arrangement is considered to be unstable or, although at rest, the chalk is in unstable equilibrium. Any number of similar examples can be imagined. Let us now consider the equivalent condition in the atmosphere.

STABLE AIR. Assume that the surface air is at a temperature of 70°F, with a lapse rate of 4°F per 1,000 feet existing. Imagine that a parcel of this 70° air is forced into vertical ascent as shown in Fig. 5 · 3. After 1,000 feet of uplift, the rising air cools to 64.5°F, while the surrounding quiet air, cooling at the vertical lapse rate, cools to only 66°F. Thus the rising air is colder than the adjacent air at the same level and must sink downward,

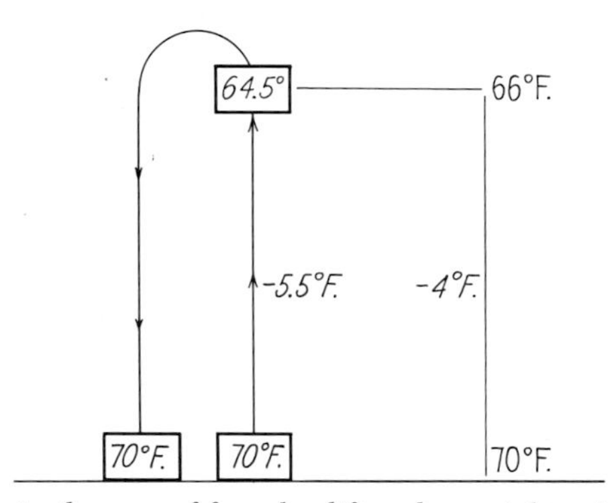

Fig. 5 · 3 Stability in the case of forced uplift such as might exist when air is forced to rise over a mountain or other topographic irregularity.

until its *temperature*, and hence its *density*, equal that of the air at some lower level. Where is this level? Obviously, it is back at the ground, where the upward motion originated and where the temperatures and densities are the same. Theoretically, this parcel of air would never start to rise at all unless forced, for the slightest ascent would cause it to become colder and heavier than the surrounding air. Such air, with adiabatic and existing lapse-rate relations such as to cause a resistance to vertical displacement, is said to be *stable*. Notice that the real or existing lapse rate in this case is the lower of the two rates.

Consider another case. On a warm summer day, with the surface air at 80°F, a parcel of air becomes relatively overheated and reaches 90°F. Assume the lapse rate to be 3.5°F per 1,000 feet on this occasion. Clearly the warmer air will rise and expand, thereby cooling adiabatically, while the surrounding quiet air cools at the current lapse rate of 3.5°F per 1,000 feet. Thus the rising air will cool faster than the surrounding air, for the lapse rate is here also less than the adiabatic rate. At some level the ascending air will therefore overcome the 10° initial difference and will then cease to rise.

This is illustrated in Fig. 5 · 4. Note that the temperature difference between the rising and the adjacent quiet air grows steadily less, until it is nil at 5,000 feet.

This case illustrates again the behavior of the atmosphere when stable equilibrium conditions prevail. Even though a local air parcel rises from its original position as a result of receiving an energy increase as its temperature rises, this air again reaches a state of stable equilibrium at a new level.

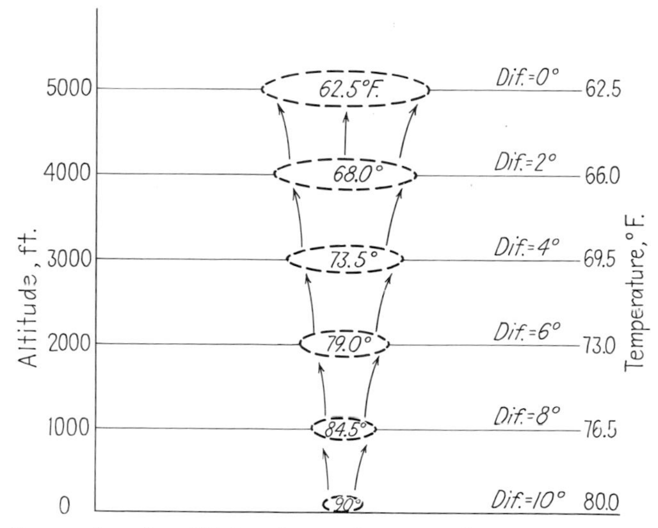

Fig. 5 · 4 Restoration of equilibrium in air when vertical motion is initiated by surface temperature differences and when stable lapse-rate conditions prevail.

As we shall see, this would not occur if the atmosphere were in an unstable equilibrium state.

From these cases we conclude that whenever the *lapse rate* is *less* than the *adiabatic rate,* the air is stable. Air that is displaced vertically for mechanical reasons (fronts or mountains) tends to resist displacement or return to its original position. Air that is displaced through temperature differences soon reaches the temperature of the surrounding air at some new level and comes to rest.

UNSTABLE AIR. We can illustrate unstable air conditions in the same manner as above. As a first case, we assume the lapse rate to be greater than the adiabatic (Fig. 5 · 5), being 7°F per 1,000 feet, with the surface air temperatures still at 70°. Suppose we now permit a parcel of the 70° air to rise and thus cool adiabatically. But the lapse rate is here greater than the adiabatic rate, causing the surrounding air to become increasingly colder with reference to the rising air at increasingly higher levels. Consequently, the rising air becomes relatively warmer at each new level, accelerating its upward velocity.

Let us also consider the case in which an initial temperature difference exists, with a parcel of air warming to 90°F, 5° more than the adjacent air. The lapse rate is again 7°F per 1,000 feet. Note that in this case the rising air temperatures do not overtake those of the surrounding atmosphere. Rather, the difference at each level becomes increasingly greater, causing the ascending air to become increasingly warmer than the surrounding air, resulting in an acceleration of the rising air (Fig. 5 · 6). Here again is an example of instability in the atmosphere, caused simply by the

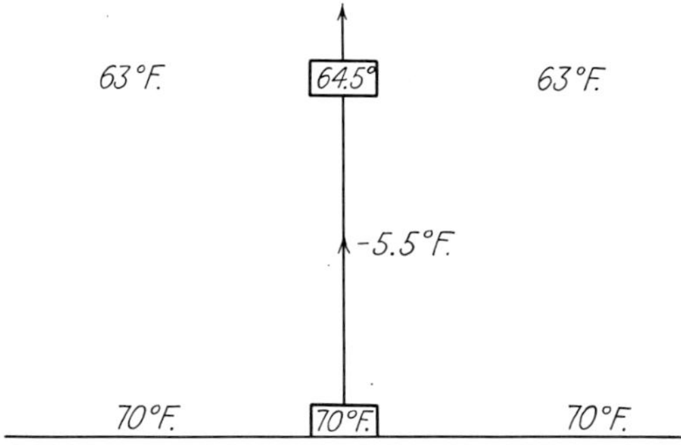

Fig. 5 · 5 Instability in the case of forced uplift such as may occur over mountains.

fact that the lapse rate is greater than the adiabatic rate. Thus, air is unstable whenever the *lapse rate* is *greater* than the *adiabatic rate.* In this case rising or falling air always shows a greater temperature difference than the adjacent quiet air at any new level, and thus cannot come to rest until a new lapse rate is encountered.

CONDITIONAL INSTABILITY. To complete our discussion of stability, we should also consider the results when air is saturated, with the moist or saturated adiabatic rate prevailing instead of the dry. The average saturated adiabatic is close to 3.2°F per 1,000 feet. In the example above for stable air, we considered the lapse rate to be 4°F per 1,000 feet, or less than the dry adiabatic of 5.5°F per 1,000 feet. This relationship yielded stable air. But for saturated air, the lapse rate now *exceeds* the adiabatic rate; hence this air is unstable. The reader should prove this to himself by developing an example similar to those used to illustrate stability and instability above.

Whenever the existing *lapse rate* lies *between* the values for the *wet* and the *dry adiabatics,* whether or not the air is stable depends on whether the air is dry or saturated. *The air is then said to be conditionally unstable.* If the air is initially dry, it is then stable, but should the air for any reason become saturated, it will immediately assume an unstable condition.

A column of rising air may be stable and unstable at different elevations as a result of condensation after sufficient uplift and cooling.

In Fig. 5 · 7 the ascending air column is stable up to 2,000 feet, for its

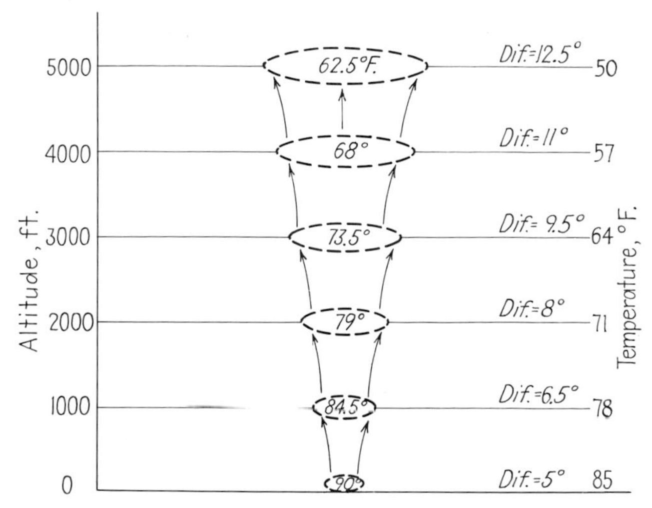

Fig. 5 · 6 Loss of equilibrium in air when vertical motion is initiated by surface temperature differences and when unstable lapse-rate conditions prevail.

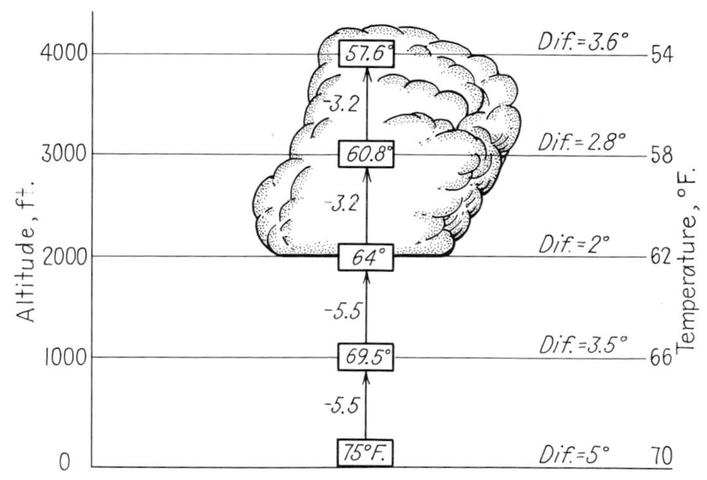

Fig. 5 · 7 Conditional instability in an air column where stability exists below the original dew-point level (2,000 feet) and instability above.

temperature is slowly overtaking the surrounding air temperature. But at 2,000 feet the air has cooled to its dew point, with condensation forming a cloud as the air rises above this level. The air now cools at the saturated adiabatic, which is here exceeded by the lapse rate, resulting in the moist air becoming unstable above 2,000 feet. This is clearly shown by the increasing temperature differences between rising and surrounding air at corresponding levels.

It is evident that whenever the lapse rate is less than the moist adiabatic, the air must be stable under all conditions, since dry air with its high adiabatic rate is certainly stable, as is moist air also, whose lower adiabatic rate is higher than the lapse rate. Such air is said to be *absolutely stable.*

When the lapse rate exceeds even the dry adiabatic, the air is said to be *absolutely unstable,* for instability will exist regardless of the humidity condition of the air.

SUMMARY. Let us summarize the conditions affecting equilibrium in the atmosphere:

In general:

Stability. Lapse rate is less than the adiabatic rate, or $LR < AR$

Instability. Lapse rate exceeds the adiabatic rate, or $LR > AR$

In particular:

Neutral Stability. Lapse rate equals dry adiabatic rate for dry air and wet adiabatic for moist air, or $LR = AR$

Conditional Instability. Lapse rate lies in between values of wet and dry adiabatic rates, or $LR \begin{matrix} > SAR \\ < DAR \end{matrix}$

Absolute Stability. Lapse rate is less than the saturated adiabatic, or $LR < SAR$

Absolute Instability. Lapse rate exceeds the dry adiabatic, or $LR > DAR$

We can also illustrate graphically the actual examples considered earlier.

1. Figure 5 · 8 shows the stability example considered previously. The coordinates of the point of intersection of the adiabatic and lapse-rate curves give the altitude and the final temperature at which the rising air comes to rest (5,000 feet and 63°).
2. Figure 5 · 9 illustrates the case of instability, wherein the curves representing the adiabatic and the lapse rates separate continuously with increasing altitude.
3. The example of conditional instability is shown graphically in Fig. 5 · 10. The dry adiabatic curve below 2,000 feet approaches the lapse-rate curve, showing stability below that altitude. Above 2,000 feet, the saturated adiabatic curve now separates from the lapse rate, showing instability.

The importance of equilibrium and stability factors of the atmosphere cannot be overlooked in explaining and interpreting many of the weather phenomena. Note well that stability and instability refer to equilibrium conditions of the air. Thus, air may be in a very unstable state of equilibrium and yet be perfectly free of vertical motion. The unstable equilib-

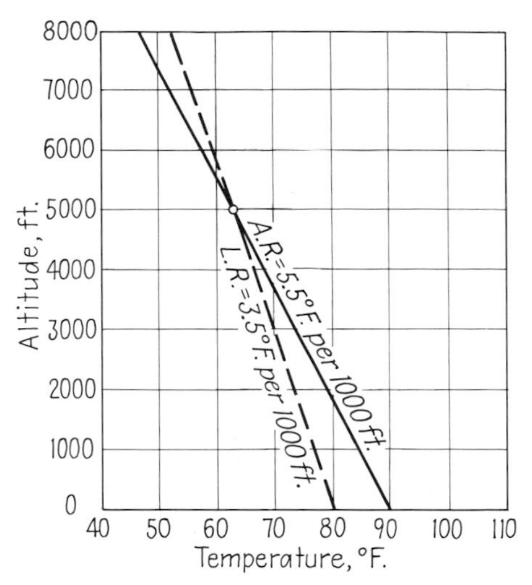

Fig. 5·8 Graphic solution of problem in which dry surface air at 90° F (rising and cooling at the dry adiabatic rate as a result of initial 10-degree temperature difference) reaches new position of equilibrium at 5,000 feet at 62.5° F when the lapse rate is 3.5° F per 1,000 feet.

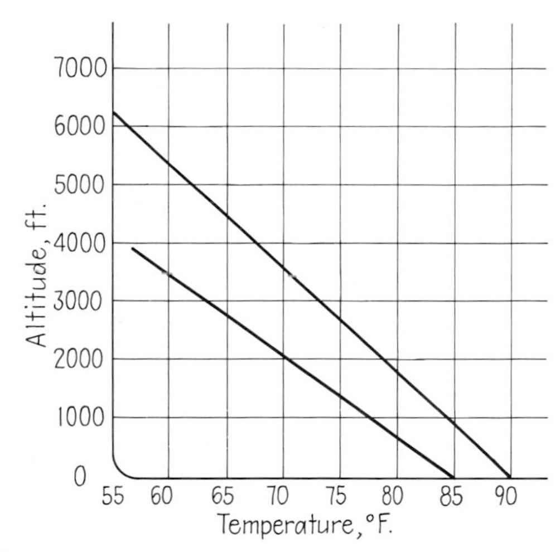

Fig. 5·9 Loss of equilibrium due to instability when the lapse rate is greater than the adiabatic rate as shown by the divergence of lapse-rate curve (left) from adiabatic curve (right).

rium conditions do not become manifest until vertical motion is initiated. Similarly, the state of stable equilibrium does not preclude the existence of vertical motion. But, should such motion be initiated by any of the factors considered heretofore, it would tend to diminish, and the air would come to rest or return to its original position.

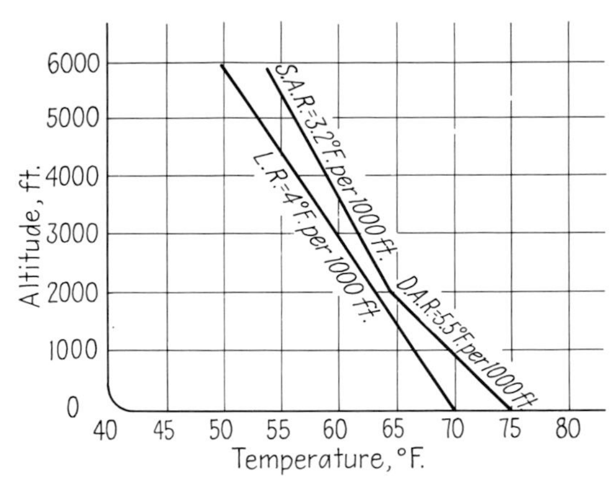

Fig. 5 · 10 Conditional instability with saturation occurring at 2,000 feet in a rising air column.

INVERSIONS AND STABILITY. Inversions in the temperature lapse rate promote stable conditions and tend to reduce vertical motion. In case 1 in Fig. 5 · 11, although the air is basically stable (lapse rate is less than the adiabatic lapse rate), initial temperature differences start air rising.

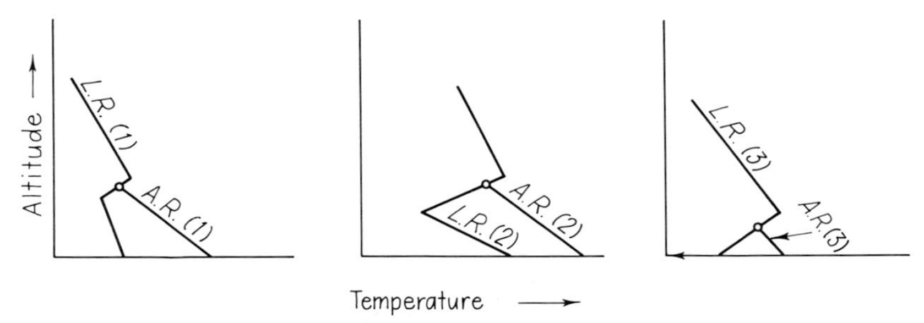

Fig. 5 · 11 Three curves showing the effects of temperature inversions in promoting stability or in limiting vertical motion to a level near the surface.

Because of the presence of the inversion, the adiabatic curve intersects the lapse-rate curve at a much lower elevation, the level of the intersection marking the height at which the rising air will come to rest. In case 2, although the air at the ground is both heated and unstable, it will come to rest at the inversion level. In case 3, the inversion starts at the ground, which restricts vertical motion, especially from thermal convections, to within a very shallow layer of air.

The condition illustrated in case 3 is fairly common over cold ocean currents (especially in summer) when air is carried in from nearby warmer land or warmer water. The cool surface produces a strong low-level inversion that greatly inhibits vertical motion, often resulting in strong marine fogs, as described in the next chapter.

The notorious smog of the Los Angeles region of California is in large part a result of low-level inversion. Particularly in summer, cool marine air often wedges in beneath the warmer continental air. The cool air becomes trapped by the high coastal mountains some miles in from shore and the overlying warm air, thus producing a very stable shallow air layer.

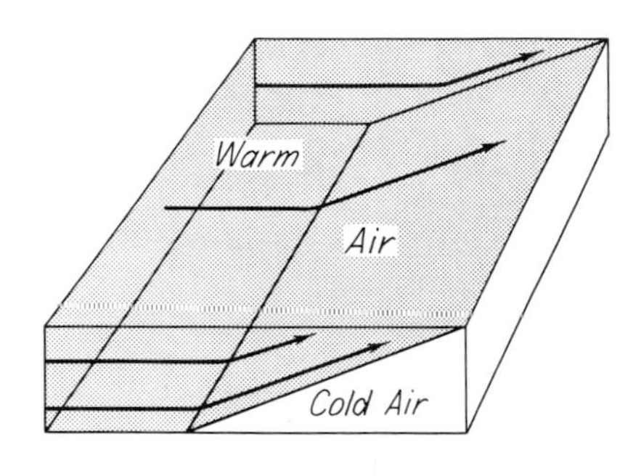

Fig. 5 · 12 Illustration of the motion of warm air rising over a cold-air wedge.

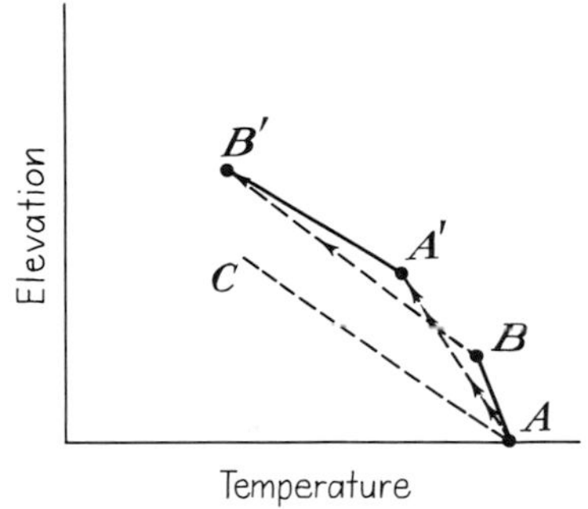

Fig. 5 · 13 Illustration of convective instability.

Fumes from factory and automobile fuel combustion become restricted to this layer.

CONVECTIVE INSTABILITY. We have previously considered instability as related to a parcel of air rising or falling through a layer of air essentially at rest along the vertical. However, a layer of air as a whole may be forced to rise, as in the case of the warm layer being lifted over the inclined cold wedge of air illustrated in Fig. 5 · 12. As the result of such uplift, air which may be initially in a state of stable equilibrium is often converted to an unstable state from the adiabatic effect on internal temperature–humidity conditions.

Convective instability is illustrated graphically in Fig. 5 · 13. Imagine a layer of air in which points *A* and *B* represent the temperature at the particular elevations. The lapse rate in this air is shown by the sloping line *AB*. If the dry adiabatic rate is indicated by the sloping line *AC* and the saturated adiabatic by the line *AA′*, then *AB* indicates absolute stability in view of its steepness. Imagine further that the lower part of the layer is saturated or very close to saturation, while the upper part is quite dry (unsaturated). If the layer is then forced to ascend, the air represented by point *A* will cool according to the saturated adiabatic rate and will travel to point *A′* following the broken (saturated adiabatic lapse-rate) curve. The air represented by point *B*, being dry, will cool along the broken curve *BB′* (parallel to *AC*), the dry adiabatic lapse rate. Thus, after uplift, the layer having the lapse rate *AB* has the lapse rate *A′B′*, which has a gentler slope and hence a steeper rate than the dry adiabatic (*BB′* or *AC*). From absolute stability, the layer has thus gone to absolute instability. Clearly, between these values, during uplift, the layer must have gone through conditional instability also. Since the lower part of the layer is saturated, this air would become unstable at this time.

The reader should work out the reverse situation in which an initially unstable layer becomes stable upon uplift. This would, of course, require that the lower portion be dry and the upper part of the layer be saturated prior to lifting. Because uplift over frontal surfaces, identical with conditions shown in Fig. 5 · 12, may often involve areas of scores to hundreds of thousands of miles, the mechanism of convective instability is quite important in the dynamics of weather conditions.

Potential and Equivalent Potential Temperature

The potential temperature and the equivalent potential temperature are two "conservative" properties of the atmosphere that are very important both in dynamic calculations of atmospheric conditions and in the study of air-mass movements.

POTENTIAL TEMPERATURE. Potential temperature is the temperature of a quantity of air at a standard pressure—the pressure of 1,000 millibars. A small quantity or parcel of air may have to be raised or lowered in the atmosphere until this pressure is reached. If the unit of air is considered to expand or contract adiabatically during this raising or lowering, its temperature at the 1,000-millibar level would be the potential temperature. As long as the air remains unsaturated during the adiabatic process, this temperature value is a constant for a particular unit of air. Let us consider an example: Assume the atmospheric pressure at a given time to be 1,000 millibars at an altitude of 400 feet; assume also that the temperature of a unit of air, measured at sea level, is 60°F. If this air is now raised 400 feet, until the pressure is 1,000 millibars, it will cool at the dry adiabatic rate (5.5°F per 1,000 feet) and become 57.8°. Should the air rise another 1,000 feet, its temperature will then be 52.3°F. If brought back to the 400-foot 1,000-millibar level, the temperature will be 57.8°. For this particular unit of air, 57.8° is thus the potential temperature. Regardless of any change in position experienced by the air, it will always have this temperature, if brought adiabatically to a pressure of 1,000 millibars, provided the air remains unsaturated.

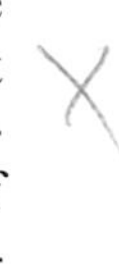

However, the potential temperature of a given sample of air may be increased if, in rising, the air becomes saturated and continues cooling at the saturated adiabatic rate. Let us assume, for example, that in the foregoing case the air rises another 1,000 feet after cooling to a temperature of 52.3 at 1,400 feet, but that it becomes saturated with resulting condensation and precipitation above this level. The temperature at 2,400 feet will be only 3.2° cooler (the average saturated adiabatic rate), or will have a

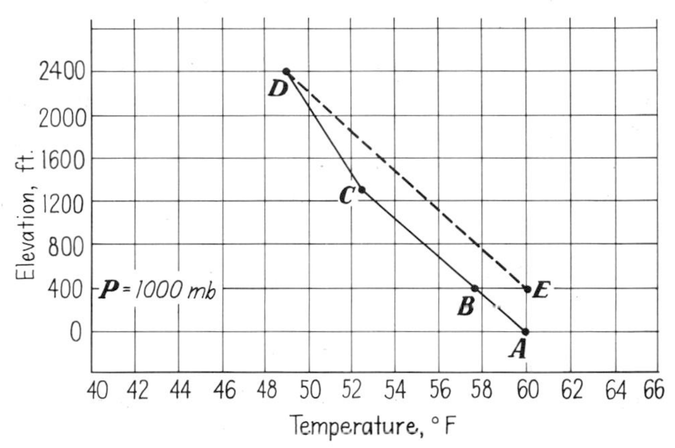

Fig. 5 · 14 Illustration of the concept of potential temperature.

temperature of 49.1°. If we now allow the air to return to the 1,000-millibar level at 400 feet, warming at the dry adiabatic rate as it descends, it will warm by 11°, thereby reaching a new potential temperature of 60.1°, not 57.8°. Had the air risen still farther after becoming saturated, its potential temperature would have increased still morc.

The explanation of potential temperature may be aided by reference to a simple graphic statement of the problem, as shown in Fig. 5 · 14. Point *A* represents a unit of air with a temperature of 60°F at sea level. Rising and cooling at the dry adiabatic rate to point *B*, where the pressure is 1,000 millibars at the 400-foot level, the potential temperature is shown to be 57.8°. Further uplift and cooling to point *C* reduces the temperature to 52.3°, but upon descending to *B*, the original potential temperature is reached. In fact any mass of air ascending or descending along the line *ABC* would have the potential temperature of 57.8°, shown at *C*. Hence,

ABC, which is the dry adiabatic lapse rate, is a line of constant potential temperature. To generalize, *all dry adiabatic lapse-rate curves are lines of constant potential temperature.*

If, in rising above 1,400 feet, the air cools according to the saturated adiabatic rate, its temperature will be 49.1°, shown at *D*. The line *CD* represents the saturated adiabatic curve. Now, in descending, the air warming at the dry adiabatic rate reaches the new potential temperature, 60.1°, at point *E*.

EQUIVALENT POTENTIAL TEMPERATURE. We have just seen that the potential temperature is fairly constant for a given air mass provided no condensation occurs; otherwise this property may vary. To overcome this defect of potential temperature, the more conservative concept of equivalent potential temperature has been developed. This is arrived at in the following manner: Consider a unit of air to be raised, cooling at the dry adiabatic rate until condensation begins; then continue the uplift of the air at the saturated adiabatic rate until all the moisture has condensed, with the air absorbing the latent heat of condensation; the air is then returned to the 1,000-millibar level, warming at the dry adiabatic rate all the way. The temperature at this level is the equivalent potential temperature and is clearly a more constant air-mass property. This feature can be calculated easily for a given unit of air, for it is seen to be the temperature of a unit of air at standard pressure with the total latent heat of condensation added. Note that once a parcel of air becomes saturated upon uplift at a given saturated adiabatic curve, its equivalent potential temperature is the same as that for any other air which becomes saturated upon crossing this curve, regardless of the level at which this takes place. Thus, *saturated adiabatic curves are lines of constant equivalent potential temperature.*

EXERCISES

5·1 (*a*) Define moist air. (*b*) When and why does rising air cool at the moist adiabatic lapse rate?

5·2 Why does the moist adiabatic lapse rate approach the dry adiabatic lapse rate with increase of elevation?

5·3 Explain how moist air can ascend a mountain and then descend to the original elevation on the lee side, but be at a warmer temperature.

5·4 Define stable and unstable air.

5·5 Describe several processes by which air, originally stable, can be made unstable.

5·6 Why is air within clouds generally more unstable than air below or above the clouds?

5·7 (*a*) Illustrate the following conditions graphically: The lapse rate is 4°F per 1,000 feet; the air temperature at sea level is 75°F; sea-level air, warmed to 85°F, rises according to the dry adiabatic rate of 5.5°F per 1,000 feet. (*b*) If the humidity is such that the rising air will cool to the dew point at 8,000 feet, determine whether the rising air will come to rest before clouds form. (*c*) If clouds should form prior to the equilibrium level for dry air, would the rising air then be stable or unstable?

5·8 How do inversions promote stability?

5·9 Set up numerical conditions in a layer of air that would result in convective instability.

5·10 (*a*) What is the difference between potential temperature and equivalent potential temperature? (*b*) Which is more conservative as a property of the atmosphere?

5·11 Explain the relationship between curves of dry and moist adiabatic lapse rates and lines of constant potential and equivalent potential temperature, respectively.

5·12 If air is in stable equilibrium, does the potential temperature increase or decrease as elevation increases? Explain your answer. (This question can be solved best by a simple graphical construction.)

CHAPTER 6

Condensation and Precipitation

In continuing our examination of the behavior of atmospheric moisture, let us note again the ceaseless hydrologic cycle. The vast expanse of oceans is the main source for the moisture evaporated into the atmosphere. This condition is as old as the atmosphere and the oceans themselves. Yet there has been no significant diminution of the volume of the seas nor increase in the water vapor of the air. How do we account for this remarkable balance between the waters of the air and the oceans?

By the processes of condensation and precipitation, huge quantities of water are returned to the seas, either directly or from streams that convey this water to the oceans. In addition to the rain, snow, etc., that fall directly on the sea surface, thousands of rivers pour their contents into the sea. The mighty Amazon, for instance, contributes so great a share that fresh water is experienced in the Atlantic 200 miles beyond its mouth! But with all this, the ocean volume is not significantly increased, for the cycle of evaporation, condensation, and precipitation is continuous and will be so as long as the sun exists to supply the heat energy necessary for the operation of this tremendous atmospheric machine.

We have previously considered one phase of this cycle—the process of evaporation and the nature and behavior of water while in the atmosphere. This was considered in relation to the all-important humidity regulator—temperature. We followed the variations in humidity until the air was saturated or the dew-point temperature was attained. In continuing our study of the cycle, we shall examine the process and forms of condensation and then of precipitation.

Process and Conditions of Condensation

Condensation has been defined earlier as the process in which water vapor is changed to liquid water. It should be noted that when this process occurs at subfreezing temperatures, the vapor often turns directly into the solid form—ice—with no intervening liquid stage. Technically, the transition from vapor to solid is known as *sublimation,* but we shall use the term *condensation* to cover either situation, for they are both a part of the same major process.

In the process of condensation in the atmosphere, three general conditions must be satisfied: (1) a sufficient water vapor content must be present, (2) cooling of the air to and below the dew-point temperature is usually required, and (3) nuclei of condensation must exist. These conditions are interrelated in their effects.

An adequate water vapor content is necessary so that saturation can occur. As noted in Chap. 4, the air can be brought to a condition of saturation with respect to water vapor through a temperature decrease until the dew point is reached (relative humidity of 100 percent) or by the addition of water vapor. With the exception of certain cases to be described later, cooling to the dew point is the most important immediate cause of saturation.

All of the water and ice particles which condense in the atmosphere do so on microscopic- or submicroscopic-sized particles called *nuclei of condensation.* It has been found that only *hygroscopic* particles—particles which are somewhat water-absorbent or have an affinity for water—can act as nuclei for condensation. Such nuclei are necessary because pure water under normal natural conditions will not form into droplets. The vapor pressure of water in a droplet is inversely proportional to the radius, so that as a droplet becomes infinitely small, the vapor pressure becomes infinitely large, and the minute droplet would evaporate as fast as it tended to form. If the hygroscopic particles are large enough, they can therefore maintain condensation of the droplet.

The most important of the hygroscopic nuclei in the atmosphere are sea salt (sodium chloride), nitric oxide, organic particles, and sulfur trioxide, the latter being especially abundant near large cities or industrial areas.

Sea salt is, of course, much more important over the seas, and decreases progressively toward the interiors of the continents. Without the presence of these nuclei, air can become supersaturated, that is, have a relative humidity greater than 100 percent. Although in laboratory experiments pure air has been brought to a relative humidity of over 400 percent, the atmosphere rarely achieves a supersaturation value much above 100 percent relative humidity. The highest relative humidity values probably occur in the very cold upper part of the troposphere where nuclei borne aloft from the earth's surface are relatively scarce. It should also be noted that although the condensation process is most rapid when the air is fully saturated with a relative humidity of 100 percent, small amounts of condensation droplets may form when the relative humidity exceeds 90 percent, particularly when large hygroscopic particles are present.

In the sublimation process, water vapor crystallizes directly into ice without an intervening liquid stage. In this process, inorganic dust particles, which are unimportant in droplet condensation, appear to play an important role. It seems now that those particles with a crystal form similar to that of ice are the most efficient in sublimation. This was the principle involved in the use of silver iodide smoke to cause increased ice sublimation in artificial rain generation, because silver iodide crystals are similar to those of ice.

Observations in cold regions and at high elevations reveal that water can exist as liquid droplets at temperatures as low as −40° (F or C). The reason for the existence of such *supercooled* water is not yet clear but may be related to the availability of crystallization nuclei.

Forms of Condensation

The forms assumed by condensed moisture in nature are varied and often beautiful in appearance. Depending on the differing conditions under which cooling occurs, the resulting condensation types may be classified under the common headings of dew, frost, clouds, and fog.

DEW. Dew is familiar to everyone. Leaves and blades of grass covered with sparkling beads of water droplets are almost normal morning features of the spring, summer, or autumn landscape. The formation of dew is readily explained. On clear, calm evenings, the earth will cool rapidly by the process of radiation and become colder than the air resting on the surface. Consequently, this surface air will cool by contact with the colder earth beneath. As this process continues, the air will become cooler and cooler until the dew point is reached. It should be carefully understood that this all takes place in a very thin layer of air in contact with the earth, perhaps a mere few inches. This cooling of the air in contact with the earth is essentially a conduction process and is restricted to a thin layer, since the air is, as explained previously, a very poor conductor of heat.

On further cooling below the dew point, the excess water vapor in the air will condense. Dark objects such as vegetation always cool the fastest, for it will be remembered that dark objects are good absorbers of heat, and good absorbers are always good radiators and lose their heat rapidly when the source is removed. Thus the moisture will condense directly on the cold dark surfaces, whether they be of natural or artificial origin.

Consequently, what is commonly known as "sweating" of pipes, cargo, or hulls is actually the formation of dew. It is obvious that *clear* and *calm* conditions must prevail. Clouds act as blankets which greatly reduce the cooling of the earth by radiation, and hence, also, the subsequent cooling of the air. Windy nights prevent the warm air from remaining in contact with the earth or cold objects long enough to cool sufficiently, even if clear skies prevail. Both clear skies and calm air are necessary conditions for dew formation.

FROST. Frost is *not* frozen dew. The conditions of formation of dew and frost are practically identical, with one exception. Dew forms when condensation occurs on cold objects above the freezing point; frost forms when condensation takes place below freezing temperatures. Under such conditions the moisture changes from the vapor state directly into the solid or ice state, skipping the liquid condition entirely. A similar but reverse phenomenon occurs with dry ice, where the solid carbon dioxide evaporates directly into a gas, with no liquid forming.

CLOUDS. Clouds are so important a topic in weather description and analysis that they will be treated separately in Chap. 7. Briefly, however, clouds form when the air at higher altitudes is cooled to the dew point, yielding water droplets which condense around the nuclei of condensation. Nearly all clouds form as a result of air that has cooled adiabatically, i.e., air that has risen, expanded, and consequently cooled. When air rises sufficiently, it will cool to the dew point. Continued uplift of the air above this level results in clouds.

Rising air currents tend to keep the clouds from falling, since the droplets composing the clouds are very small and light. Many a cloud often does fall, but as its base falls to an altitude where the air temperature is above the dew point, the small droplets readily evaporate. If the clouds form at sufficiently high altitudes, the temperatures are below freezing and, as in the case of frost, the condensation often yields ice crystals, with no water-droplet stage. Thus high clouds are composed not of water but of ice crystals.

FOG. Fog is one of the common forms of condensation that occur when the air near the earth's surface falls below the original dew-point temperature. Physically, there is very little difference between a fog and a cloud. They are both composed of minute water droplets suspended in the air. Fog, however, forms in the air near the earth's surface whereas clouds are features of much higher altitudes. Essentially, then, the difference between fog and clouds is one of method and place of formation rather than of structure or appearance.

Clouds form when the air cools adiabatically through rising and expanding. Fog forms through cooling of the air by contact and mixing or on occasions through saturation of the air by increasing the water content. Frequently a continuous gradation exists from thick fogs into low-lying clouds, there being no definite distinction in appearance. The type of fog that forms depends on existing conditions and falls into four recognized categories: *radiation fog, advection fog, frontal fog,* and *upslope fog.*

In general, then, if surface air (the air near the earth's surface) is close to and is approaching the dew point (as determined by the psychrometer), fog formation can be anticipated. If the temperature should increase after the fog has formed, the dispersal of the fog may be expected. The thickness of the fog depends on various factors of humidity, temperature, wind, nuclei, etc.

Fogs are usually classified according to their effect on visibility. These conditions are given in Table 6 · 1.

Table 6 · 1 Fog-visibility Table

	OBJECTS NOT VISIBLE AT
Dense fog	50 yds
Thick fog	200 yds
Fog	500 yds
Moderate fog	½ nautical mile
Thin fog	1 nautical mile

Radiation Fog. Radiation fogs are of less importance than the other types. They form under the same conditions, though further advanced, that formed dew and frost. On clear nights, with very slight wind, the earth and consequently the air above it will cool rapidly. If the air is cooled to a greater depth than the slight layer necessary for dew and reaches its dew point, the resulting condensation will form not only on the ground but also on the nuclei of condensation in the air. This yields minute droplets suspended in the air which constitute fog. Since the cooling of the air depends on the earth beneath cooling rapidly by radiation, it is known as *radiation fog.*

Obviously, there will always be dew or frost when a radiation fog forms. Since clear nights are always necessary, radiation fog, in addition to dew or frost, usually indicates fair weather for 12 hours at least. A slight wind is necessary in order to stir the cold air in contact with the ground and scatter it sufficiently so that fog forms. Further, the fog forms first and is thickest in the low areas or depressions since cool, relatively heavy air flows to the lowest points. Thus, radiation fog is sometimes known as *ground fog* owing to this tendency to "hug" the ground.

Harbors, especially those surrounded by forested hills, are often shrouded in radiation fog during the night and early morning. Then, as the sun comes up over the horizon the fog is said to "lift" or "burn off." Here again we have an excellent example of the earth's influence in heating the

Fig. 6 · 1 Advection fog bank. (Humphreys, U.S. Weather Bureau)

air. The fog is not simply blown away and does not actually rise. Rather, it evaporates as the air is warmed. But does it evaporate from the top down? On the contrary, the rising sun warms the earth, which then warms the surface air first. Hence the fog evaporates from the bottom up, or lifts. The last remnant of a thick radiation fog may therefore look like a low white cloud extending outward from the hilltops or other high spots.

In certain areas where inversions are frequent, the cold, heavier air is trapped beneath a lid of overlying warm air. Consequently during nights favorable to radiation, pronounced cooling occurs in this cold-air layer which may have considerable thickness and yield thick, high fogs that dispel slowly even after the sun rises.

Advection Fog. It will be recalled that convection refers to the transpor-

Fig. 6 • 2 Fog-shrouded ships in the North Atlantic Ocean.

Fig. 6 • 3 Advection fog bank seen from mountaintop above the fog.

tation of air in a vertical direction, the warm air rising and the cold air falling. *Advection* refers to air in horizontal motion. Thus, advection fogs are those which result from air being cooled following horizontal movement.

There are essentially two processes responsible for the production of this type of fog:

1. Cold air may, as a result of its horizontal motion, pass across a warmer sea surface and at the same time mix with the warmer air prevailing there.
2. Warm moist air may blow across a cold surface and become chilled by contact and by mixing with the cold air associated with the cold surface beneath.

In the first case, the warm vapors evaporating from the water will immediately condense and form a very shallow *steam fog.* This is encountered very frequently when cold air blows across a warm ocean current, such as the Gulf Stream. As a consequence, the water vapor rising from the warm current and meeting the cold air appears to give the water a steaming or smoking appearance as it condenses. This feature is also very prevalent over the northern portion of the warm Japan Current. In general, the waters in the Arctic areas are warmer than their continental surroundings in the winter. Here again the passage of cold air across the warm waters will cause this advection fog to form, giving rise to the name *arctic smoke* owing to its appearance.

The second type of advection fog, in which warm, humid air passes across a cold surface, is by far the most important kind of marine fog encountered. It has been estimated that about four-fifths of all maritime fogs owe their origin to this process. To a great extent, their occurrence depends on a definite contrast between air and water temperatures. Further, the development of these fogs is greatly aided by the presence of ocean currents whose waters differ markedly in temperature. Then, if the winds are such as to transport the air of the warm current across the cold sea, the warm humid air will be chilled by contact with the cold sea surface and by some mixing with the colder air.

Clearly then, fog will develop in these instances over the cold surface and, when the wind direction changes so that the warm air no longer meets the cold current, the fog will be dispelled.

Usually, a wind force of about 4 to 15 miles an hour is required for the proper development of this type of fog. If the wind is less, the fog will form but will be limited to a very shallow layer of the surface air, the effect of the above wind being to disseminate the fog to a greater thickness of air. Too strong a wind will lift the fog to form a low overcast condition.

Typical examples of this important kind of fog are very common. The warm Gulf Stream flowing to the northeast encounters the cold, southward-flowing Labrador Current east of Newfoundland in the vicinity of the Grand Banks. There is an abrupt change in temperature across the line of separation of these currents which is common not only to the water but to the air above as well. Whenever the wind is southerly to easterly, large quantities of warm, humid air cross this temperature break, with the consequent thick fog formation so common to the Grand Banks and Newfoundland waters. This forms with a frequency of one out of three days in winter months, but is more common in summer, with an occurrence of two out of three days.

Further eastward, the passage of the warm Gulf air across the colder British Isles in wintertime produces some of the thickest fogs to be encountered, including the famous "pea-soup" fogs of England.

The air traveling toward the northeast with the Japan Current in the North Pacific forms frequent and dense fogs on encountering the cold air and water flowing south through the Bering Straits. Thick, common fog predominating over the Aleutians owes its origin to this same condition. Warm, moist air is often blown onto the coast of California by the prevailing westerly winds in that area, which have picked up much moisture from the Pacific. When meeting the cold shore currents off California, or on striking the chilled coast itself in wintertime, the well-known California fog develops. The distribution and occurrence of fog over the oceans is shown in Figs. 6 · 4 and 6 · 5.

Precipitation and Frontal Fogs. These are probably the most generally important continental forms of fog. Actually, this type forms not so much by cooling as by saturation through increasing the water content. We have already considered the importance of a cold-air wedge in forcing warm air to rise. The line of meeting of the warm and the cold air was defined as a *front.* As the warm air rises over the cold wedge, it cools adiabatically, and ultimately clouds will form *in the warmer air.* The cold-air wedge itself may be very humid and near the dew point. Should the clouds in the warm air yield rain, this rain will partly evaporate in falling through the humid but yet unsaturated cold air beneath. Ultimately saturation of the cold air will result, and any added water vapor or temperature decrease, or a combination of the two, will cause fog to form in the cold air which may extend from the ground up to the layer of clouds in the warmer air. Although a more or less continuous mass of condensed water droplets will then exist, from fog to clouds, there will still be a definite distinction as regards method of formation (Fig. 6 · 6).

Upslope Fog. This form of fog develops when relatively humid air ascends a gradually sloping plain. In the course of this motion, adiabatic cooling causes a slow decrease in the air temperature. Ultimately, the dew point may be reached, whereupon an extensive fog layer usually forms. Fog of this type is a rather common occurrence in the interior plains of the

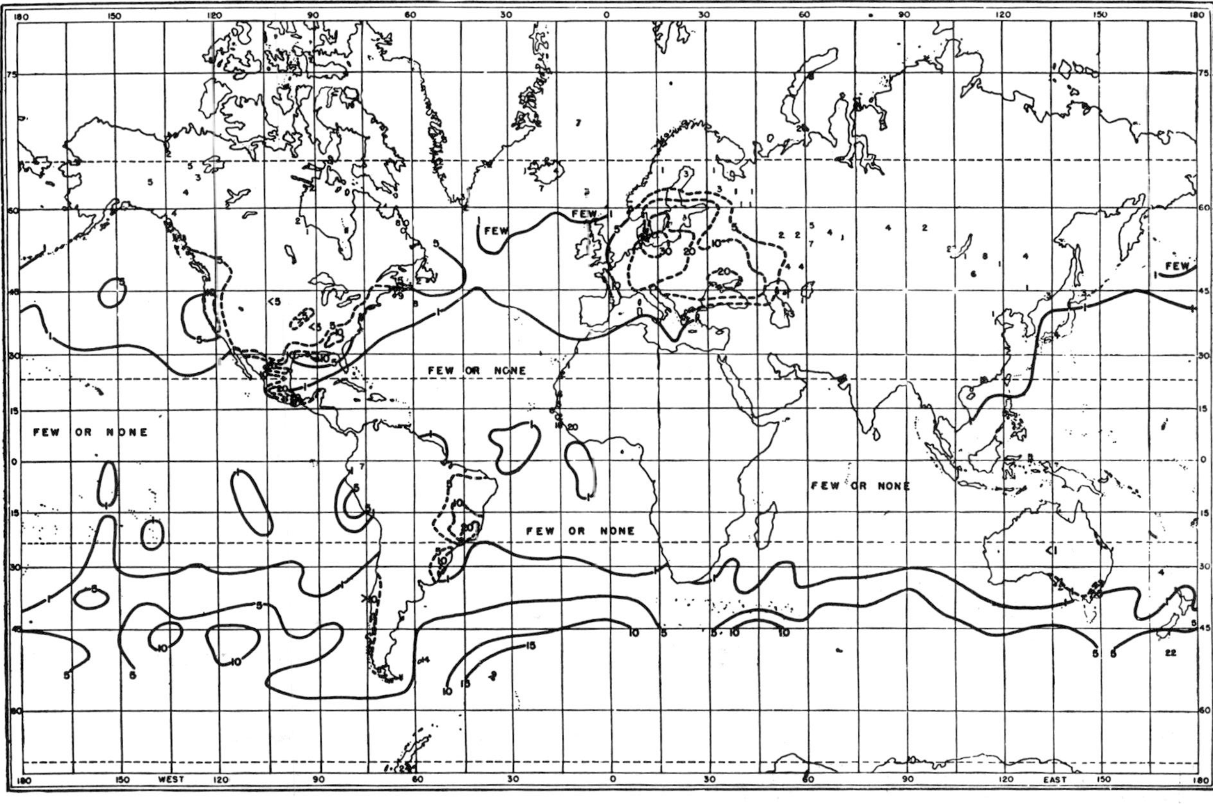

Fig. 6·4 Average frequency of fog for December to February. Solid lines, percentage of all ship observations. Broken lines and isolated numbers, number of days with fog at land stations. (Haurwitz and Austin, Climatology)

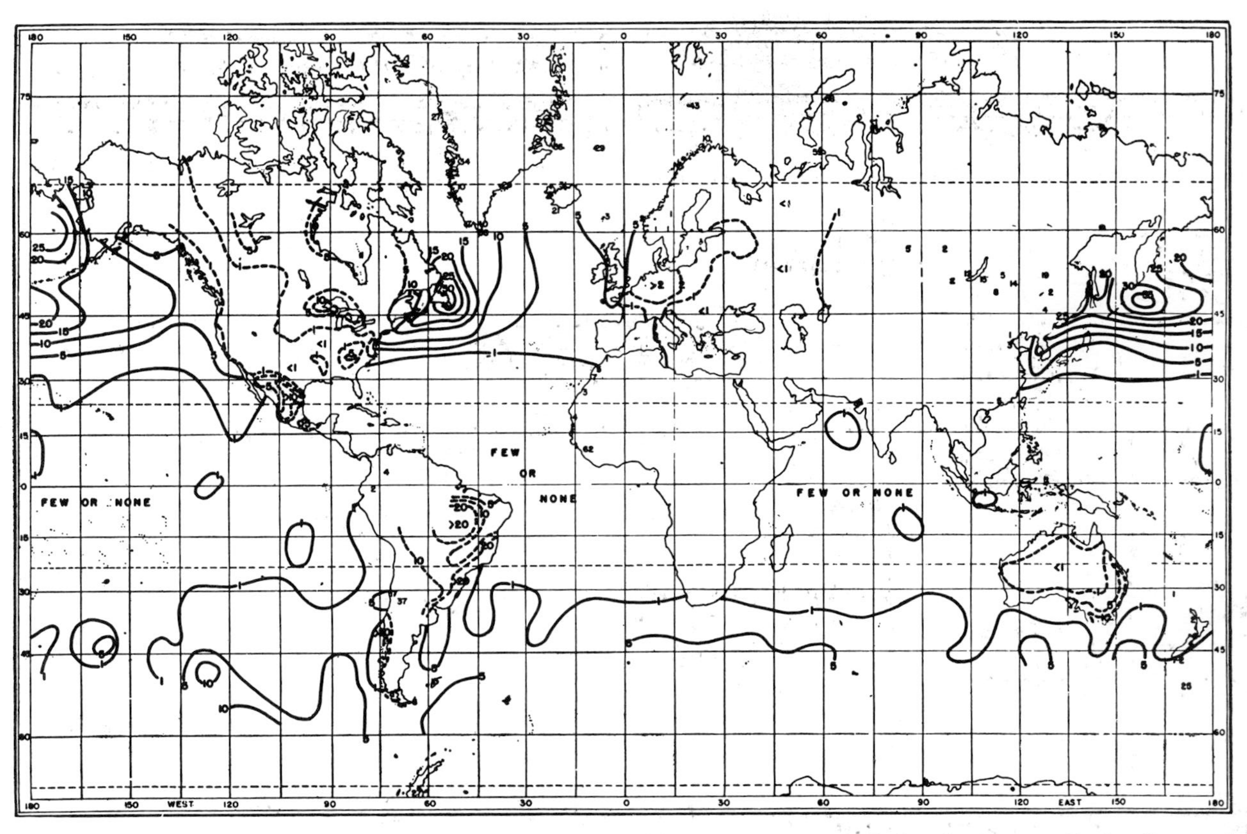

Fig. 6·5 Average frequency of fog for June to August. Solid lines, percentage of all ship observations. Broken lines and isolated numbers, number of days with fog at land stations. (Haurwitz and Austin, Climatology)

United States. It is particularly frequent when humid south or southeast winds, originating in the Gulf of Mexico, follow the slope of the gently rising plains west of the Mississippi, which reach an altitude of more than 2,000 feet.

Commonly in winter, a combination of upslope and advection fog occurs in this area as a result of the cold land surface beneath the warm air approaching from the Gulf. Such conditions may yield thick and persistent fog.

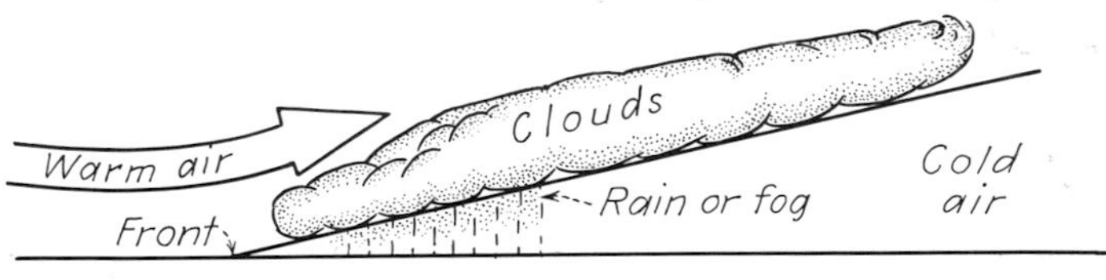

Fig. 6 • 6 The formation of frontal fog.

Precipitation

For purposes of definition we may consider precipitation to be any condensed moisture that falls to the ground surface. The process of condensation must thus precede precipitation. For the most part, precipitation falls from clouds of one type or another. Yet most clouds do not yield precipitation. Only when droplets, ice pellets, or crystals become large enough to overcome the normal buoyancy and updrafts in the atmosphere does precipitation occur.

When a comparison is made between the droplet sizes of clouds which do not yield precipitation and those that do, it becomes clear that some special process or processes not yet fully understood must operate in a cloud from which rain falls. For example, the average size of a cloud droplet, which is estimated to grow from condensation in about 100 seconds, is 0.04 millimeter (0.00016 inch), and the maximum size of a condensation droplet is 0.2 millimeter (0.008 inch). However, ordinary raindrops vary from 0.5 to 4 millimeters (0.02 to 0.16 inch).

It has been suggested that one method of growth of droplets from condensation to precipitation size is by the process of coalescence of many droplets. Why this process, if it occurs, should be restricted to only some clouds is not yet too clear. Another proposal, which seems to have considerable merit, is the growth of droplets from the presence of water and ice particles in the same cloud which lies, in part at least, above the freezing level. Since water droplets have a greater vapor pressure (evaporate faster) than ice particles, water will be transferred from the former to the latter when both water and ice exist close together in a cloud. The particle consisting of an ice nucleus and a water mantle can thus exceed the size of a straight condensation droplet and will grow heavy enough to fall, increasing still further through contact with other droplets.

It is at present believed that most precipitation in middle and higher latitudes takes place from this or a closely related process since clouds in these latitudes usually grow to elevations above the freezing level before precipitation begins. However, it has also been established that rain does fall from "warm clouds," particularly in lower latitudes. The growth of droplets by transfer of water from those at higher to those at lower temperatures has been suggested for these situations. We must at present regard the problem as incompletely solved.

Forms of Precipitation

All forms of precipitation, regardless of appearance, are often termed collectively *hydrometeors.* The U.S. Weather Bureau has classified these hydrometeors into about fifty specific types, of which only the more common will be examined here.

RAIN. Rain is obviously the most familiar example of precipitation. It is so common that in ordinary conversation it usually replaces the more general word. Although the definition of rain is simple, the explanation of its origin is complex, as just noted. Briefly, rain is precipitation in the form of liquid drops. Clouds are the source of rain; yet all or most clouds do not yield any precipitation, as we also noted above. Much rain is the direct result of condensation of water droplets in clouds followed by growth to a size necessary to overcome the buoyant effect of the air. However, a considerable amount of rain is actually melted snow, especially in high latitudes or during the winter in the mid-latitudes.

Regardless of the size of the drops in the cloud, their size upon reaching the ground surface depends on two main factors: (1) the amount of evaporation undergone by the drop in its descent and (2) the effect of the friction of the air on the falling drop. In relatively quiet air, friction imposes a limiting size on raindrops, the effect being to shatter them if they grow too large. Should the air in the precipitation zone also be descending (which exists in certain parts of thunderstorms), drops of maximum size may occur.

Thus, both the size of raindrops and the intensity of the rainfall show considerable variation. In general, the conditions responsible for heavy rains (as will be studied later) are of relatively brief duration, whereas light and moderate rainfalls are usually associated with widespread weather patterns, of much longer duration. Thus the duration of rain is, in general, inversely proportional to its intensity.

Rain of very light intensity, composed of fine droplets barely reaching the ground, is called *drizzle.* If the droplets completely evaporate before reaching the ground, we refer to the condition as *mist.*

SNOW. When condensation occurs in rising air that has cooled to subfreezing temperatures, typical hexagonal ice crystals tend to form, instead of liquid droplets. It will be remembered that condensation at the earth's surface yields frost when the dew point is below 32°F. Snow crystals may exist in isolated form or may coalesce to form snowflakes of varying sizes

Fig. 6 · 7 Some types of snow crystals. (U.S. Weather Bureau)

and shapes. As a result of the union of hexagonal ice crystals, beautiful snowflake patterns in a tremendous variety occur. It is possible, and often common, for the lower section of a cloud to consist of water droplets, while the upper portion, above the freezing level, consists of snowflakes.

SLEET. Sleet is often true frozen rain. If the waterdrops falling from the clouds encounter a layer of air with freezing temperature, they solidify into small, hard, clear ice pellets. Sleet thus indicates a temperature inversion even though it may be a very slight one. The temperature in the clouds may be very close to freezing and, if the air below the clouds is somewhat colder, sleet may result. Sleet may also develop from the freezing of melted snow as the latter falls through a colder layer of air near the ground.

HAIL. Hail is a product of the violent convection found in a thunderstorm (discussed in Chap. 7) and occurs only in connection with a thunderstorm. In the thundercloud the strong vertical air swirls the raindrops above and below the freezing level. As a result, the drop freezes when it is carried to a height above the freezing line and grows by the accumulation of snow and water at the different levels. Consequently, when a hailstone is cut apart,

Fig. 6 · 8 Results of ice storm and freezing rain, New York, January, 1943. (Robidean Studios, U.S. Weather Bureau)

it shows a series of concentric shells formed by the successive passages above and below the freezing level.

GLAZE. When rain falls on objects or on ground having subfreezing temperatures, it freezes into a sheet or coating of ice, known as *glaze* or freezing rain. If this coating becomes thick, if often has a destructive effect owing to its heavy weight (Fig. 6 · 8).

RIME. Rime is a freezing fog. It forms as a thick, frosty deposit when objects with subfreezing temperatures encounter a fog. In such a case the minute fog droplets freeze and adhere to the cold surface. Rime is thicker on the windward sides of objects, particularly when forming on hulls,

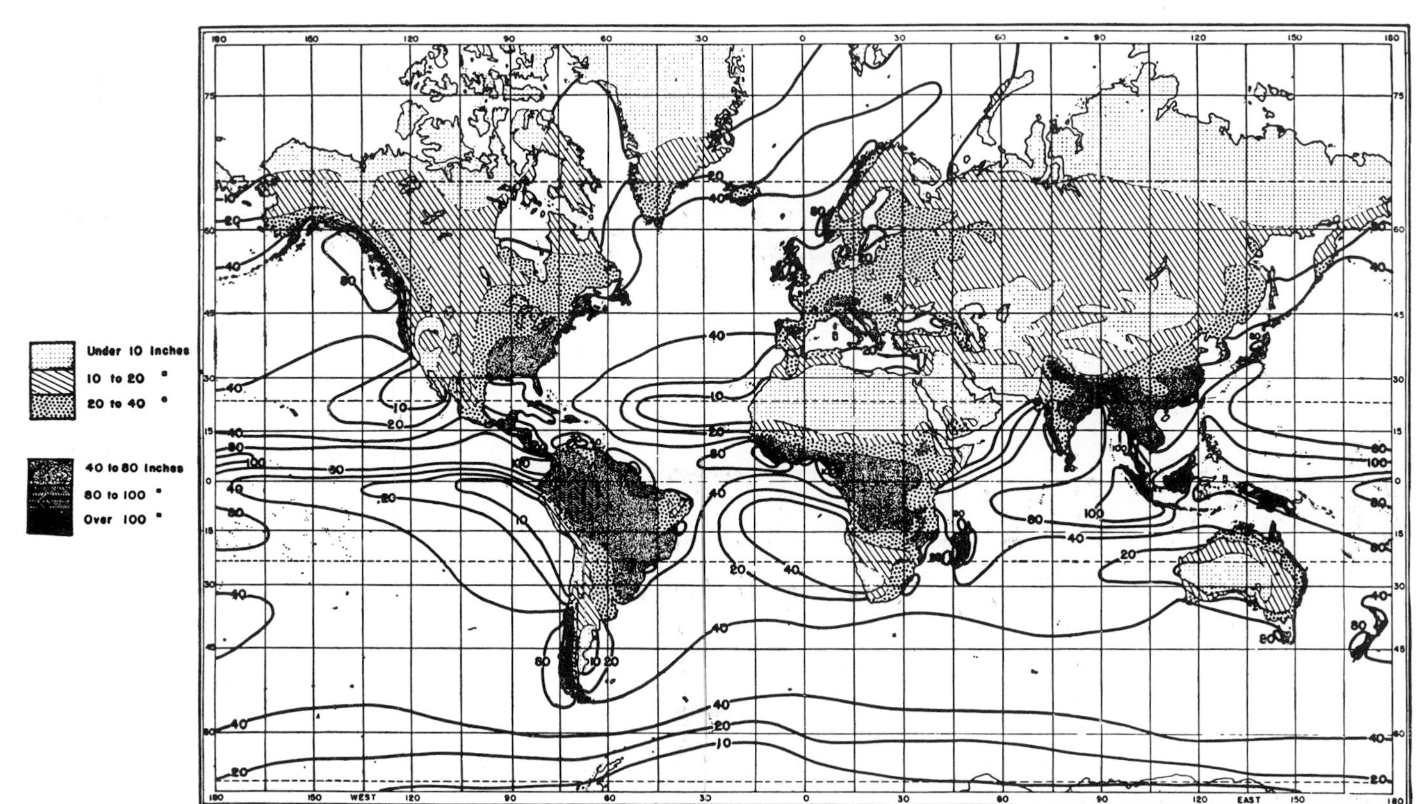

Fig. 6·9 Annual precipitation. (Haurwitz and Austin, Climatology)

masts, or bulkheads of vessels sailing through fog in cold weather. It also occurs on leading edges of airplanes flying through certain types of clouds.

Measurement of Precipitation

It is equally important to determine the amount of precipitation, as well as the kind and duration. Nearly all the measurable precipitation occurs as rain or snow, depending on the latitude and the season. A rather

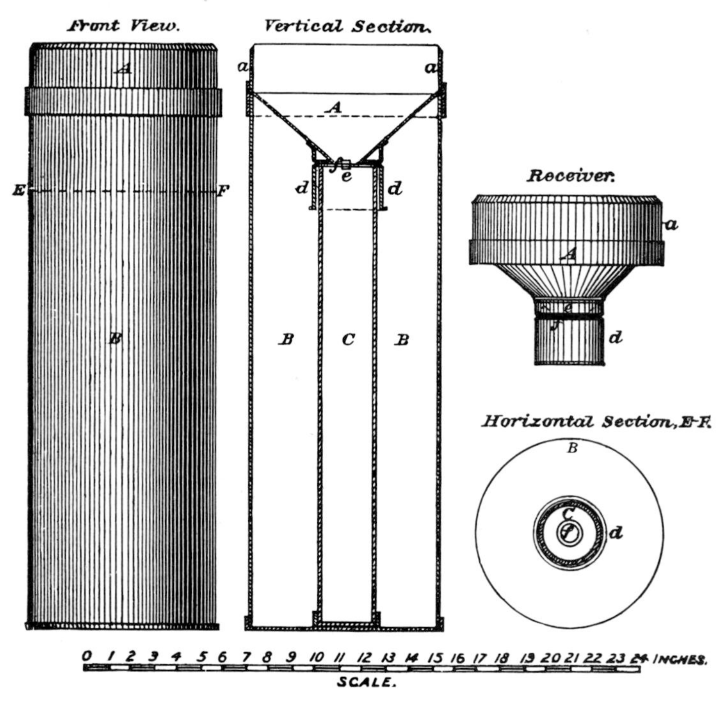

Fig. 6 · 10 Standard rain gauge.

simple instrument known as the *rain and snow gauge* is employed in their quantitative measurement.

RAIN GAUGE. The common type of rain gauge is illustrated in Fig. 6 · 10. *A* is the receiver; *B* is the overflow compartment; *C* is the inner measuring tube; *d* is the sleeve of the funnel-shaped receiver *A* and fits over the measuring tube *C*.

The gauge is supported in a vertical position by means of a tripod or other suitable stand. The area of the receiver *A* is ten times the area of the measuring tube. Therefore as water collects in the inner tube *C*, its height is exaggerated tenfold, permitting more precise measurement of the rainfall.

Since the inner tube is usually 2 inches high, it actually can contain only 2 inches of rainwater. The depth of precipitation is determined by inserting a special hardwood measuring stick through the receiving funnel *A* and into the measuring tube. When withdrawn, the moist portion of the stick shows the depth of water within the tube. The measuring stick is so calibrated as to allow for the vertical exaggeration of the gauge. Thus, 10 inches on the stick is marked as 1 inch; 1 inch is marked as ⅒ inch; and ⅒ inch is marked as 1⁄100 inch. After measurement, the water within the gauge is poured out. Should more than 2 inches of rain occur between observations, the excess water drains into the overflow tube. After the water in the inner tube is measured and disposed of, the excess in the larger container should be poured carefully into the measuring tube and measured in the regular manner. The sum of these two values is the total precipitation in the interval between observations.

The gauge obviously should be so situated as to give the most valid indication of the actual amount of precipitation. No obstructions should be near it. It should be placed where the least wind eddies may occur. If mounted on some elevated surface, it should be as close to the center as possible.

Somewhat more complicated automatic recording gauges are in use in the more elaborate weather observation stations.

In the type known as the *tipping bucket* rain gauge, a balance consisting of two small buckets at opposite ends of a beam seesaws as the buckets alternately fill and drain. Automatic records are obtained from electrical contacts made with each excursion of the balance, which usually corresponds to 0.01 inch of rain. Another automatic type of gauge responds to the direct weight of the water accumulated.

MEASURING THE DEPTH OF SNOW. It is well known that a mass of snow has a much greater volume than an equivalent mass of water. On the average, 10 inches of snow equal 1 inch of rain, or an average ratio exists between snow and rain of 10:1. However, this ratio varies considerably depending on the condition of the snow. Consequently it is preferable to express the depth of snow according to the depth of rain it represents.

If the actual snow depth is also desired, care should be taken to measure this where uniform, level snow surfaces exist and not in drifts.

To determine the water equivalent of snow, the snow collected in the overflow container, *without the use of the inner tube* or *receiving funnel,* is simply melted and then poured into the measuring tube to be measured as rain. A better method consists of adding a known quantity of warm water to the overflow container to melt the snow collected therein. The difference between the water added and the resulting melted snow, as measured by pouring into the measuring tube, represents the water equivalent of the existing snow depth.

EXERCISES

6·1 Distinguish between condensation and sublimation, and the types of nuclei that are required in the atmosphere for each of these processes.

6·2 How are dew, frost, and radiation fog related?

6·3 Which of the different types of fog requires adiabatic cooling for its formation? Explain.

6·4 Explain the fundamental differences between the location and mechanisms of formation of clouds and fog.

6·5 What is the origin of the fog over the Grand Banks and the British Isles? How do these differ as regards the season of greatest frequency?

6·6 Which are the two regions of highest fog frequency, and what similarities exist between them?

6·7 Which of the fog types is probably the greatest hazard in aviation?

6·8 (*a*) What are the average- and the maximum-size cloud droplets, and what is the range of sizes of resulting precipitation droplets? (*b*) How do the latter grow from the relatively small cloud droplets?

6·9 What is the difference in conditions within clouds yielding rain and snow? Is there any fundamental difference in the meteorology of such clouds?

CHAPTER 7

Clouds and Thunderstorms

The varying cloud forms are among the greatest beauties of nature. At times they exhibit the most delicate of appearances, when composed of thin, feathery, curling wisps; or they may present a dappled sky formed of numberless individual scalelike or woolpack masses. At other times the clouds build up in towering majestic masses, with huge billowing domes extending upward thousands of feet.

To the poet and the artist, clouds are mere things of beauty. To the weather-wise observer they are one of the most important of the local weather features. Remember that all clouds are formed of minute water droplets, or ice crystals a few thousandths of an inch across, that actually float or remain suspended in the atmosphere at varying distances above the ground. So small are these particles that the slightest movement of the air is usually sufficient to keep them aloft. Precipitation results only when these particles increase greatly in size.

Clouds indicate the prevailing and past conditions in the air and, more important, the probable future atmospheric conditions. For making short-period forecasts from local observations alone, clouds are one of the two most important criteria available. We will therefore examine the observational features and then the physical causes of clouds.

Amount of Cloudiness

Normally the amount of cloudiness is either stated in the number of tenths of the sky obscured by clouds or described verbally. The use of specific decimal fractions to indicate cloud coverage will of course give a more specific indication of this amount. The decimal point is usually omitted when writing numerical designations. Thus 7 indicates that seven-tenths of the sky is covered.

If described verbally, two sets of descriptive terms are often used. These are given in Table 7 · 1.

Table 7 · 1 States of the Sky

DESIGNATION		AMOUNT OF CLOUDINESS
Clear	Clear	Less than 1 tenth, or 1
Scattered	Partly cloudy	1 to 5 tenths inclusive, or 1 to 5
Broken	Cloudy	6 to 9 tenths inclusive, or 6 to 9
Overcast	Overcast	More than 9 tenths, or more than 9

The term *overcast with breaks* is used for a sky more than nine-tenths covered but with slivers of clear sky showing through.

CEILING. The term *ceiling,* although common mainly to aviation, is so familiar that we shall consider it briefly. Ceiling is defined as the height above the ground at which the cloud cover appears broken or overcast. Thus, if less than six-tenths of the sky is covered, the ceiling is considered "unlimited," regardless of the cloud heights. If broken or overcast conditions exist, but above 20,000 feet, the ceiling is "unlimited." A cloud ceiling above 10,000 feet is reported in weather messages as "unlimited."

It is a matter of common observation that clouds always appear thicker, darker, and closer together toward the horizon than they do overhead. This is merely an illusion and not a coincidence for any particular observer.

If breaks exist between clouds, the angular opening between the clouds appears to close up when seen at a distance. This perspective effect is seen in Fig. 7 · 1. The clouds may appear darker since several clouds, or a greater thickness of cloud, may be in the observer's line of sight near the horizon, allowing less light to penetrate.

MEASUREMENT OF CEILING. A knowledge of the height of the cloud ceiling is often of great value. This factor can be determined quite readily whether it be day or night. In the daytime the ceiling height is found by releasing balloons that have been inflated so as to have a known rate of ascent, usually 400 feet per minute. After release, the balloon is kept under observation until it rises and disappears into the cloud base. The time consumed from release to disappearance, multiplied by 400, gives the height of the cloud base in feet.

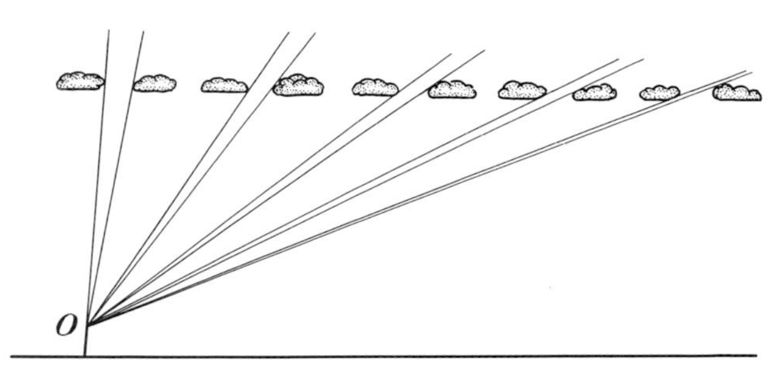

Fig. 7 • 1 Perspective effect causing clouds to appear more closely together when near the horizon.

At night, this height is determined by means of a *ceiling light.* This is a powerful projection lamp, usually arranged to shine vertically upward. The observer stands at some known distance from the light and measures the elevation angle at which the light beam is reflected by the cloud base.

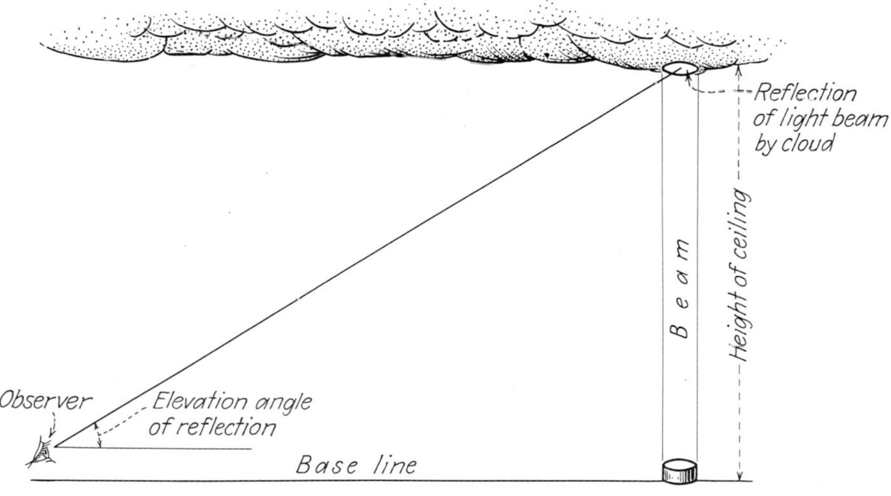

Fig. 7 • 2 Use of the ceiling light.

The angle of elevation may be determined by means of the clinometer, a simple device used for measuring vertical angles, or a sextant, or any other instrument capable of measuring elevation angles. Knowing this angle and the base line of the observer, one can calculate by simple trigonometry the ceiling height. If this base line is constant for such measurements and only the angle varies, a table yielding the required height may be used (Fig. 7 • 2). The *ceilometer,* consisting of an ultraviolet beam and a photoelectric scanner, is often used for daytime measurements.

Aboard ship, the ceiling light may be at one end of the vessel, with the observer at the other end. The length of the ship will then be the base line.

Classification of Clouds

When clouds are first examined with a view toward recognition, they may well appear to be confused chaotic masses streaming across the sky in more or less disorder. However, most of this confusion vanishes once the cloud classification is understood and the description of the common forms is known. Frequently more than one type of cloud is present in the sky, but when the existence of two or more types is recognized, the identification of each is greatly simplified.

In accordance with the 1956 *International Cloud Atlas* of the World Meteorological Organization, clouds are classified into 10 characteristic forms or genera which are mutually exclusive. Variations within most genera are recognized, but will be omitted from the descriptions that are given here.

Basic Cloud Forms

Three basic cloud forms are recognized: *cirrus, cumulus,* and *stratus.* All the other standard types are either these pure forms or modifications and combinations of them at different elevations, where varying air and moisture conditions are responsible for their form. Much more water vapor is available at lower than at higher levels. Clearly then, the higher a cloud, the thinner it will usually be; the lower the cloud, the denser and darker will it appear.

If a basic cloud form (with the exception of cirrus) occurs above its normal level, the cloud will be thin and the prefix *alto* precedes the name. If any cloud is associated with precipitation, the word *nimbus* (Latin for *rain*) is often introduced in combination with the name.

Cirrus. This form embraces very high, thin, separated, or detached clouds that develop delicate patches or long extended fibers, frequently with a feathery appearance and always white in color.

Cumulus. This form always exhibits flat-based individual cloud masses, with a pronounced vertical doming, and frequently a cauliflower-like structure.

Stratus. This form implies an extended sheet or layerlike cloud covering all or large portions of the sky. The stratus type is usually a continuous cloud deck and may show minor rifts, but no definite individual cloud units.

CLOUD DESCRIPTIONS. Let us consider a brief description of the foregoing cloud types. This, together with good cloud photographs, should aid measurably in cloud identification. U.S. Weather Bureau publications and cloud charts will prove very profitable for the earnest cloud observer.

Cirrus (*Ci*). Cirrus are about the highest of all the clouds, usually forming above 30,000 feet. They are detached, delicate, white cloud units appearing in all seasons. Often they are feathery, fibrous, or tufted in appearance, indicating the well-known mare's tails. Owing to their great height and the consequent low temperatures prevailing there, cirrus clouds are composed of thin crystals or needles of ice and not droplets of water.

Cirrostratus (*Cs*). These clouds form typically as a thin whitish veil or sheet, often covering all or a good portion of the sky. They may be very thin, giving the sky a slight milky white or veiled appearance, or they may form a definite white sheet. Cirrostratus clouds are responsible for the halos often occurring about the sun or moon. In fact, the presence of such features usually indicates the presence of cirrostratus. The very thin appearance of these clouds indicates the great height at which they commonly occur, which is the same as for cirrus. Hence these, too, are formed of ice spicules or needles.

Cirrocumulus (*Cc*). These clouds form as small, white, flaky or scaly globular masses covering small or large portions of the sky and have no shading. The delicate groups of cirrocumulus often appear to be rippled in appearance, or they may be arranged in bands crossing the sky. It is this banded arrangement of the delicate white cirrocumulus packs that has resulted in the application of *mackerel sky* to these clouds. They are the least common of the cloud types, often forming from the degeneration of original cirrus or cirrostratus clouds with which they must be associated in the sky. Being at heights equivalent to the other high clouds, they are also usually formed of ice. The individual cloudlets are less than one degree across.

Altostratus (*As*). Altostratus clouds are uniform bluish or grayish-white cloud sheets, covering all or large portions of the sky and sometimes occurring in uniform broad bands. The sun may be totally obscured or may shine through in a weak, watery condition. The typical watery sun is characteristic of altostratus. Very frequently there is a complete or nearly complete absence of shadows associated with this weak sun, for the general illumination of the clouds is sufficient to offset the shadows cast by the weakened sun. Just how thick these clouds are depends on the height at which they form. If very high, they may grade into cirrostratus. The lower they form, the heavier and denser they become. Altostratus yield a large percentage of precipitation, particularly in the middle and high latitudes, being composed of both water and ice particles.

Altocumulus (*Ac*). These clouds form as elliptical globular units occurring individually or in groups. When in groups, altocumulus may form as confused, and more or less closely grouped, masses or in definite bands, with clear sky alternating. Altocumulus may have gray shading on their undersurfaces. Individual altocumulus clouds are frequently elongated elliptical or lenticular units distinguishable from the cumulus (to be studied later) by their height and absence of vertical doming. The wavy or parallel bands of altocumulus, mentioned above, are particularly characteristic of this cloud type. The well-known "sheep clouds" or woolpack clouds are examples of high globular altocumulus groups.

Stratus (*St*). This is a uniform gray cloud sheet or layer. Stratus clouds have no particular form or structure and usually completely cover the sky. The uniform cloud sheet may sometimes appear partly broken into elongated patches. The stratus sheet is normally thicker and darker than the higher altostratus which may be overlying. It is often difficult or impossible to distinguish low stratus formations from high fog, for an almost continuous graduation exists. Thus, warm, humid air, flowing across cooler regions, may yield very thick advection fogs. During the day, much of the lower portion of the fog may evaporate and leave a high fog or stratus above the ground. Stratus clouds frequently become broken and wind-blown, being more or less formless ragged patches which are then called *fractostratus* or *scud* clouds.

Nimbostratus (*Ns*). Nimbostratus are thick, dark gray, shapeless cloud sheets with irregular broken clouds beneath and surrounding them. They are the common associates of steady precipitation, whether rain or snow. Nimbostratus have a poorly defined "wet" undersurface in contrast to the "dry" undersurface of stratus clouds, and frequently are underlain by ragged fractostratus with which they may merge.

Stratocumulus (*Sc*). These form large, heavy rolls or elongated globular masses arranged in long, gray parallel bands that usually cover all or most of the sky. They grade in appearance from definite cloud rolls that are simply close together to a more or less continuous sheet broken into irregular parallel bands. They often form from the flattening of cumulus clouds which may be arranged in bands, or may develop as a continuation of altocumulus occurring at low altitudes. In this latter case, the stratocumulus will appear darker, lower, and heavier than the related altocumulus.

Cumulus (*Cu*). Cumulus clouds are the majestic, billowing, white clouds so prominent in the summertime. However, they may occur at any season. These clouds are typically flat-based, with a pronounced vertical thickness which extends upward as a domed or cauliflower or turreted mass.

Cumulus are for the most part fair-weather clouds. Frequently after a storm has passed, a continuous train of small flat cumulus, with relatively small vertical doming exhibited, will float across the sky. Their flat bottoms, if extended toward each other, form a nearly perfect plane surface—the dew-point level. Irregular wind-torn patches of cumulus, formed by wind action on larger cumulus clouds, are called *fractocumulus.*

Cumulonimbus (*Cb*). Cumulonimbus arise from cumulus that have developed into tremendous towering clouds with a vertical range, from base to top, of 2 to 5 miles. When grown to this height, such clouds form the well-known thunderstorms, the cloud itself being called a *thunderhead.* Such clouds are marked by the turrets which are everchanging in form and shape as the cloud builds up higher and higher. In the well-developed thunderhead or cumulonimbus, the top becomes flattened and drawn out in the direction of motion, resulting in the anvil shape or top.

Conditions of Cloud Formation

Nearly all clouds are the products of temperature and humidity changes in rising and adiabatically cooling air. Which cloud genera form depends on the method by which the air is set in vertical motion and on the height at which the air cools to the dew point. We noted earlier that air may be forced to ascend in response to:

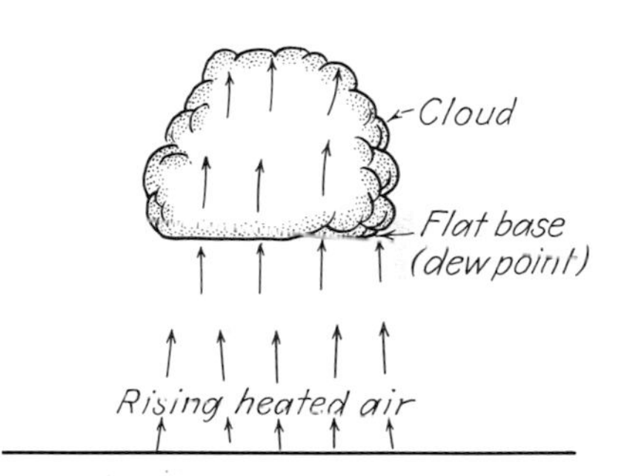

Fig. 7 · 3 Production of cumulus cloud by isolated, heated, rising air column.

1. Local heating, or direct convection
2. The effect of topography
3. The effect of fronts or cold-air wedges
4. Convergence

Surface heating results in air rising vertically from the heated surface, usually over a rather limited area. Topography is important mostly in cases of steep slopes, which also cause steeply rising air. Fronts can produce either gently or steeply rising air depending on the frontal type, while convergence usually results in air rising steeply over a broad area.

EFFECT OF STEEPLY RISING AIR. Whenever air is brought into vertical motion, in response to any of the foregoing causes, the temperature will decrease at the adiabatic rate. Such motion is usually restricted to independent columns of air, which may be widely scattered or relatively close. In any case the temperature throughout such air columns is the same horizontally. Consequently the dew point of these rising air bodies will be reached at the same level and produce clouds having flat, even bases, all at the same elevation. The vertical thickness of the cloud depends upon the height to which the air continues to rise above the dew-point level, and the available moisture supply. Figure 7 · 6 shows strikingly the effect of a

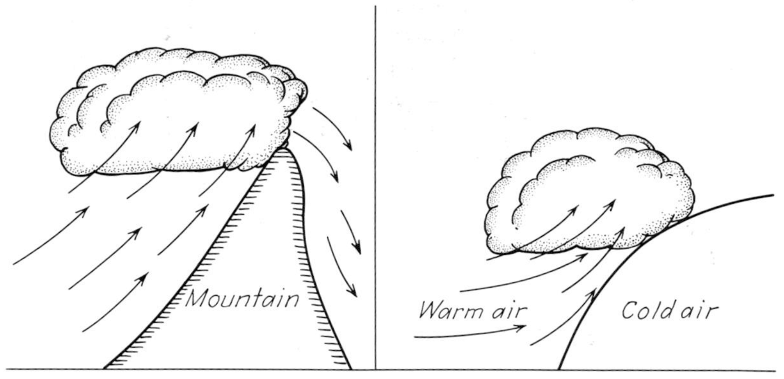

Fig. 7 · 4 Production of cumulus clouds by air rising over a steep mountain (left) and a steeply sloping cold-air wedge (right).

Fig. 7 · 5 View looking south along the eastern face of the Sierra Nevada Mountains showing the Sierra wave cloud. This cloud forms from the steep reascent of air following the initial steep descent along the lee side of the range. The region between the cloud and the mountain is characterized by descending air. (Symons Flying Service, Bishop, California)

warm land area in the production of cumulus. Clearly this process will be responsible for the formation of cumulus clouds and, with very exaggerated uplift, cumulonimbus clouds.

The height of a cumulus base may be very easily calculated from a

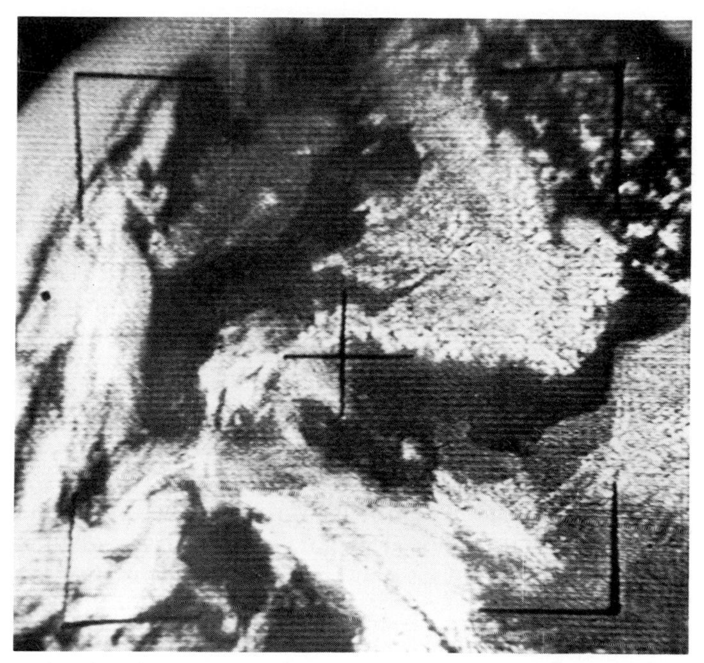

Fig. 7·6 Tiros IV *photograph, showing in particular England outlined by the development of cumulus clouds over the entire land area.* (U.S. Weather Bureau)

formula derived by means of the adiabatic rate and dew-point decrease for rising air. This formula is

$$H = \frac{T - DP}{4.5} \times 1{,}000$$

where H is the height of the cloud base, T is the air temperature at the ground (from observation), and DP is the dew point at the ground (from observation). If Fahrenheit temperature units are used, the results will be in feet. Approximately the same results are obtained from the formula

$$H = 222(T - DP)$$

Cumulus clouds will therefore form in locally isolated air columns or in adjacent columns along a steeply sloping mountain range or cold-air wedge.

EFFECT OF FRONTS ON RISING AIR. Whenever warm air meets a cold air mass, the cold air, it will be recalled, wedges beneath the warm. Note that the warm-air layer overlying the cold may have a considerable horizontal extent. As the sheet of warm air rises over the sloping upper surface of the

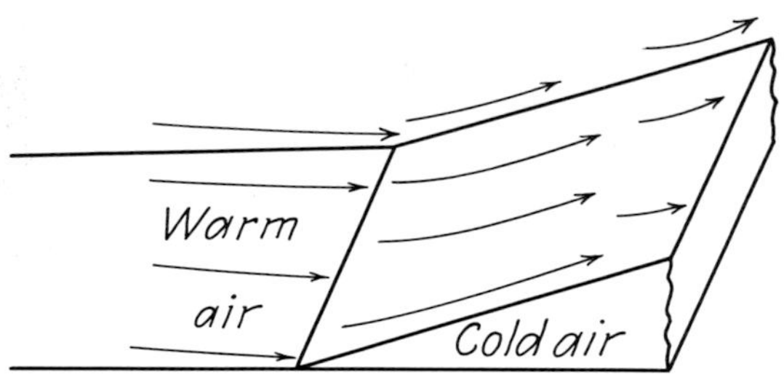

Fig. 7 · 7 Production of an extended sheet of rising warm air by a gently sloping cold-air wedge.

colder air, the temperature decreases adiabatically, causing cloud condensation to begin at the dew-point level and continue as the air rises still farther. This process occurs more or less uniformly throughout the sheet of warm air and produces a cloud blanket of the stratus type.

The cloud deck thus formed in the warm air will be thickest nearest the ground or bottom of the slope, for more moisture is available at lower altitudes; also, the lower clouds will here be covered by middle and high clouds directly above.

The cross section in Fig. 7 · 8 indicates this process, showing the sequence of formation of the stratus cloud.

It is clear that clouds must become thinner and generally lighter in color the higher the air ascends. Far up near the head of this warm, rising

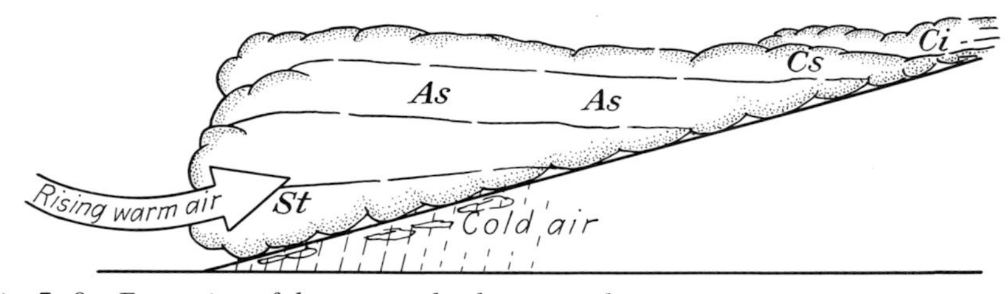

Fig. 7 · 8 Formation of the stratus cloud sequence by air rising over gently sloping cold-air wedge.

sheet the clouds will be scattered and detached, occurring at great heights. These are the cirrus clouds. Following the cirrus the main cloud mass is encountered. The thinnest and highest portion of this sheet will of course be the cirrostratus. Then, descending the wedge, the series passes into the thicker altostratus with its watery sun, and stratus or nimbostratus, with associated precipitation.

The significance of an approaching cloud sequence of this type, with the cirrus and cirrostratus appearing first, is therefore very often a good indicator of the approaching rain, snow, and general bad weather. This cloud picture will later be related to the weather pattern as a whole, and its significance will then become still greater.

CONVERGENCE. Horizontal convergence, shown earlier in Figs. 3 · 12 and 3 · 13, usually produces rising air over the entire region involved. Under such conditions stratus-type clouds often develop. However, if the air layer has an initially unstable lapse rate or is in a state of convective instability, cumulus and cumulonimbus clouds and related thunderstorms can develop.

Reporting and Identification of Clouds

Although the fundamental classification of clouds depends on form and appearance, further criteria are used in the reporting of clouds in view of the important relationship between clouds and weather structure. Height categories are used which are based on the elevation of the *bases* or lower surfaces of clouds. All of the genera described previously can be included in the categories of *high, middle,* and *low* clouds as shown in Table 7 · 2.

Table 7 · 2

	LOW CLOUDS	MIDDLE CLOUDS	HIGH CLOUDS
POLAR REGIONS	0–6,500 ft	6,500–13,000 ft	10,000–25,000 ft
TEMPERATE REGIONS	0–6,500 ft	6,500–23,000 ft	16,000–45,000 ft
TROPICAL REGIONS	0–6,500 ft	6,500–25,000 ft	20,000–60,000 ft
	Stratus Cumulus Cumulonimbus Stratocumulus	Altostratus Nimbostratus Altocumulus	Cirrus Cirrostratus Cirrocumulus

Note that the upper portion of the middle cloud range overlaps the lower portion of the high cloud range. The average heights of clouds vary with latitude owing to the great range of temperature and water vapor content of the air from the equator to the poles.

Before clouds can be used by the observer for the purpose of forecasting from local observations or for reporting to the Weather Bureau for the purpose of preparing the synoptic weather maps necessary for official forecasts, the correct identification is necessary. Only experience in observation and reference to a cloud atlas will make positive identification possible. Both the World Meteorological Organization and the U.S.

Weather Bureau provide (the former at some cost) cloud atlases and codes to use in the reporting of cloud observations from sea or land stations. The drawings and photographs reproduced here can help considerably in the recognition of clouds.

Figure 7 · 9 shows graphic cloud diagrams taken from the *International Cloud Atlas,* Volume 1, of the World Meteorological Organization. These drawings are very helpful because they tend to remove some of the irregularities and atypical features common to cloud photographs and at the same time they emphasize the characteristics important in recognition and reporting. To help in this, a large number of cloud photographs are included at the end of this chapter. Figures 7 · 41 to 7 · 69 represent an abridged group taken from the U.S. Weather Bureau Circular on Clouds and indicate the appropriate classification to be used in coded reports together with the symbol shown on weather map station models.

In the identification of clouds for reporting purposes, it may be of interest and importance to note some special features. Gray wisps or trails are often suspended from a variety of cloud types. This feature, known as *virga,* may be composed of light rain or snow precipitation which is scattered and evaporated long before reaching the ground. Another striking form is the *mammatus,* which consists of bulbous or festoon-shaped, rounded masses covering the undersurfaces of cumulonimbus and less frequently stratocumulus clouds. These features are usually associated with very severe turbulence in the lower parts of these clouds, which become known, for example, as *mammato-cumulonimbus* (Fig. 7 · 33). *Incus* is the term applied to the marked anvil tops of mature cumulonimbus.

Thunderstorms

Thunderstorms are among the most violent displays of nature. The violent and gusty winds which accompany the storm are deadly for small surface craft, and the extremely severe turbulence within the storm cloud is even more deadly for aircraft. All of the storm features—heavy showers, hail, severe gusty winds, and frequent lightning and thunder—are the products of a rather localized giant convection cell in the atmosphere. The most striking visible result of this convection is the towering dark cumulonimbus cloud which usually begins as a simple cumulus cloud. The combination of adequate water vapor and steeply ascending air produces a strong vertical development of the original cumulus into the billowing cumulonimbus cloud whose top may be miles above its base.

THUNDERSTORM DEVELOPMENT. Although the development of a thunderstorm is a continuous process, the events in this genesis can be grouped conveniently into three stages: the cumulus stage, the mature stage, and the dissipation stage. The most complete study so far has been that of the

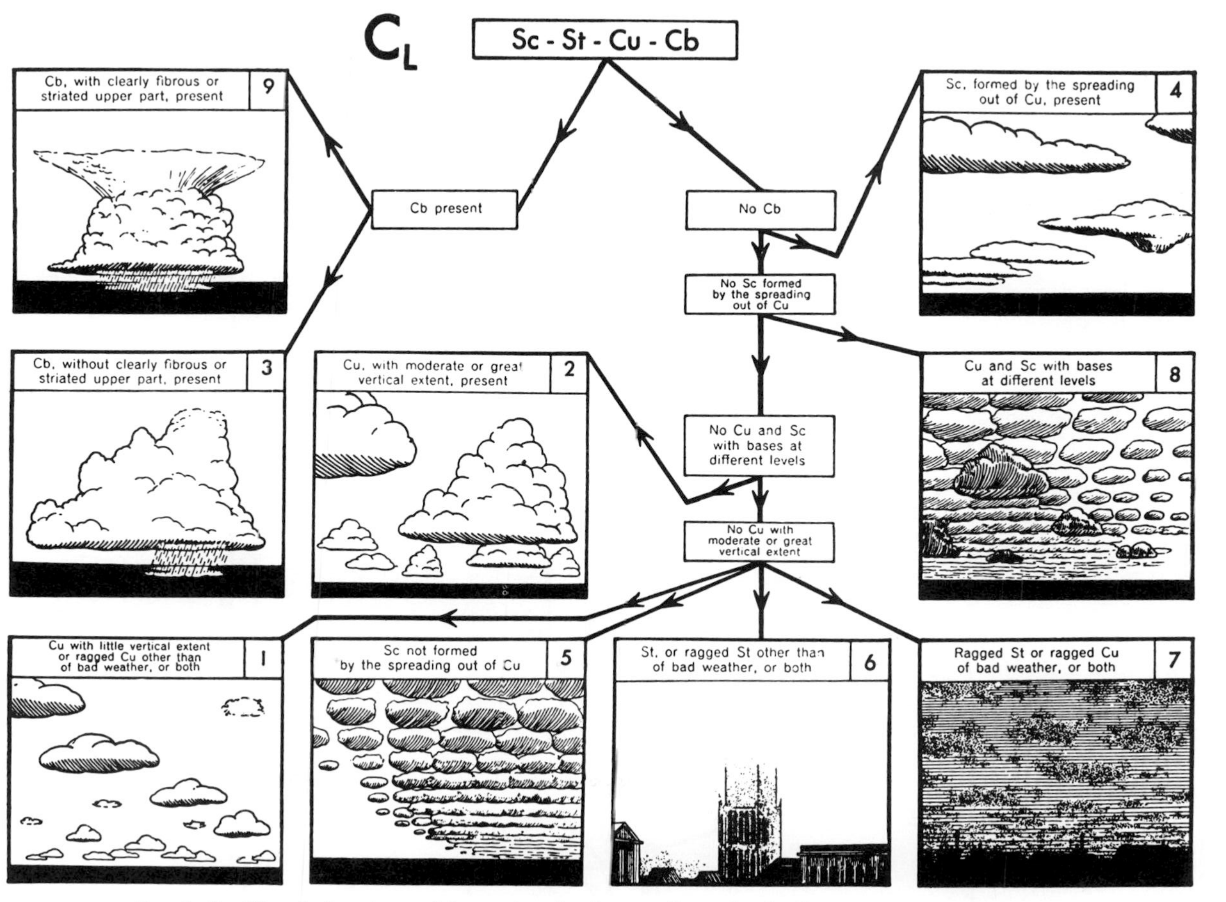

Fig. 7·9 Sketch drawings of the major cloud types. Low clouds. (International Cloud Atlas)

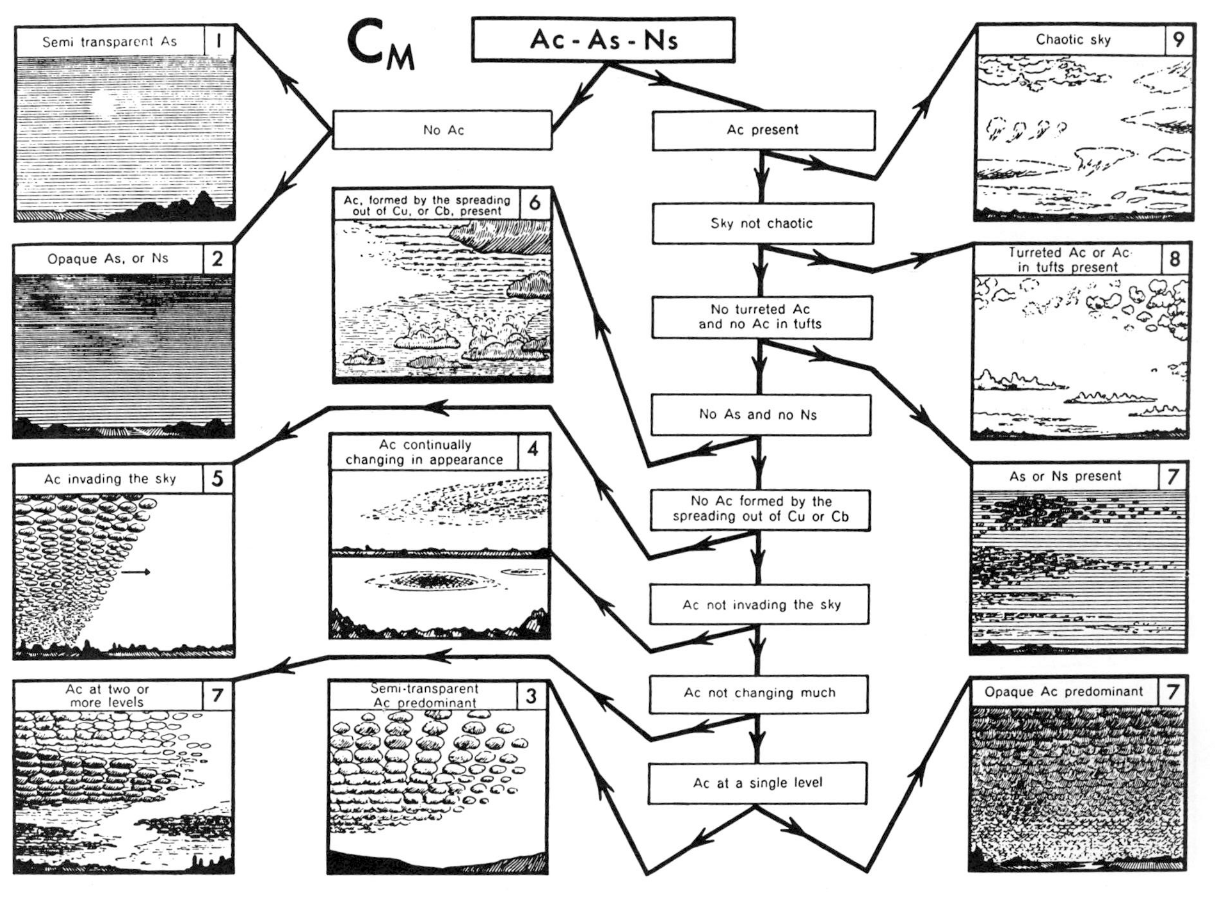

Fig. 7·9 (Continued) *Middle clouds.*

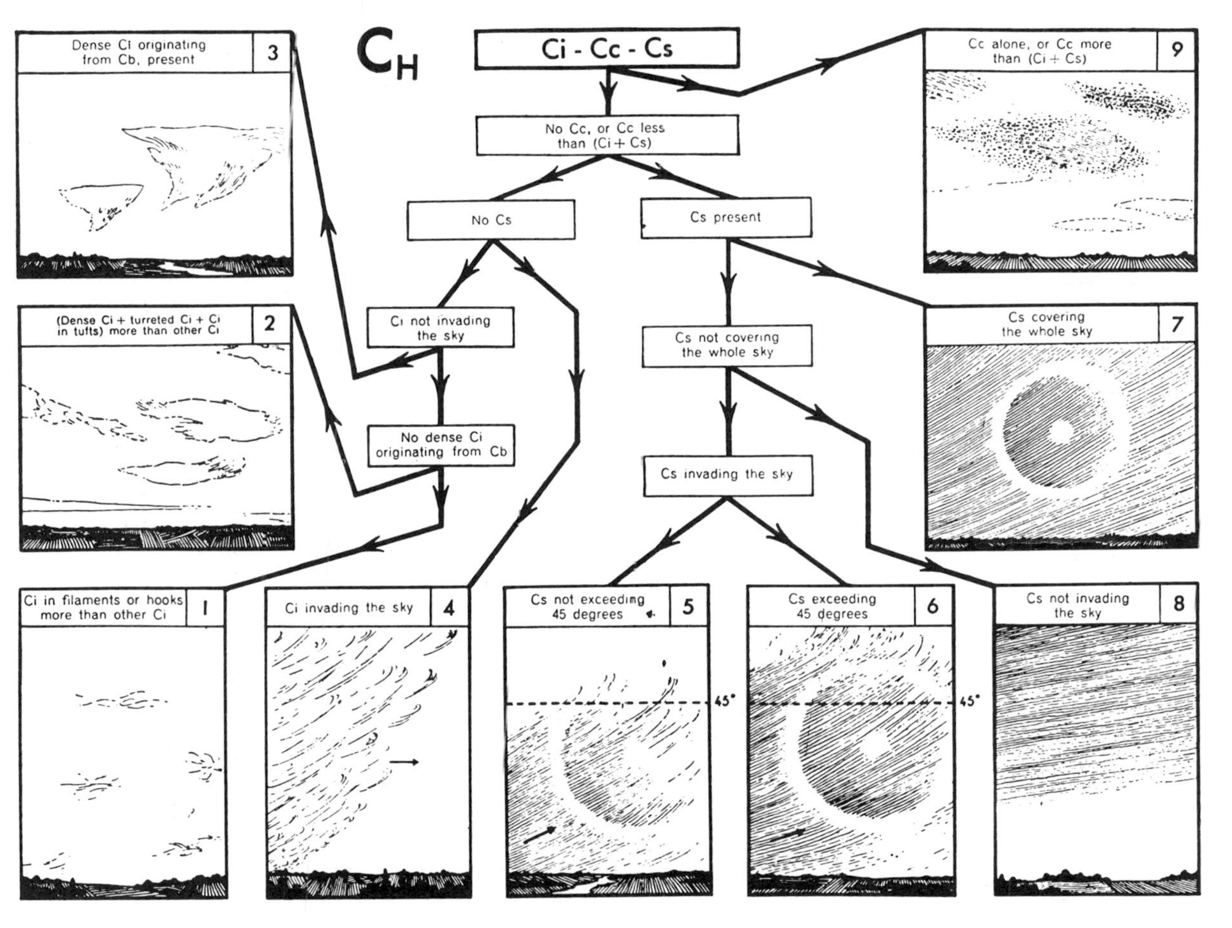

Fig. 7·9 (Continued)   High clouds.

Thunderstorm Project of the U.S. Weather Bureau, U.S. Navy, U.S. Air Force, and the former National Advisory Committee for Aeronautics (now the National Aeronautics and Space Administration). Although the three stages to be described are based on observations of Florida storms, they are probably applicable to most thunderstorms.

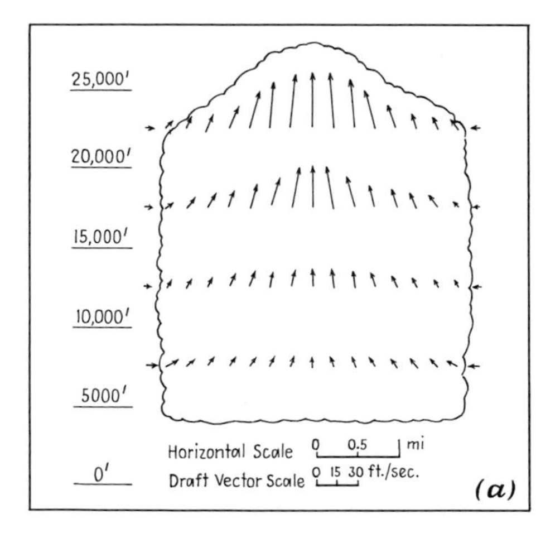

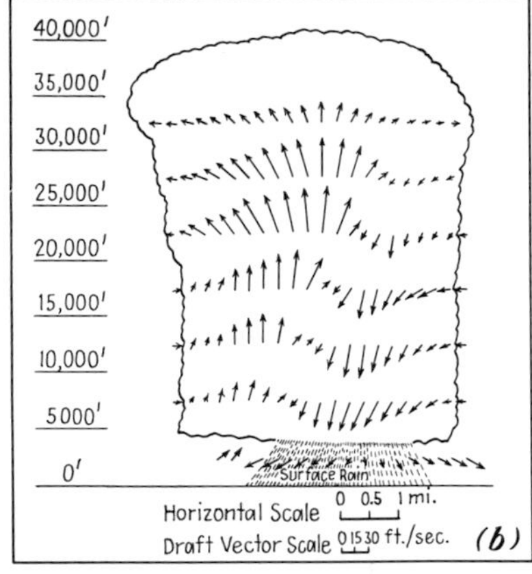

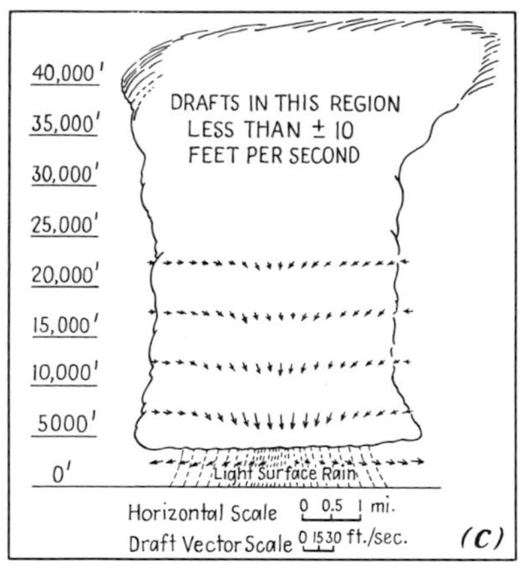

Fig. 7 · 10 Three stages in the development of a thunderstorm from original cumulus.

The cumulus stage is characterized by updrafts throughout the cell (as indicated by Fig. 7 · 10*a*) which become stronger toward the upper portion of the cloud where speeds up to 35 miles per hour occur. Although much weaker, the updraft actually begins at the ground. Air also enters the cumulus from the sides of the cloud, this process being called *entrainment.* This stage may be considered to begin when the cumulus grows above the freezing level, after which water and ice particles grow large enough to give a radar echo. The cloud remains in the cumulus stage for about 15 minutes, during which interval it can grow to 25,000 or 30,000 feet (somewhat lower in high latitudes).

The mature stage (Fig. 7 · 10*b*) begins with the initial appearance of rain at the surface after drops have grown beyond the size supportable by the updrafts. Updraft speed still increases with elevation and appears to be at a maximum in the early mature stage when speeds of 70 miles per hour may occur, while strongest downward speeds are a little more than half this. The region of principal organized downdrafts occurs in the central forward portion of the cloud, downdrafts being initiated by the drag of

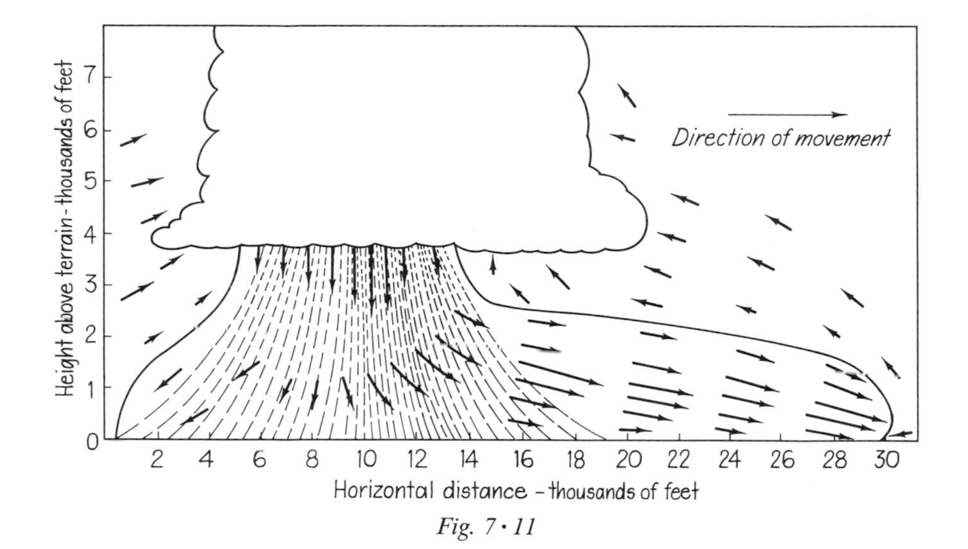

Fig. 7 · 11

falling water and ice particles, together with the effects of cool air entrainment from the sides. When the downdraft reaches the surface, the air spreads horizontally, with the strongest motion being in the direction of storm travel—usually eastward. Associated with the downdraft are the heaviest showers of the storm, considerable gustiness, a fall in temperature, and an abrupt increase in pressure—all of which are experienced by a ground or sea-level observer. The phenomena below the cloud are shown in Fig. 7 · 11. Hail occurs during the mature stage in many but not all storms and grows in concentric zones from particles being carried cyclically above and below the freezing level. During the mature stage, about 15 to 30 minutes, the thunderstorm cell reaches its maximum vertical extent—30,000 to 60,000 feet. Upper westerlies may flatten out the upper portion of the cloud into the well-known *anvil top*, from which thin *pseudocirrus* may be blown.

The dissipation stage (Fig. 7 · 10*c*) begins when the region of downdraft has spread over the entire cell in the lower levels. During this stage the downdraft region spreads vertically as well as horizontally until the entire cell becomes a region of either weak downdraft or air with no motion.

LIGHTNING AND THUNDER. In fair weather the earth normally has a negative charge with respect to the air, the potential gradient being directed downward at approximately 100 volts per meter. The intense friction of air on hydrometeors within the cumulonimbus builds very high charges in

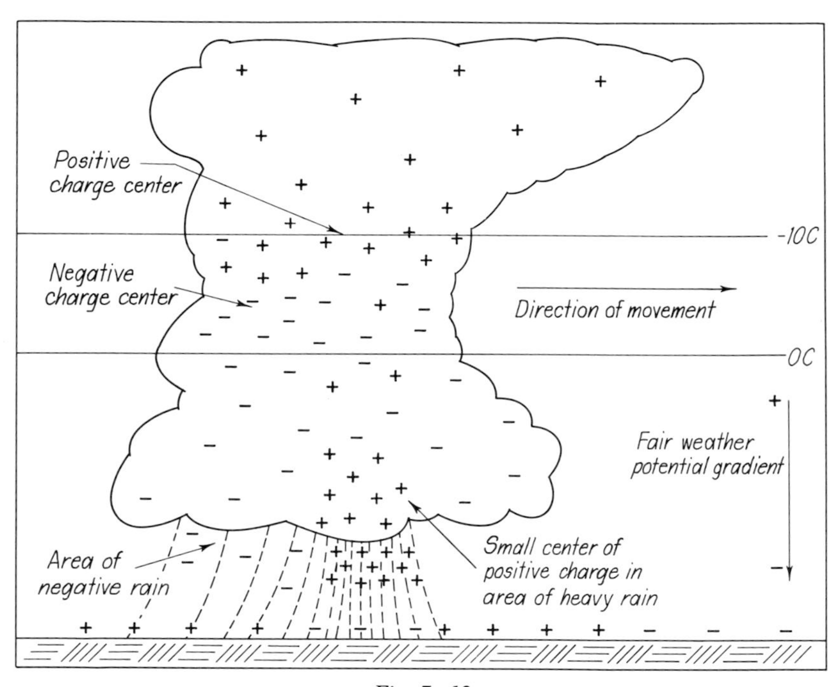

Fig. 7 · 12

such clouds, with a clustering of positive charges in the upper portion and negative charges in the lower portion of the cloud; the latter results in a positive charge in the surface beneath, reversing the fair-weather occurrence as shown in Fig. 7 · 12. When the potential difference becomes high enough, discharges (lightning) may occur from cloud to ground, cloud to cloud, or within the same cloud. The rapid expansion and contraction of highly heated air along the discharge path produces the almost explosive sound called *thunder,* which may reverberate by successive reflection from neighboring clouds. Sheet lightning is simply the general illumination of the sky produced by a lightning streak obscured by clouds or below the horizon.

Frequently in thundery weather at sea, a high voltage or potential difference is built up between a ship and the air or clouds above. This concentration of static electricity on the vessel tends to leap from all the pointed objects such as the ends of masts and spars. A purplish or bluish spray of light results, called *St. Elmo's fire.* Balls of lightning have been observed rolling along the masts and rigging of ships and disappearing with a sudden explosion.

TYPES OF THUNDERSTORMS. Most thunderstorms can be classified into types which depend on their mode of origin. The steeply ascending air required for initiation of a storm is usually provided by one of several

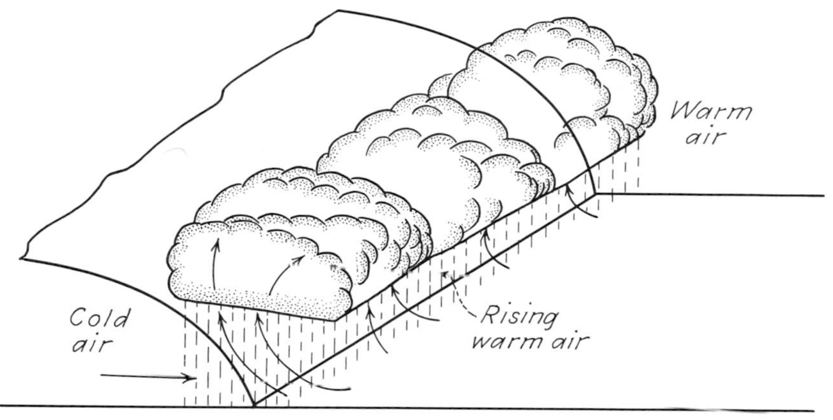

Fig. 7 · 13 Formation of a band of storms along a steeply sloping front.

factors: pronounced local heating, orographic uplift (on windward sides of mountains), steep frontal surfaces, or advection.

Air-mass Thunderstorms. These include storms resulting from the first two of the above causes. Strong local heating resulting in convection usually produces isolated storms within a single air mass which are known variously as *local* or *convection* storms. *Orographic,* or mountain, storms result from a pronounced uplift of air on the windward sides of steep mountains when the lapse rate is superadiabatic (unstable).

Local thunderstorms of marine origin are most frequent in the early morning hours. On a clear night the air temperature may fall considerably at high levels, while the sea itself, as explained earlier, retains the heat of the day for long periods. Consequently, the air adjacent to the sea remains warm. During the night the temperature contrast between warm sea-level air and high-altitude air becomes more pronounced, developing a high lapse rate, necessary to all thunderstorms. Finally, any surface temperature inequality causes a local rising air column to form, culminating in a storm.

Over land, local thunderstorms are most common in the late afternoon, when the effect of solar heating of the land surface is greatest although nocturnal thunderstorms are quite common in the northern plains states of the United States.

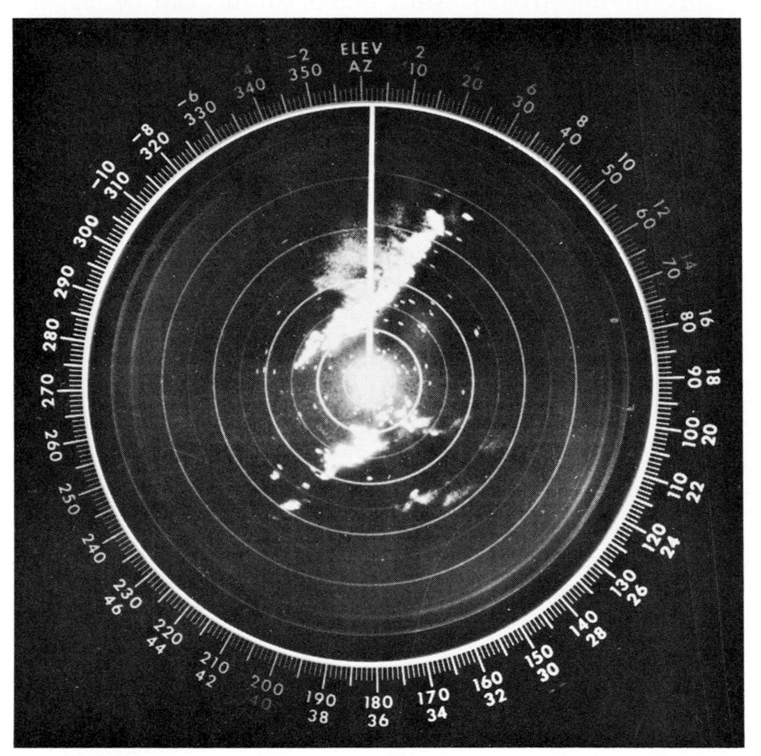

Fig. 7 • 14 Thunderstorms observed on radarscope at Washington National Airport, June 17, 1960. The range is 250 nautical miles. (U.S. Weather Bureau)

Frontal Storms. These disturbances occur when the warm air of one air mass rises over the underrunning steep boundary of a colder air mass, forming huge cumulonimbus clouds along the length of this cold-air boundary. Frontal thunderstorms thus occur as a long belt or band of storms moving progressively across country, with the forward movement of the cold air mass.

Many thunderstorms result from a third process. The lapse rate may be increased through the advection (horizontal movement) of warm air near the ground, and/or cold air at high levels. The presence of a steep lapse rate is conducive to the production of rising air columns which may culminate in thunderstorm formation. Some "trigger action" is still required to initiate the vertical air motion.

Radar Observation of Thunderstorms

Radar, both ground-based and airborne, has become a valuable tool in the observation and analysis of thunderstorms. Because radar waves are best reflected by relatively large-size hydrometeors, it is of greatest value in the study of the mature stage of the storm.

Radar is of value in outlining the horizontal and vertical dimensions of mature cumulonimbus clouds and of tracking the movement of the storms, thus permitting better forecasting. Figure 7 · 14 shows a photograph of the radarscope at Washington National Airport and indicates storms both north and south of the station.

EXERCISES

7 · 1 On an overcast night, the reflection of the ceiling light at an airport subtends an angle of 60° for an observer 4,000 feet from the light. How high is the base of the overcast?

7 · 2 Compare the nature of the air motion responsible for the cooling involved in the generation of cumulus clouds and stratus-type clouds.

7 · 3 What is the significance of the sharp base of fair-weather cumulus clouds?

7 · 4 If the ground temperature is 85°F and the dew point is 74°F, at what elevation will cumulus clouds begin to form—or what will be the level of the cloud bases?

7 · 5 Give a summary description in one sentence of the 10 principal cloud types.

7 · 6 From which clouds may precipitation be expected?

7 · 7 Give four mechanisms that produce the uplift necessary for cloud formation and the associated types of clouds.

7 · 8 Describe briefly the stages in the development of a thunderstorm.

7 · 9 Compare the normal atmospheric electrical field with the field generated by a thunderstorm.

Fig. 7 · 15 Cirrus. (Ellerman, U.S. Weather Bureau)

Fig. 7 · 16 Cirrus bands. (Davis, U.S. Weather Bureau)

Fig. 7 · 17 Cirrus—mare's tails. (U.S. Weather Bureau)

Fig. 7 · 18 Cirrostratus, terminating in cirrus filaments. (Weed, U.S. Weather Bureau)

Fig. 7·19 Cirrocumulus—degenerating from cirrostratus, also visible. (U.S. Army Photographic Section, U.S. Weather Bureau)

Fig. 7·20 High cirrus illuminated by rising sun. Dawn patrol over Iceland—1 A.M. (Official U.S. Navy photograph)

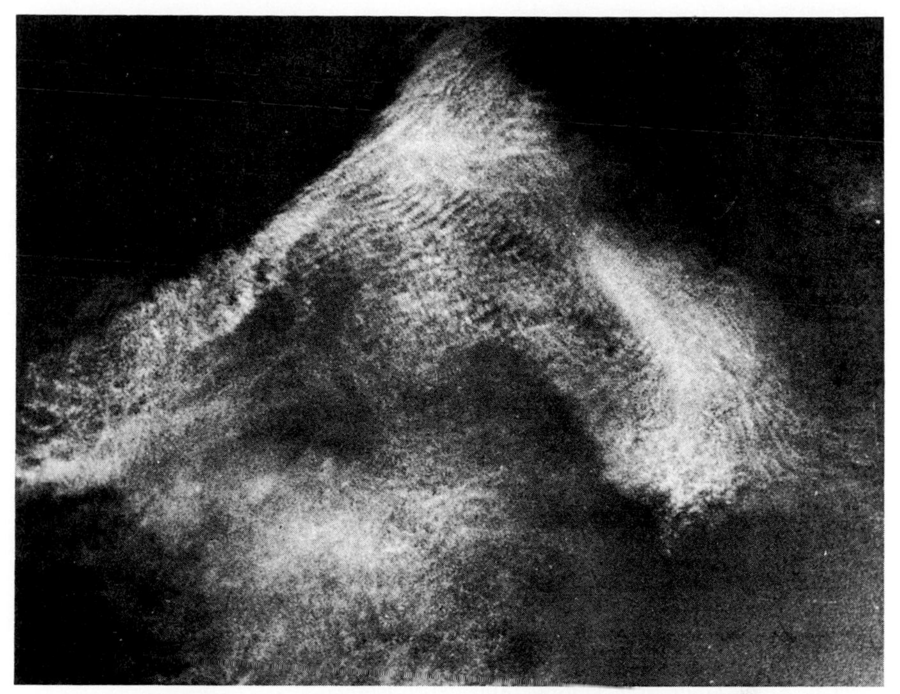

Fig. 7 · 21 *Cirrocumulus.* (Barnard, U.S. Weather Bureau)

Fig. 7 · 22 *Altocumulus bands.* (U.S. Weather Bureau)

Fig. 7 • 23 Altocumulus groups with some thicker, lower stratocumulus in right background, Kings Point, N.Y. (Photograph by Motowski)

Fig. 7 • 24 Altostratus, showing watery sun with cumulus and fractocumulus beneath. (Official U.S. Navy photograph)

Fig. 7·25 Lenticular altocumulus. (I. Tannehill, U.S. Weather Bureau)

Fig. 7·26 Thick altostratus with some altocumulus at a lower level. (U.S. Weather Bureau)

Fig. 7 · 27 Low stratus moving over a small rocky island. (U.S. Weather Bureau)

Fig. 7 · 28 Stratocumulus rolls. (Davis, U.S. Weather Bureau)

Fig. 7 · 29 Broken stratocumulus. (U.S. Weather Bureau)

Fig. 7 · 30 Cumulus clouds—Levanter cloud—hovering over Gibraltar. (Great Britain Meteorology Office, U.S. Weather Bureau)

Fig. 7 · 31 Typical cumulus and fractocumulus of fair weather. (U.S. Weather Bureau)

Fig. 7 · 32 Low cumulus and fractocumulus seen in evening with illuminated bands of cirrus visible at high levels. (Official U.S. Navy photograph)

Fig. 7 · 33 Mammato cumulus clouds, which often precede and may follow severe thunderstorm or tornado conditions. (U.S. Weather Bureau)

Fig. 7 · 34 Cumulus clouds at low level overlain by high-level cirrus. (Official U.S. Navy photograph)

Fig. 7 · 35 Fractocumulus showing pronounced wind streaming. (Official U.S. Navy photograph)

Fig. 7 · 36 Cumulus and fractocumulus at sunset.

Fig. 7 · 37 Cumulonimbus east of Pensacola, Fla., showing precipitation beneath central portion of cloud. (U.S. Weather Bureau)

Fig. 7 · 38 Thunderstorm at sea, with heavy precipitation beneath forward half of cumulonimbus cloud. (Official U.S. Navy photograph)

Fig. 7 · 39 Nose of a thunderstorm advancing along a line of frontal storms. (H. L. Crutchen, U.S. Weather Bureau)

Fig. 7 · 40 Lightning striking the Empire State Building, New York, N.Y. (Gary, U.S. Weather Bureau)

CLOUD FORMS ARRANGED IN ACCORDANCE WITH INTERNATIONAL CODE TABLES AND STATION MODEL SYMBOLS[1]

Modifying Cloud Terms Are Included in Titles

LOW CLOUDS

C_L 1

Fig. 7 · 41 Cumulus (cumulus of fair weather).

C_L 2

Fig. 7 · 42 Cumulus (heavy and swelling without anvil top).

[1] The cloud illustrations in this section have been furnished through the courtesy of the U.S. Weather Bureau.

C_L 3

Fig. 7 · 43 Cumulonimbus (without anvil top).

C_L 4

Fig. 7 · 44 Stratocumulus (formed by flattening of cumulus clouds).

C_L 5

Fig. 7 · 45 Stratocumulus (layer of stratocumulus).

C_L 5

Fig. 7 · 46 Stratocumulus (layer of stratocumulus).

C_L 6

Fig. 7 · 47 Stratus (layer of stratus).

C_L 7

Fig. 7 · 48 Low, broken clouds of bad weather: scud.

Fig. 7 · 49 Cumulus and stratocumulus—above (fair-weather cumulus and stratocumulus).

MIDDLE CLOUDS

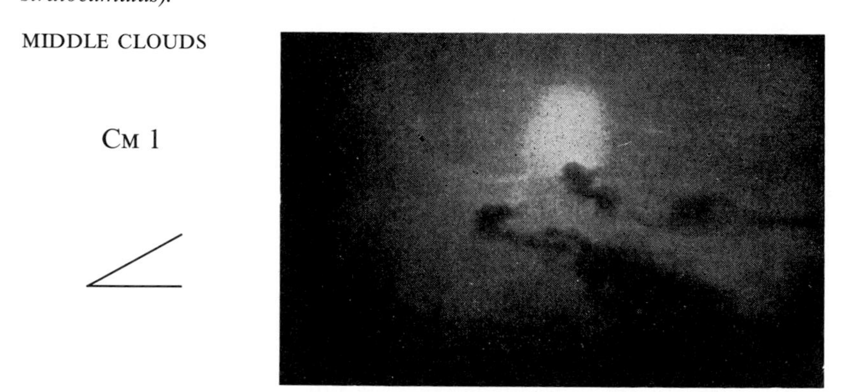

Fig. 7 · 50 Altostratus translucidus (typical thin altostratus).

Fig. 7 · 51 Altostratus opacus (typical thick altostratus).

Fig. 7 · 52 Altocumulus (altocumulus sheet at one level only).

Fig. 7 · 53 Altocumulus (altocumulus in isolated patches and more or less lenticular in shape).

Fig. 7 · 54 Altocumulus patches—lenticular.

C_M 5

Fig. 7·55 Altocumulus (altocumulus arranged in more or less parallel bands or an ordered layer advancing across the sky).

C_M 6

Fig. 7·56 Altocumulus (formed by the spreading out of tops of cumulus).

C_M 7

Fig. 7·57 Altocumulus (associated with altostratus or altostratus having an altocumulus character).

CM 7

Fig. 7 · 58 Altocumulus (see description Fig. 7 · 57).

CM 8

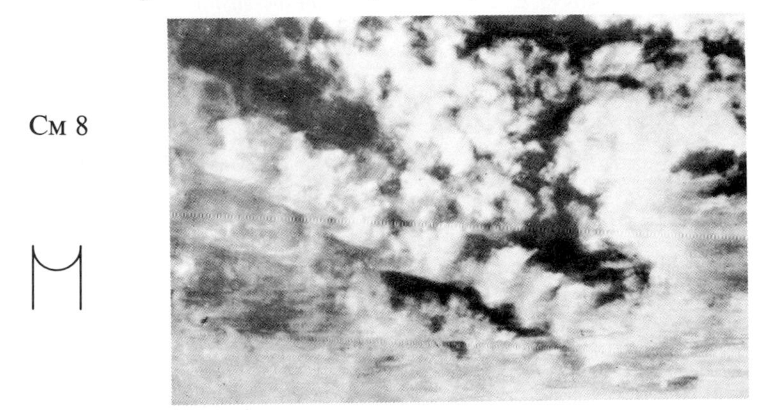

Fig. 7 · 59 Altocumulus (scattered cumuliform tufts).

CM 9

Fig. 7 · 60 Chaotic altocumulus. (Chaotic sky appearance with altocumulus in several sheets at different levels, associated with fibrous cloud veils.)

Fig. 7 · 61 Cirrus filosus (scarce delicate cirrus, not increasing, scattered and isolated).

Fig. 7 · 62 Cirrus filosus (abundant delicate cirrus but not forming a continuous layer).

Fig. 7 · 63 Cirrus (dense white cirrus).

CH 4

Fig. 7·64 Cirrus (cirrus increasing, generally in the form of hooks ending in a point or a small tuft).

CH 5

Fig. 7·65 Cirrus below 45° altitude. (Cirrus in bands, or cirrostratus, advancing over the sky but not more than 45° above the horizon.)

CH 6

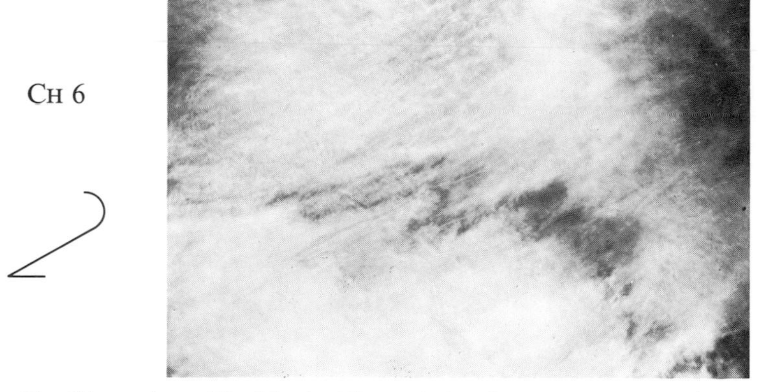

Fig. 7·66 Cirrus above 45° altitude. (Cirrus, in bands, or cirrostratus, advancing over the sky and more than 45° above the horizon.)

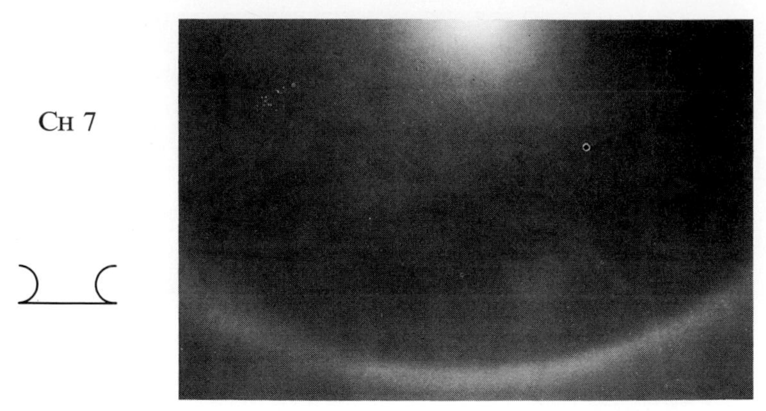

Fig. 7 · 67 Cirrostratus filosus (veil of cirrostratus covering the whole sky with cumulus below).

Fig. 7 · 68 Cirrostratus (cirrostratus layer not increasing and not covering the whole sky).

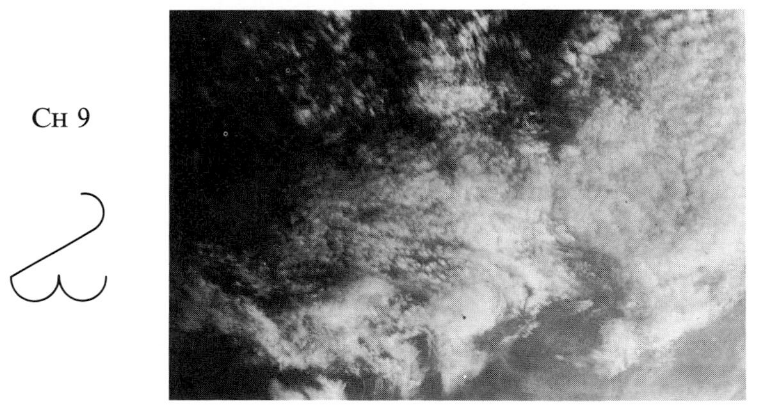

Fig. 7 · 69 Cirrocumulus. (Cirrocumulus predominating, associated with small amounts of cirrus or cirrostratus.)

CHAPTER 8

Atmospheric Pressure

Meaning of Atmospheric Pressure

The detectable atmosphere extends many hundreds of miles above the earth's surface, being held by the attraction of the earth's gravity. Historically, air pressure has usually been described in terms of certain equivalent values, such as the mass that could be balanced by the atmosphere or the height to which a column of mercury would rise in a barometer in response to the pressure. For example, the average weight (that is, mass) that is balanced by the pressure of the air on a square inch of the earth's surface at sea level is 14.7 pounds; or under the same conditions, mercury stands at a height of 29.92 inches (or 76 centimeters) in a standard barometer.

But neither of these adequately describes the *pressure* of the atmosphere. Technically, pressure is defined as the force per unit area. In the case of the atmosphere, this force is really determined by the number and speed of air molecules which strike a given surface, usually taken as a square inch or square centimeter, depending on the system of units used. Normally, when air pressure increases, more molecules are present in a given volume; when air pressure decreases, fewer molecules are present.

To clarify this a bit more, imagine a given pressure to exist at a given time and place. This pressure is determined by the total amount of air from the place measured to the top of the atmosphere. If the amount of air in this column changes at any place, the measured pressure will change owing to the gravitational effect of overlying air on the air below. Further, with the conditions referred to above, imagine that a completely sealed box of air is removed and isolated from the rest of the atmosphere—with the box surrounded by a vacuum. If a pressure-measuring instrument is enclosed in the box, it would be observed to indicate the same pressure as was shown by the entire column of air prior to the removal and isolation.

Although technically pressure must be given in terms of a force, this force, as noted earlier, can be equated with a unit of length, for example, the height to which it will support a column of mercury in a barometer, or it may be given in terms of the mass of mercury supported in the tube.

The Laws of Gases

There are a number of physical relationships so fundamental to meteorology, and so important to anyone pursuing the subject further, that brief attention must be given them.

Historically, the first of these, relating pressure and volume, was discovered by Robert Boyle. According to this relationship, known as *Boyle's law,* pressure and volume are inversely related such that, if the temperature remains constant, a twofold pressure increase on a gas results in a decrease to one-half the original volume. A threefold pressure increase decreases the volume to one-third the original value, etc. Thus, the product of the pressure and volume must be a constant, or

$$pV = K$$

A later discovery, known as *Charles's law,* which relates temperature and pressure, was developed by Jacques Charles. According to this law, if the volume remains constant, the pressure of a gas varies directly with the temperature. More specifically, the pressure increases with increase of temperature at the rate of 1/273 of the value at 0°C for each 1°C change in temperature, or

$$p = p_o + t\tfrac{1}{273}p_o$$

where p is the pressure at any time, p_o is the pressure at 0°C and t is the change of temperature in °C. By factoring p_o, the more common expression of the law is found as

$$p = p_o(1 + \tfrac{1}{273}t)$$

The law of *Gay-Lussac* relates temperature and volume, in the same manner as above, but with the pressure remaining constant. The volume of a gas increases with increase of temperature at the rate of 1/273 of the volume at 0°C for each 1° increase in temperature, or

$$V = V_o + t\tfrac{1}{273}V_o$$

or

$$V = V_o(1 + \tfrac{1}{273}t)$$

where V is the volume at any time, V_o is the volume at 0°C, and t is the change in temperature in °C.

By combining these relationships, it is seen that pressure and volume vary directly with temperature as shown by the expression

$$pV = KT$$

where T is the absolute temperature in °K, and K is a constant of proportionality. If $\mathbf{v}$ is taken to be the volume of a unit mass (one gram-molecular weight of the gas), then the more usual expression is found as

$$p\mathbf{v} = RT$$

where R is the universal gas constant (8.314 joules).

The last equation, known as the equation of state for ideal gases, is more suitable to application in laboratory study than in meteorology. The fundamental gas equation in meteorology, modified directly from the above, is

$$p = \frac{dRT}{m}$$

where p is the pressure of the air, d is the density, R is the gas constant, T is the absolute temperature, and m is the gram-molecular weight of the air.

Pressure Units

Traditionally, units of length have been used to express air pressure, based on long use of the mercurial barometer in which fluctuations of the height of the mercury column are a function of changing pressure in the atmosphere. The "height of the glass" which has been given in either inches or millimeters is still used frequently to express atmospheric pressure. To meet the need for a pressure unit involving force rather than length, the *millibar* has come into meteorological use and stems from the *bar,* which is the conventional engineering unit of pressure.

The pressure exerted by the entire atmosphere on one square centimeter is approximately one bar (a bar being further defined as a force of one million dynes per square centimeter). Since fluctuations of atmospheric pressure are very small compared to this large value, the millibar, which is a thousandth of a bar, is used in meteorology, thus avoiding pressure expressions involving many decimal places.

NORMAL PRESSURE. Although atmospheric pressure varies continuously over a relatively small range, the average of these fluctuations is very close

Table 8·1 Conversion of Inches to Millibars

In	Mb	In	Mb	In	Mb
27.32	925	28.64	970	29.97	1,015
27.34	926	28.67	971	30.00	1,016
27.37	927	28.70	972	30.03	1,017
27.40	928	28.73	973	30.06	1,018
27.43	929	28.76	974	30.09	1,019
27.46	930	28.79	975	30.12	1,020
27.49	931	28.82	976	30.15	1,021
27.52	932	28.85	977	30.18	1,022
27.55	933	28.88	978	30.21	1,023
27.58	934	28.91	979	30.24	1,024
27.61	935	28.94	980	30.27	1,025
27.64	936	28.97	981	30.30	1,026
27.67	937	29.00	982	30.33	1,027
27.70	938	29.03	983	30.36	1,028
27.73	939	29.06	984	30.39	1,029
27.76	940	29.09	985	30.42	1,030
27.79	941	29.12	986	30.45	1,031
27.82	942	29.15	987	30.47	1,032
27.85	943	29.18	988	30.50	1,033
27.88	944	29.21	989	30.53	1,034
27.91	945	29.23	990	30.56	1,035
27.94	946	29.26	991	30.59	1,036
27.96	947	29.29	992	30.62	1,037
27.99	948	29.32	993	30.65	1,038
28.02	949	29.35	994	30.68	1,039
28.05	950	29.38	995	30.71	1,040
28.08	951	29.41	996	30.74	1,041
28.11	952	29.44	997	30.77	1,042
28.14	953	29.47	998	30.80	1,043
28.17	954	29.50	999	30.83	1,044
28.20	955	29.53	1,000	30.86	1,045
28.23	956	29.56	1,001	30.89	1,046
28.26	957	29.59	1,002	30.92	1,047
28.29	958	29.62	1,003	30.95	1,048
28.32	959	29.65	1,004	30.98	1,049
28.35	960	29.68	1,005	31.01	1,050
28.38	961	29.71	1,006	31.04	1,051
28.41	962	29.74	1,007	31.07	1,052
28.44	963	29.77	1,008	31.10	1,053
28.47	964	29.80	1,009	31.13	1,054
28.50	965	29.83	1,010	31.15	1,055
28.53	966	29.85	1,011	31.18	1,056
28.56	967	29.88	1,012	31.21	1,057
28.58	968	29.91	1,013	31.24	1,058
28.61	969	29.94	1,014	31.27	1,059

to a value adopted for certain standard conditions, defined as the *standard atmosphere.* At a temperature of 15°C and a latitude of 45° the *normal pressure* is given as 1,013.2 millibars. This latter value is equivalent to 29.92 inches and 760 millimeters. Thus, 1 inch is equivalent to 33.86 millibars and 25.40 millimeters; 1 millibar is about the same as 0.03 inch and 0.75 millimeter. Most United States ships still record and report observations in inches, while most European vessels use the metric (millimeter) units in reporting, although in time the millibar will become standard. Conversion tables for both English and metric units are given in Tables 8 · 1 and 8 · 2, respectively.

Table 8 · 2 Conversion of Millimeters to Millibars

Mm	Mb	Mm	Mb	Mm	Mb
696	927.9	726	967.9	756	1,007.9
697	929.3	727	969.3	757	1,009.3
698	930.6	728	970.6	758	1,010.6
699	931.9	729	971.9	759	1,011.9
700	933.3	730	973.3	760	1,013.3
701	934.6	731	974.6	761	1,014.6
702	935.9	732	975.9	762	1,015.9
703	937.3	733	977.3	763	1,017.2
704	938.6	734	978.6	764	1,018.6
705	939.9	735	979.9	765	1,019.9
706	941.3	736	981.3	766	1,021.2
707	942.6	737	982.6	767	1,022.6
708	943.9	738	983.9	768	1,023.9
709	945.3	739	985.3	769	1,025.2
710	946.6	740	986.6	770	1,026.6
711	947.9	741	987.9	771	1,027.9
712	949.3	742	989.3	772	1,029.2
713	950.6	743	990.6	773	1,030.6
714	951.9	744	991.9	774	1,031.9
715	953.3	745	993.3	775	1,033.2
716	954.6	746	994.6	776	1,034.6
717	955.9	747	995.9	777	1,035.9
718	957.3	748	997.3	778	1,037.2
719	958.6	749	998.6	779	1,038.6
720	959.9	750	999.9	780	1,039.9
721	961.3	751	1,001.3	781	1,041.2
722	962.2	752	1,002.6	782	1,042.6
723	963.6	753	1,003.9	783	1,043.9
724	965.3	754	1,005.3	784	1,045.2
725	966.6	755	1,006.6	785	1,046.6

Pressure Instruments

As previously noted, atmospheric pressure varies from place to place and from time to time at the same place. To understand and interpret weather conditions it is necessary to measure this pressure and its changes. The standard instruments for this purpose are the *mercurial barometer,* the *aneroid barometer,* and the *barograph.*

MERCURIAL BAROMETER. The principle of the mercurial barometer is very simple. Basically it consists of a long, hollow glass tube from which the air is evacuated. This tube is placed with its open end down, immersed in a receptacle called a *cistern,* which is filled with mercury. The pressure of the outside air then forces the mercury in the cistern upward into the vacuum chamber within the tube. The mercury will rise until its weight in the tube just balances the pressure of the air on an area of the mercury in the cistern just equal to the interior cross sectional area of the tube. Thus if the tube area is 1 square inch and the air pressure is normal, the mercury will rise until its weight within the tube is 14.7 pounds. As the atmospheric pressure changes, the height of the mercury will fluctuate accordingly.

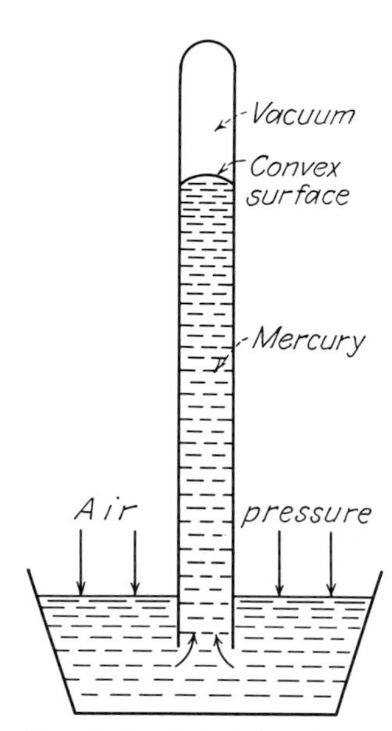

Fig. 8·1 Principle of mercurial barometer.

Regardless of the area of the tube, the mercury will always rise to the same height at a given time. If the tube is narrow, as it commonly is, the mercury within simply balances the air pressure on a smaller area in the cistern. Under normal pressure conditions the mercury will rise to a height of 29.92 inches (1,013.2 millibars) in the barometer tube.

If the barometer is calibrated in inches, there will be a fixed scale near the top of the barometer divided into inches and tenths of inches. A sliding vernier scale, much the same as that on a sextant, permits readings to a tenth of the smallest fixed scale division, or to a hundredth of an inch.

When in a narrow tube, mercury always exhibits a curved or convex surface as a result of surface tension. As the top of the curve represents the actual mercury height, readings are taken for the uppermost part of the curving mercury surface.

Marine mercurial barometers have been devised for use on shipboard which minimize agitation of the mercury caused by movements of the vessel. They are also mounted suspended in gimbals in order to remain as nearly vertical as possible. When not in use the barometer is locked in a protecting wooden case.

CORRECTIONS TO MERCURIAL BAROMETRIC READINGS. For one to attempt to foretell weather conditions, it is normally necessary to observe only the *change* in pressure and not so much the actual reading. For local interpre-

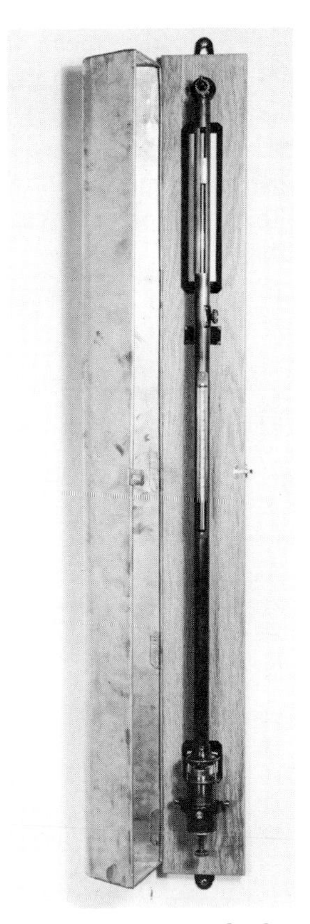

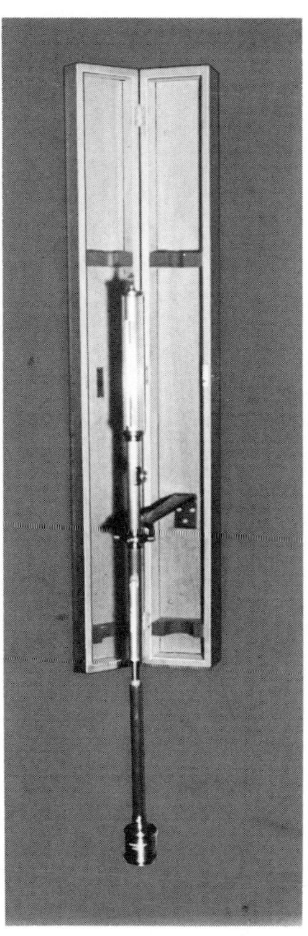

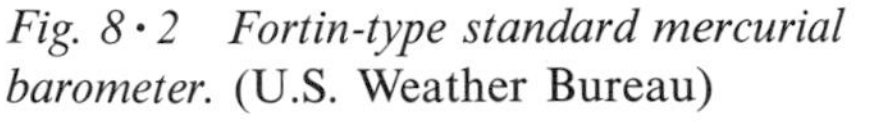

Fig. 8·2 Fortin-type standard mercurial barometer. (U.S. Weather Bureau)

Fig. 8·3 Marine mercurial barometer shown in position for use.

tation, whether the "glass" is rising or falling is of more importance than the actual value. However, when data are sent to a central station, it is important to transmit pressure readings, as these are needed for general weather analysis. If barometer readings are to be sent from many points to a central station, these readings must show true differences in atmospheric pressure and not differences resulting from varied circumstances under which the readings are taken. In other words, all readings must be reduced to normal conditions of sea level, latitude, and temperature. Accordingly the following corrections are necessary after the barometer is read: elevation correction, temperature correction, latitude or gravity correction, and instrument correction. The *station pressure* is the observed pressure including all corrections except elevation.

Elevation Correction. Unless a barometer is located at sea level, it will naturally show a lower reading than at that level, since pressure decreases with elevation. All barometer readings should be reduced to mean sea level by *adding* the proper correction for elevation. Normal pressure is approximately 30 inches. At 900 feet it would be one-thirtieth less or approximately 29 inches. A barometer on a ship's bridge 30 feet above sea level would thus have a correction of 0.03 inch. Since the barometer is read to a hundredth of an inch, this is a significant difference. Note that this correction is always positive. Complete corrections are given in Table 8 · 3.

Table 8 · 3. Reduction of Barometric Reading to Mean Sea Level (Reading, 30 in. The correction is always to be added.)

Height, ft	Temperature of air (dry bulb), °F									
	0°	10°	20°	30°	40°	50°	60°	70°	80°	90°
5	0.01	0.01	0.01	0.01	0.01	0.01	0.01			
10	0.01	0.01	0.01	0.01	0.01	0.01	0.01	0.01	0.01	0.01
15	0.02	0.02	0.02	0.02	0.02	0.02	0.02	0.02	0.02	0.02
20	0.02	0.02	0.02	0.02	0.02	0.02	0.02	0.02	0.02	0.02
25	0.03	0.03	0.03	0.03	0.03	0.03	0.03	0.03	0.03	0.03
30	0.04	0.04	0.04	0.04	0.03	0.03	0.03	0.03	0.03	0.03
35	0.04	0.04	0.04	0.04	0.04	0.04	0.04	0.04	0.04	0.04
40	0.05	0.05	0.05	0.05	0.04	0.04	0.04	0.04	0.04	0.04
45	0.06	0.05	0.05	0.05	0.05	0.05	0.05	0.05	0.05	0.05
50	0.06	0.06	0.06	0.06	0.06	0.06	0.05	0.05	0.05	0.05
55	0.07	0.07	0.06	0.06	0.06	0.06	0.06	0.06	0.06	0.06
60	0.07	0.07	0.07	0.07	0.07	0.07	0.06	0.06	0.06	0.06
65	0.08	0.08	0.08	0.08	0.07	0.07	0.07	0.07	0.07	0.07
70	0.09	0.08	0.08	0.08	0.08	0.08	0.08	0.07	0.07	0.07
75	0.09	0.09	0.09	0.09	0.08	0.08	0.08	0.08	0.08	0.08
80	0.10	0.10	0.09	0.09	0.09	0.09	0.09	0.08	0.08	0.08
85	0.10	0.10	0.10	0.10	0.10	0.10	0.09	0.09	0.09	0.09
90	0.11	0.11	0.11	0.10	0.10	0.10	0.10	0.10	0.09	0.09
95	0.12	0.11	0.11	0.11	0.11	0.11	0.10	0.10	0.10	0.10
100	0.12	0.12	0.12	0.12	0.11	0.11	0.11	0.11	0.10	0.10

Table 8·4 Correction of Mercurial Barometer for Temperature (English Measures)

Add

Temperature, °F	Observed reading, in.					Temperature, °F	Observed reading, in.				
	28.5	29.0	29.5	30.0	30.5		28.5	29.0	29.5	30.0	30.5
0	0.07	0.08	0.08	0.08	0.08	16	0.03	0.03	0.03	0.03	0.04
1	0.07	0.07	0.07	0.08	0.08	17	0.03	0.03	0.03	0.03	0.03
2	0.07	0.07	0.07	0.07	0.07	18	0.03	0.03	0.03	0.03	0.03
3	0.07	0.07	0.07	0.07	0.07	19	0.02	0.02	0.03	0.03	0.03
4	0.06	0.06	0.07	0.07	0.07	20	0.02	0.02	0.02	0.02	0.02
5	0.06	0.06	0.06	0.06	0.07	21	0.02	0.02	0.02	0.02	0.02
6	0.06	0.06	0.06	0.06	0.06	22	0.02	0.02	0.02	0.02	0.02
7	0.06	0.06	0.06	0.06	0.06	23	0.02	0.02	0.02	0.02	0.02
8	0.05	0.05	0.06	0.06	0.06	24	0.01	0.01	0.01	0.01	0.01
9	0.05	0.05	0.05	0.05	0.05	25	0.01	0.01	0.01	0.01	0.01
10	0.05	0.05	0.05	0.05	0.05	26	0.01	0.01	0.01	0.01	0.01
11	0.05	0.05	0.05	0.05	0.05	27					
12	0.04	0.04	0.04	0.04	0.05	28					
13	0.04	0.04	0.04	0.04	0.04	29					
14	0.04	0.04	0.04	0.04	0.04	30					
15	0.04	0.04	0.04	0.04	0.04						

Subtract

Temperature, °F	28.5	29.0	29.5	30.0	30.5	Temperature, °F	28.5	29.0	29.5	30.0	30.5
31	0.01	0.01	0.01	0.01	0.01	66	0.10	0.10	0.10	0.10	0.10
32	0.01	0.01	0.01	0.01	0.01	67	0.10	0.10	0.10	0.10	0.11
33	0.01	0.01	0.01	0.01	0.01	68	0.10	0.10	0.10	0.11	0.11
34	0.01	0.01	0.01	0.02	0.02	69	0.10	0.11	0.11	0.11	0.11
35	0.02	0.02	0.02	0.02	0.02	70	0.11	0.11	0.11	0.11	0.11
36	0.02	0.02	0.02	0.02	0.02	71	0.11	0.11	0.11	0.12	0.12
37	0.02	0.02	0.02	0.02	0.02	72	0.11	0.11	0.12	0.12	0.12
38	0.02	0.02	0.02	0.03	0.03	73	0.11	0.12	0.12	0.12	0.12
39	0.03	0.03	0.03	0.03	0.03	74	0.12	0.12	0.12	0.12	0.12
40	0.03	0.03	0.03	0.03	0.03	75	0.12	0.12	0.12	0.13	0.13
41	0.03	0.03	0.03	0.03	0.03	76	0.12	0.12	0.13	0.13	0.13
42	0.04	0.04	0.04	0.04	0.04	77	0.12	0.13	0.13	0.13	0.13
43	0.04	0.04	0.04	0.04	0.04	78	0.13	0.13	0.13	0.13	0.14
44	0.04	0.04	0.04	0.04	0.04	79	0.13	0.13	0.14	0.14	0.14
45	0.04	0.04	0.04	0.04	0.04	80	0.13	0.14	0.14	0.14	0.14
46	0.04	0.05	0.05	0.05	0.05	81	0.14	0.14	0.14	0.14	0.14
47	0.05	0.05	0.05	0.05	0.05	82	0.14	0.14	0.14	0.14	0.15
48	0.05	0.05	0.05	0.05	0.05	83	0.14	0.14	0.14	0.15	0.15
49	0.05	0.05	0.05	0.06	0.06	84	0.14	0.14	0.15	0.15	0.15
50	0.06	0.06	0.06	0.06	0.06	85	0.15	0.15	0.15	0.15	0.16
51	0.06	0.06	0.06	0.06	0.06	86	0.15	0.15	0.15	0.16	0.16
52	0.06	0.06	0.06	0.06	0.06	87	0.15	0.15	0.16	0.16	0.16
53	0.06	0.06	0.06	0.07	0.07	88	0.15	0.16	0.16	0.16	0.16
54	0.06	0.07	0.07	0.07	0.07	89	0.16	0.16	0.16	0.16	0.17
55	0.07	0.07	0.07	0.07	0.07	90	0.16	0.16	0.16	0.17	0.17
56	0.07	0.07	0.07	0.07	0.08	91	0.16	0.16	0.17	0.17	0.17
57	0.07	0.08	0.08	0.08	0.08	92	0.16	0.17	0.17	0.17	0.18
58	0.08	0.08	0.08	0.08	0.08	93	0.17	0.17	0.17	0.17	0.18
59	0.08	0.08	0.08	0.08	0.08	94	0.17	0.17	0.17	0.18	0.18
60	0.08	0.08	0.08	0.08	0.09	95	0.17	0.17	0.18	0.18	0.18
61	0.08	0.08	0.09	0.09	0.09	96	0.17	0.18	0.18	0.18	0.19
62	0.09	0.09	0.09	0.09	0.09	97	0.18	0.18	0.18	0.18	0.19
63	0.09	0.09	0.09	0.09	0.10	98	0.18	0.18	0.18	0.19	0.19
64	0.09	0.09	0.10	0.10	0.10	99	0.18	0.18	0.19	0.19	0.19
65	0.09	0.10	0.10	0.10	0.10	100	0.18	0.19	0.19	0.19	0.20

Temperature Correction. The mercury in a barometer will expand or contract just as in a thermometer. Thus an arbitrary reference level must be taken; for the mercurial barometer this is 32°F or 0°C. Hence, when the temperature is above freezing, the mercury stands too high in the tube, and the correction is negative and must be subtracted to lower the reading of the mercury column to normal. If below freezing, the correction is added in order to raise the reading of the now contracted mercury column to normal. At freezing, the correction is zero. All mercurial barometers have an attached thermometer to indicate the temperature. The correction for any temperature can be found in Table 8 · 4.

Latitude or Gravity Correction. Since the earth is flattened at the poles and bulges at the equator, there will be a greater pull on a mercury column near the poles and a lesser pull at the equator, producing increased and decreased densities, respectively. This correction is therefore positive above 45° latitude, negative below, and zero at 45°. These correction figures are found in Table 8 · 5.

Table 8 · 5 Reduction of the Mercurial Barometer to Standard Gravity (45°) (30 in.)

LAT.,°	CORR., IN.	LAT.,°	CORR., IN.	LAT.,°	CORR., IN.	LAT.,°	CORR., IN.
0	—0.08	25	—0.05	45	0.00	70	+0.06
5	—.08	30	—.04	50	+.01	75	+.07
10	—.08	35	—.03	55	+.03	80	+.08
15	—.07	40	—.01	60	+.04	85	+.08
20	—.06	45	0.00	65	+.05	90	+.08

Instrument Correction. This varies with the particular instrument and is found by comparison with a standard barometer. The U.S. Weather Bureau will make such comparisons.

The following example is given, the corrections being those found in the tables. A barometer on a ship's bridge is 35 feet above sea level; temperature, 90°F; latitude, 10°N or S.

		Inches
Barometer as read		29.97
Elevation (sea-level) correction	+0.04	
Temperature correction	−0.17	
Latitude or gravity correction	−0.08	
Instrument correction	+0.03	
Total correction	−0.18 inch	
Corrected barometer reading		29.79

In standard weather bureau practice, the corrected value in inches is converted to millibars for transmission and entry on charts.

ANEROID BAROMETER. The aneroid barometer is a mechanical device which registers pressure. Briefly, it contains a cylindrical vacuum chamber with corrugated tops and sides known as a sylphon chamber. A strong spring within the chamber prevents it from collapsing under the air pressure. As the outside pressure changes, the chamber will either expand or contract. By means of an intricate system of levers and pulleys this change is magnified and conveyed to a pointer. The pointer swings around a dial that is calibrated in the same units as in a mercurial barometer. The whole is enclosed in a protective brass case with a glass window. A movable indi-

Fig. 8·4 Aneroid barometer with scales in both inches and millibars. (Friez Instruments Division of Bendix Aviation Corporation)

cator can be set over the barometer needle to show the change in pressure since the last reading.

Owing to the tendency of the mechanical parts to stick, this instrument should be tapped prior to each reading and should be read with the eye directly in front of the needle pointer to avoid parallax error.

The aneroid is met with much more frequently than is the mercurial, owing to its compactness and ease of mounting and reading. However, of the two, the mercurial is the standard instrument, being more reliable and accurate.

CORRECTIONS APPLIED TO ANEROID READINGS. The aneroid has essentially two corrections: an elevation correction, which is similar to that of the mercurial, and an instrument correction. All good aneroids are compensated for temperature and obviously show no gravity effect. Should the instrument error become too large, the pointer can be reset after comparison by means of a setscrew on the back of the case. Such barometers should be checked every three months. This can be done by taking the barometer to the nearest U.S. Weather Bureau office or, if not convenient, by sending to the Bureau a series of readings on special Weather Bureau cards. The necessary correction will be returned by mail.

Fig. 8·5 Microbarograph. (Friez Instruments Division of Bendix Aviation Corporation)

BAROGRAPH. The barograph is simply a recording aneroid barometer. The effect of changes of atmospheric pressure on the vacuum chamber is communicated to a long pen arm which exaggerates this movement. The pen rests on a rotating drum similar to that of the thermograph and yields a continuous trace on the coordinate paper surrounding the drum. This sheet is calibrated vertically in pressure units and horizontally in time, with a line for every 2 hours. The barograph requires the same corrections as the aneroid: elevation and instrument corrections. A setscrew permits adjustment of the barograph to standard pressure after comparison with a more accurate instrument.

A sensitive barograph has been developed which permits pressure readings directly to one-hundredth of an inch (one-tenth of a millibar). This is known as a *microbarograph* and has come into more common use than the simple barograph.

THE MICROBAROVARIOGRAPH. This is an extremely sensitive atmospheric pressure recorder which is of value in the investigation of certain subtle but important features of the atmosphere related to very small changes in pressure—changes of the magnitude of hundredths to thousandths of a millibar. Such devices are so sensitive, in order to record these minute pressure variations, that ordinary daily changes would be so great as to

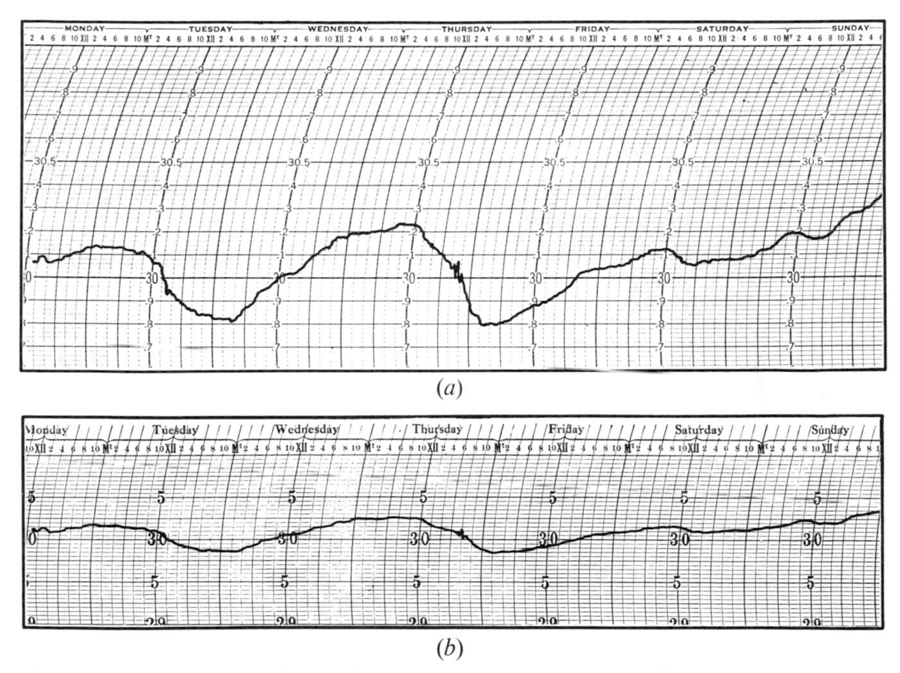

Fig. 8·6 Comparison of microbarogram (a) with barogram (b) for the same week, showing the vertical exaggeration of the former. The chart shows passage of two lows with rising pressures at the close of the week. The irregularity of the barograph tracing on Thursday evening marks the abrupt minor pressure fluctuations attending the passage of a severe thunderstorm.

throw the recorder far off scale. Since the small variations are usually of very short duration—seconds to minutes, it is possible to construct the sensitive recorders to filter out the large variations because these have a much longer duration—of a day or more. The sensitive microbarovariograph can easily detect the minute pressure fluctuations produced by the passage of waves through the air and permit the study of these waves

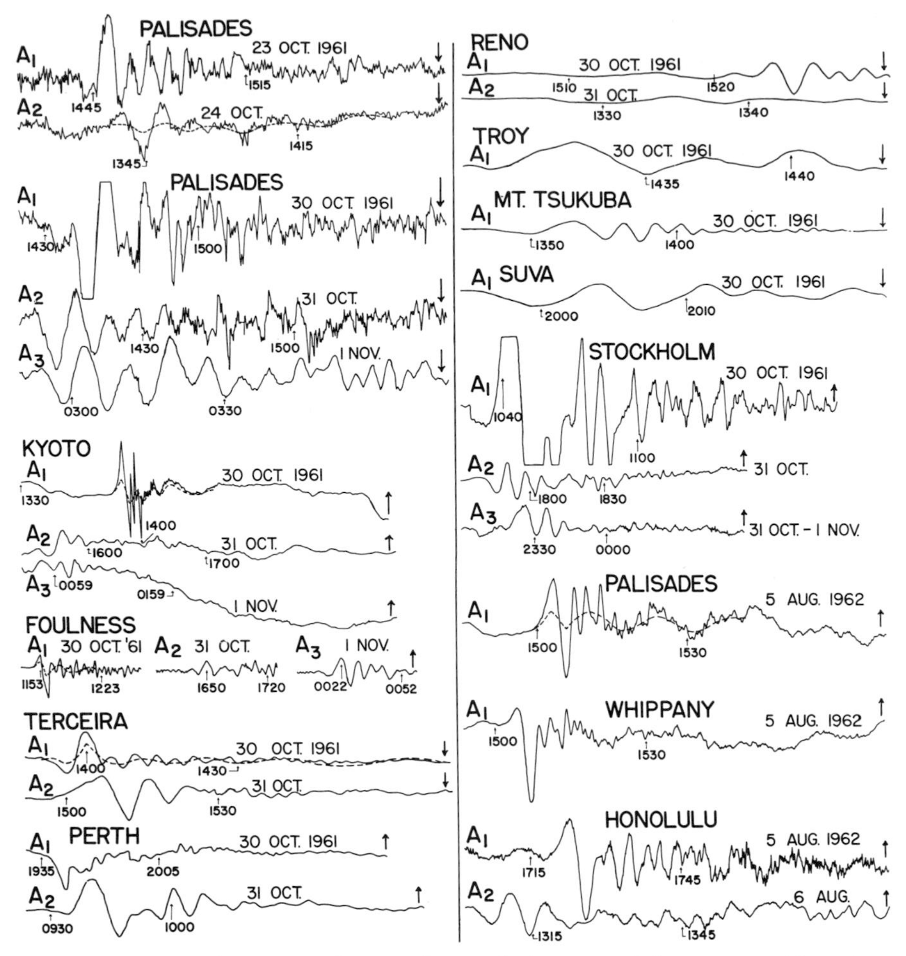

Fig. 8 · 7 Atmospheric waves from nuclear explosions recorded on very sensitive micro-barographs. A_1 *is the train of direct waves;* A_2, *the waves traveling the opposite direction around the earth;* A_3, *the second passage of the direct waves. The greatest amplitudes are about 0.5 millibar. Arrows indicate the direction of increasing pressure.*

whose behavior is very revealing about many important scientific aspects of the atmosphere. Figure 8 · 7 illustrates some examples of waves recorded from large nuclear explosions in the atmosphere and shows details not available from conventional instruments.

Pressure Variations

As in the case of temperature, atmospheric pressure also exhibits important variations vertically, horizontally, and periodically.

VERTICAL VARIATIONS. As noted earlier, the atmosphere is a mixture of gases held to the earth by gravity. Because gases are compressible, the pressure of overlying air compresses the lower air considerably and thereby greatly increases the density of the air near the ground. This compression accounts for the rapid decrease in density or pressure with increase in elevation pointed out in Chap. 1. In the lower part of the atmosphere the rate of pressure decrease is quite uniform, the decrease being one-thirtieth of the previous value for each 900-foot increase in elevation.

The pressure at any point in the atmosphere is given by the simple hydrostatic formula common to fluids in general,

$$p = hdg$$

where h is the height of the air column above the point, d the average density, and g the acceleration of gravity. The difference in pressure between two points with a relatively small vertical separation is thus given by

$$p_1 - p_2 = (h_1 - h_2)dg$$

or the change in pressure $\Delta P = \Delta hdg$, where Δh is the change in elevation. Airplane altimeters which are essentially aneroid barometers have their scales calibrated to read elevation on the basis of such vertical pressure-change relationships.

HORIZONTAL VARIATIONS. The nonuniform solar heating of the earth's surface, and consequently of the air above, results in a variety of pressure variations over the earth's surface. The horizontal pressure patterns are of two broad types: those whose average position is roughly the same and those whose position moves with time. These patterns are so closely interrelated with the physical processes in the atmosphere associated with weather changes that much of the remainder of this book will be devoted to them in one way or another.

PERIODIC PRESSURE VARIATIONS. The atmosphere is not a static body. It is a dynamic medium. At a given station the pressure changes continuously as areas of higher or lower pressure approach.

It has been observed that a periodic pressure change occurs daily. The air pressure shows two high and two low points each day, the pressure peaks occurring at 10 A.M. and 10 P.M., while the low points are midway between, at 4 P.M. and 4 A.M. This effect, which is known as the solar, or thermal, tide, is not yet fully understood.

For reasons to be explained later, this small daily pressure variation is generally overshadowed by other, greater changes in the middle latitudes. However, in the tropics and subtropics the daily change is characteristic. Figure 8·8 illustrates this characteristic variation on a Friez microbarogram recorded in tropical latitudes.

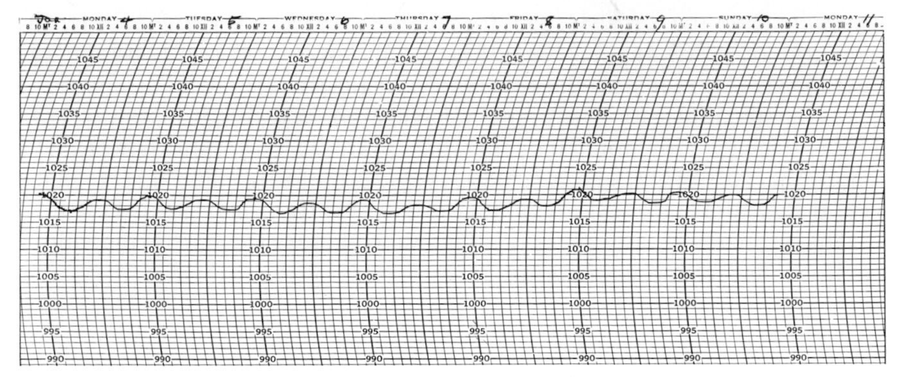

Fig. 8 • 8 Friez microbarogram showing the nature of diurnal variation in the tropics (Island of Barbados).

Isobars and the Pressure Gradient

ISOBARS. Isobars are lines drawn through points of equal pressure. The values of air pressure are obtained by the use of barometers at local observation stations on land and from ships at sea. The corrected pressure readings, reduced to sea level, are entered on maps, and points on the map showing the same pressure readings are connected with lines (isobars). Obviously all points on a map with the same pressure cannot be so joined, as an infinite number of lines would result. Hence isobars are usually drawn at pressure intervals of either 3, 4, or 5 millibars. Thus successive isobars might be drawn through points having readings of 1,008, 1,011, 1,014 millibars, etc. On U.S. Weather Bureau weather maps, millibars have replaced inches in designating the pressure.

RELATION BETWEEN ISOBARS AND THE PRESSURE GRADIENT. If isobars are lines of equal pressure, then adjacent isobars indicate a change in pressure from one isobar to the next. In Fig. 8 • 9, a continuous pressure

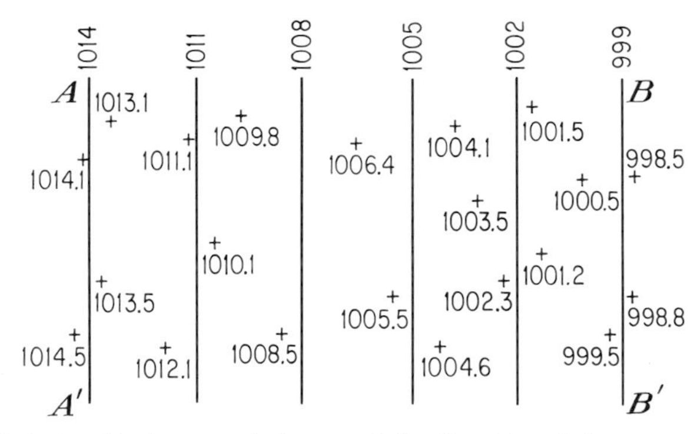

Fig. 8 • 9 Relationship between isobars and the direction of the pressure gradient.

change exists between the lines *AA′* and *BB′*, for the pressure over this distance falls from 1014 to 999 millibars. This change in pressure with horizontal distance is known as the *pressure gradient.* To see more clearly the meaning of the pressure gradient, let us consider its relationship to the isobars as regards (1) the direction of the gradient and (2) the steepness of the gradient, or the amount of change of pressure.

1. *Directional Relationship of Isobars and the Pressure Gradient.* The pressure gradient is measured along the line of greatest pressure change. This is shown in Fig. 8 · 9. From *A* to *B* there is a pressure change of 15 millibars. The same pressure change exists between *A* and *B′*, or between *A* and any other point on the 999 isobar. Clearly the gradient should be measured over the shortest distance between isobars or in a direction perpendicular to them. This is indicated by the arrow.

   Let us compare the pressure gradient to a sloping surface. Actually, the term is borrowed from the surveyor and refers to a slope of the ground surface (Fig. 8 · 10). The gradient of the surface *ABCD* is measured directly down its slope. In the same way then, the pressure gradient is measured directly down the pressure change. The dotted lines in the figure represent lines of equal altitude on the sloping surface. Just as the gradient here is perpendicular to the line of equal altitude, so the pressure gradient is perpendicular to lines of equal pressure (isobars).

   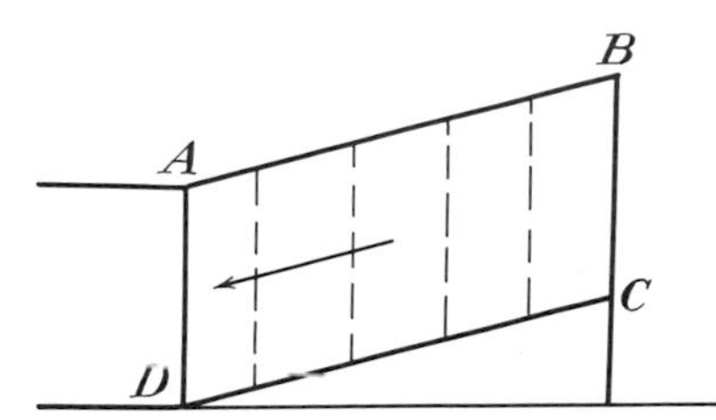

   Fig. 8 · 10 Relationship between the direction of slope of a surface and lines of equal altitude on the surface.

2. *Steepness of Pressure Gradient.* Borrowing another surveyor's term, we note that the amount of pressure change between two points is described as *steepness* of the pressure gradient. In Fig. 8 · 11, the first surface slopes up to 50 feet while the second slopes up to 100 feet over the same horizontal distance. If we draw lines of equal altitude on the surfaces with a vertical interval of 10 feet between them, it is clear that these lines drawn on the surface must be much closer together and also be more numerous in the second case. This is the steeper gradient.

   Similarly in the case of atmospheric pressure, the steeper the pressure gradient between the points, the more numerous and more closely spaced will be the isobars. Isobars relatively far apart show a very slight pressure difference or gradient. Thus the gradient between *D* and *E* is 0.5 inch in the first case and 1 inch in the second

(Fig. 8 · 12). Hence the number of isobars is doubled, and they are much closer. A barometer carried over the horizontal distance *DE* would fall twice as fast in the second as in the first case.

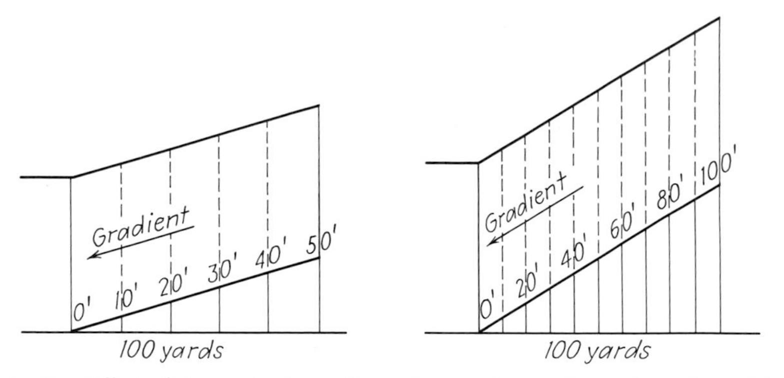

Fig. 8 · 11 Effect of change in slope of a surface on the spacing and number of lines of equal altitude.

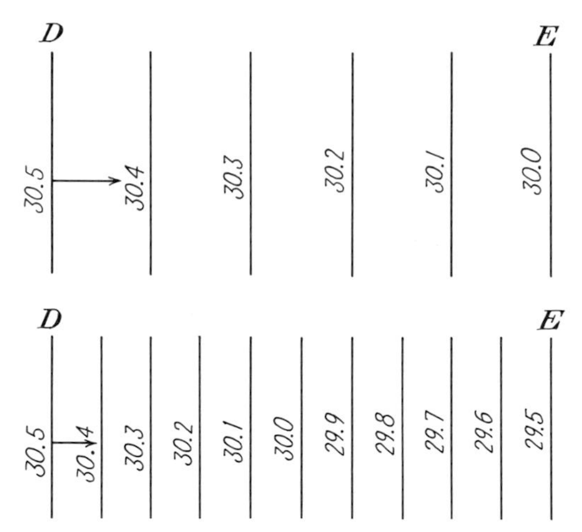

Fig. 8 · 12 Effect of change in pressure gradient on the spacing and number of isobars.

TYPES AND FORMS OF ISOBARS. If an isobar is extended sufficiently, it must always close up on itself and become a continuous curved line. However, in depicting atmospheric pressure distribution, isobars may assume various shapes and patterns for any particular part of their length. This is particularly true in the middle latitudes, between 30° and 60°, where continuous horizontal pressure variations occur. Some typical forms are shown in Fig. 8 · 13.

ISALLOBARS. We noted earlier that one group of pressure patterns moves through the atmosphere. As a result, rather continuous changes in pressure occur at stations experiencing such changes. The magnitude of the pressure change experienced depends upon the pressure gradient within the moving system, the speed of the system, and the orientation of the isobars with respect to the observer and the line of motion. Since these factors may be different for different stations, it is often more instructive to know the magnitude of the pressure change rather than simple pressure over a particular region. The change in pressure in the 3 hours prior to the observation recording time is called the *pressure tendency* and is a net value because the change may be continuous in one direction or may involve a rise followed by a fall, or the reverse. Pressure tendency is reported as a weather

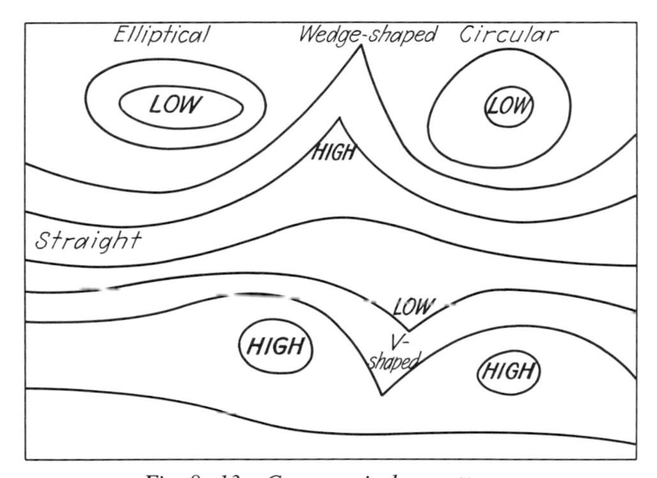

Fig. 8 · 13 Common isobar patterns.

observation item along with the other weather elements. Lines drawn connecting places having equal pressure tendencies are called *isallobars* and constitute an important tool in weather analysis.

High- and Low-pressure Areas

In the middle latitudes isobars display characteristic shapes indicating relatively large areas of alternately high and low pressure. These areas as shown by isobars are rudely circular or elliptical in shape and usually cover tens of thousands of square miles. A particular pressure configuration, whether high or low, may cover a few states or one-half of the entire country. Figure 8 · 14 shows a typical middle-latitude pressure pattern as indicated by isobars which are labeled in millibars.

In Fig. 8 · 14 the lower pressure is within the 999 isobar but never reaches

996, or another isobar would be drawn. The word *low* is simply written within the last isobar, and the whole is known as a *low-pressure area* or *low* or *depression.*

The highest pressure is within the 1,017 isobar, and that whole area is known as a *high-pressure area* or *high.* It should be emphasized that there is no particular point where the low ends and the high begins. There is rather a continuous increase in pressure from the center of the low to the center of the high, and then a steady decrease to the next low, and so on.

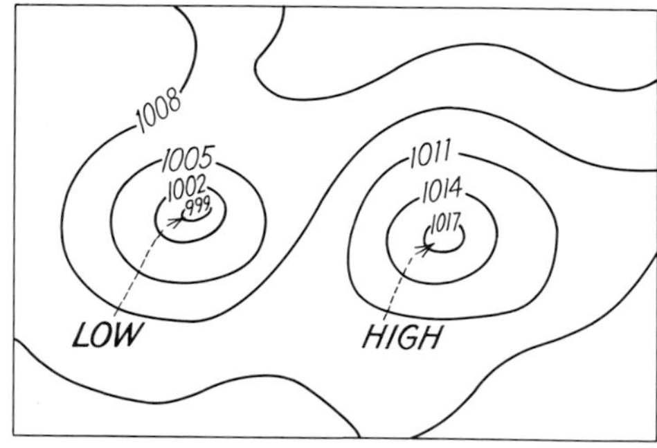

Fig. 8 · 14 A common middle latitude pressure pattern showing a low and a high.

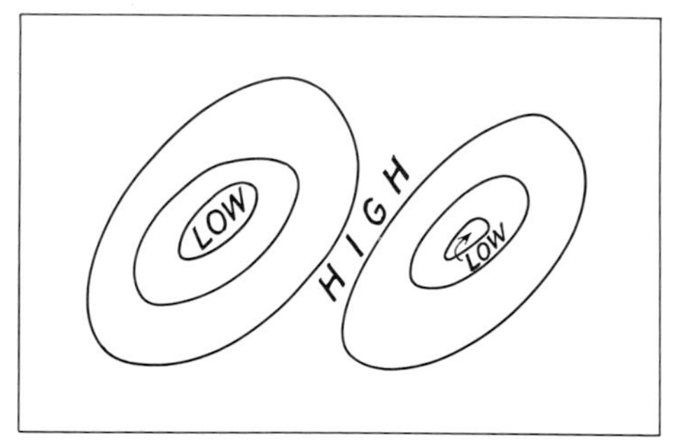

Fig. 8 · 15 Areas of low pressure separated by a ridge of high pressure.

The highs and lows are not always equally well developed. One may be much stronger than the other, and their circular or elliptical shapes may be considerably distorted at times. Thus, two well-developed low-pressure areas may exist on a map adjacent to each other, in which case they are separated by a narrow *ridge* of higher pressure (Fig. 8 · 15). The reverse may be true, with two strong highs separated by a narrow *trough* of low pressure. Whatever the case, the uniformity or irregularity of the pressure gradient will be indicated by the shape and spacing of the isobars, as explained earlier.

A continuous series of highs and lows exists in the atmosphere of the middle latitudes, which may be likened to water waves whose crests and troughs follow each other continuously. In the same way these low- and high-pressure areas (waves) move continuously through the atmosphere of the middle latitudes, more or less in the direction of the earth's rotation, or approximately west to east. The passing of the highs and lows is associated with the weather changes experienced in these latitudes.

The rise or fall in a particular barometer is now seen to herald approaching highs or lows. We know roughly from experience that a falling barometer heralds bad weather and a rising one, good weather. A partial reason for this can now be seen. In order for a low-pressure area to be maintained, air must rise within it. Since rising air cools adiabatically, clouds and precipitation may result. For a high-pressure area to be maintained, air must descend within it, thereby warming adiabatically and tending to become dry and clear. A more complete picture of the wind and weather structure of lows and highs will be developed in ensuing chapters.

Isobaric Surfaces

Although we have considered pressure variation and distribution in the horizontal dimensions, it must be emphasized that the atmosphere is three-dimensional. Pressure patterns at the surface are but the low-level expression of pressure effects that extend well up into the troposphere.

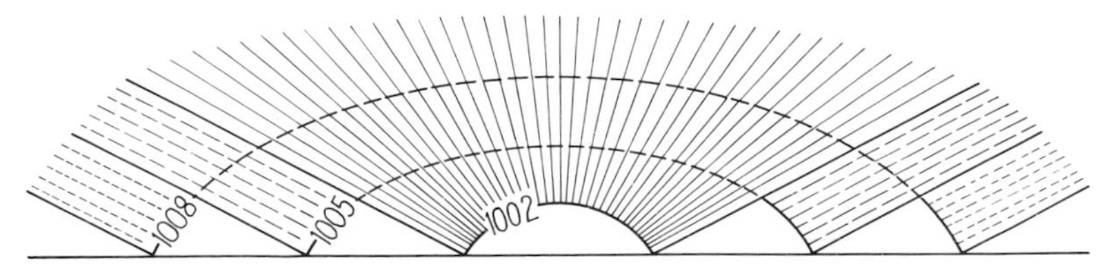

Fig. 8 · 16 Isobaric surfaces in a low-pressure area.

Just as we can draw lines of equal pressure on the two-dimensional map at the ground level, we can imagine *surfaces* of equal pressure extending from the ground up into the atmosphere. Thus, the isobars of a circular low-pressure area are simply the lines of contact between surfaces of equal pressure—isobaric surfaces—and the ground. An attempt to illustrate this three-dimensional relationship for both low- and high-pressure areas is shown in Figs. 8 · 16 and 8 · 17, respectively. The slopes of the isobaric surfaces which are shaded are, of course, exaggerated for purposes of illustration.

These drawings illustrate one of the hazards of flying by pressure altimeter alone, as occurred particularly during the early days of aviation. When flying and maintaining altitude by the use of the altimeter only, without either frequent correction or absolute elevation indicators, the plane will tend to fly on an isobaric surface. When heading toward a region of lower pressure, a slow decrease in elevation will occur. At times of poor visibility this presented a real hazard before there were additional means of obtaining elevation.

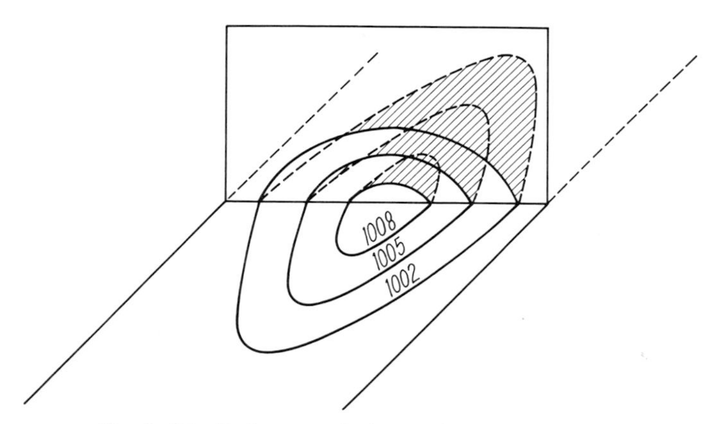

Fig. 8 · 17 Isobaric surfaces in a high-pressure area.

The true appearance of isobaric surfaces within lows and highs also depends upon comparative temperatures within each. If the center of the low is relatively warm, the effect of the warmed, lifted air will be to decrease the vertical pressure gradient, causing the isobaric surfaces to flatten with increased elevation. If the effect is strong enough, the slope may reverse, resulting in a high above the low-pressure center. A warm high-pressure center, by similar reasoning, will increase the slopes of the isobaric surfaces, thus intensifying the high with elevation.

EXERCISES

8 · 1 What is normal or standard sea-level pressure expressed in inches and millimeters of mercury, and in millibars?

8 · 2 Calculate (*a*) the number of millibars equivalent to 1 inch of mercury and (*b*) the number of millimeters equivalent to 1 inch of mercury.

8 · 3 If the ocean acts like an inverted barometer to changes in air pressure (sea level rises when pressure falls, and the reverse), and if normal air pressure can support a column of water 34 feet high,

determine the change of sea level in millimeters and inches that corresponds to a change of one millibar of air pressure.

8·4 Using the appropriate correction tables, determine the corrected reading for a mercurial barometer that indicates a pressure of 30.08 inches in a ship's wheelhouse 45 feet above sea level. The vessel is in latitude 18°N and the temperature is 85°F. The instrument correction is known to be −0.04 inch.

8·5 Determine the pressure gradient in millibars per 100 miles for the cyclone in Fig. 14·6. (Note: One degree of latitude equals 60 nautical miles.)

8·6 If the pressure of a gas at 0°C is 1,010 millibars, find the pressure at 30°C if the volume is unchanged.

8·7 If the volume of a gas at 0°C is 5 cubic meters, find the volume at 20°C assuming no change in pressure occurs.

8·8 Explain the relationship between isobar spacing and the horizontal pressure gradient.

8·9 Make a schematic diagram showing isobaric surfaces at a 3-millibar interval in a cross section through two low-pressure systems and an intervening high. Assume the highest isobaric value is 1,014 millibars and the lowest, 1,002 millibars. (Draw a cross section rather than a three-dimensional diagram.)

CHAPTER 9

Wind—Observation and Theory

Wind is air in horizontal motion. Such motion tends to equalize lateral differences in temperature, humidity, and pressure that may exist in the atmosphere. Although this equalization is never attained, because new differences are created continuously, the wind does maintain an approximate average state for these differences. Wind is thus an extremely important regulator of the atmosphere.

Characteristics of Wind Motion

Wind refers to air moving horizontally or approximately so. Vertically moving air columns are called *currents* and give the effect known in aviation as "bumpy air." An *air pocket* is simply a current whose descent is rapid enough to create the illusion of an absence of air to support the aircraft. Although important, especially in the development of certain clouds and resulting precipitation, the amount of air moving vertically is negligible compared to that transported as wind.

Air motion at low levels is strongly affected by friction with the underlying surface and by thermal convection if the surface is warmer than the air. The smooth or sheetlike flow of air (called *laminar flow*) which prevails aloft is converted into turbulent or eddy motion by these effects, producing the familiar gusts and lulls. Wind turbulence increases with the roughness of the terrain as well as with the degree of temperature contrast between the warm surface and the cooler air above. Turbulence also increases as the wind speed increases. In Fig. 9 · 1, a comparison is shown for winds at the surface and those aloft.

Since wind-eddy action varies with the irregularity in both the relief and temperature of the surface, it follows that winds should be more uniform over water than over land owing to the greater roughness and temperature variations of the latter. As these effects also reduce the effective wind speed, winds are stronger at sea than on land for the same meteorological conditions.

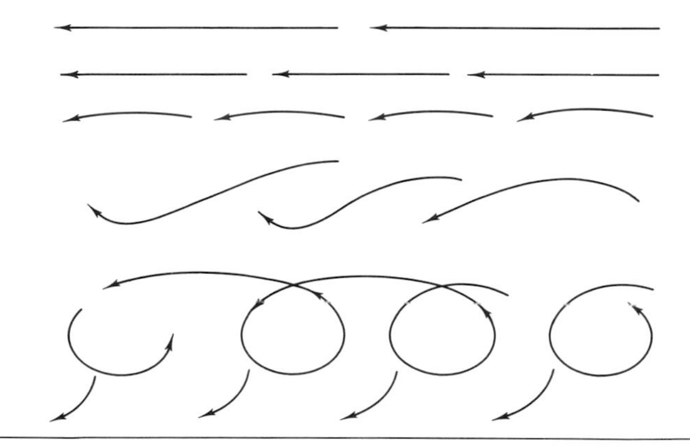

Fig. 9 · 1 Turbulent wind motion at the surface with more laminar (straight-line) flow shown in the overlying air, where friction is much less.

The results of surface friction and convection decrease with altitude, with a consequent decrease of turbulence and increase in the steadiness and velocity of the wind. After a very rapid increase in which the ground velocity is doubled in a few tens of feet, the velocity increases more slowly on the average until an altitude of 1,500 to 2,000 feet. However in rugged, mountainous country turbulence produced by the exaggerated relief may extend to many thousands of feet, producing much of the roughness experienced even by modern high-level jet aircraft. But such conditions are limited to well-known mountainous locations. High-level turbulence of a more widespread occurrence results from the *shearing* effect of wind streams having strong contrasts in velocity.

Although the wind may be quite variable in direction and speed for short periods as a consequence of eddy action, the main flow of air usually is steady on the average from a given direction. The variable gusty quality of winds should not be confused with the true shifting of winds. A *wind shift* is a progressive change in direction. The causes of such shifting will be studied later.

A wind may shift in either a clockwise or a counterclockwise manner. Thus, a wind whose direction changes from east, through south, to west, has shifted clockwise through the compass; such a shifting wind is called a *veering* wind, or is said to *veer*. A wind that shifts in a counterclockwise

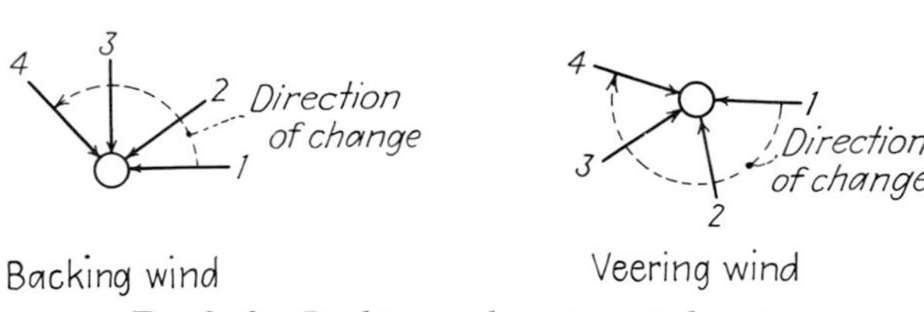

Fig. 9 · 2 Backing and veering wind motion.

manner is said to *back* and is termed a *backing* wind. Note that the wind is not blowing in a circular manner whether clockwise or counterclockwise. The *directions* of the shifting wind describe this circular motion.

This is illustrated in Fig. 9 · 2. The terms veering and backing therefore indicate immediately the compass points through which the wind direction has changed, without actual mention being made of these points, i.e., a wind backing from west to southeast must pass through all the intervening directions.

Wind Observations and Measurement

Wind direction and velocity can be measured and recorded accurately by means of instruments, or they may be obtained through estimation from certain observations. The means that are used naturally depend on the equipment available and the purpose for which the observations are taken. In placing wind instruments, care should be taken that no wind obstructions are near them. Otherwise the influence of eddies described above will yield false readings.

WIND DIRECTION. Instrumentally, wind direction is most commonly indicated by the familiar *wind vane*, of which there are several types. The vane always points *into* the wind. Wind vanes are usually constructed so that the fluctuations in wind direction are communicated instantaneously to some form of indicator or automatic recorder by means of proper electrical connections.

The wind sock, a typical airport device to show the wind direction, always flies *with* the wind.

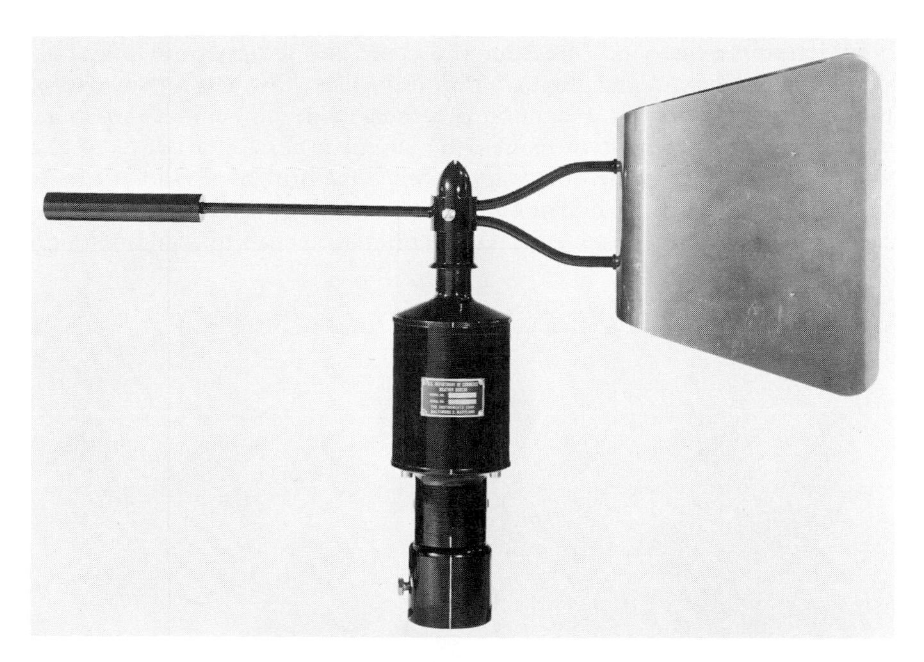

Fig. 9 · 3 Wind direction transmitter. The electrical signal generated by this instrument is used to activate both dial indicator and continuous chart recorders. (Belfort Instrument Company)

Fig. 9 · 4 Wind-speed (left) and wind-direction (right) dial indicators. (Belfort Instrument Company)

To determine the wind direction and speed in the upper air, pilot balloons are utilized. When inflated properly, they have a known rate of ascent. Observations on a balloon are then made at 1-minute intervals with a *theodolite,* which measures the angular horizontal drift of the balloon and also its angle of elevation. Since the drift of a balloon equals the wind speed at the particular elevation, it is a simple matter to compute these values from the observed data. A lantern attached to a pilot balloon

Fig. 9 · 5 Pilot balloon being released from a flight deck.

provides for nocturnal observations. A knowledge of upper-air conditions is of great importance in modern weather analysis. Pilot balloon observations are therefore very valuable.

From fixed or stationary positions, determination of the wind direction by ordinary observation is usually fairly simple. Any object that is bent or swayed or blown by the wind may serve as a useful direction indicator. Thus, smoke, flags, or pennants are normally reliable wind guides.

On a moving vessel, wind direction is best indicated by the surface of the sea. The wind will be nearly perpendicular to the crest line of the ripples, waves, or whitecaps. The direction of the spume blowing from the white caps will coincide with the wind direction. Care should be taken not to confuse waves with any prevailing swell. Swells are long undulating waves which are produced by winds at a distant source and which have since outrun the storm that caused them. They bear no particular relationship to local winds.

Clouds are excellent guides to the wind direction prevailing in the upper free air. Low clouds usually conform in their motion to the surface winds, with some variation to be explained later. The higher clouds, however, may be under the influence of an entirely different air stream.

WIND VELOCITY. Usually wind velocity can be determined quite accurately by means of the *anemometer*. The common cup-type anemometer

Fig. 9 · 6 Keuffel and Esser shipboard theodolite. (U.S. Weather Bureau)

consists essentially of three or more hemispherical cups extending on horizontal arms from a vertical shaft or spindle. The higher the wind velocity, the faster will the cups rotate the movable spindle. By means of a magnetogenerator arrangement, or a gear system with proper electrical contacts, this spinning motion is translated to show the wind speed on remote instruments. This equipment may indicate the instantaneous velocity directly by means of dials or buzzers, or it may record instantaneous or average velocities for reference purposes.

There are several types of wind transmitting and indicating systems, all of which are completely described in various government manuals on the subject.

In order to obtain more accurate and detailed systematic observations of all the weather elements, the National Bureau of Standards has developed a remote-weather-transmitting system of which the *Nomad I* weather station shown in Fig. 9 · 9 is an example. The more complete use of such marine weather stations will very greatly improve weather service at sea in addition to the understanding of weather in general.

When no instruments are available, wind velocity cannot be determined so easily. A moving mass of air, or wind, has a certain effect on objects in its path. The stronger the wind, the greater the force exerted. Once the relation between wind force and its effect is carefully determined, future

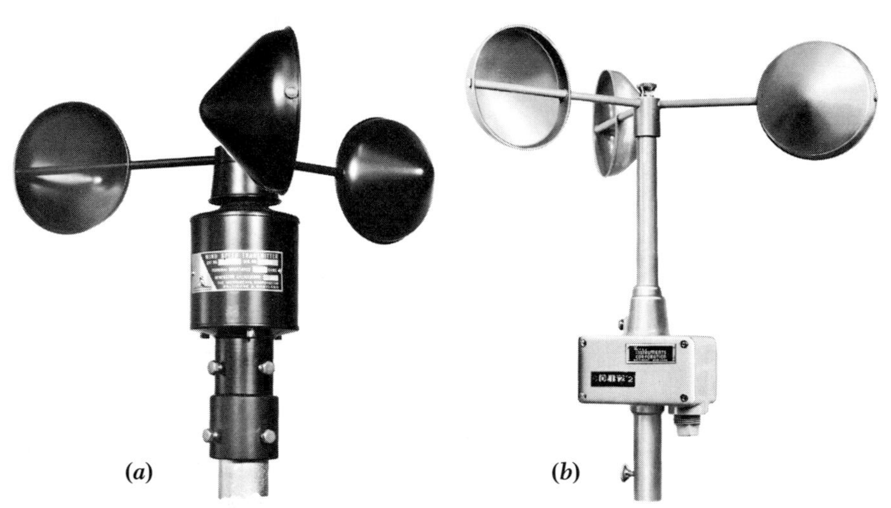

Fig. 9 · 7 Types of three-cup anemometers wind-speed transmitters. (a) Indicates gusts and instantaneous wind-speed changes by generating an electrical voltage proportional to speed. (b) Totalizing anemometer in which electrical contacts are closed after a preset unit of air has passed, permitting determination of average speed and actual amount of air passage. (Belfort Instrument Company)

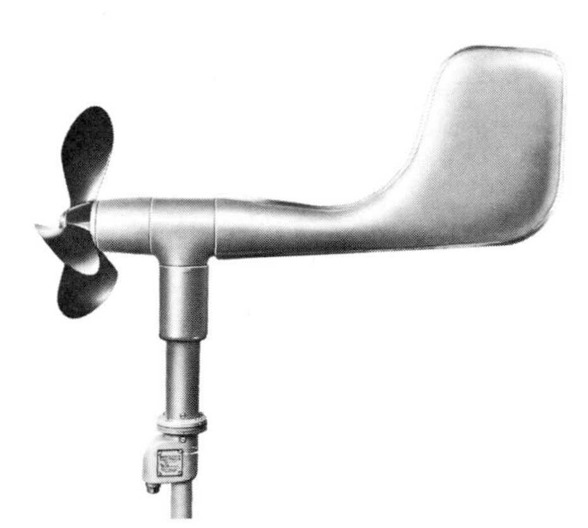

Fig. 9 · 8 Propeller-type anemometer—combines speed and direction transmitters in a single unit for general purpose use. (Friez Instruments Division of Bendix Aviation Corporation)

estimates of wind velocity may be made by noting the reactions of obstacles to the wind. Obviously, determinations to within a mile an hour cannot be obtained by this rough method.

The pressure exerted by a wind is proportional to the square of its velocity. Thus

$$P \propto V^2$$

or

$$P = KV^2$$

Fig. 9·9 The Nomad I *marine weather station installed in position. The direction of the weather vane shows the station is now pointed into the wind. The wind chargers are turning. The tail appendage at the stern points away from the station in the direction of the ocean current. The floats in the tail appendage, which are nearest the station, serve to keep a water temperature indicator 18 inches below the surface. Between the last two floats is a partially immersed experimental solar battery, which is contained in a plastic housing and makes experimental measurements of solar radiation.* (National Bureau of Standards)

If the pressure is to be found in pounds per square foot, then the velocity should be in miles (statute) per hour. The constant K then equals 0.004. For this formula to hold without correction, the surface exposed to the wind must be perpendicular to the path. Note that the effect of the wind on any exposed object is determined actually by the direct impact of the air on the windward side, together with the resulting suction on the leeward side.

In the early part of the nineteenth century, Admiral Beaufort of the Royal Navy developed the well-known scale of wind velocity that bears his name. The modern Beaufort system employs a series of numbers from 0 to 17, each number standing for a wind velocity between certain limits in miles per hour. Table 9 · 1 shows the relationship between the Beaufort scale and the force in miles per hour. This is followed by a brief wind description for land stations (see Table 9 · 2). Criteria for determining the Beaufort force at *sea* are given in Table 9 · 3.

Table 9 · 3 of sea and wind relations is to be used only on the open sea; *it is not applicable to inland or restricted waters.* Clearly, the effect of the wind on the sea depends, in addition to its velocity, on the *fetch,* or distance it has blown over the water. Also, the motion of currents adds to or detracts from the wind effects. If the current opposes the wind, the sea surface indicates a higher velocity than exists. If the current runs with the wind, a lower velocity is apparent. Thus whitecaps are produced when a tidal current of 3 or 4 knots, opposing a wind with a force of but 2, would indicate a force of 3.

*Table 9 · 1 Table of Comparative Wind-velocity Terminology**

BEAUFORT FORCE	VELOCITY		SEAMAN'S DESCRIPTION OF WIND	U.S. WEATHER BUREAU TERMINOLOGY
	MPH	KNOTS		
0	1	1	Calm	Light
1	1–3	1–3	Light air	Light
2	4–7	4–6	Light breeze	Light
3	8–12	7–10	Gentle breeze	Gentle
4	13–18	11–16	Moderate breeze	Moderate
5	19–24	17–21	Fresh breeze	Fresh
6	25–31	22–27	Strong breeze	Strong
7	32–38	28–33	Moderate gale	Strong
8	39–46	34–40	Fresh gale	Gale
9	47–54	41–47	Strong gale	Gale
10	55–63	48–55	Whole gale	Whole gale
11	64–72	56–63	Whole gale	Hurricane
12	73–82	64–71	Hurricane	Hurricane
13	83–92	72–80	Hurricane	Hurricane
14	93–103	81–89	Hurricane	Hurricane
15	104–114	90–99	Hurricane	Hurricane
16	115–125	100–108	Hurricane	Hurricane
17	126–136	109–118	Hurricane	Hurricane

* *American Practical Navigator,* rev. 1958, U.S. Navy Oceanographic Office Publication 9.

To aid further in the estimation of wind speed at sea, photographs of sea conditions corresponding to Beaufort numbers from 1 to 12 are illustrated in Fig. 9 · 12*a* to *m*. These scenes, which are used to illustrate the revised

Table 9 · 2 Criteria for determining Beaufort wind force

BEAUFORT FORCE	SPECIFICATIONS FOR USE ON LAND
0	Calm; smoke rises vertically
1	Direction of wind shown by smoke drift, but not by wind vanes
2	Wind felt on face; leaves rustle; ordinary vane moved by wind
3	Leaves and small twigs in constant motion; wind extends light flag
4	Raises dust and loose paper; small branches are moved
5	Small trees in leaf begin to sway; crested wavelets form on inland waters
6	Large branches in motion; whistling heard in telegraph wires; umbrellas used with difficulty
7	Whole trees in motion; resistance felt in walking against wind
8	Breaks twigs off trees; generally impedes progress
9	Slight structural damage occurs (chimney pots and slate removed)
10	Seldom experienced inland; trees uprooted; considerable structural damage occurs
11	Very rarely experienced; accompanied by widespread damage
12–17	Maximum wind damage

Table 9 · 3 Criteria for determining Beaufort wind force

BEAUFORT FORCE	SPECIFICATIONS FOR USE AT SEA
0	Sea like mirror
1	Ripples with scaly appearance; no foam crests
2	Small wavelets, crests of glassy appearance and not breaking
3	Large wavelets with crests beginning to break, scattered whitecaps
4	Small waves growing larger, numerous whitecaps
5	Moderate waves with greater length, many whitecaps with some spray
6	Larger waves, whitecaps very numerous, more spray
7	Sea tends to heap up, streaks of foam blown from breaking waves
8	Fairly high waves of greater length, well-marked streaks of foam
9	High waves with sea beginning to roll, dense streaks of foam with spray blown higher into air—may cut visibility
10	Very high waves with overhanging crests, sea is white with foam, heavy rolling and reduced visibility
11	Waves exceptionally high, sea covered with foam, visibility further reduced
12–17	Sea completely covered with spray, air filled with foam, greatly reducing visibility

edition of *American Practical Navigator* (Hydrographic Office Publication 9), have been provided through the courtesy of the U.S. Naval Oceanographic Office.

UPPER-LEVEL WINDS. Although wind direction aloft can be determined by cloud motion, the more complete and precise information required for weather analysis is obtained through the use of pilot balloons and radiosondes (Fig. 9 · 10).

When properly inflated, pilot balloons have a known rate of ascent (400

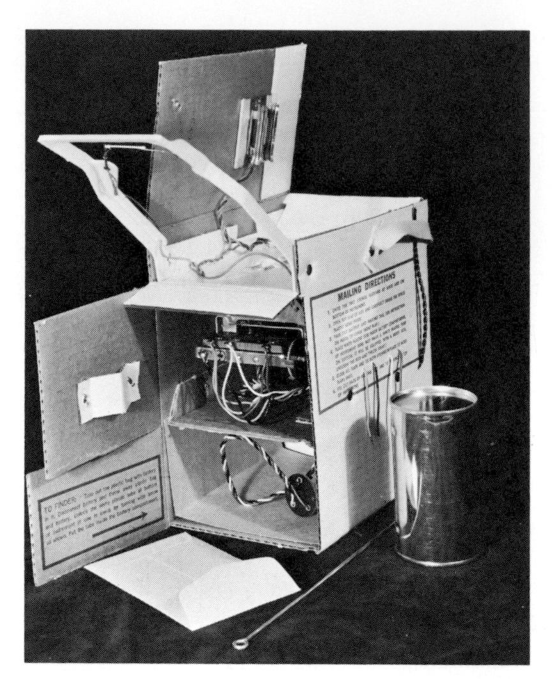

Fig. 9·10 Radiosonde. This instrument, which is carried aloft by a balloon, automatically transmits radio messages of temperature, pressure, and humidity conditions aloft. (Friez Instruments Division of Bendix Aviation Corporation)

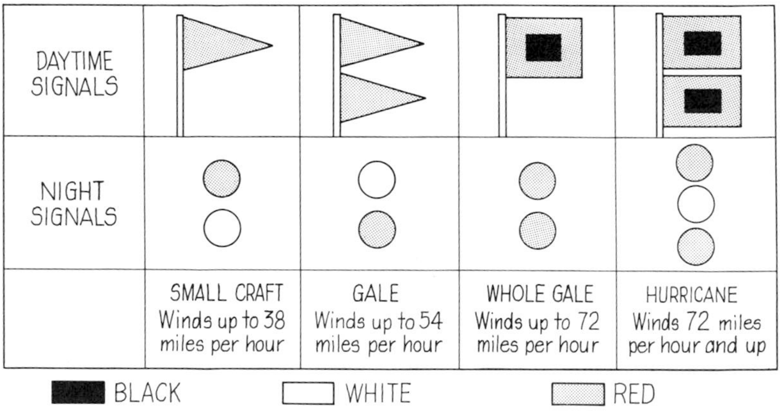

DAYTIME SIGNALS				
NIGHT SIGNALS				
	SMALL CRAFT Winds up to 38 miles per hour	GALE Winds up to 54 miles per hour	WHOLE GALE Winds up to 72 miles per hour	HURRICANE Winds 72 miles per hour and up

BLACK WHITE RED

Fig. 9·11 Explanation of small craft, gale, whole gale, and hurricane warnings by means of signal flags and lights.

Fig. 9 • 12a
Beaufort scale 0

Fig. 9 • 12b
Beaufort scale 1

Fig. 9 • 12c
Beaufort scale 2

Fig. 9 • 12d
Beaufort scale 3

Fig. 9 • 12e
Beaufort scale 4

Fig. 9 • 12f
Beaufort scale 5

Fig. 9 • 12g
Beaufort scale 6

Fig. 9 • 12h
Beaufort scale 7

Fig. 9 • 12i
Beaufort scale 8

Fig. 9 • 12j
Beaufort scale 9

Fig. 9 • 12k
Beaufort scale 10

Fig. 9 • 12l
Beaufort scale 11

Fig. 9 · 12m
Beaufort scale 12

feet per minute). Observations are made at 1-minute intervals, after release, with a *theodolite* which measures both the angular horizontal drift and the angle of elevation. The direction of drift gives the wind direction at each observation level, and a simple trigonometric calculation gives the wind speed from level to level. Nocturnal observations can be made using an illuminating device attached to the balloon. The wind determinations made from pilot balloon observations are known as *pibals*.

N

Fig. 9 · 13

The *radiosonde*, a more sophisticated instrument carried aloft by a pilot balloon, automatically transmits radio messages at predetermined eleva-

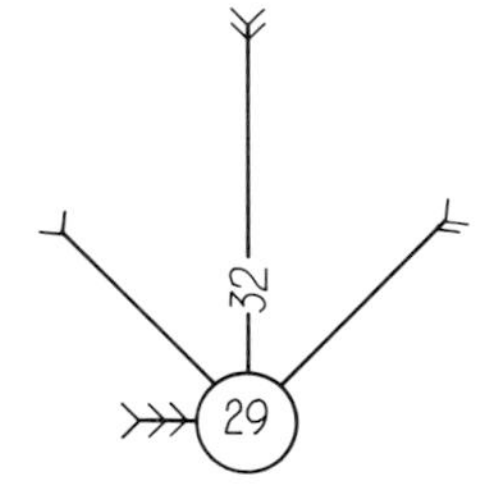

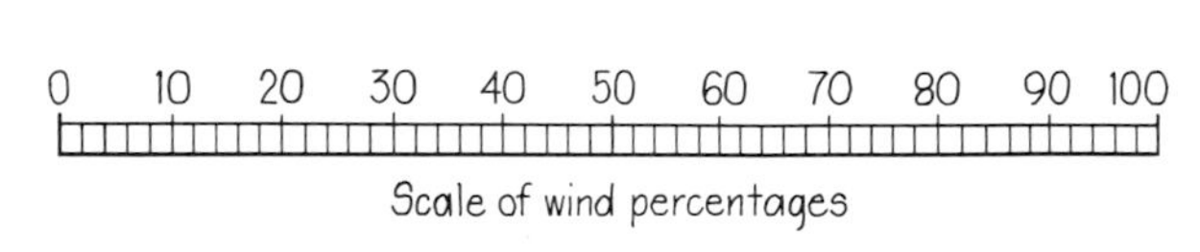

Fig. 9 · 14 Representation of average wind conditions by means of wind roses. The wind percentages are concentrated upon eight points. The arrows fly with the wind. The length of the arrow, measured from the outside of the circle on the attached scale, gives the percent of the total number of observations in which the wind has blown from or near the given point. The number of feathers shows the average of the wind on the Beaufort scale. When arrow is too short, feathers are shown beyond its end. The figure in the center of the circle gives the percentage of calms, light airs, and variable winds. When the arrow is too long to be shown conveniently, the shaft is broken and the percentage is indicated by numerals.

Example: The attached wind rose should be read thus: In the recorded observations the wind has averaged as follows: from N, 32 percent, force 4; from NE, 20 percent, force 3; from W, 1 percent, force 6; from NW, 18 percent, force 2; calms, light airs, and variables, 29 percent.

tions from which determinations of temperature, pressure, and humidity are made. Radiosondes can be followed by either radar units or radio direction-finding instruments which thus give successive locations of the airborne instrument from which the direction and speed of the winds aloft can be determined. Such data are known as *rawins.* Radiosonde observations of temperature, pressure, humidity, and winds are taken twice daily at the same time all over the world by stations specially instrumented for this purpose. The resulting simultaneous synoptic weather picture of conditions aloft provides the basis of modern weather analysis.

WIND REPRESENTATION. Winds are usually described by compass direction and speed in miles per hour, knots, or Beaufort force, or they may be shown diagrammatically as on weather maps and pilot charts. On either of these maps the wind is indicated by an arrow which flies with the wind, thereby indicating the direction. The number and length of the tails or barbs on the arrows indicate the wind speed in knots. Each half-barb represents the wind to the nearest 5 knots. Hence Fig. 9 · 13 shows a southeast wind of 25 knots (23 to 27 knots). The small circle at the head of the arrow represents the station at which the observation is made.

To show the average wind conditions in a given locality, the wind rose, illustrated in Fig. 9 · 14, is used.

True and Apparent Wind Relationship

The wind experienced on a ship under way is the result of two variable components: the wind created by the ship's motion and the true wind. The resultant wind is called the *apparent wind.* The direction and velocity of this apparent wind, as experienced on the deck of a vessel, depend on the force and direction of the true wind and that of the wind caused by the forward motion of the ship. For convenience, the latter wind, which results from the vessel's headway, will hereafter be referred to as the *ship's wind.*

The apparent wind, being the resultant of these forces, must always lie *between* the true and the ship's wind, except when the true wind is dead ahead or dead astern. Thus, if the velocity and direction of the true and ship's winds are known and plotted to scale on a chart, the apparent wind can be determined by completing the typical diagram of parallelogram of vectors. This is indicated in Fig. 9 · 15. *SW* stands for ship's wind, *TW* for true wind, and *AW* for apparent wind.

It is the *true* and not the apparent wind which is of significance for weather purposes. When conditions of visibility make observation of the sea's surface difficult or impossible, the true wind may be determined by this relationship. The direction and force of the ship's wind are easily determined from the course and speed of the vessel. The apparent wind force and direction can be estimated by observation, or measured accurately if instruments are aboard. With this known, the true wind can be obtained by application of the diagram method described. Some cases are considered in Fig. 9 · 16. The true wind *TW* is found by drawing a line connecting the tails of the arrows representing the apparent wind *AW* and the ship's wind *SW*. The length and direction of this line represent the true wind direction and velocity. For clarity in the diagram, the opposite side of the parallelogram is labeled *TW*.

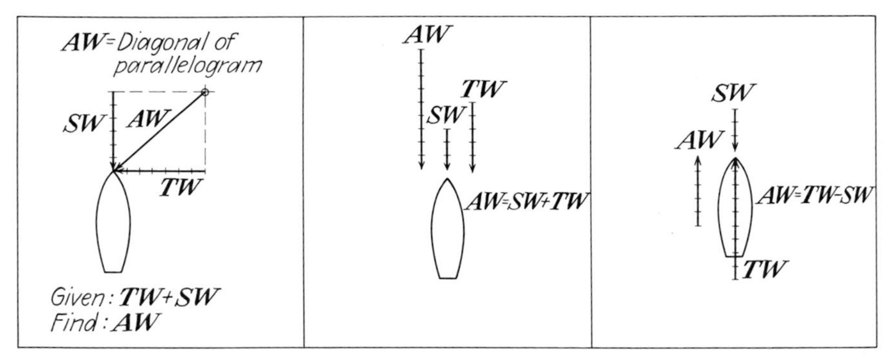

Fig. 9 · 15 Determination of apparent wind when the true wind and the ship's wind are known.

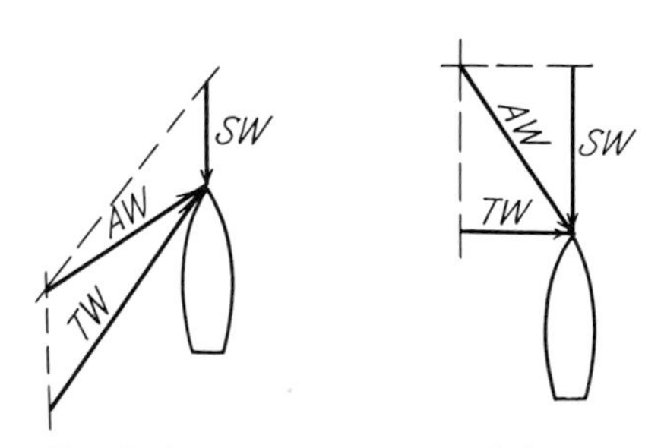

Fig. 9 · 16 Determination of the true wind when the apparent wind and ship's wind are known.

Any number of other possibilities exist and may be worked by this method. Many generalities may also be drawn:

1. The true wind direction is always farther from the bow than the apparent wind direction.
2. When the apparent wind is abaft the beam, the true wind is stronger and farther aft than the apparent wind.
3. When the apparent wind is forward of the beam, the true wind is less than the apparent.
4. If the apparent wind is dead astern, the true wind is dead astern and is stronger than the wind created by the ship.

Table 9 · 4 permits the determination of the true wind when the apparent wind force and direction and the ship's heading are known. Obviously the table gives approximate results, its accuracy being ordinarily limited.

Table 9·4 Table for Obtaining the True Direction and Force of the Wind from the Deck of a Moving Vessel

Apparent force of the wind, Beaufort scale	Speed of the vessel, knots	Apparent direction of the wind (points off the bow): 0		1		2		3		4		5		6		7		8		9		10		11		12		13		14		15		16	
		True direction, points off the bow	True force, Beaufort scale	True direction, points off the bow	True force, Beaufort scale	True direction, points off the bow	True force, Beaufort scale	True direction, points off the bow	True force, Beaufort scale	True direction, points off the bow	True force, Beaufort scale	True direction, points off the bow	True force, Beaufort scale	True direction, points off the bow	True force, Beaufort scale	True direction, points off the bow	True force, Beaufort scale	True direction, points off the bow	True force, Beaufort scale	True direction, points off the bow	True force, Beaufort scale	True direction, points off the bow	True force, Beaufort scale	True direction, points off the bow	True force, Beaufort scale	True direction, points off the bow	True force, Beaufort scale	True direction, points off the bow	True force, Beaufort scale	True direction, points off the bow	True force, Beaufort scale	True direction, points off the bow	True force, Beaufort scale	True direction, points off the bow	True force, Beaufort scale
0	10	16	3	16	3	16	3	16	3	16	3	16	3	16	3	16	3	16	3	16	3	16	3	16	3	16	3	16	3	16	3	16	3	16	3
	15	16	4	16	4	16	4	16	4	16	4	16	4	16	4	16	4	16	4	16	4	16	4	16	4	16	4	16	4	16	4	16	4	16	4
	20	16	5	16	5	16	5	16	5	16	5	16	5	16	5	16	5	16	5	16	5	16	5	16	5	16	5	16	5	16	5	16	5	16	5
1	10	16	3	16	3	16	3	15	3	15	3	15	3	15	3	15	3	15	3	15	3	15	3	15	4	15	4	16	4	16	4	16	4	16	4
	15	16	4	16	4	16	4	16	4	15	4	15	4	15	4	15	4	15	4	15	4	15	4	15	4	16	4	16	4	16	5	16	5	16	5
	20	16	5	16	5	16	5	16	5	16	5	16	5	16	5	15	5	15	5	16	5	16	5	16	5	16	5	16	5	16	6	16	6	16	6
2	10	16	2	15	2	14	2	14	2	13	3	13	3	13	3	13	3	14	4	14	4	14	4	14	4	15	4	15	4	15	4	16	4	16	4
	15	16	3	16	3	15	3	15	4	14	4	14	4	14	4	14	4	14	4	14	5	15	5	15	5	15	5	15	5	16	5	16	5	16	5
	20	16	4	16	4	15	4	15	4	15	5	15	5	15	5	15	5	15	5	15	5	15	6	15	6	15	6	15	6	16	6	16	6	16	6
3	10	16	1	12	1	11	2	11	2	11	3	11	3	12	3	12	4	12	4	13	4	13	4	14	4	14	5	15	5	15	5	10	5	16	5
	15	16	2	15	3	14	3	13	3	13	3	13	4	13	4	13	4	13	5	14	5	14	5	14	5	15	6	15	6	15	6	16	6	16	6
	20	16	4	15	4	15	4	14	4	14	4	14	5	14	5	14	5	14	5	14	6	14	6	15	6	15	6	15	6	15	7	16	7	16	7
4	10	0	1	3	2	6	2	7	3	8	3	9	4	10	4	11	4	11	5	12	5	12	5	13	5	14	6	14	6	15	6	15	6	16	6
	15	16	1	11	1	10	2	10	3	11	4	11	4	11	4	12	5	12	5	13	6	13	6	14	6	14	6	15	6	15	7	16	7	16	7
	20	16	2	14	3	13	3	12	4	12	4	12	5	12	5	13	5	13	6	13	6	14	7	14	7	14	7	15	7	15	7	16	7	16	7
5	10	0	3	2	3	4	3	5	4	7	4	8	4	9	5	10	5	10	5	11	6	12	6	13	6	13	6	14	7	15	7	15	7	16	7
	15	0	2	4	2	6	3	8	3	9	4	9	5	10	5	11	5	11	6	12	6	13	7	13	7	14	7	14	7	15	7	15	8	16	8
	20	16	1	10	2	10	3	10	4	10	4	11	5	11	5	12	6	12	6	13	7	13	7	14	8	14	8	15	8	15	8	16	8	16	8
6	10	0	4	2	4	3	4	5	5	6	5	7	5	8	6	9	6	10	6	11	7	12	7	12	7	13	7	14	7	15	8	15	8	16	8
	15	0	3	2	3	5	4	6	4	7	5	8	5	9	6	10	6	11	7	12	7	12	7	13	8	13	8	14	8	15	8	15	8	16	8
	20	0	2	4	2	7	3	8	4	9	5	10	5	10	6	11	7	11	7	12	8	13	8	13	8	14	9	14	9	15	9	15	9	16	9
7	10	0	5	1	5	3	5	4	6	5	6	7	6	8	7	9	7	10	7	11	7	11	8	12	8	13	8	14	8	14	8	15	8	16	8
	15	0	4	2	4	4	5	5	5	6	6	8	6	9	7	9	7	10	7	11	8	12	8	13	8	13	9	14	9	15	9	15	9	16	9
	20	0	3	3	4	5	4	6	5	8	5	9	6	9	7	10	7	11	8	12	8	12	9	13	9	14	9	14	10	15	10	15	10	16	10
8	10	0	6	1	6	3	7	4	7	5	7	6	7	7	8	8	8	9	8	10	8	11	9	12	9	13	9	14	9	14	9	15	9	16	9
	15	0	6	2	6	3	6	5	6	6	7	7	7	8	8	9	8	10	8	11	9	12	9	12	9	13	10	14	10	15	10	15	10	16	10
	20	0	5	2	5	4	5	6	6	7	6	8	7	9	8	10	8	11	9	11	9	12	10	13	10	13	10	14	10	15	11	15	11	16	11
9	10	0	8	1	8	3	8	4	8	5	8	6	8	7	9	8	9	9	9	10	9	11	10	12	10	13	10	14	10	14	10	15	10	16	10
	15	0	7	2	7	3	7	4	7	6	8	7	8	8	8	9	9	10	9	11	10	11	10	12	10	13	11	14	11	14	11	15	11	16	11
	20	0	6	2	6	3	6	5	7	6	7	7	8	8	8	9	9	10	10	11	10	12	10	12	11	13	11	14	11	15	11	15	11	16	11
10	10	0	9	1	9	2	9	4	9	5	9	6	9	7	10	8	10	9	10	10	10	11	11	12	11	13	11	13	11	14	11	15	11	16	11
	15	0	8	1	8	3	8	4	8	5	9	6	9	7	9	8	10	9	10	10	10	11	11	12	11	13	11	14	12	14	12	15	12	16	12
	20	0	7	2	7	3	7	5	8	6	8	7	9	8	9	9	10	10	10	11	11	12	11	12	12	13	12	14	12	15	12	15	12	16	12
11	10	0	10	1	10	2	10	4	10	5	10	6	10	7	11	8	11	9	11	10	11	11	12	12	12	13	12	13	12	14	12	15	12	16	12
	15	0	9	1	9	3	9	4	10	5	10	6	10	7	11	8	11	9	11	10	12	11	12	12	12	13	12	14	12	14	12	15	12	16	12
	20	0	9	1	8	3	9	4	9	6	10	7	10	8	11	9	11	10	11	10	12	11	12	12	12	13	12	14	12	14	12	15	12	16	12
12	10	0	10	1	10	2	10	3	11	5	11	6	11	7	11	8	11	9	12	10	12	11	12	12	12	12	12	13	12	14	12	15	12	16	12
	15	0	10	1	10	3	10	4	10	5	11	6	11	7	11	8	11	9	12	10	12	11	12	12	12	13	12	14	12	14	12	15	12	16	12
	20	0	9	1	9	3	9	4	10	5	10	7	11	8	11	9	11	10	12	10	12	11	12	12	12	13	12	14	12	14	12	15	12	16	12

First figure column indicates speed of the vessel, knots. Second column gives direction, points off the bow. Third column, true force, Beaufort scale. Proper allowance should be made for compass variation.

Factors Affecting Wind Motion

The wind motion of the atmosphere is a continuous and cyclical tendency on the part of the atmosphere to stabilize itself and to reach equilibrium. The process has been in operation since the birth of the atmosphere and will continue to operate as long as inequalities in temperature exist over the earth's surface. As Humphreys has stated it, "Atmospheric circulation is a gravitational phenomenon, induced and maintained by temperature differences."

The scale of wind motion is extremely variable from winds affecting very localized and restricted areas to those which are planetary in scope. Although small-scale wind activity is often the result of a simple convection or modified convection cell produced by local temperature differences, as in Fig. 9 · 17, the origin of large-scale wind activity is usually more complicated. In the latter case, although temperature inequalities are still at the root of the air motion, the complete dynamics involve other factors that will be developed in the course of subsequent chapters.

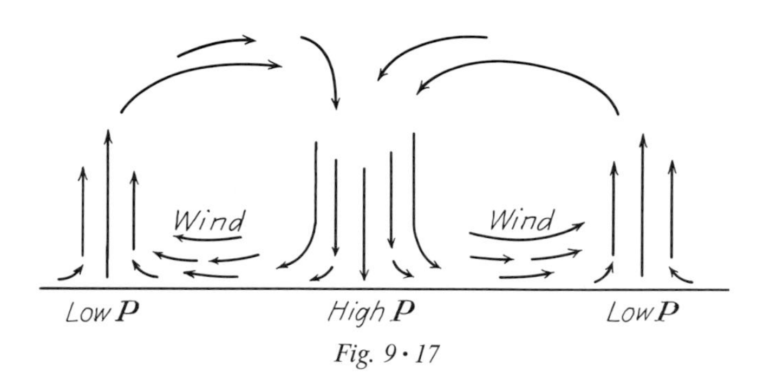

Fig. 9 · 17

Regardless of the mechanism of wind generation, the motion of winds is governed by a number of factors that can usually be evaluated numerically in order to predict the velocity. These are pressure gradient, friction, rotation of the earth (Coriolis effect), and the centrifugal effect (for winds with circular paths).

HORIZONTAL PRESSURE GRADIENT. The horizontal pressure gradient, which was defined in the preceding chapter, provides the horizontal force determining the speed and *initial* direction of wind motion. The greater the difference in pressure between two points, the steeper is the pressure gradient or the greater is the force on the air, and the higher is the wind speed. Since the direction of the force is from higher to lower pressure, and perpendicular to the isobars (lines of equal pressure), the initial tendency of the wind is to blow parallel to the gradient and perpendicular to the isobars. However, as soon as wind motion is initiated, a deflective force is generated which alters the direction of motion as described below. It is noteworthy that this force does not exist until the air is set in motion and that it increases as the wind speed increases, with a consequent effect on the motion.

FRICTIONAL FORCES. Moving air behaves like any moving fluid with respect to any boundary surface. For example, a river flows fastest at the center of its channel and slowest along the banks and bottom of the channel, owing to the frictional retardation produced by contact with the solid surface. Also, the river runs slower when the channel floor and sides are irregular or rough.

In the same way, the wind nearest the underlying surface has a distinctly lower speed than the air aloft. And the more irregular the surface (or the greater the surface *roughness*), the slower is the speed. Because the slowed layers impart a drag on the air just above, the frictional drag continues with decreasing effect until approximately 2,000 feet above the surface. In general this effect is least over the oceans, whose surfaces are, on the average, much smoother than those of land areas which are roughened by hills, valleys, trees, buildings, etc. The relatively smooth oceans slow the surface wind by about 20 percent, whereas the continental effect may be much greater.

ROTATION OF THE EARTH. The earth is a spinning globe whose rotational velocity of about 1,000 miles per hour at the equator decreases continuously with increase in latitude, becoming zero at the poles. As winds travel horizontally over this rotating sphere in response to pressure differences, the surface tends to turn beneath the moving air. To an observer on the surface, and turning with it, the wind *seems* to suffer a continuous deflection as the surface turns beneath. To an observer situated so as to be independent of the earth's rotation, the wind will show no such deflection but the earth's surface will be seen to rotate. Thus the winds undergo an apparent deflection which is real enough to observers on the earth. Some simple physical considerations may aid in understanding this effect.

Rotational Deflection: Coriolis Force. The apparent force resulting from rotation, which causes wind deflection, is known as the *Coriolis force.* An exact quantitative treatment and explanation of this force is beyond the scope of this chapter. However, a simple treatment of the problem can aid in understanding the cause of deflection—the Coriolis force—which is so important in the behavior of all material moving freely over the earth's surface.

It might be well to examine the explanation from two viewpoints: (1) by considering the special case of the behavior of winds moving in a north-south direction in the vicinity of the North Pole and (2) by extending the results of the special case to the general case of the deflection of winds moving in any direction and in any latitude. For convenience, we shall first substitute the motion of a bullet for the motion of the wind, since a bullet, being a more tangible object, may help to simplify the explanation.

1. A relatively small area of the earth around the poles can be considered as a plane surface. Figure 9 · 18 shows the North Pole at the center of the projection with the perimeter of the circle rotating as shown by the arrows.

   Assume that *PX* is the direction of an observer's meridian at a given instant. Assume also that a bullet is shot from *P*, along the meridian *PX*, and that it takes 1 hour to reach *X*. During this interval, however, the meridian originally located along the direction *PX* has

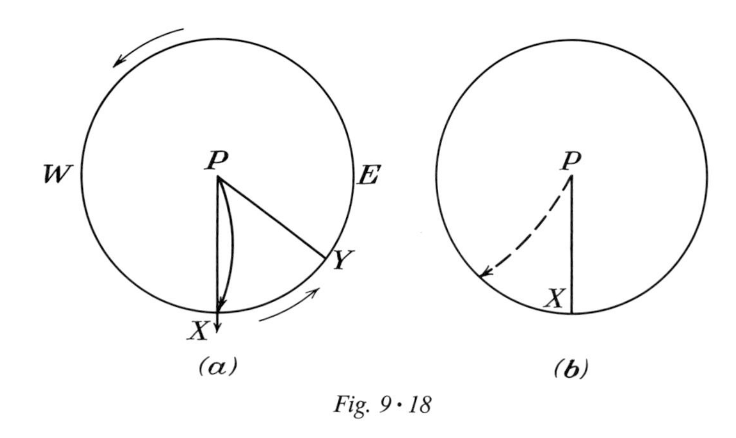

Fig. 9 · 18

   rotated 15° eastward and is now at *PY*. The bullet, which does not partake of this rotation, continues along the line *PX* and appears, to the observer firing the gun, to be deflected farther and farther to the right, describing the curved path shown by the curved line *PX*. Actually, any observer on the earth is never aware of rotation, so that the bullet seems to describe the curving path indicated by the broken line in (*b*).

   Any observer situated in a relatively fixed position off in space will see the bullet trace the straight path from *P* to *X* in 1 hour.

   Note that the direction of deflection will in every such case be to the *right* of the original direction of motion.

   In the case of a bullet moving toward the North Pole from point *X*, we can see the results again in a similar diagram, Fig. 9 · 19.

   Now the bullet is shot from *X* toward *P*. Again *X* describes in that time an angle of 15° to the east. As the bullet moves from *X* to *P*, in accordance with the law of inertia, it partakes of the eastward rotational spin in addition to its direct forward motion, and after 1 hour reaches a point to the right of *P*, appearing again to be deflected to the right of its original direction of motion along the meridian.

We need now only substitute moving air for this grand-scale shooting, and we arrive at the behavior of winds under the influence of

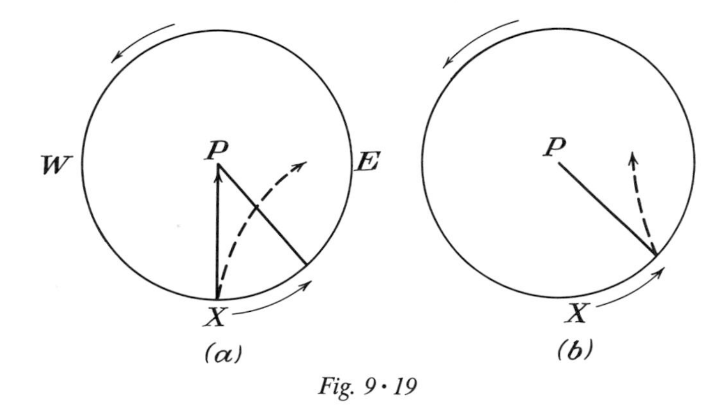

Fig. 9 · 19

rotational deflection. It is of interest to note that corrections for this deflection are required for long-range rifle fire. It is negligible over short distances.

2. Having examined the special case of deflection at the poles, let us now generalize the effect for any wind direction at any latitude. Con-

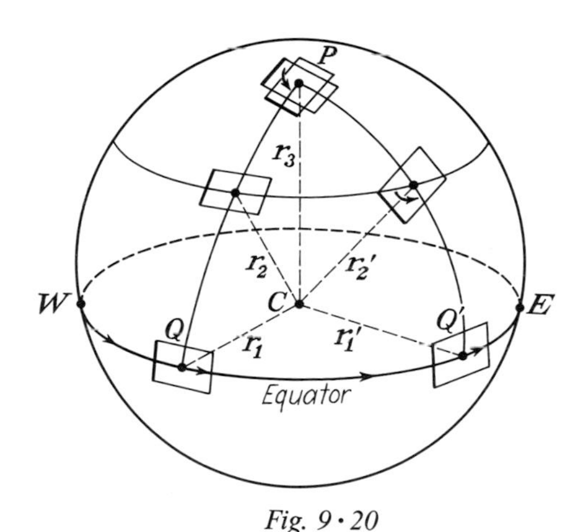

Fig. 9 · 20

sider the earth's surface at any point to be a horizontal plane perpendicular to a radial axis from the center (Fig. 9 · 20). Only at the poles does this axis coincide with the earth's axis of rotation. Clearly, the polar horizontal plane describes a full rotation of 360° in one day about its radial axis. But all other planes will describe only some component of a full rotation about their radial axes in one day, the amount of this rotation decreasing with distance from the poles, becoming zero at the equator.

The rotational motion of these planes is indicated in Fig. 9 · 20. As the meridian *PQ* travels to *PQ′*, it is apparent that the equatorial plane shows no horizontal rotation about the axis *r*, and the mid-latitude plane shows less rotation about r_2 than the polar plane shows about r_3. Hence, anywhere but at the equator, the rotation of the earth's surface shows a component of rotation about a radial axis, and the amount of this motion increases with the latitude (quantitatively with the sine of the latitude angle).

The rotational deflection effect described under number 1, above, is thus applicable to any point on the earth's surface with the exception that the effect becomes less as the rotational component of the surface about a radial axis becomes less with decrease in latitude.

It is not difficult to conceive of the rotation of the polar plane and the lack of rotation of the equatorial plane. But the movement of a plane between the equator and the poles requires a bit more concentration. The fact that this rotation actually exists has been proved by means of the Foucault pendulum experiment, which it might be well to review. Once a long, freely swinging pendulum is set in motion, it will continue to oscillate in a *fixed* plane—fixed as regards some object in space, such as a star. If such a pendulum is set swinging directly over the North Pole, the surface beneath will slowly rotate toward the left, about the fixed plane of vibration, and make a complete rotation in one day. At the equator, the surface beneath the pendulum will remain unchanged with respect to the plane of vibration. Between the equator and the poles, the surface beneath the swinging pendulum shows a partial rotation about the pendulum, the amount of this rotation increasing with increase in latitude. Actually, to the observer who partakes of the rotation of the earth, and thus of the rotation of any plane at the earth's surface, it is the plane vibration of the pendulum that *appears* to be displaced (toward the right) rather than the surface beneath! Wind deflection results from the same process.

Any number of arbitrary planes can be imagined on the earth's surface, all of which have some rotation about an axis perpendicular to the plane. Because, as we saw above, this rotation varies from zero at the equator to one full rotation per day at the poles, it must vary as the sine of the latitude (0 at 0°, 1 at 90°). Since, at any instant, the wind anywhere on the earth's surface can be considered as moving toward, or away from, the axis of an arbitrary plane, the problem can then be likened to that developed in Figs. 9 · 18 and 9 · 19 by remembering to include the latitude effect. A formula expressing Coriolis force which can be developed from fairly simple physical reasoning is

$$F = 2mv\omega \sin \Phi$$

where m is the mass of the wind; v its velocity; ω the angular rotation of the earth (radians per second); and Φ the latitude angle.

Since the direction of rotation of the earth is clockwise in southern latitudes, the resulting deflection of wind in the Southern Hemisphere is counterclockwise, or to the *left* of the path of motion. Figure 9 · 21 illustrates the fact that, although both hemispheres rotate from west to east as a unit, the Northern Hemisphere appears to have a counterclockwise, and

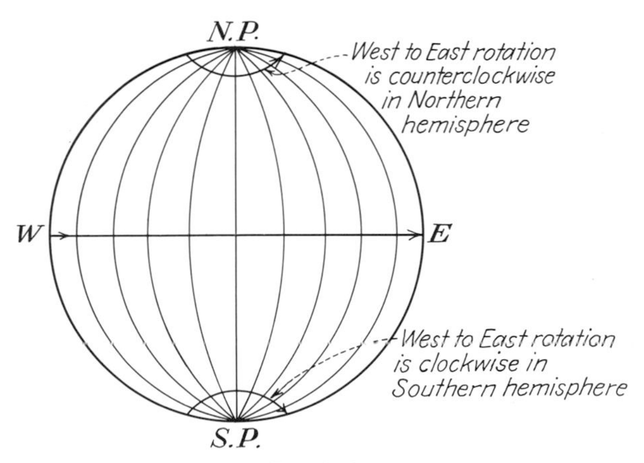

Fig. 9 · 21

the Southern a clockwise, motion. Although the direction of the wind is altered by the influence of rotation, the velocity of the wind is entirely unaffected. Only the pressure gradient and frictional effects have any noteworthy influence on the wind velocity.

The Geostrophic Wind

Once wind motion becomes established, the Coriolis effect causes continuous deflection further to the right of the pressure gradient. Deflection must cease when the wind direction becomes parallel to the isobars (perpendicular to the gradient); otherwise it will blow toward higher pressure or against the pressure gradient (which would be like water running uphill from the effect of gravity alone). Since the Coriolis force acts 90° to the right of the wind, it must act in a direction opposite to and thus balance the pressure gradient at the time of maximum deflection. This relationship is shown in Fig. 9 · 22 in which the pressure gradient force acts inwards toward the low (double arrow), and the Coriolis force acts away from the low (single arrow), with the resulting wind blowing parallel to the isobars (dashed arrows). The wind that results from this balance between gradient and Coriolis forces *in the case of straight or broadly curved isobars* is called the *geostrophic wind.*

Near the ground or water surface, the frictional force slows the wind, thereby decreasing the deflection so that the wind actually crosses the isobars at a small angle and is not geostrophic. However, at an elevation of a few thousand feet, frictional effects are sufficiently small so that maxi-

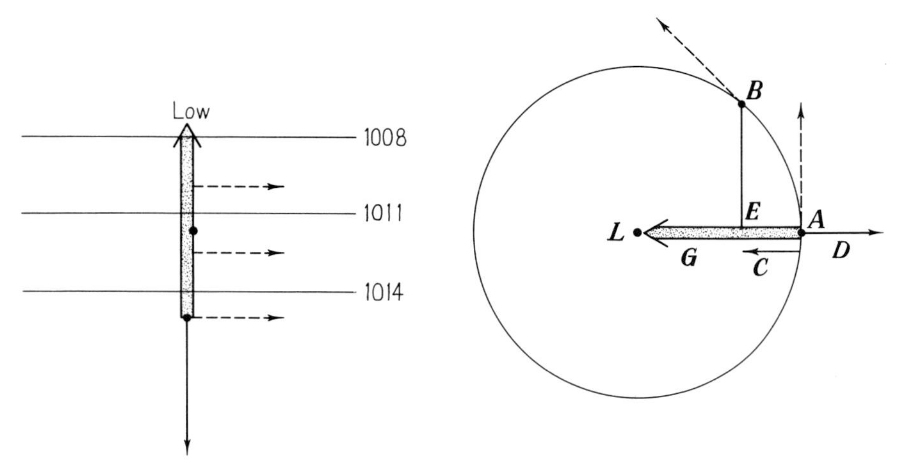

Fig. 9 · 22 Illustration of the balance between Coriolis and pressure gradient forces for the geostrophic wind.

Fig. 9 · 23 Explanation of forces on a wind with a circular path about a low.

mum Coriolis deflection occurs with a nearly complete balance between the gradient and the deflective force, resulting in the geostrophic wind. The value of the geostrophic wind is usually obtained from graphs based on the isobar spacing and the latitude.

The Gradient Wind

In the case of winds following a curved path in a circular pressure field shown by strongly curved isobars, the simple balance between the pressure gradient and the Coriolis force that was developed for the geostrophic wind no longer holds.

Consider the wind moving about a circular low-pressure system as at point A in Fig. 9 · 23. Again the gradient force is indicated by the double arrow (G) and the Coriolis deflective force by the opposing single arrow

(D). If these forces were in balance, as for the geostrophic wind, the resulting motion would be that indicated by the dashed arrow. Although the instantaneous motion of the wind at A is in the direction of the dashed arrow, the forces G and D cannot be in balance because the real wind follows the curved path toward B. In so doing, the wind has actually moved (been accelerated) toward the center L by the amount AE. Hence, when at point A, a third force must be considered which also acts toward the center, as does the gradient force G. This force is known as the *centripetal force* (C), which together with the gradient force overbalances the Coriolis or deflective force in the development of circular motion. The relationship can be described algebraically as

$$G - C = D$$

which shows that the gradient force exceeds the deflective force by the value of the centripetal force.

These forces are usually considered from the point of view of the balance between them for the instantaneous wind, as at A, arranged so that no net inward or outward forces exist on the wind. This is accomplished by rewriting the above equation as

$$G = D + C$$

According to this, the gradient force can be balanced by the sum of the deflective (Coriolis) force and an outward force equivalent to the centripetal force, for the instantaneous motion of a wind following a strongly curved path. This outward force is known as the *centrifugal force,* and the wind existing when this balance occurs is known as the gradient wind.

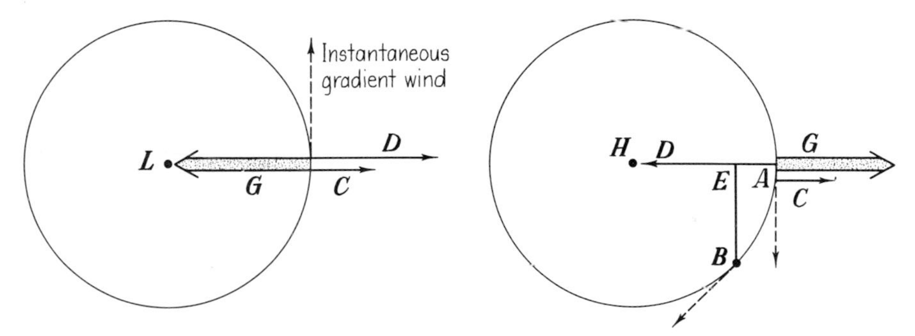

Fig. 9 · 24 The gradient wind and related forces in a low-pressure area.

Fig. 9 · 25 The gradient wind and related forces in a high-pressure system.

For a low-pressure area, the balance just described involves the gradient force acting inward and the deflective and centrifugal forces outward, as in Fig. 9 · 24. For a high-pressure area, the balance of forces for a gradient wind is different. The gradient force is directed outward from H and the deflective force inward (to the right of the wind). Now, however, the centripetal force necessary to curve the wind from A toward B acts opposite to the gradient force, so that the deflective force exceeds the gradient force by the centripetal force, or

$$D - C = G$$

But to express this as a balance so that there are no net inward or outward forces on the instantaneous wind, the equation is written as

$$D = G + C$$

where C, which now acts outward from the high and in the direction of the gradient force, is again called centrifugal force.

The same force relationship can be established for any point on the circle of motion, as for example, the wind at point B in Figs. 9·23 and 9·25. The real gradient wind is thus the combined motion of all of the instantaneous winds considered in relationship to the force balance just described.

EXERCISES

9·1 Give examples of the changes in wind direction experienced in the case of (*a*) a veering wind and (*b*) a backing wind.

9·2 Find the total pressure acting on a sail of 300 square feet held perpendicular to a wind of 30 statute miles per hour.

9·3 Describe the procedures used to determine the speed and direction of winds aloft.

9·4 Find the true wind at sea using the graphical vector method when a ship is heading NW at 20 knots and the apparent wind is NE at 30 knots. Compare the result with information obtainable from Table 9·4.

9·5 What is the relationship between wind speed and the horizontal pressure gradient?

9·6 Describe the frictional effect on wind motion, noting its variation with ground conditions and elevation.

9·7 Briefly explain the Coriolis force and its variation with latitude.

9·8 What is the relation between the direction of surface winds and that of the low clouds? How are both related to the trend of the sea-level isobars?

9·9 Distinguish between the geostrophic and the gradient wind.

CHAPTER 10

Terrestrial or Planetary Winds

Winds have blown over the earth's surface as long as horizontal temperature inequalities have existed. In some cases the air motion is simply and directly a result of air circulation within a convectional cell (as in Fig. 9 · 17) and in others the motion is part of a more complicated atmospheric pattern that can be classified into two broad groups. One group consists of wind patterns that prevail over large regions, planetary in size, and are, on the average, fairly constant in direction. The second broad group comprises winds that involve smaller areas, are changeable in speed and direction, and may be quite restricted and local in occurrence or may be part of a traveling wind pattern. The former, large-scale pattern is referred to as the *terrestrial* or *planetary* wind system and is controlled by the *general circulation* of the atmosphere. The latter, more variable type may be referred to as *secondary winds,* and although they originate from a variety of processes, the most important type is associated with moving low- and high-pressure areas, referred to in Chap. 8. Originating directly or indirectly from temperature differences, wind systems—the large-scale ones in particular—play an extremely important part in the redistribution of heat over the earth's surface.

Ideal Planetary Wind System

Although somewhat schematic, the idealized view in Fig. 10 · 1 is useful in giving a general picture of the terrestrial pressure-wind distribution. In some large areas or at some seasons, the actual pattern differs strongly from this view, the difference being mostly a consequence of the irregular heating of the earth's surface in both time and position. Note that winds converge on the low-pressure belts, which must be characterized by vertical ascending air motion, and diverge from high-pressure belts, which must therefore be characterized by vertically descending air motion.

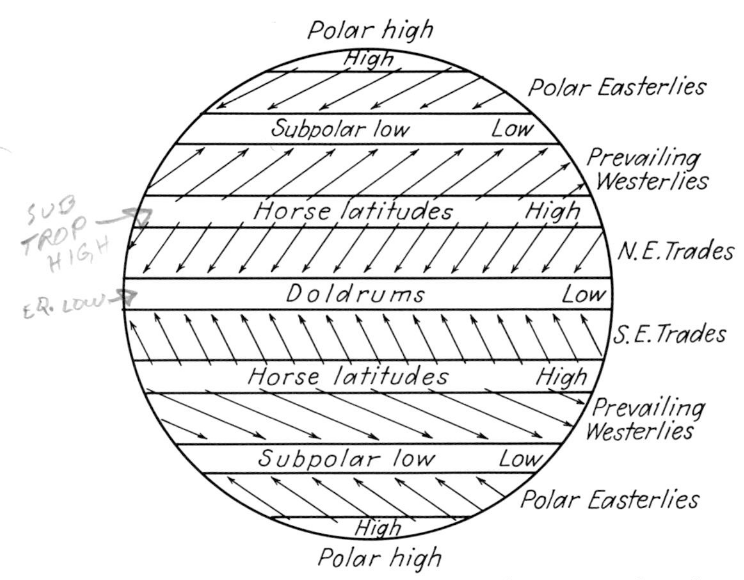

Fig. 10 · 1 The ideal primary or terrestrial pressure and wind systems.

The planetary wind motion is an aspect of the general circulation of the atmosphere. The concept of a relatively simple global convectional pattern which was generally supported until about the middle of the twentieth century has been overthrown by more complete observational data. Although the new theory of circulation is not yet fully developed, it already seems a more valid explanation of the planetary wind patterns. The new and old ideas will be examined briefly after a better background in wind knowledge is obtained in the following chapters.

Owing to the inhomogeneous nature of the earth's surface to which we have referred so often, the schematic pattern shown in Fig. 10 · 1 may suffer considerable modification. The major modifications that occur show well in the charts reproduced in Figs. 10 · 2 and 10 · 3, which show average

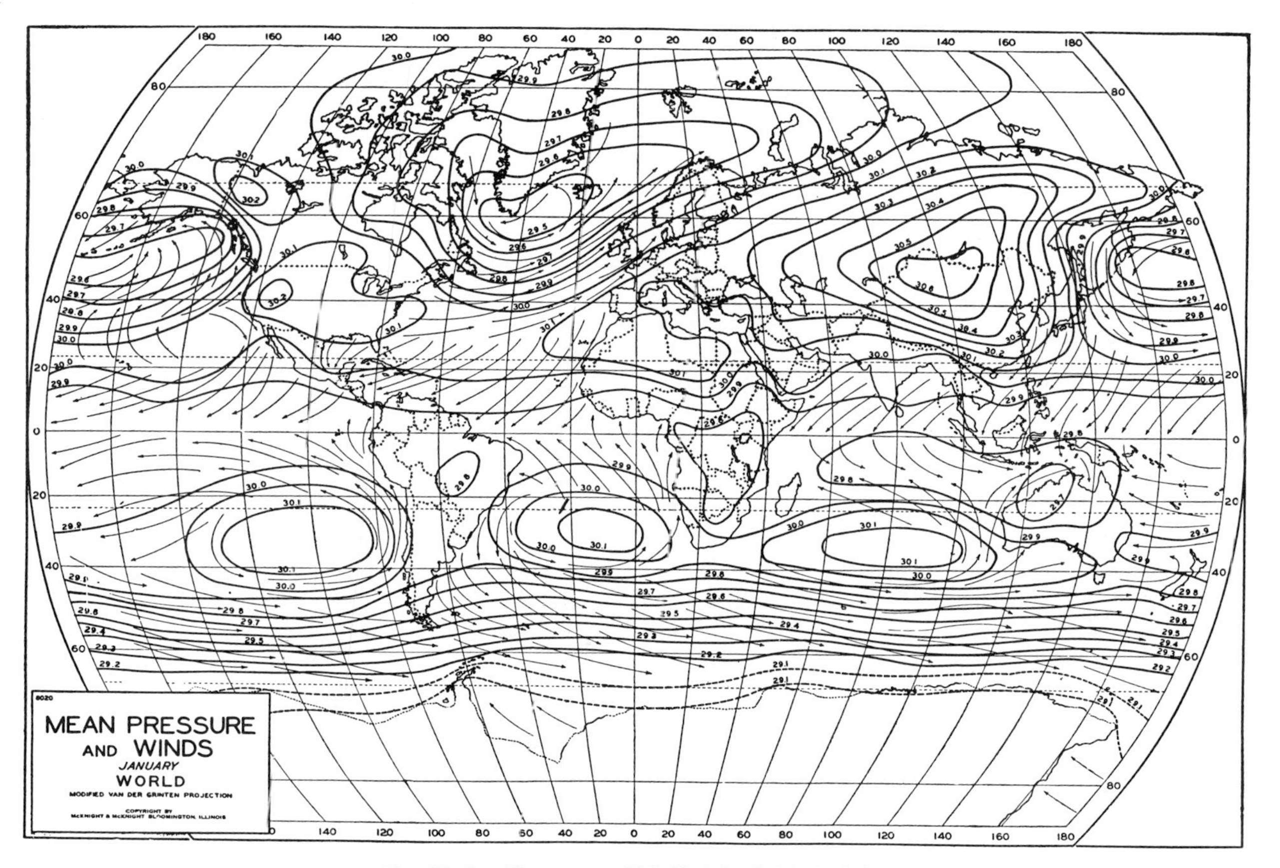

Fig. 10 · 2 (Courtesy of McKnight & McKnight)

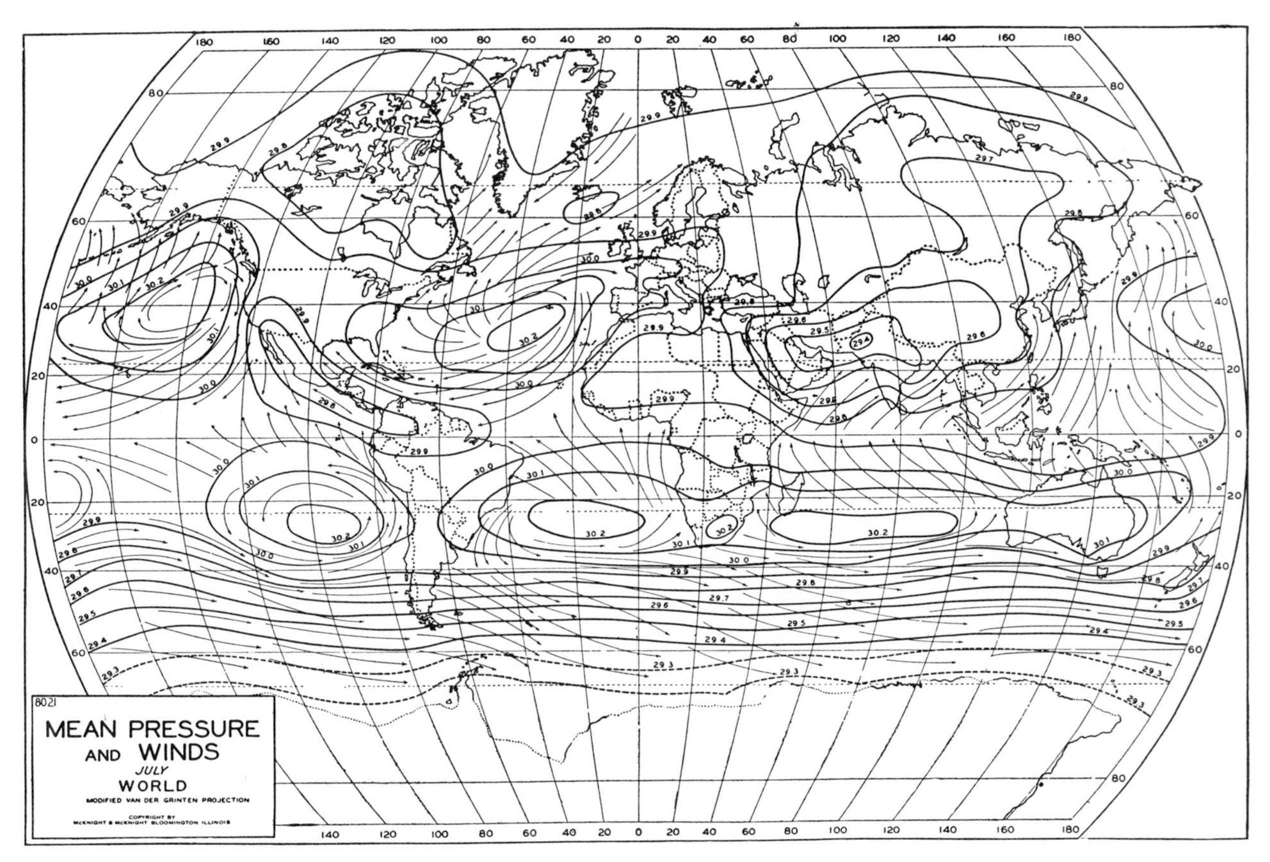

Fig. 10·3 (Courtesy of McKnight & McKnight)

global pressure (by means of global isobars) and winds for the entire months of January and July, respectively. It is important to realize that these charts show average conditions. In some of the pressure and wind belts to be described, day-to-day conditions may be quite different. It may be helpful to consider the pressure and wind belts by referring to both the ideal and the real mean conditions. Note that these descriptions, which follow immediately, refer to the lower troposphere. The overlying circulation, whose description comes later, is uncomplicated by temperature variations of the earth's surface and is schematically simpler in structure.

The Planetary Pressure Belts

THE EQUATORIAL CALMS (DOLDRUMS). Throughout the year a belt of low pressure surrounds the earth in the equatorial zone as a result of the average overheating of the globe in this region. Late afternoon showers are usual from the strong convection and resulting adiabatic cooling at this time of highest diurnal temperature. Most of the air motion here is thus *vertical* with but light and variable winds, having in general a slight westward drift. Thus, the region is known as the *belt of equatorial calms*. The oppressive, hot, sticky atmosphere with calm winds and slick glassy seas was named *doldrums* in the days of sailing ships.

The following extract is from the log of Commodore Arthur Sinclair, aboard the frigate *Congress* on a cruise to South America in 1817–1818, and describes the murky atmosphere beneath the equatorial cloud band. "This is certainly one of the most unpleasant regions in our globe. A dense, close atmosphere, except for a few hours after a thunderstorm, during which time torrents of rain fall, when the air becomes a little refreshed; but a hot glowing sun soon reheats it again, and but for your awnings, and the little air put in circulation by the continual flapping of the ship's sails, it would be most insufferable. No person who has not crossed this region can form an adequate idea of its unpleasant effects. You feel a degree of lassitude unconquerable, which not even the sea bathing, which everywhere else proves so salutary and renovating, can dispel. Except when in actual danger of shipwreck I never spent twelve more disagreeable days in the professional part of my life than in these calm latitudes."

During the Northern Hemisphere's winter, as shown in Fig. 10 · 2, the doldrums approaches the ideal condition (Fig. 10 · 1) straddling the equator except in the vicinity of the eastern Indian Ocean-East Indies-Australia region. Here the equatorial low is displaced southward, mostly from the warming effects associated with the Australian continent in the southern summer. However, during the northern summer when the sun is north of the equator (Fig. 10 · 3), a considerable northern displacement of the low-pressure belt occurs owing to the heating of the large continental areas. At this time of the year, the low-pressure belt near the Americas becomes centered over and oriented along the Central Americas. And in eastern longitudes an even more pronounced displacement to southern Asia takes place. It is important to note that the mean annual position of the doldrums generally lies north of the equator owing to displacement during the northern summer.

BELT OF SUBTROPICAL CALMS (HORSE LATITUDES). In the ideal picture, two belts characterized by high pressure (also called *subtropical highs*) and relatively light winds or calms occur symmetrically about the equator at 30°N and 30°S. The descending air that maintains the high-pressure pattern is warmed adiabatically and therefore develops a low relative humidity and clear skies. The dryness of this descending air is responsible for the prevalence of the world's great deserts either in or adjacent to the horse latitudes. The latter name supposedly stems from the historic days of sailing vessels when the horse cargo was often set adrift to lessen the load and conserve water after the ship was becalmed in this zone.

In the Southern Hemisphere the horse latitudes are mostly over water, so that conditions are fairly uniform throughout the year. The annual configuration shown in Figs. 10 · 2 and 10 · 3 closely resembles the ideal pattern (Fig. 10 · 1), except for the continental breaks in the high-pressure ridge. These breaks become less pronounced in the southern winter (July) owing to the cooling of the lands, thereby increasing air subsidence and enhancing the high-pressure belt.

In the Northern Hemisphere, a more drastic annual modification of the idealized pattern occurs following the pronounced temperature variation of the large land areas relative to that in the oceans. During the northern winter a high-pressure belt roughly encircles the earth although the position over the continents is displaced to the north and over the oceans, to the south of the 30th parallel. Also, the high pressure is on the average greatly strengthened over the continents, particularly over Asia where the *Siberian high* is quite intense as a consequence of the pronounced refrigeration of this large land mass.

In summer (Fig. 10 · 3) there is a partial reversal of pressure over North America and a very strong reversal over Asia. At the same time an intensification of the high-pressure belt takes place over the oceans because they become cool relative to the continents. The high-pressure region west of the United States is known as the *Pacific high,* while that over the Atlantic Ocean, although not shown as such here, is often a doublet known as the *Bermuda high* and *Azores high.*

THE SUBPOLAR LOW-PRESSURE BELTS. Although observations in high oceanic latitudes of the Southern Hemisphere are relatively sparse, there seem to be enough to indicate that in the average views little change appears to occur from summer to winter. This would be expected on the basis of the global encirclement by the southern oceans in these latitudes. But in the Northern Hemisphere strong annual changes occur in this zone as pronounced temperature contrasts develop between land and water. In January, the lows reverse to highs over the lands (to form the Canadian and Siberian highs) but become extremely intense and stormy low-pressure areas over the relatively warm North Atlantic and North Pacific Oceans, where the designations *Iceland low* and *Aleutian low* are used, respectively.

POLAR HIGH-PRESSURE CAPS. On the average, high-pressure areas exist over both polar regions. However, both the intensities and locations of the centers of these highs are known to shift, with the latter aspect being only rarely centered on the geographic poles. When completely evaluated, the results of polar explorations during the International Geophysical Year 1957–1958 and the International Geophysical Cooperation 1958–1959 should yield a much more complete understanding of the meteorology of the polar zones.

The Planetary Wind Systems

Since the planetary winds are physically a part of a global pattern involving the pressure belts just described, they too often vary considerably from the idealized pattern. Again this variation is most pronounced where zonal contrasts in land and water produce parallel contrasts in temperature distribution and temperature variation.

THE TRADE WINDS (TROPICAL EASTERLIES). The winds blowing from the horse latitudes to the doldrums are among the most constant of the planetary system. The name *trade* as applied to winds is derived from the expression "to blow trade" which meant to blow constant. These winds, which are best developed over the Atlantic and the much broader Pacific, away from the pressure perturbations of the continents, tend to follow the pressure gradient directed toward the equator, but become deflected to the right and left in each hemisphere, becoming the *northeast and southeast trades,* respectively. The trades have an average speed of 10 to 15 knots and are strongest, particularly over the Northern Hemisphere oceans, during the summer when the subtropical high is strongest.

Since the equatorial low is often north of the equator, the southeast trades suffer a deflection to the right, becoming *hooked trades* in order to reach the low. This is particularly noticeable in Fig. 10·3 for the Indian and eastern Atlantic Oceans. Owing to the trade-wind convergence on the equatorial low-pressure belt, or doldrums, this zone is also known as the *intertropical convergence,* the meteorological significance of which will be discussed in a later chapter on weather structures.

It is important to realize again that the wind systems and latitudinal zones are symmetrically disposed about the equator. Hence the area within latitudinal zones diminishes from a maximum in the equatorial region to zero at the poles. The trade-wind system is thus area-wise by far the largest of the planetary wind groups which become progressively smaller with increasing latitude.

THE PREVAILING WESTERLIES. In both hemispheres a system of winds is directed poleward from the subtropical high-pressure belt. These winds become southwesterly to westerly in both hemispheres as a result of the Coriolis deflection and are known generally as the prevailing westerlies. In the Southern Hemisphere, as was noted, conditions are very uniform owing to the broad ocean areas, and little variation occurs.

Owing to this uniformity, a relatively strong and uniform pressure gradient is normally present in the Southern Hemisphere between the subtropical high and the subpolar low. This is striking in both Figs. 10 · 2 and 10 · 3, where the parallel, east-west isobars show a uniform and fairly strong gradient from about 30°S to 60°S. The westerlies associated are quite steady and strong throughout the year with an average force of 5 to 6 Beaufort (17 to 27 knots). Owing to the wind strength which often increases to gale force, particularly with increasing southerly latitude, the progressive zones have often been designated as "roaring forties, furious fifties, and screaming sixties."

In the Northern Hemisphere, the prevailing westerlies are very variable and are often either masked by the more prominent circulation about moving low- and high-pressure areas or completely reversed as reversals in seasonal pressure of the related horse latitude and subpolar lows occur.

MONSOON WINDS. Monsoon winds are winds whose direction reverses with the seasons. Such a reversal in wind direction obviously requires a reversal of the pressure gradient. There are many areas on the earth where such pressure variations occur, following changes in land temperature from winter to summer. By far the most notable and widespread pressure change is that responsible for the Indian monsoon.

We noted that from winter to summer over Asia there is a reversal of pressure from high to low. This is the direct cause of the famous Indian monsoon. During the winter the winds blow normally from the northeast, originating in the high over Asia. But in the summer, as seen in Fig. 10 · 3, pressure is low over Asia and the winds have completely reversed, arriving from the southwest after hooking across the equator. This air motion begins in the southern horse latitudes and follows the continuous pressure gradient into central Asia, thus combining trade and monsoon winds.

After blowing across the open tropical ocean, this southwest summer monsoon is extremely humid and warm. Upon striking the high plateaus and mountains of India, the air is forced to ascend and cools adiabatically as it does so. This cooling lowers the air temperature beneath the dew point, resulting in the well-known rainy season of the Indian summer.

The winter monsoon, blowing from the northeast in central Asia, is cold and dry owing to its continental origin. The monsoon winds become light and variable during the period of change—spring and autumn.

Another strong, though not so widespread, monsoon exists west and southwest of the African bulge, in the Atlantic. Here, again owing to the migration of the doldrums with change of seasons, this area is covered by the northeast trades in winter and the hooked south to southwest trades in summer. This is clearly evinced on the world wind maps. Somewhat less-pronounced monsoons can also be shown to occur in other regions.

Winds in the Upper Troposphere

So far we have considered the wind motion in the lower troposphere, sometimes referred to as the *surface circulation.* However, any complete description of the planetary wind system must also consider the fundamentally more important airflow patterns in the upper part of the troposphere, which together with the surface circulation constitute the general circulation of the atmosphere. Note again that we are involved here with average or mean air flow and that the day-to-day values of wind speed and direction, on which the means are based, may vary considerably from the average values.

In the upper troposphere, an average westerly flow of air prevails beyond about 15 to 20 north and south latitudes with the exception of weak easterly flows in higher latitudes. (Observations are not yet fully reported for high latitudes.) Within the tropical zone, an easterly flow prevails with winds that increase from the surface to a maximum near the top of the troposphere. These upper easterlies have speeds distinctly lower than the westerlies of the middle latitudes.

A summary of average conditions throughout the troposphere of the Northern Hemisphere is shown in Fig. 10 • 4 for the months of January and July. Positive numbers refer to winds whose major component of motion is from west to east; negative numbers refer to easterly winds. Elevations are given in terms of pressure values (centibars; multiply by ten for millibars) which can be converted to approximate elevations by means of the scale included.

On the average, maximum troposphere winds in July are seen to be westerly at about 20 meters per second (39 knots) at about latitude 42°N and an elevation of 200 millibars (about 38,000 feet); in January, the strong westerly current is displaced southward to about 27°N with an average speed of about 40 meters per second (78 knots). Maximum easterlies in the low latitudes, which are not too well shown here, reach 10 to 15 meters per second (19 to 29 knots).

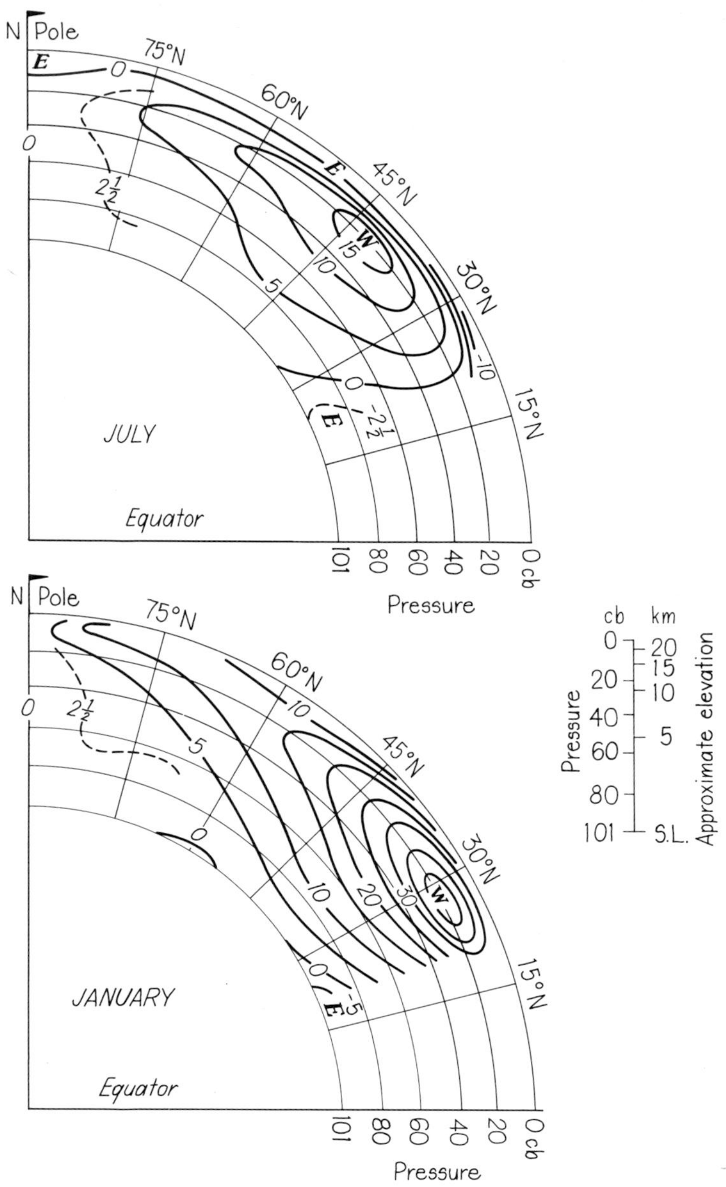

Fig. 10 · 4 Cross section through the troposphere of the Northern Hemisphere, from the equator to the north pole, showing average wind velocity. Wind speed contours are given in meters per second (mps) in intervals of 5 (twice the values given is nearly the wind speed in knots). The elevation scale is given both in units of pressure, centibars (cb) and in kilometers (km). (B. Bolin, after Y. Mintz and G. Dean, courtesy Academic Press, Inc.)

When the actual motion of the upper-level winds are examined, they are seen to follow a broad hemispheric pattern of waves rather than a simple linear motion. The distribution of pressure and winds associated with these waves in the upper westerlies (known as Rossby waves after C. G. Rossby who formulated their quantitative characteristics of motion) is shown in the example in Fig. 10 · 5. Note that the solid lines (equivalent to isobars) are quite parallel to the wind motion at these levels.

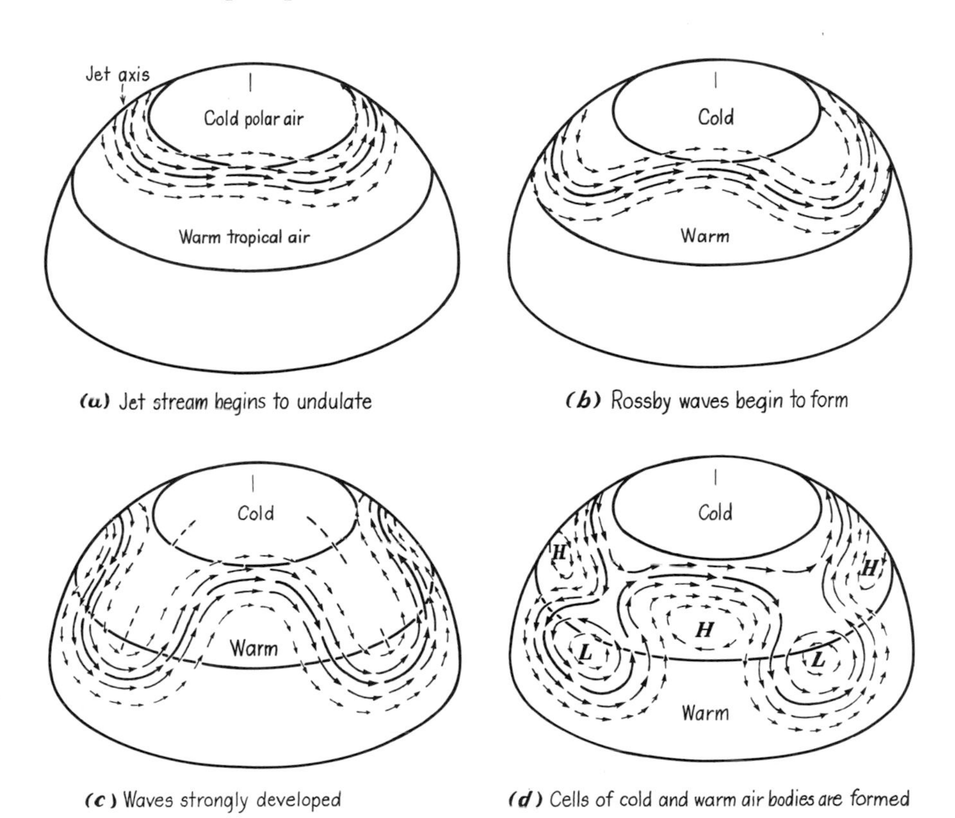

Fig. 10 · 5 Waves in the upper westerlies. (A. Strahler, *The Earth Sciences,* after G. T. Trewartha)

Rossby waves follow growth cycles of a few days to a week or so in length, so that the upper wind pattern changes frequently in detail over short periods despite the average values shown in Fig. 10 · 4. The relationship between surface pressure and winds and upper-level winds at 500 millibars (18,000 to 19,000 feet) for the Northern Hemisphere is shown in Figs. 10 · 6 and 10 · 7, respectively. We can see in these illustrations that the surface wind, particularly in the middle latitudes, is controlled by low- and high-pressure areas in a manner to be described in the next chapter and tends to describe a closed circulation about these pressure cells. The upper winds, however, are more uniform and simpler in pattern, showing a generally zonal (latitudinal) flow. The speed is also much greater. In fact the upper westerlies always show a high-speed axis (also obvious in the mean winds of Fig. 10 · 4) known as the *jet stream.*

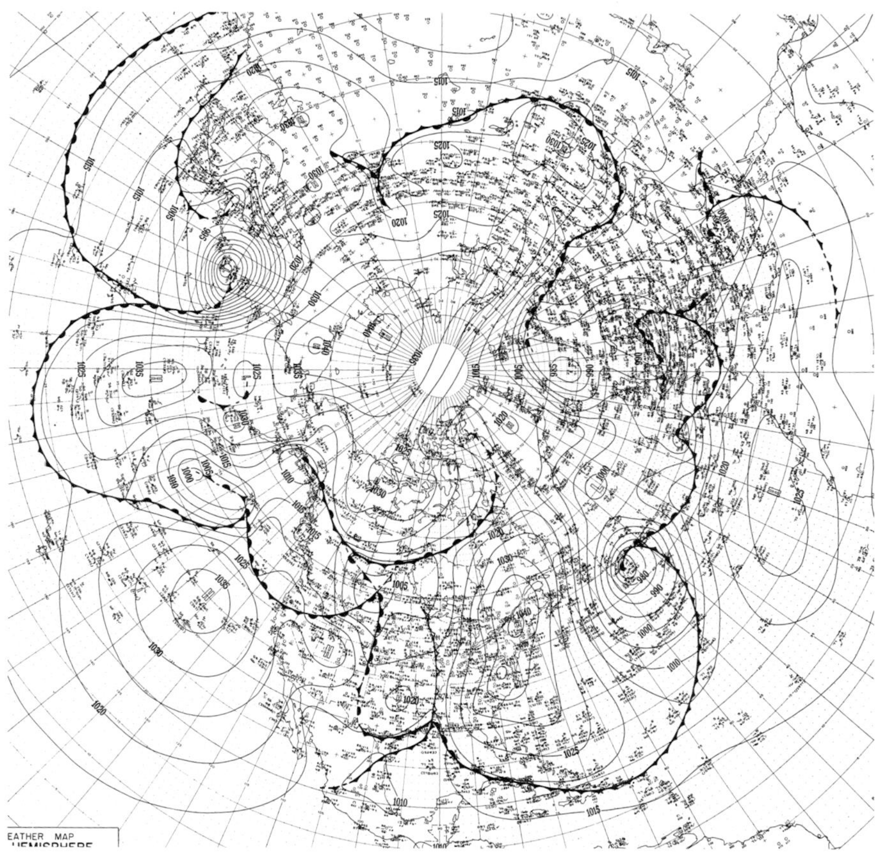

Fig. 10 · 6 Northern Hemisphere surface chart showing pressure and wind distribution (February 5, 1955).

JET STREAMS. During World War II, high-level military aircraft capable of speeds of 300 miles per hour developed ground speeds which, depending on direction of flight, approached either 0 (westward flight) or 600 miles per hour (eastward flight). The explanation was soon discovered to be in the existence of a high-elevation, narrow, elongated core of very high winds—the jet stream. Although discovered during the mid-1940s, the jet stream's existence was suspected earlier, particularly by C. G. Rossby who deduced that the atmosphere should have high-speed currents similar to the Gulf Stream, which is a confined, relatively high-speed ocean current embedded in the slower-moving water of the Atlantic Ocean.

An atmospheric jet stream is described as a westerly air current in the

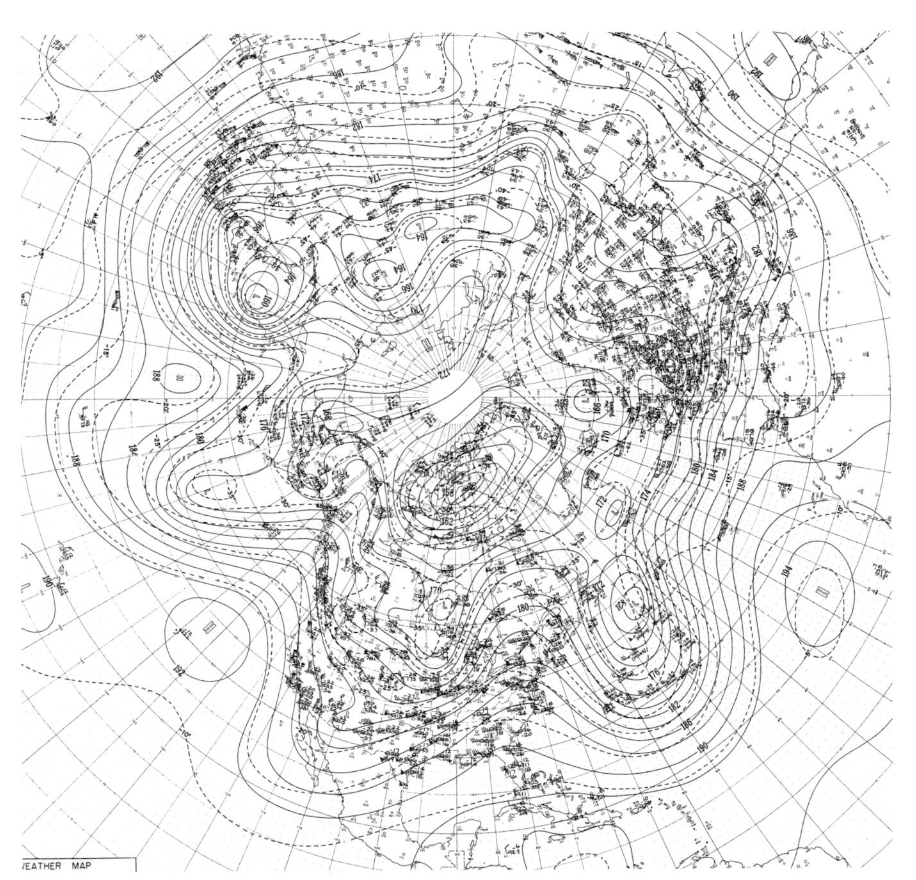

Fig. 10 · 7 500 millibar constant pressure chart of the Northern Hemisphere showing related wind motion (February 5, 1955).

form of a flattened, narrow core or tube, thousands of miles in length, a hundred or more miles in width, and one or more miles in vertical thickness. Although reports of maximum winds in the center of the core reach 300 knots, 100 to 200 knots is more typical of the maximum jet winds encountered. These maximum winds usually occur between 35,000 and 40,000 feet, but jet winds of lesser magnitude may be experienced at times above 20,000 feet. The rapid expansion of upper-level observations to the tropopause and higher which took place in the 1950s resulted in the discovery of two basic jet streams in the troposphere—the *subtropical* and the *polar-front* jet streams. The mean positions of these jets for winter is shown in Fig. 10 · 8 on a chart of the Northern Hemisphere centered on the North Pole.

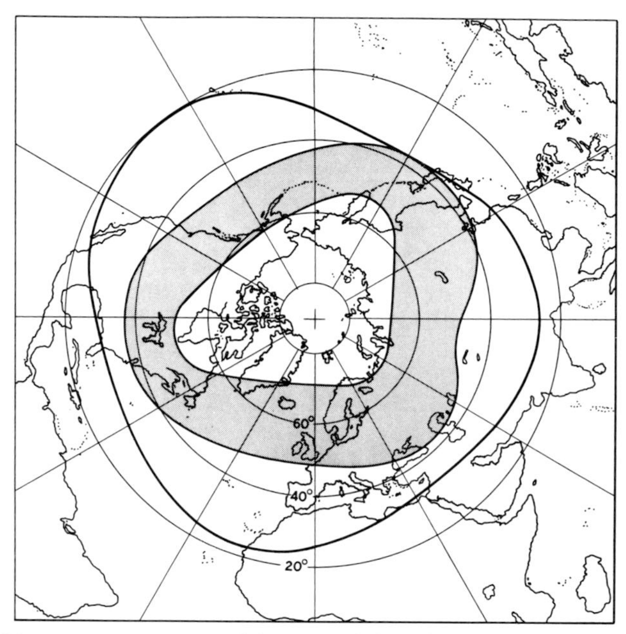

Fig. 10 · 8 Mean winter position of the axis of the subtropical jet stream (heavy line) and mean area of principal winter activity of the polar front jet stream (shaded). (H. Riehl, Jet Streams of the Atmosphere)

During the winter the subtropical jet is relatively constant and continuous in position and in time, as indicated by the rather narrow axis in Fig. 10 · 8. This jet marks the northern boundary of the trade-wind cell in the upper troposphere and, although extremely prominent in the winter, it becomes broken and weakened during the summer, when its importance as a strong global current is lost. When well developed, winds of more than 150 knots are not uncommon, as illustrated in Fig. 10 · 9 which shows the subtropical jet on four successive days at the 200-millibar level (38,000 to 39,000 feet). The pattern of long, fairly symmetrical waves which characterizes this current is very noticeable. The waves are composed of ridges (northward bends) and troughs (southward bends).

The polar-front jet stream is much more variable in location, continuity,

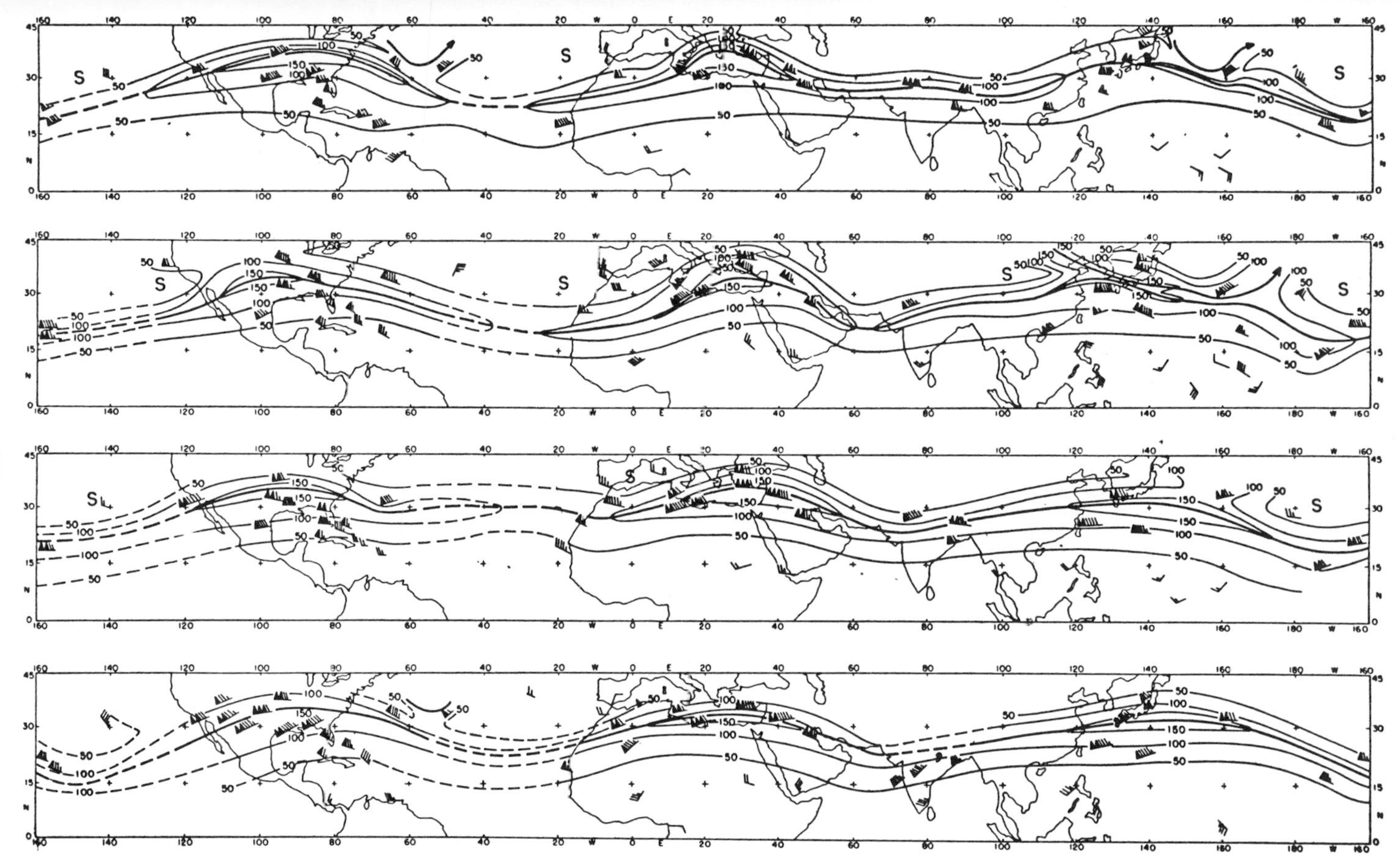

Fig. 10 • 9 The Northern Hemisphere subtropical jet stream on February 25 to 28, 1956 (top to bottom). Arrows fly with the wind—each barb represents 10 knots; each flag is 50 knots and each rectangle 100 knots. The heavy line is the jet axis and the finer lines are isotacs, *lines of equal velocity.* (H. Riehl, Jet Streams of the Atmosphere)

wind speed, and elevation. Some of this is indicated by the broad zone over which this jet is active, as shown in Fig. 10 · 8. In general, the polar-front jet is farthest south in midwinter and farthest north in summer, and its elevation decreases as it migrates northward. The polar-front jet is intimately associated with the location and motion of the low- and high-pressure areas that mostly control the weather of the mid-latitudes. It represents the high-speed core of the prevailing westerlies in the upper troposphere and takes on a meandering or serpentine pattern conforming to the shape of the Rossby waves referred to in the previous section. When this jet is located in the lower middle latitudes in winter, a ridge or ridges of the subtropical jet and a trough or troughs of the polar jet may temporarily coalesce into a single jet, at which time important exchanges of low- and high-latitude air occur.

EXERCISES

10 · 1 What are the two primary causes of the disturbances of an ideal or uniform planetary wind system?

10 · 2 Describe the average weather of the belts of equatorial and subtropical calms.

10 · 3 Describe the planetary surface wind systems.

10 · 4 What seasonal changes occur in (*a*) the Iceland and Aleutian lows and (*b*) the Bermuda, Azores, and Pacific highs?

10 · 5 Explain the pressure and wind variations associated with the Indian monsoon.

10 · 6 Compare the surface wind systems with the patterns of the upper troposphere.

10 · 7 Describe Rossby waves and their relation to the upper westerlies.

10 · 8 Give a concise description of jet streams and their seasonal migration.

CHAPTER 11

Cyclones, Anticyclones, and Secondary Winds

Although the winds of the trade wind zone are fairly constant and uniform, those of the prevailing westerlies zone are far more variable, especially in the lower portion of the troposphere. The wind in this region is associated with areas of low and high pressure that travel eastward through the troposphere as closed systems near the ground and as waves in the upper westerlies. These systems are known as *cyclones* and *anticyclones,* respectively, and the related winds as *cyclonic* and *anticyclonic* winds.

Cyclones and Cyclonic Winds

Although a more complete analysis of the structure of a cyclone will be given in Chap. 15, we may consider for the present that a cyclone is a roughly circular low-pressure area whose diameter may be from hundreds to a thousand or so miles. Atmospheric pressure is always lowest in the center of this region and increases radially outward as described in Chap. 8.

In Fig. 11 · 1, the pressure gradient is shown directed inward by the broken arrows. However, the winds in the low are under the influence of the deflective Coriolis, centrifugal, and frictional forces, as well as the pressure gradient, as described in Chap. 9. Instead of blowing directly inward, parallel to the gradient, the wind blows across the isobars, at a high angle to the pressure gradient.

In the Northern Hemisphere the deflective force causes the wind to blow to the right of the pressure gradient, resulting in a counterclockwise, spiral, inward motion for the air near the surface. At an elevation of thousands of feet above the ground, as noted in Chap. 9, frictional forces are very low, so that the wind deflection is greater, giving motion parallel to the isobars. Such winds, that move parallel to the isobars as the result of balance between the pressure gradient and the deflective forces, were defined earlier (Chap. 9) as gradient when motion is curved and geostrophic when it is more or less linear.

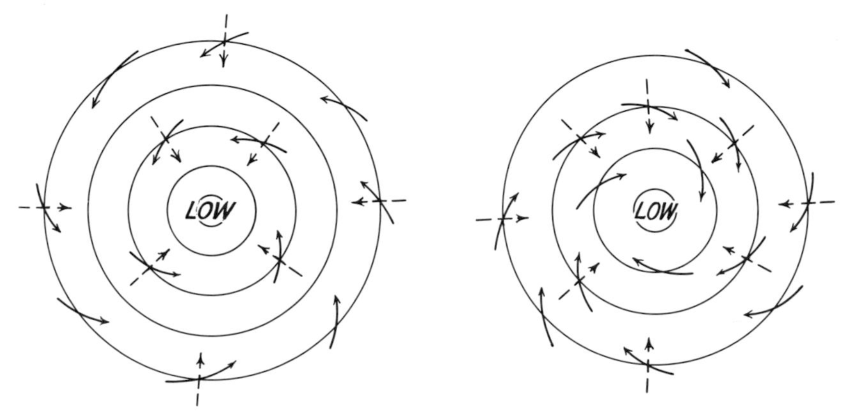

Fig. 11 · 1 Cyclonic circulation in the Northern Hemisphere.

Fig. 11 · 2 Cyclonic circulation in Southern Hemisphere.

This counterclockwise spiral to circular motion in the Northern Hemisphere is known as *cyclonic circulation.* Note that the term *cyclone* thus implies no particularly dangerous or destructive storm, but is one of the typical and common weather patterns of the middle latitudes. Although steep pressure gradients, with consequent gale winds, often develop in middle-latitude low-pressure areas, the latter should not be confused with the more violent but much smaller *tropical cyclones* or hurricanes, to be studied in Chap. 15. In distinction to these, the mid-latitude lows are known technically as *extratropical cyclones.*

Recall that the deflective force is to the left of the pressure gradient in the Southern Hemisphere. Consequently the wind motion takes the pattern shown in Fig. 11 · 2, becoming clockwise in circulation about a cyclone in the southern latitudes. Figs. 11 · 3 and 11 · 4 are striking satellite photographs of cloud distribution in the Northern and Southern Hemispheres, respectively. The former, a *Tiros* VI photograph, clearly shows the counterclockwise, inward, wind-spiral motion by means of the cloud streaks in the upper right quadrant. The latter indicates clockwise motion for the Southern Hemisphere cyclonic disturbance and shows progressive changes in this storm during an eight-day period. It is instructive to compare the cloud pattern of the cyclone in Fig. 11 · 3 with the abridged marine

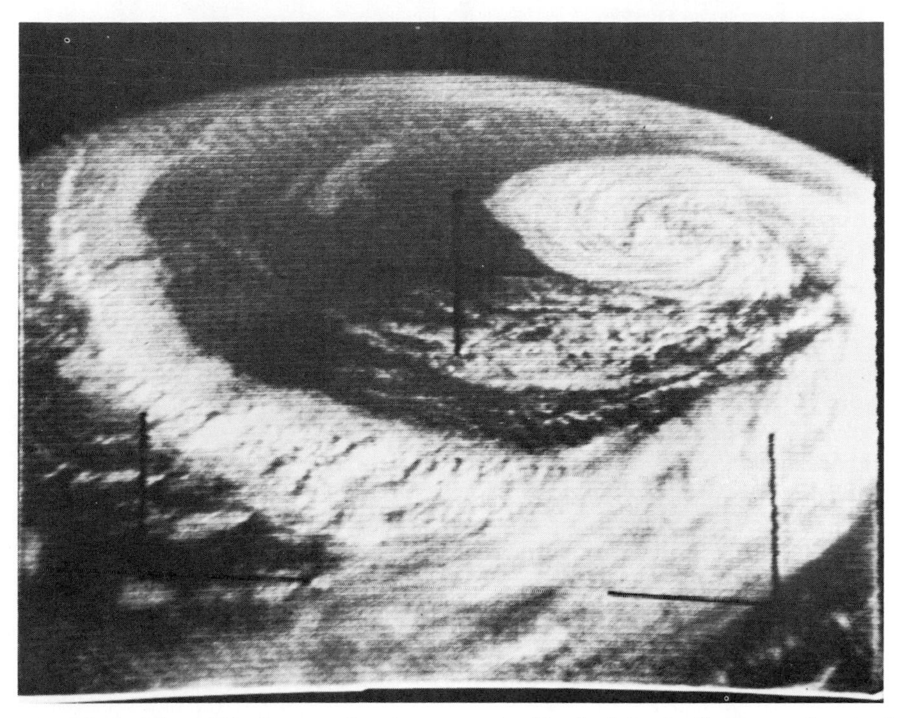

Fig. 11 · 3 Tiros VI *photograph taken over the North Atlantic Ocean southeast of Nova Scotia, May 29, 1963. The camera was looking toward the northwest (top of picture). This is an outstanding picture of the spiral cloud associated with cyclonic vortices.* (U.S. Weather Bureau)

weather map of the time, as shown in Fig. 11 · 5, which shows the isobaric pattern of lows and highs analyzed on the basis of ships' reports. The prominent cloud in the lower half of the photograph is associated with the frontal system shown by the heavy line with black semicircular and triangular symbols. The description and relationship of fronts to cyclones will be described more completely in Chaps. 13 and 14.

Extratropical cyclonic winds are invariably more intense during the cold portion of the year and often create hazardous conditions for navigation over the oceans. A typical storm series for the North Atlantic Ocean

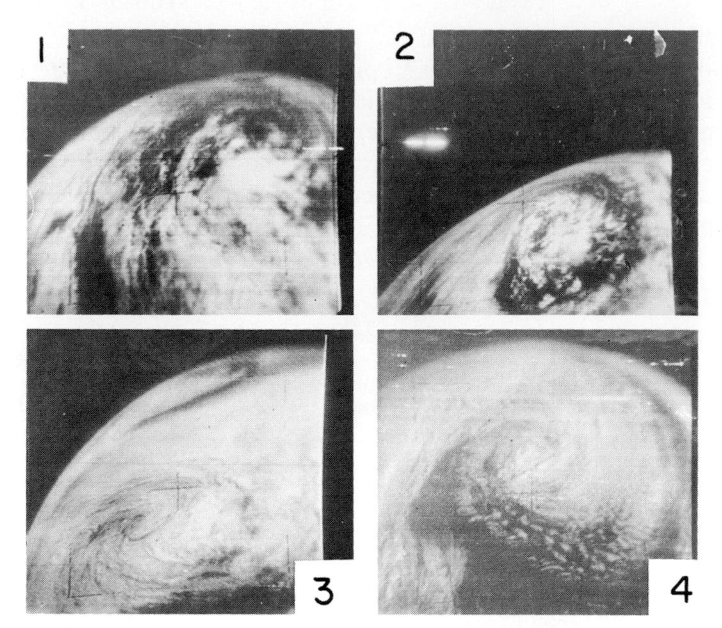

Fig. 11 · 4 Tiros *satellite photographs of cyclones in the Southern Hemisphere showing clearly the clockwise circulation as revealed by the cloud pattern. Number 1 is a tropical cyclone in the Indian Ocean on April 20, 1960. Number 2 is in the Indian Ocean on April 25, 1960. Number 3 is in the Indian Ocean, April 30, 1960. Number 4 is in the South Atlantic Ocean on April 28, 1960.* (U.S. Weather Bureau)

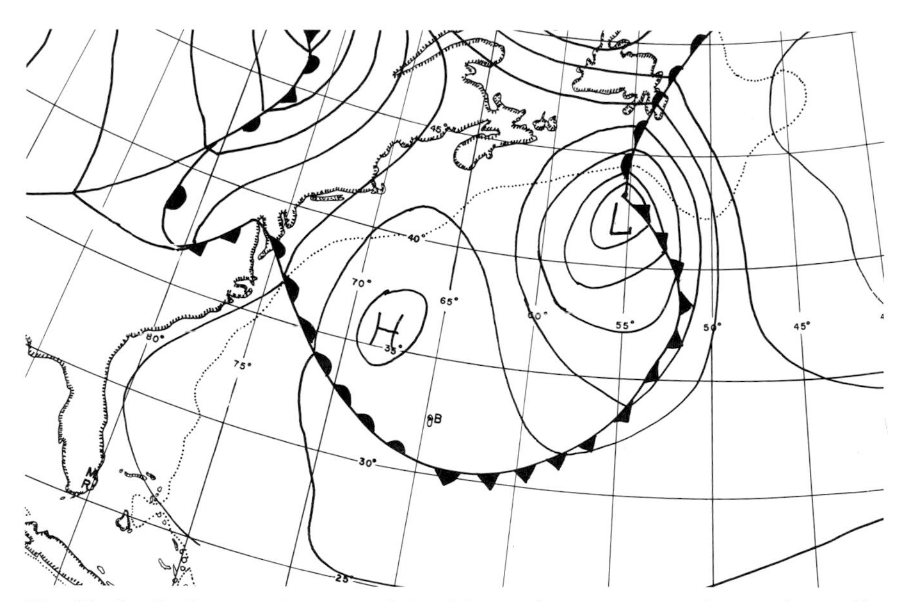

Fig. 11 · 5 Surface weather map of the Atlantic Ocean corresponding to the satellite photograph shown in Fig. 11 · 3.

is shown in Figure 11 · 6. This traces a storm as it moves in a common east-northeast path from a weak low off the Virginia Capes to a very intense cyclone showing observed winds up to 50 knots off Newfoundland. Again, the frontal structure is included in the diagrams, whose explanation is reserved for later chapters. Storms of the type illustrated in Fig. 11 · 6 are quite common in all but the summer months. Because they cross the western end of the main North Atlantic shipping routes, such storms have been a threat to safe navigation and cargo stowage.

Anticyclones

The roughly circular high-pressure areas described in Chap. 9 give rise to winds whose motion with respect to the center is essentially opposite to that developed in low-pressure areas. Under the influence of both the pressure gradient and the deflective forces, the winds, which tend to move directly out from the center of a high-pressure system, develop a clockwise outward spiral in the Northern Hemisphere and a counterclockwise motion in the Southern Hemisphere. These motions are illustrated for each hemisphere, respectively, in Figs. 11 · 7 and 11 · 8, in which the former indicates the direction of the pressure gradient by dashed arrows. Because the wind circulation in high-pressure systems is opposite that in cyclones (low-pressure systems), the motion is called *anticyclonic* and the system is called an *anticyclone.*

As in the case of cyclones, frictional forces decrease vertically upward. Consequently the higher-speed winds at levels several thousand feet above the ground suffer the maximum deflection possible, which is a direction normal to the pressure gradient. Hence the wind well above the ground is influenced by the geostrophic or gradient conditions and tends to blow parallel to the isobars, rather than across them at a small angle as near the surface.

Although we usually associate storms and strong winds with cyclonic activity, very strong anticyclones are commonly associated with strong outbreaks of dense cold air that surge from the arctic to lower latitudes. Central United States and the Gulf of Mexico often experience this effect during the cold months of the year. At such times, severe cold winds pour southward along the eastern half of an elongated anticyclone, bringing winds which may reach whole gale force to the interior United States and the Gulf of Mexico. Such strong cold winds are known as *northers.* An example of a pressure pattern over the United States associated with northers is shown in Fig. 11 · 9.

Occurrence of Gales and Strong Winds at Sea

For the most part, gales and strong winds at sea are associated with well-developed pressure gradients of strong cyclones and anticyclones

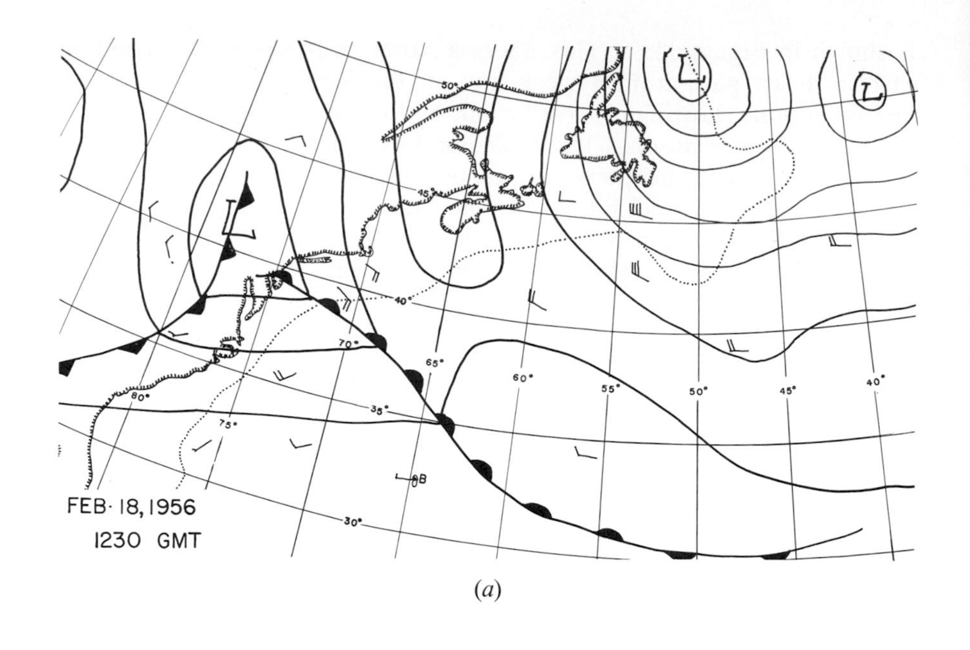

(*a*)

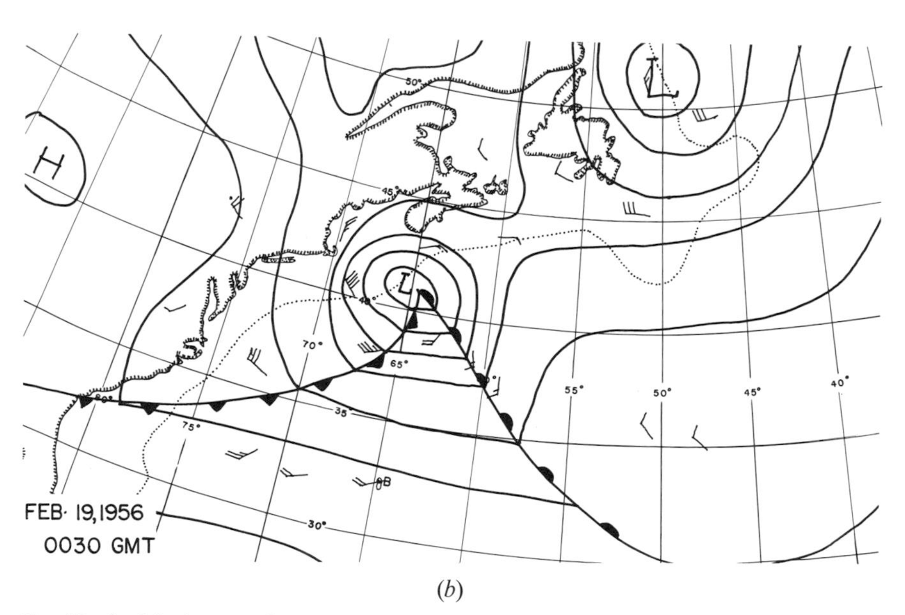

(*b*)

Fig. 11 · 6 Marine weather map series of the North Atlantic Ocean. Isobars are drawn at intervals of 6 millibars.

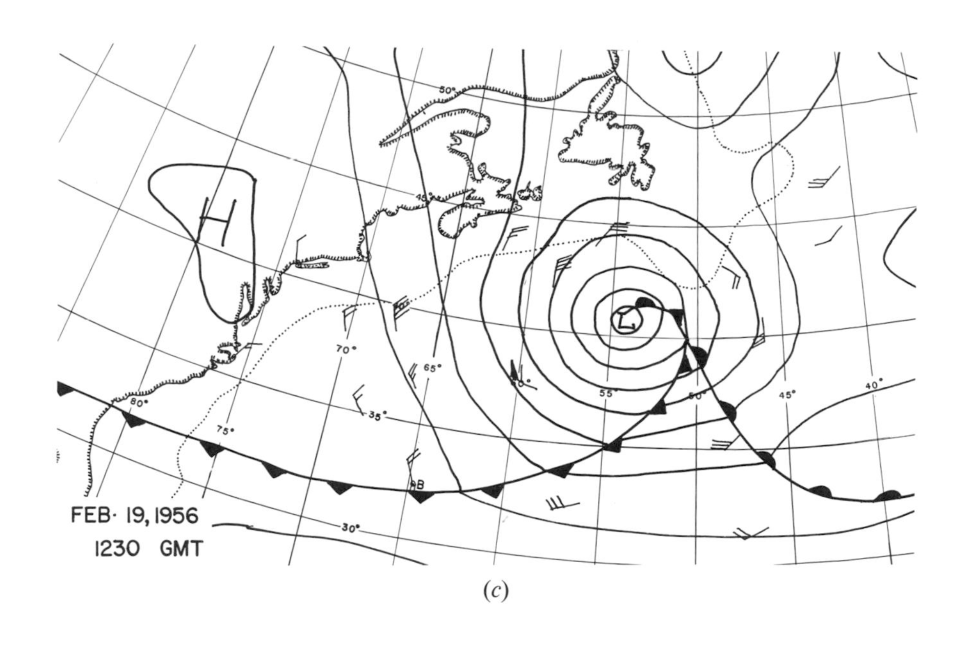
FEB. 19, 1956
1230 GMT
H
L
50°
45°
70°
65°
60°
55°
50°
45°
40°
35°
80°
75°
30°

(*c*)

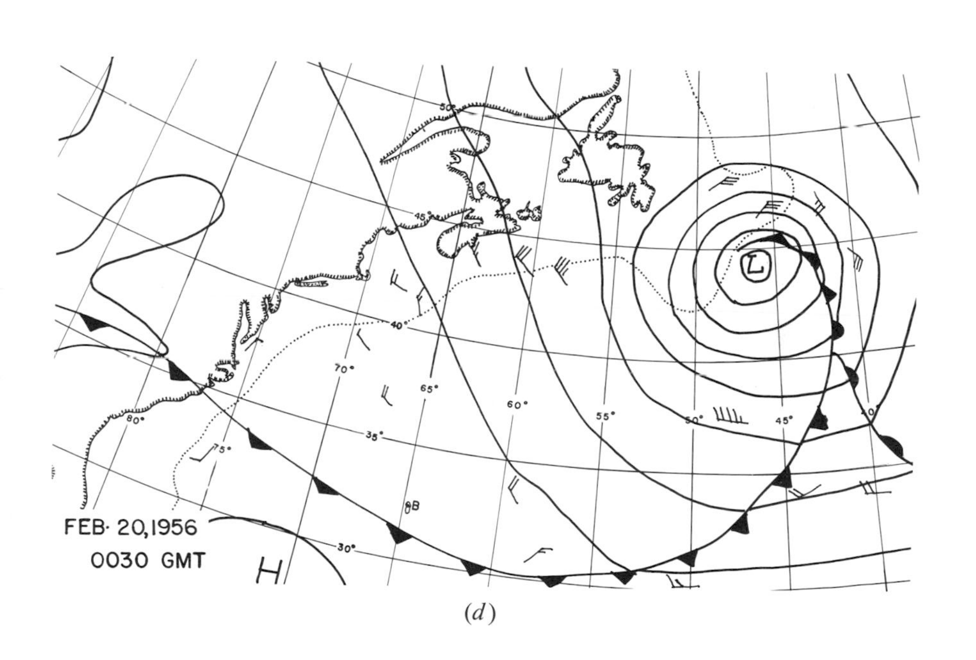
FEB. 20, 1956
0030 GMT
H
L
50°
45°
40°
70°
65°
60°
55°
50°
45°
80°
75°
35°
30°

(*d*)

rather than with the average flow of the planetary winds described in the previous chapter. In general, high winds are far more characteristic of the upper middle latitudes of both hemispheres than of the lower latitudes, except for the tropical cyclones (hurricanes) to be described in a later

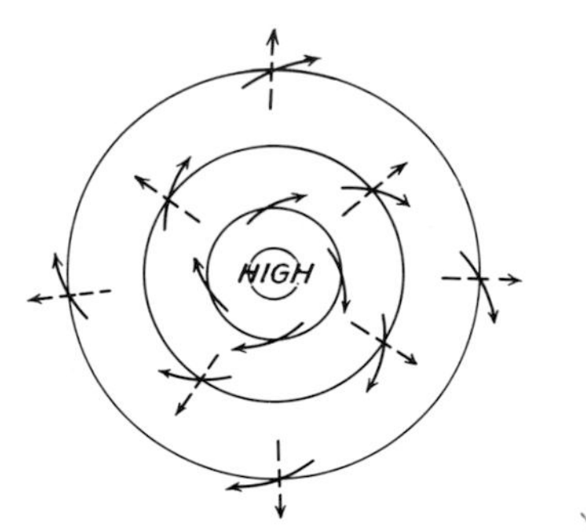

Fig. 11 · 7 Anticyclonic circulation in Northern Hemisphere.

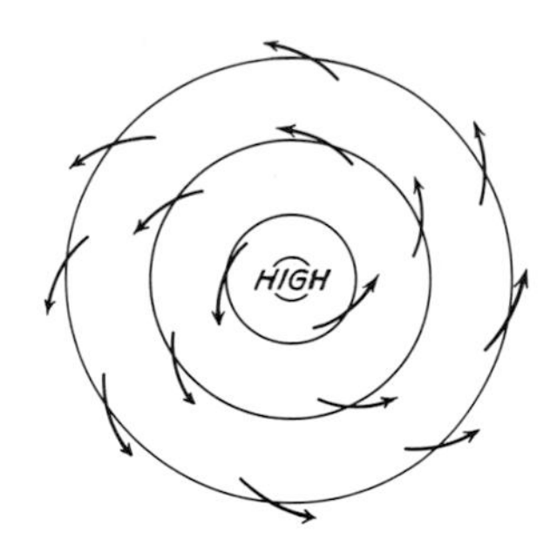

Fig. 11 · 8 Anticyclonic circulation in Southern Hemisphere.

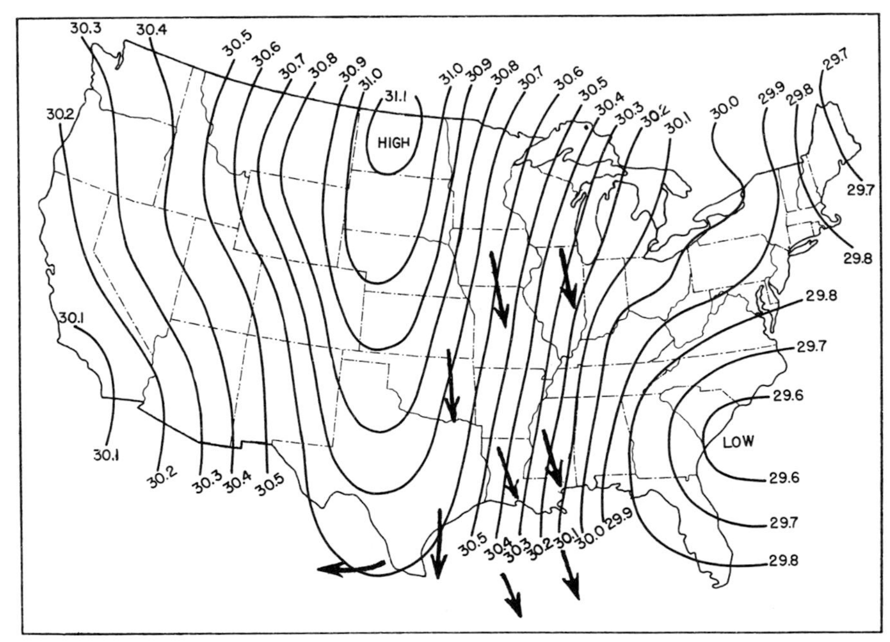

Fig. 11 · 9 Pressure distribution giving rise to a strong norther of anticyclonic origin in the Gulf of Mexico. (U.S. Weather Bureau)

chapter. And in the higher latitudes, storminess in much greater during the cold portion of the year and in the western rather than the eastern portions of the oceans.

Figures 11 · 10 to 11 · 15 give a valuable quantitative summary of the

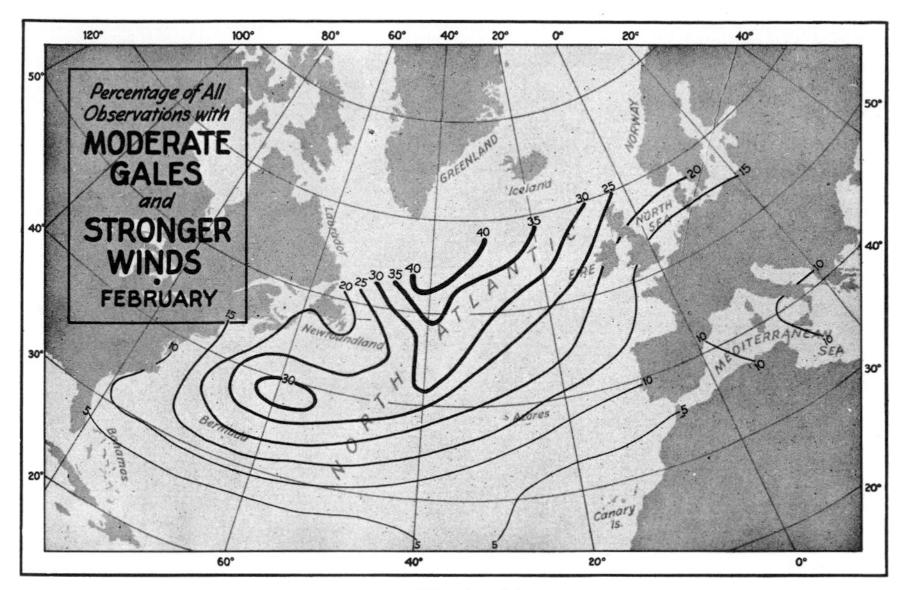

Fig. 11 · 10

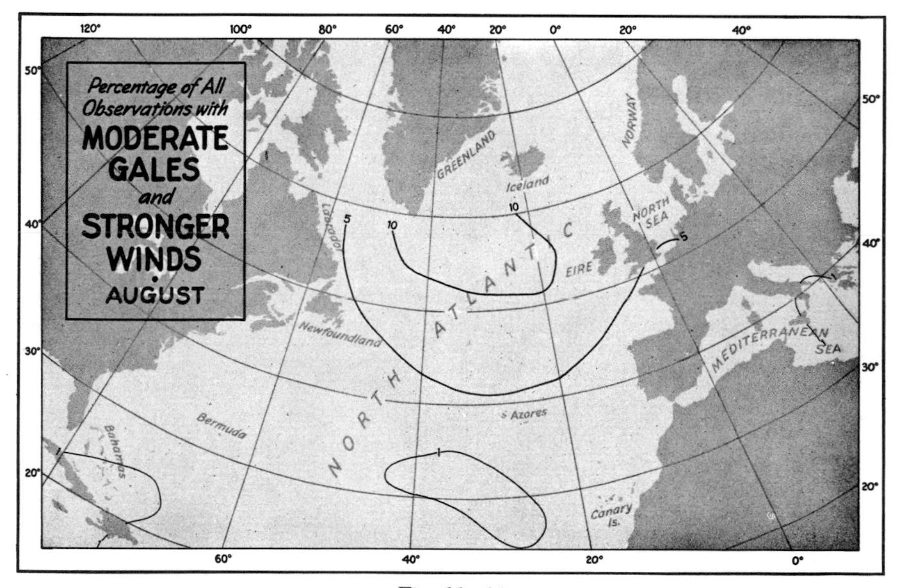

Fig. 11 · 11

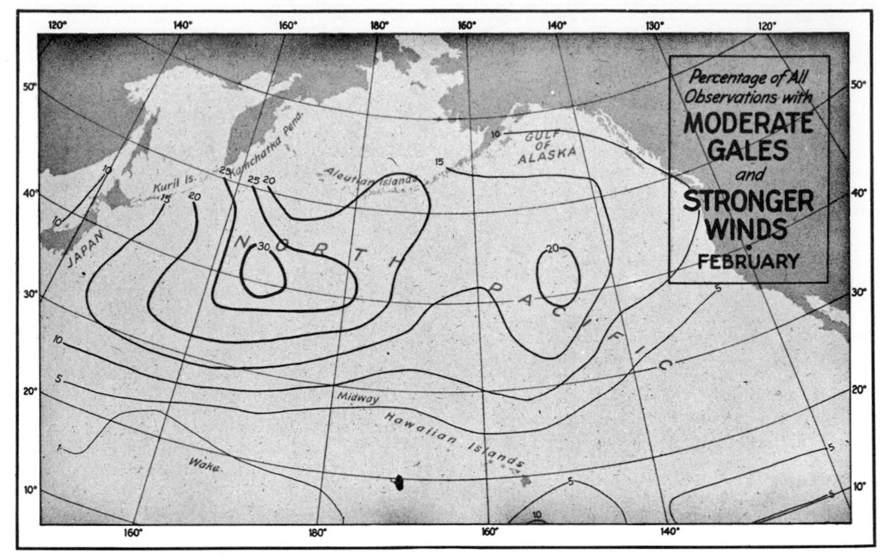

Fig. 11 · 12

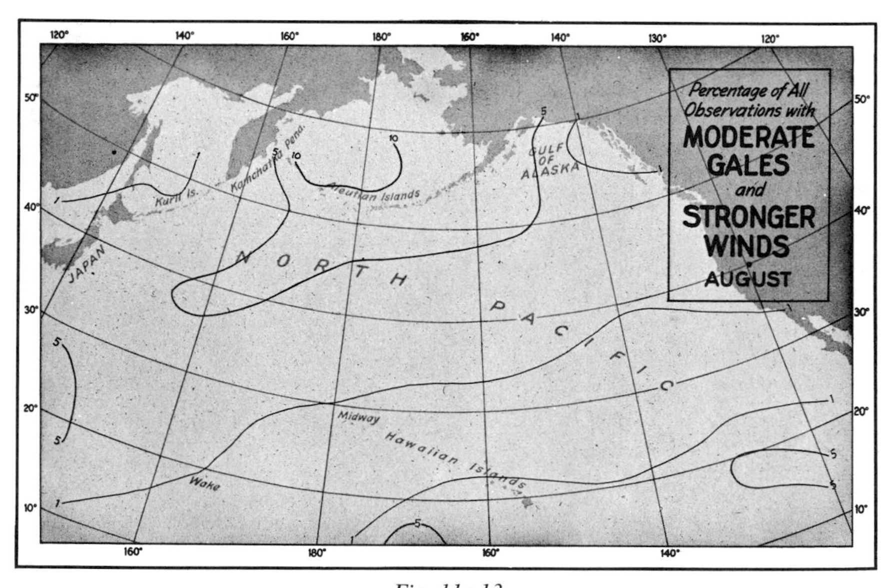

Fig. 11 · 13

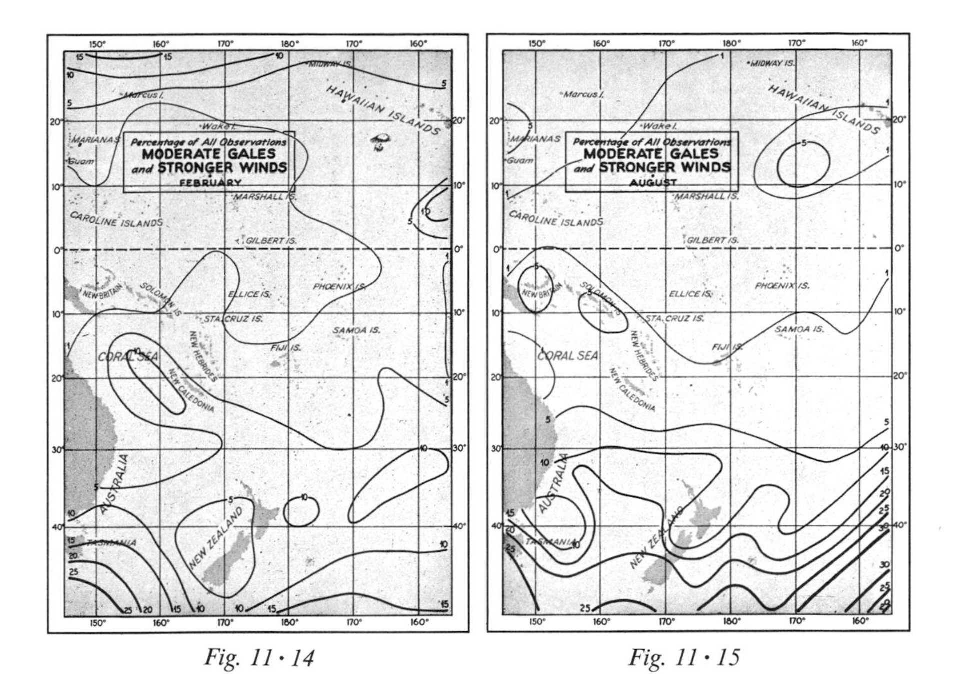

Fig. 11 · 14 *Fig. 11 · 15*

frequency of strong wind and gale occurrence over most of the oceans. The continuous lines in these illustrations connect points estimated to have the same percentage frequency of storm occurrence.

Although only immediate surface weather maps can indicate the presence of actual storms, the combination of information in Figs. 11 · 10 to 11 · 15 and in Figs. 11 · 16 and 11 · 17 can aid in estimations of the probability of storm encounters. The latter two illustrations give the tracks of storms for the North Atlantic and North Pacific Oceans for the months of temperature extremes. Primary storm tracks for February are shown by solid arrows and those for August by broken arrows. The concentration of tracks on the western side of the oceans explains the reason for the greater frequency of strong winds in these regions as noted above.

Much more detailed monthly data relating to storm occurrence for all parts of the world oceans can be found in the series *Climatological and Oceanographic Atlas for Mariners.*

Effect of Moving Cyclones and Anticyclones

We mentioned earlier that the periodic pressure variations, with maximums at 10 A.M. and 10 P.M. and minimums at 4 P.M. and 4 A.M., are overshadowed by pressure changes of greater magnitude in the middle latitudes

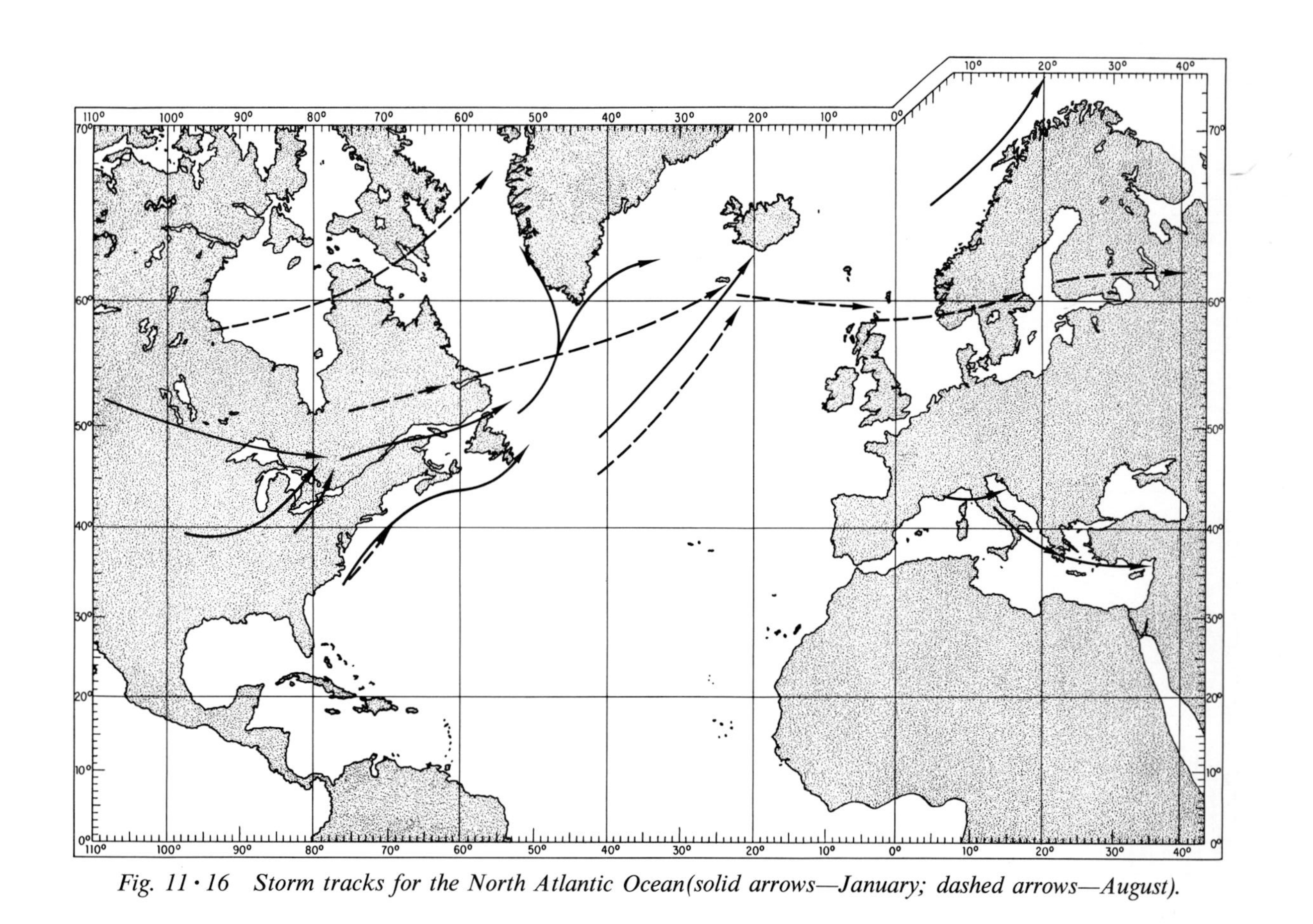

Fig. 11·16 Storm tracks for the North Atlantic Ocean(solid arrows—January; dashed arrows—August).

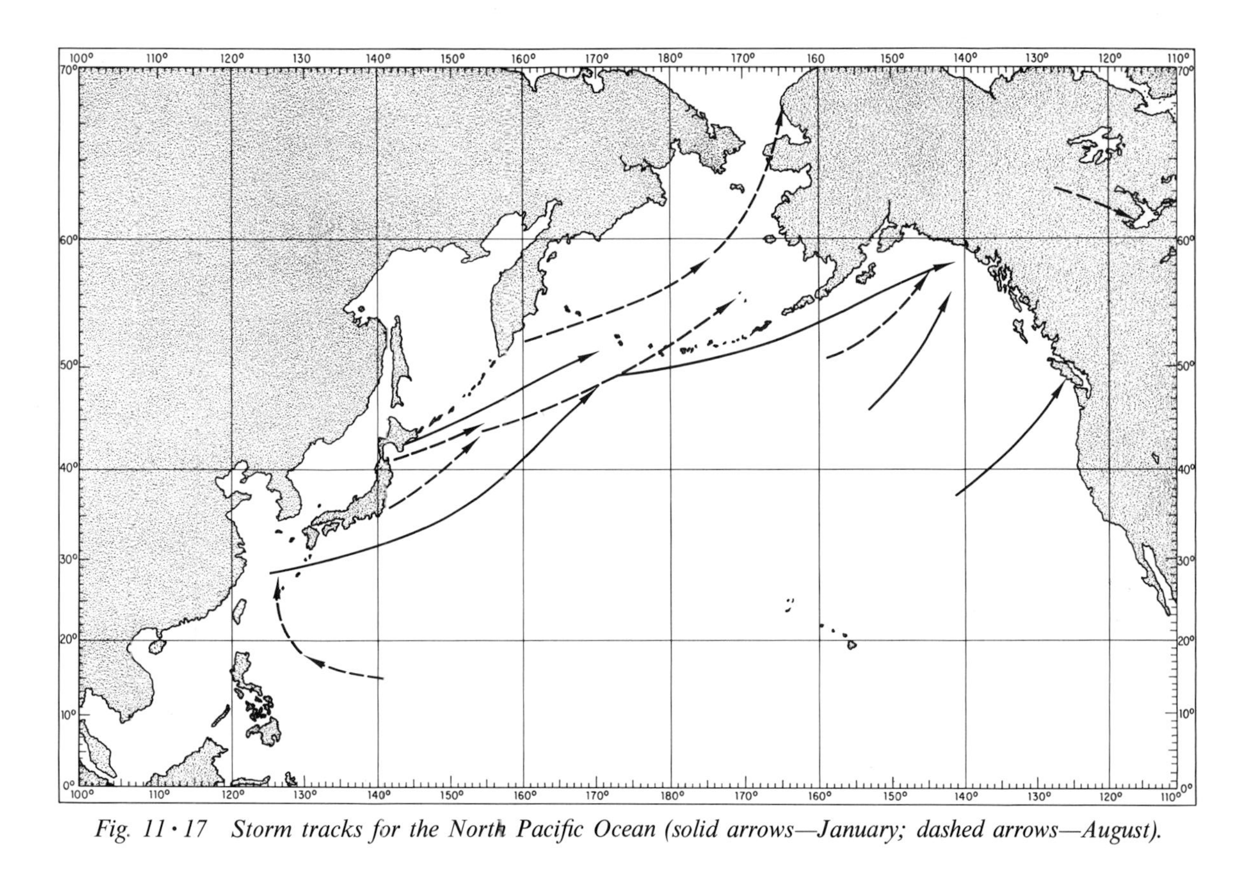

Fig. 11 · 17 Storm tracks for the North Pacific Ocean (solid arrows—January; dashed arrows—August).

(between 30° and 60°). The reason for this overshadowing can now be made more clear.

The highs and lows just considered travel through the atmosphere as huge pressure waves with a motion generally from *west to east* and with an average velocity of 20 to 30 miles an hour. The effect of low-pressure areas usually *disappears* above heights of 2½ to 4 miles, where the planetary circulation is reasserted. The complete weather conditions attending the passage of these highs and lows will be studied in more detail in a later chapter. However, we can now examine some of the pressure and wind changes attending the passage of these atmospheric disturbances.

The continuous and irregular pressure variations experienced in middle

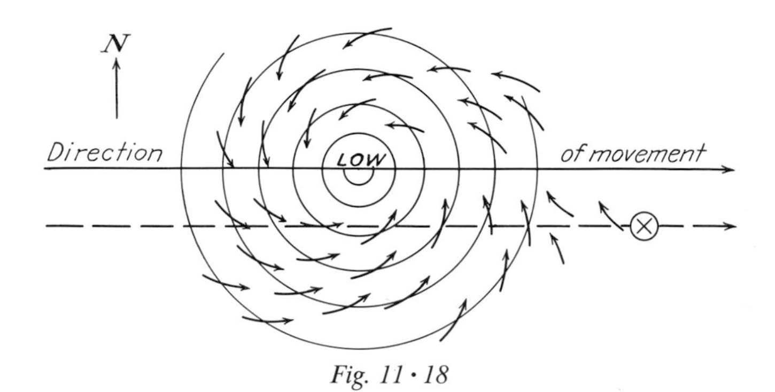

Fig. 11 · 18

latitudes are the result of the successive passage of highs and lows, which are rarely of uniform size, shape, and pressure gradient. Moreover, they travel with varying velocities. Consequently the amount and duration of these pressure variations, as registered on recording instruments, vary considerably from day to day. The barometer falls as a low or cyclone approaches. The rate at which the barometer rises or falls depends upon the pressure gradient and the velocity of the pressure areas approaching from the west. Cloudiness, precipitation, and bad weather usually accompany a low, and fair weather, a high.

Noticeable and significant wind shifts also attend the passage of these pressure areas. Let us consider a low or cyclone in the Northern Hemisphere, moving from west to east, as in Fig. 11 · 18.

As the area moves, the part of the low indicated by the dashed line will pass over the observer at *X*. With a falling barometer, the observer will experience winds setting in from the southeast and *veering* to southwest, west, and northwest as the low passes. If the *center* passes over the observer, the wind will *reverse* in direction, approximately from southeast to northwest.

In the same way, it is clear that, if the low or cyclone center passes *south* of the observer, a *backing* wind will be experienced, with the winds shifting from an easterly direction to north or northwest.

LAW OF BUYS-BALLOT. This is also known as the *law of storms* and enables the observer to obtain a fairly accurate approximation of the bearing of the lowest pressure, or storm center.

Owing to its counterclockwise motion, the wind blows inward at a slight angle with the isobars in a cyclone. Thus, when facing the wind the observer in the Northern Hemisphere will always find the lowest pressure on his right. In the Southern Hemisphere, as a consequence of the clockwise cyclonic circulation in those latitudes, the low will be on his left.

More specifically, in the Northern Hemisphere, when facing the wind, the observer will have the low or storm center bearing 8 to 10 points (90° to 120°) on his right, and in the Southern Hemisphere, an equivalent distance on his left.

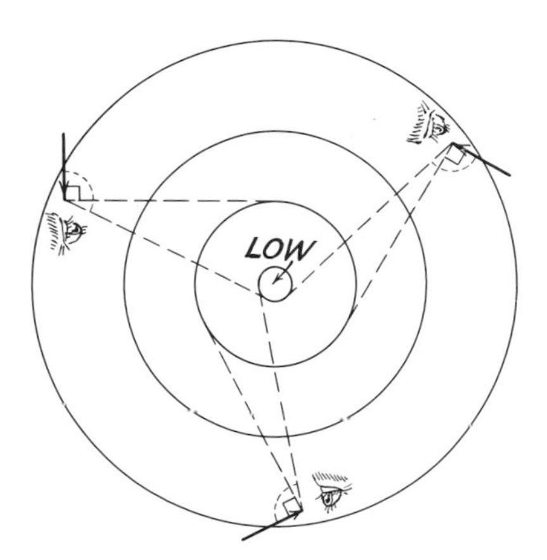

Fig. 11 · 19 The law of Buys–Ballot in Northern Hemisphere.

Since the winds at higher altitudes, being unhampered by friction, achieve higher velocities, they are, as explained previously, deflected more and have paths parallel to the isobars. Consequently, a discrepancy usually exists between the surface wind direction and the direction exhibited by the movement of the *lower clouds*. The storm center will be very nearly 8 points (90°) to the right when the observer faces into the cloud motion.

Vertical Temperature and Pressure Structure of Cyclones and Anticyclones

Although most studies of the atmosphere are limited to two-dimensional diagrams, such as maps or charts of the surface or upper levels, the three-dimensional nature of the atmosphere must again be emphasized. Surface weather behavior can only be understood when integrated into the conditions of the overlying air. Since cyclones and anticyclones are so important to the weather of the middle latitudes, we should examine the relation of the basic elements within these systems both at the ground and aloft.

THE THERMAL EFFECT. Consider a situation in which a uniform pressure distribution exists as in Fig. 11 · 20*a*, which shows the isobars (or edges of isobaric surfaces) in a vertical section of the atmosphere from the surface upward. Then, if the air in the center of the section is warmed, the expanded air will rise, so that the central air column becomes thicker than the side columns. Although the atmospheric pressure remains the same at the surface, the pressure along any surface above the ground will be greater for the central column, since this column is expanded, causing more air to be present above any level in the expanded zone than in the central column. This is shown in Fig. 11 · 20*b* where an actual pressure gradient now exists outward from the central column. In response to such a gradient aloft, wind motion tends to occur outward from the upper portion of the rising air and typical convectional circulation develops.

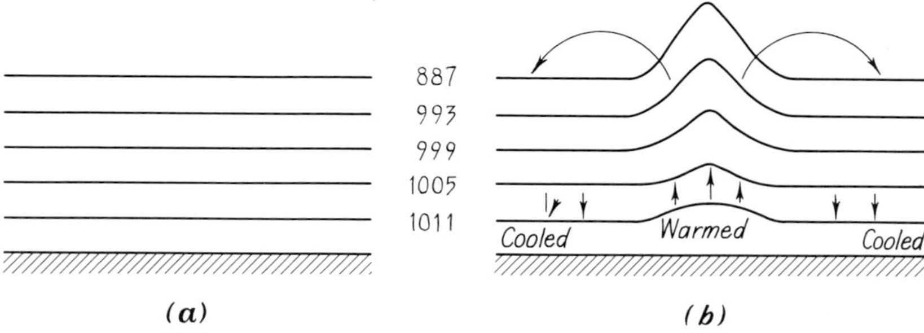

Fig. 11 · 20 Effect (exaggerated) of warming or cooling a stratified air layer.

In this model, we can just as well imagine that the center remained unchanged and the side columns were cooled, causing air to become denser and subside, thus forming thinner side layers. Then, at any level above the surface, pressure would be less in the subsiding side columns compared to the central zone. We see therefore that lateral pressure differences can occur above the surface from the simple warming and cooling of adjacent air columns. Other things being equal, we can also deduce from Fig. 11 · 20*b* that the pressure along the vertical changes more slowly when temperature is high (as in the central column) and more rapidly when temperature is low (as in the side columns). The thermal effect just described produces modifications of the vertical pressure structure of cyclones and anticyclones with resulting changes in cyclonic and anticyclonic winds aloft.

WARM AND COLD CYCLONES. A warm cyclone is defined as one whose central or core region is warmer than the outer portion at any level in the air; a cold cyclone is one whose central region is colder than the outer portion. From the discussion just given, it appears that the pressure gradient along the vertical must be less in the warm central zone than in the surrounding region. The pressure gradient in a vertical plane through a warm cyclone is shown in Fig. 11 · 21*a*, where it is clear that the low-pressure center at the ground disappears rapidly with height until the level where no gradient occurs. The gentler central gradient is indicated by

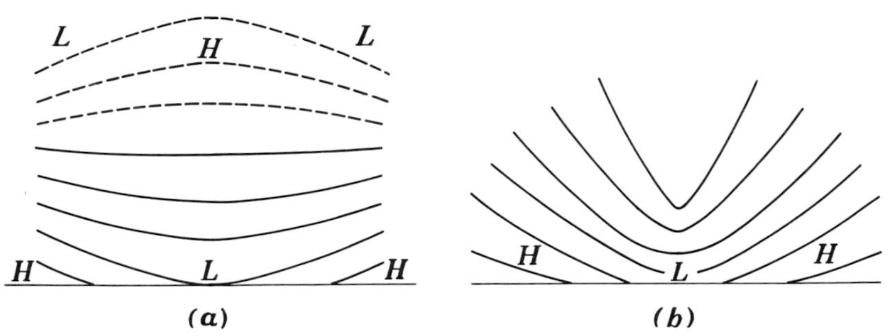

Fig. 11 · 21 (a) Vertical pressure gradient in a warm cyclone; (b) Vertical pressure gradient in a cold cyclone.

the increased isobar spacing. Above the level of no pressure gradient, the thermal effect can be identical with that in Fig. 11 · 20*b*, with the development of a high-pressure region above the low-pressure center, as shown by the broken lines.

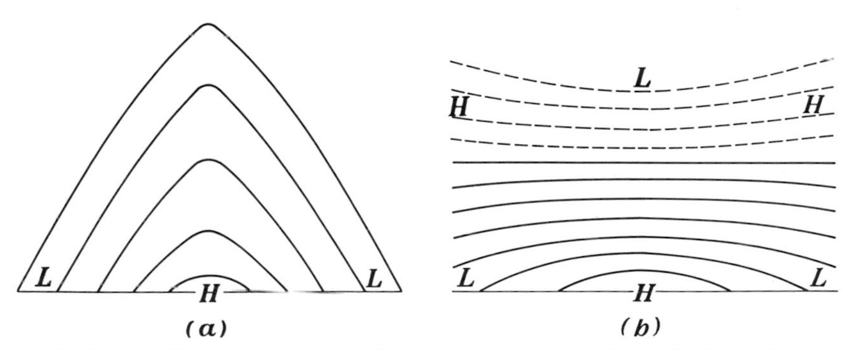

Fig. 11 · 22 (a) Vertical pressure gradient in a warm anticyclone; (b) Vertical pressure gradient in a cold anticyclone.

In a cold cyclone, the increasing pressure gradient in the colder central portion causes an intensification with height as shown by the increasing slope of the isobars in Fig. 11 · 21*b*. (Note that the gradients are exaggerated in these diagrams for emphasis in the presentation.)

WARM AND COLD ANTICYCLONES. The thermal effect on the vertical pressure structure of an anticyclone is quite analogous to that on a cyclone except that the effect is reversed. Thus, a warm anticyclone increases in intensity with height, and a cold anticyclone decreases in intensity with height and soon disappears. These effects are shown in Fig. 11 · 22*a* and *b*. In the former the isobars become closer in the peripheral region as elevation increases, thus increasing the slope of the isobaric surfaces and hence the pressure gradient. In the latter case, the isobars are closer in the center, causing the gradient to disappear quickly with height. Above the level of no pressure change, the gradient actually reverses, producing a cyclonic center above the high.

Some Secondary Wind Types

All winds are of course basically the result of differences in temperature. There are, however, certain important local small-scale wind types that are caused directly by temperature distinctions resulting either from variations in composition or topography of the earth's surface or both.

LAND AND SEA BREEZES. Shore and coastal areas frequently exhibit winds whose direction reverses from day to night. They can be considered as a kind of daily monsoon since their origin is similar to that of monsoon winds. They differ in that monsoons result from large *seasonal* pressure changes, whereas land and sea breezes are a consequence of lesser *daily* changes in pressure. In any case, changes in temperature are responsible for all such winds.

During the daytime at nearly all seasons in the tropics and during the summer in temperate latitudes, the difference in the rate of warming between land and water results in a distinct temperature contrast by midmorning. The expansion and rising of air over the relatively warmer land causes a partial convection cell to form in which air moves horizontally from sea to land (the sea breeze) after descending over the adjacent ocean.

The sea breeze, which can be detected by 0900 local time, increases in strength until mid to late afternoon. During this time, the *front* or landward margin of the sea breeze can progress many miles inland. Since the front is the region of maximum upward motion, it is often marked by a zone of cumulus clouds which grow progressively higher during the life of the breeze, particularly in tropical and semitropical regions. Heavy shower activity may form if the cumulonimbus stage is reached. The landward penetration and development of clouds in the marginal area of the sea breeze are illustrated in Fig. 11 · 23. The sea breeze has a definite moderating influence on the climate of the narrow shore area, invoked particularly from the reduction of afternoon temperatures by the relatively cool marine air.

A relatively important navigational aid results from this daily heating coupled with the sea breeze. As the warm sea air reaches the heated island during the daytime, it partakes of the convectional air motion over the island and rises. The rising air produces adiabatic cooling and yields cumulus clouds as it cools to the dew point. Such clouds are daily characteristics of these small islands and can be seen from great distances when the low land beneath is completely invisible. They are sometimes called

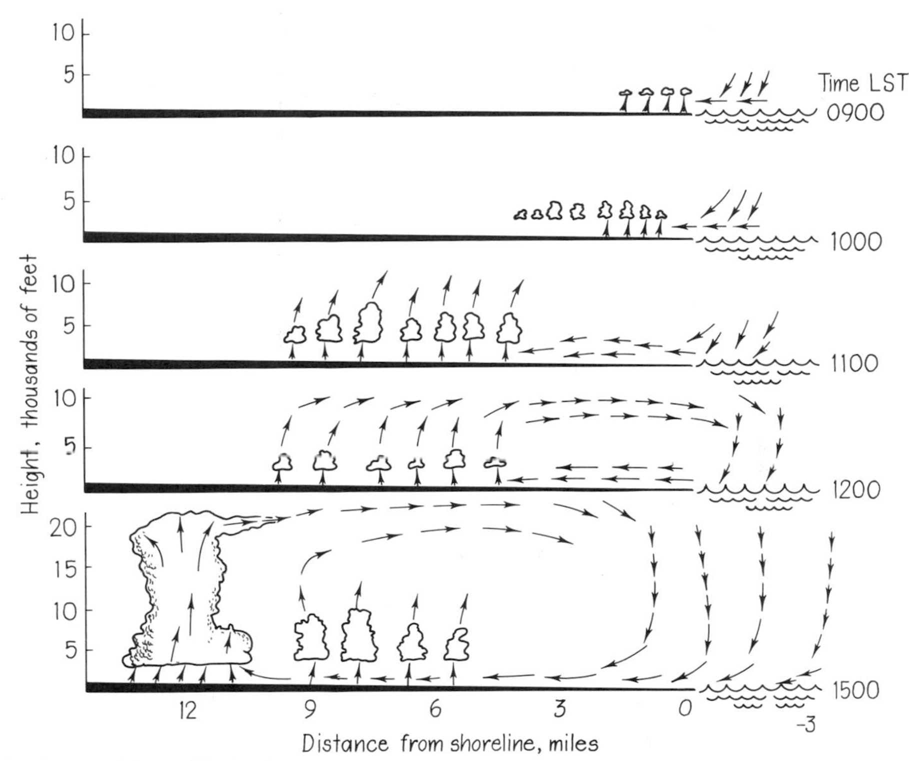

Fig. 11 · 23 The daily landward penetration of the sea breeze with related air motion and cloud development. (Modified from L. Crow and G. Cobb, NSF Report NSF C-184)

atoll clouds owing to their prevalence over the coral atolls of tropical waters. Brief, heavy showers may develop if the heating is sufficient to cause towering cumulonimbus clouds to form from the original cumulus.

Land breezes are *offshore night winds* that develop as the land cools by radiation at night. We have learned previously that land cools more rapidly than water. Consequently the land, which was warmer during the day, at night becomes colder than the adjoining water. The air over the land gradually becomes colder and therefore denser than that offshore. Hence, the pressure gradient is now reversed, with pressure higher over the land, yielding the offshore winds known as land breezes. Land breezes are experienced to within only a few miles offshore, although they are frequently very strong along actual beach areas.

Land and sea breezes are often particularly well developed over tropical and subtropical islands, such as those in the West Indies and the Pacific. Marked temperature differences form between these islands and the surrounding oceans during sunny days, which differences reverse at night. The sea breeze dies down at or shortly after sunset as the heating effects of the sun disappear. This often happens with extreme regularity.

FOEHN AND CHINOOK WINDS. These are warm *dry* winds descending the leeward slopes of mountains. The air, in falling to the floor of an adjacent

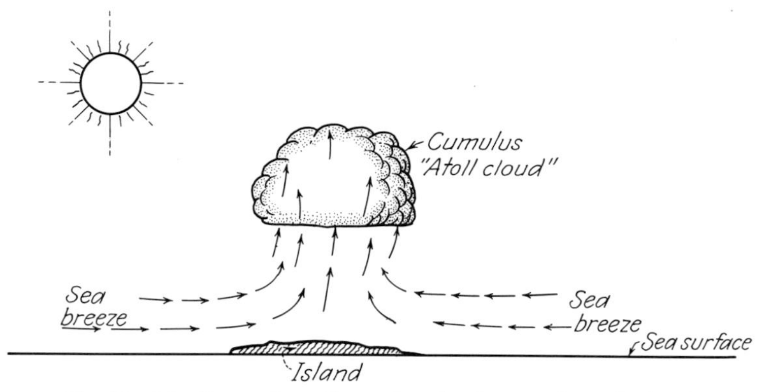

Fig. 11 · 24 Formation of cumulus cloud over a heated island.

valley or plain after coming across the mountaintop, warms adiabatically at the rate of 1°F for each 185 feet of descent. Thus if the wind descends the side of a mountain several thousands of feet high, it will be considerably warmer and hence drier than the air prevailing in that area. In the winter and early spring such warm dry winds are responsible for rapid melting of the snow, clearing the soil for spring farming. In Europe the name *foehn* is applied to this wind, while in the United States and Canada along the eastern or lee slopes of the Rockies, the Indian name *chinook* is used.

The chinook may originate from winds that ascend the windward slopes and descend the leeward slopes of a mountain range. If the air in ascending cools sufficiently to reach the dew point, further rising will cause cooling according to the reduced moist adiabatic rate. But on descending the lee slopes, the dry air now warms only according to the dry rate and may reach a much higher temperature than the air at the same level on the approach side.

MOUNTAIN AND VALLEY BREEZES. As implied by the name, these breezes exist in mountainous areas, or areas of strong topographic relief. During daytime heating, valley areas become relatively overheated. Consequently, the warmed air rises up the sides of the valley (or mountain) during periods of sunshine. These breezes may be weak or strong, depending on the nature of the topography. Such air movements, originating at the valley floor, are called *valley breezes.*

During the night in the same areas, the air on the mountain sides, in contact with the cooling slopes, cools faster than the surrounding air or that in the sheltered valley below. Hence the air along the slopes settles downward, being more pronounced on nights when good radiation weather prevails and permits rapid cooling of the mountain sides and consequently of the adjacent air. Again, local topography may favor the development of rather strong nocturnal mountain breezes.

Both valley and mountain breezes are sometimes slowed considerably by the effect of adiabatic cooling in the rising valley breeze, and adiabatic heating in the case of the settling mountain breeze.

Other Local Wind Types

All winds, regardless of their location, can be explained by the facts and principles developed in this and the preceding chapter. At the root of the whole matter lie the basic temperature differences followed by pressure irregularities. Many local winds have such marked and definite characteristics in the localities of their occurrence that colloquial names are ascribed to them. Brief definitions (from the U.S. Weather Bureau) of these local winds are given since they are so frequently encountered.

BORA. A cold wind of the northern Adriatic blowing down from the high plateaus in the north. A similar wind occurs on the northeastern coast of the Black Sea.

CAT'S-PAW. A slight and local breeze which shows itself by the rippling of the sea surface.

CHUBASCO. A violent squall on the west coast of tropical and subtropical North America.

CORDONAZO. A hurricane wind blowing from a southerly quadrant on the west coast of Mexico as a result of a hurricane passing off the coast.

ETECIANS. Northerly winds blowing in summer over the eastern Mediterranean.

GREGALE. A stormy northeast wind on the Mediterranean.

HARMATTAN. A dry, dusty wind of the west coast of Africa, blowing from the deserts.

KATABATIC. The cold, sometimes gale-force wind which may blow outward from a cold ice cap as a result of intense cooling, subsidence, and spreading of the air.

KHAMSIN. A hot, dry, southerly wind occurring in Egypt in springtime.

LESTE. A hot, dry, easterly wind of the Madeira and Canary Islands.

LEVANTER. A strong easterly wind of the Mediterranean, especially in the Straits of Gibraltar, where damp, foggy weather also attends the wind.

MISTRAL. A stormy, cold, northerly wind blowing down from the mountains of the interior along the Mediterranean Coast from the mouth of the Ebro to the Gulf of Genoa.

NORTHER. A stormy northerly wind of sudden onset occurring during the colder half of the year over the region from Texas southward across the Gulf of Mexico and the western Caribbean. These winds are dependent on the strengthening of the cold-weather high-pressure area prevailing over the southern United States.

PAMPERO. A northwest squall blowing over or from the pampas of South America. Off the coast of Argentina it is most prevalent from July to September.

PAPAGAYO. A strong to violent northeast wind blowing during the colder months in the Gulf of Papagayo, on the northwest coast of Costa Rica, and in adjacent Pacific coastal waters.

SHAMAL. A northeast wind of Mesopotamia and the Persian Gulf.

SIMOON. An intensely hot and dry wind of Asiatic and African deserts.

SIROCCO. This is applied to various warm winds in the Mediterranean area, particularly in North Africa.

TEHUANTEPECER. A strong to violent northerly wind of Pacific waters off southern Mexico and northern Central America, confined mostly to the Gulf of Tehuantepec and occurring during the colder months.

Tornadoes and Waterspouts

Tornadoes and waterspouts are closely related phenomena. When the motion of a tornado carries it out to sea, it is called a *waterspout.* However, not all waterspouts are true tornadoes but may result from other causes as well. As such, we shall consider them under separate headings.

TORNADOES. Tornadoes are by far the most violent and destructive manifestations of all nature. Nothing else is comparable to their fury. Fortunately the path of this "atmospheric monster" is so narrow that the total damage left in its wake is not nearly as great as that of less violent but larger storms.

Tornadoes are common mostly to the lower middle latitudes of both

Fig. 11 · 25 Four stages in the development of a tornado, Gothenberg, Neb., June 24, 1930. (U.S. Weather Bureau)

hemispheres, although even here, their occurrence is limited in location. The storm itself is a gigantic, whirling funnel of air extending earthward from heavy, black cumulonimbus clouds. The air motion is characterized by a cyclonic upward spiral, causing rapid expansion, cooling, and condensation, which forms the dark cloud of the tornado funnel. A downpour of rain and hail is the common associate of the tornado, occurring just before and after its passage. The accompanying lightning and thunder are related more directly to the parent cumulonimbus cloud than to the tornado itself.

The width of the storm at the ground level averages about 300 yards, although it may vary anywhere from 20 or 30 yards to a mile. The forward velocity of motion varies from 25 to 40 miles per hour. Curiously, it often travels in an erratic, skipping path, so that the tapering end of the funnel may hop over one area and descend to wreak further havoc, often exhibiting a writhing, serpentine appearance as it does so.

The pressure and winds of a tornado can rarely be measured directly but must be inferred from observation. The extremely high velocity of the whirling air causes a marked decrease in pressure at the center. A steep pressure gradient results owing to the short horizontal distance over which the fall in pressure occurs. Thus, in many cases the destructive effect of the storm has been ascribed to the explosive action of air under normal pressure within a building, as the outside pressure falls rapidly with the passage of the storm.

The wind velocity is excessive, not only in a horizontal direction but vertically as well. Thus, very heavy objects, in some cases a team of horses and a wagon, have been lifted gently and carried some distance, being held aloft by rising air, and then deposited, just as gently. These cases lead to estimates of up to 200 miles per hour for the velocity of the vertically rising air. The horizontal winds are in excess of this. Dry straws have been driven through wooden telegraph poles under the impact of the driving winds. Theoretical calculations show that wind speeds from 300 to 500 miles per hour may be required for this result.

In the Northern Hemisphere, tornadoes occur in the warm air in the southern section of a cyclone (northern section in the Southern Hemisphere), being more common in the spring and early summer than in any other period. After formation, tornadoes usually travel east to northeastward, or in accordance with the direction of motion of the low-pressure area.

It is not accidental that tornadoes usually form when and where they do. Rather, the conditions of formation are opportune at that time and place. The conditions necessary for tornado formation are (1) warm, humid air flowing into the cyclone from the south or southwest and (2) a mass of cold air advancing from the west, into the warm air, at a level above the ground.

Without entering into an involved explanation, we may note that the tornado whirl originates in the warm air at the level of the overrunning cold upper-air layer, as a result of the contrasting properties and the contrasting directions of motion. Typical violent thundersqualls develop along this upper cold-air boundary at the same time, and it is from the base of one of these thunderclouds that the dark vortex descends.

Clearly, tornado conditions are not localized but may exist at many places along the upper cold-air line, which may at times be some 50 miles ahead of the cold-air line at the ground. Consequently, it is not uncommon to have many tornadoes traveling simultaneously across country along a fairly well-defined line.

WATERSPOUTS. Waterspouts are among the most curious of marine phenomena and have given rise to speculations and legends of all sorts in order to explain their origin and nature. They are actually of two types, depending on their origin. One type occurs, as explained above, when a tornado moves beyond the confines of the continent; in this case it has all the violence and other features of the true tornado. The more common type of spout is a simpler convectional feature of the atmosphere that may occur at almost any time or place, in temperate or tropical latitudes. Further, it may be associated with fair weather or foul.

The tornado and the tornado-type waterspout always exhibit cyclonic rotation owing to the nature of their origin. However, the air in the convection-type spout may rotate in either cyclonic or anticyclonic motion, depending on the manner of formation. Having examined the features of the tornado, we shall consider now the features of the waterspout of the second type.

Individual waterspouts may vary considerably as to features of origin and structure, although certain typical characteristics can be given. The spout is most frequently associated with a heavy cloud of the cumulus family, resulting from local convection. From the cloud base, the spout tapers downward in the form of a narrow, dark cone. Running through the center of this cone is a long hollow tube of relatively low pressure which is responsible for a certain amount of seawater being sucked into the spout for a short vertical distance. The rapidly rotating winds also carry seawater aloft in the form of spray picked up from the surface. However, the dark appearance of the spout is a result of neither seawater carried upward nor the cloud base protruding downward. It is rather the result of condensation in the moist air, whirling into and around the low-pressure center as it spirals upward and expands.

Waterspouts are commonly observed to originate in two sections. The upper part is seen to develop as a small protuberance, funneling downward from the cloud base in a halting, hesitating fashion. Directly below, at this time, an agitation of the water surface occurs, characterized by a boiling and tossing of spray, together with the formation of a low mound of water rising into the center of the yet invisible lower vortex. Then, from near the center of this disturbance, a second, dark, vapor-laden cone forms, but this one funnels upward, being an inverted counterpart of the one above. The tapering ends of the two cones continue to extend and finally unite, resembling a drawn-out hourglass at first and then a continuous, elongated, whirling tube.

The spout follows in the direction taken by the overlying parent cloud mass. It rarely lasts longer than 30 to 60 minutes, the end coming rather abruptly. The tube may thin out and separate into two sections, the upper retreating into the cloud and the lower falling back to the water surface. At times, the entire spout may appear to rise or roll directly up into the cloud. Thunderstorm conditions often attend the spout, being associated with the cumulonimbus cloud above.

In the violence exhibited, waterspouts are extremely variable. Often they are no stronger than common dust whirls or whirlwinds frequently observed over land surfaces. Occasionally the spout may contain wind motion of great strength, sufficient to destroy small craft and create much damage on larger vessels. It is doubtful whether any modern ocean-going vessel need have much to fear from waterspouts of ordinary convectional origin. There is, in fact, no record of any modern ship suffering disaster as a result of such an encounter.

EXERCISES

11 · 1 Describe the nature of cyclonic and anticyclonic wind motion in the Northern and Southern Hemispheres.

11 · 2 How will this motion, with respect to the isobars, differ between surface wind and the wind at the low-cloud level?

11 · 3 Describe the changes in wind direction experienced by an observer in the Northern Hemisphere as a cyclone approaches and passes (*a*) with the center north of the observer and (*b*) with the center south of the observer.

11 · 4 Describe the motion of a low relative to an observer if the winds experienced remain steady from the northeast and then reverse to the southwest.

11 · 5 Describe the motion of the low-pressure area illustrated in Fig. 11 · 6, giving the average speed of the center per 6 hours and the change in pressure gradient in millibars per 100 nautical miles along a line due east from the center. (Note: Each latitude degree is equal to 60 nautical miles.)

11 · 6 Using Figs. 11 · 16 and 11 · 17, describe the typical cyclone paths for the North Atlantic and North Pacific Oceans.

11 · 7 Explain the vertical changes in warm and cold cyclones.
11 · 8 Explain the vertical changes in warm and cold anticyclones.
11 · 9 Show how a chinook wind can develop from a 10,000-foot mountain originally having an air temperature of 40°F on its windward side. Calculate the air temperature at the base of the mountain on the lee side using the dry and saturated adiabatic lapse rates, as appropriate.
11 · 10 How are land and sea breezes and mountain and valley breezes related to the time of day?

CHAPTER 12

The General Circulation of the Atmosphere

No modern course in atmospheric science would be complete without at least a brief treatment of the nature and origin of the general circulation of the atmosphere. *General circulation* refers to the description of the fundamental air-flow patterns over the globe and their explanation. The planetary winds for the surface and upper troposphere (Chap. 10) together with the cyclonic and anticyclonic winds (Chap. 11) constitute essential features of the general circulation. At any given time, the actual motion observed may result from the superposition of motion of different scales and origins. When the observed motions are averaged over time intervals of a few days, the major features of the general circulation are revealed. Many of the details of this were already described in Chap. 10 for the surface and upper troposphere.

Because the knowledge of upper-level wind motion was not established until well into the twentieth century, the explanation of the origin of the general circulation has been delayed and still remains as one of the major problems in atmospheric science. A good way to approach the problem is to consider first the classical theory based on convection and then proceed from the historical treatment to more recent views.

The Tricellular Model of the General Circulation

In this scheme the planetary pressure and wind belts described in Chap. 10 are organized into a three-celled circulation pattern when viewed in a meridional (north-south) vertical plane. The low-latitude cell, from about the equator to 30° north or south, would consist of the meridional component of the trades or tropical easterlies, the equatorial calms (doldrums) with low pressure and rising air, the subtropical calms (horse latitudes) with high pressure and descending air, and a return (northward) flow of air above the trades. The mid-latitude cell in each hemisphere from about 30° to 60° would consist of the surface westerlies and a return flow aloft bounded by the horse latitudes on the equatorial side and the subpolar or arctic low-pressure belt on the polar side. The polar cell would consist of the polar easterlies at the surface with the opposite flow aloft, bounded by the subpolar lows and the polar high-pressure cells.

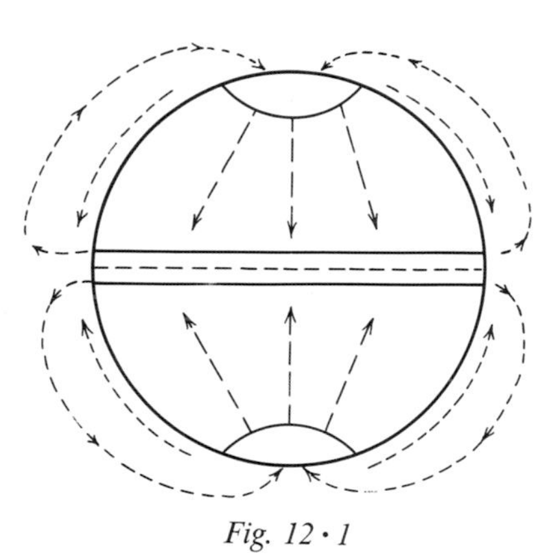

Fig. 12 · 1

These cells were explained primarily on the basis of convection and Coriolis force. If we imagine the earth to be nonrotating, the pronounced temperature gradient between the equator and the poles would tend to generate a major convection cell in each hemisphere, as shown in Fig. 12 · 1. Surface winds would move toward the equator, and winds aloft would travel poleward. However, on a rotating earth, the resulting Coriolis force would deflect the winds so strongly that they would soon tend to travel parallel to the isobars as described for geostrophic winds in Chap. 9. For example, the air traveling poleward after rising in the doldrums would be deflected so strongly as not to get farther than the upper level of the horse latitudes, where high-level radiation would cool the air and result in subsidence. The result of this deflection on the entire extent of the surface and upper air flow would be to modify the single cell in each hemisphere (Fig. 12 · 1) into the tricellular model of Fig. 12 · 2. This shows the three-cell view in the vertical meridional plane as well as the related surface winds and appears to account for most of the observations described in the first part of this section.

Limitations of the Cellular Model

With the tremendous increase in upper tropospheric observations and in the refinement of surface observations toward the middle of the 1900s, it became clear that wind motions at upper levels did not correspond to the three-cell model described above. Other defects of both observational and theoretical aspects were also discovered, resulting in the development of a new model to explain the general circulation. Before considering this, it would be instructive to review first some of the limitations of the three-cell model.

1. In the prevailing westerlies wind zone, from about 30° to 60° in latitude, instead of reversing aloft, as required by the convection-cell scheme, the westerlies actually become stronger and better established, reaching velocities up to 300 knots in the jet streams.

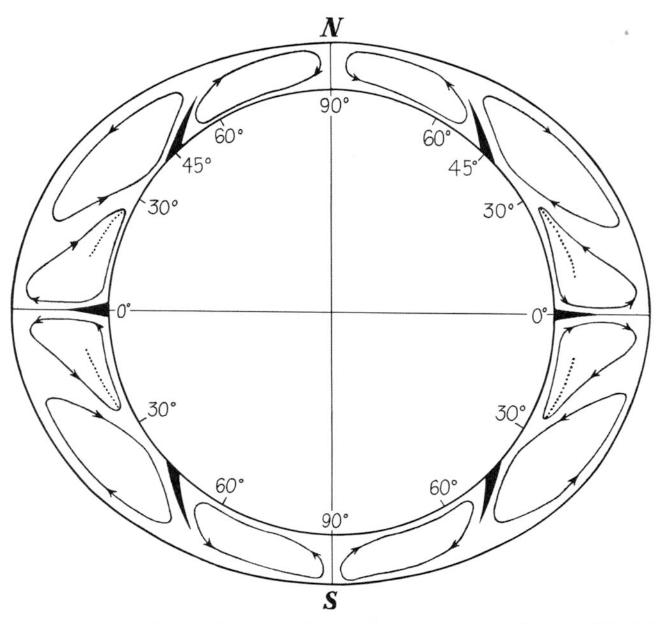

Fig. 12 · 2 Illustration of tricellular model of the general circulation. The black wedges at 45° represent polar fronts. (After Bergeron and Rossby)

2. In the trade wind zone, the antitrades, aloft, do show the required reversal but are often very weak and may be absent, despite the presence of vigorous surface trades.
3. The strong Coriolis deflection, described above, prevents the meridional circulation from transporting as much energy poleward as is put into the atmosphere in lower latitudes. This effect should increase tropical zone temperatures and produce a much higher temperature gradient from low to high latitudes than is observed. Apparently some other process must exist to distribute energy from low to high latitudes.

4. The horizontal temperature gradients in the middle latitudes (westerlies wind belt) are much stronger than those in the tropical or polar cells. This suggests that the circulation in the westerlies belt may be more important than tropical and polar convection in controlling the general circulation.
5. The observed average flow of air from low to high latitudes in the meridional plane amounts to only a few centimeters per second, but the flow necessary to transport the heat required by observations is about 2.5 meters per second. This indicates that the tricellular circulation model is insufficient to explain the necessary heat distribution and that a different circulation mechanism is required.
6. The energy of motion of the large eddies—the cyclonic and anticyclonic movements—of the middle latitudes actually exceeds the energy that is calculated for the meridional circulation. Such results suggest that the eddies are fundamental in the general circulation.
7. Recent theoretical results appear to show quite definitely that with the observed heating and rotation of the earth, the symmetrical convection of the tricellular type is not possible.

The Eddy Theory

In view of the inadequacy of simple meridional convection plus Coriolis modifications to explain the features of the general circulation, a new approach has been made, so far with considerable success. The problem has been to determine a natural mechanism that could maintain the observed surface and upper troposphere—especially the jet stream—windflow patterns without resorting to convectionally driven meridional circulation. Before considering the proposed model, we must examine one basic feature which is very important to the new theory of the general circulation, namely the transmission of *angular momentum.*

THE PROBLEM OF ANGULAR MOMENTUM. Ordinary *momentum* is simply the product of the mass and velocity (*mv*) of an object moving in a straight line. If an object moves in a circular path, the factor of the radius of the circle is also used in the product (*mvr*) and the result is called *angular momentum.* According to a very basic and elementary principle of physics, developed from Newton's laws of motion, the angular momentum of an object once set in circular motion is a constant unless a new force or forces are imposed. Thus, if a body has a given angular momentum or its *mvr* is a constant, a change in either one of the terms, velocity *v* or radius *r*, must produce an immediate and opposite change in the other of these, to keep the product unchanged. The mass term is not so affected since it cannot change in response to velocity or radius variations.

The earth's rate of rotation is such that the equatorial velocity, where the radius of rotation is greatest, is about 1,000 miles per hour. This velocity decreases steadily (as the cosine of the latitude) and becomes zero at the poles, as a consequence of the decreasing radius of rotation. Now imagine the atmosphere to be calm relative to the earth's surface, so that the atmosphere and solid earth rotate about the earth's axis as a single unit (which they do, approximately). The angular momentum of air particles as well as of the earth's surface would then decrease with increasing latitude as both the radius and the velocity about the axis decrease; but remember that the angular momentum of calm air at any latitude must be constant, or have a particular value for *mvr*. As long ago as the late 1920s, the distinguished British theoretical geophysicist Sir Harold Jeffreys noted that certain important consequences must result from the angular momentum effects of air not motionless, but moving, relative to the earth's surface.

Because of their low-latitude position, the trades involve a larger area than do wind systems of higher latitudes and are therefore of great significance in any scheme of the general circulation. Since they move westward relative to the surface, although the air involved is nevertheless eastward in rotation about the axis, they have a slightly lower angular momentum than the surface beneath. Further, since the trades move relatively opposite to the surface, frictional drag must occur on this air, tending to slow the winds and thereby increasing the eastward velocity (about the axis) and hence the angular momentum. However, the required transfer of angular momentum from surface to air is not apparent, because the trades maintain a nearly uniform velocity appropriate to the pressure gradient. The maintenance of the trade winds thus requires the transfer of angular momentum away from this zone—or else slowing must occur.

In the middle latitude or westerlies zone, the winds, whose motion is eastward, have a higher eastward velocity and hence angular momentum than the surface beneath. Friction between the air and the slower surface must result in the transfer of angular momentum to the earth. Such a frictional effect must therefore slow the westerlies. As this does not occur, the angular momentum lost by the air through friction must be replaced from the outside in order to maintain the westerlies.

Reasoning similar to that for the trades shows that the polar easterlies must also gain angular momentum from friction with the earth. Again the momentum gain must be transferred away from this zone in order to maintain the easterlies.

It can be concluded from this discussion that the easterlies of the tropical and polar zones gain angular momentum from the earth and that this excess is transferred to the westerlies in amounts which compensate for their loss of momentum to the earth. Also, the very nature of the deflection involved in the three-cell model completely opposes the transportation of angular momentum across the 30th parallel, just where the maximum transport is required. Hence, the next requirement for a model of the general circulation is a mechanism that will convey angular momentum meridionally and thus maintain the zonal flow of westerlies and trades. The great eddies already referred to seem to provide the best means of accomplishing this transfer.

IMPORTANCE AND ORIGIN OF THE EDDIES. The eddies referred to here are the large cyclonic and anticyclonic whirls in the lower troposphere of the middle latitudes and the waves in the westerlies of the upper part of the troposphere. These have already been described in preceding chapters. A careful study of the eddies has been made for different levels in the troposphere as more complete data became available. It has been shown from this investigation that the pattern and orientation of the eddies produce a wind flow which makes possible the transfer of angular momentum away from the easterlies and into the westerlies zone. By this transfer, the eddies meet the requirement of maintaining these planetary wind systems through the appropriate exchange of angular momentum. Actual determinations of the momentum transfer which have been made along meridional planes from the equator to the poles to an elevation of nearly 30,000 feet show definitely that the required amounts of angular momentum are being transmitted in the required directions. Also, the transfer is greatest at the higher elevations, a fact of considerable importance in understanding the high velocities attained by the upper westerlies.

The problem of the origin of the eddies and their relation to the general circulation has been attacked from the three viewpoints of theory, experiment, and observation, the results of which come to a single focus. In the eddy model, the driving force is still the differential heating between equator and poles. The immediate effect is the development of a simple convectional circulation, as noted earlier. The effects of rotation tend to suppress such circulation and thus prevent the poleward transportation of a supply of heat adequate to maintain a proper radiation balance, thereby tending to increase the meridional temperature gradient. According to well-established principles of hydrodynamics, at a critical temperature gradient the motion of a uniformly rotating fluid becomes unstable, developing wave patterns that distort the original flow.

In the atmosphere these disturbances take the form of the lower cyclones and anticyclones and of the waves in the upper westerlies. From the standpoint of the general circulation, the upper waves are the more important. The poleward transport of heat energy and water vapor by the eddy or wave systems is of the amount required for the proper insolation and thermal balance in the atmosphere.

A fairly successful attempt has been made to construct an experimental model of the eddy mechanism by using a relatively simple "dishpan" containing a liquid in which metallic powder is suspended to indicate fluid motion. The center of the dishpan is cooled and is analogous to the poles, while the warmer rim becomes equivalent to the equator. The temperature gradient between rim and center is then analogous to that from the equator to the poles. Coriolis force is introduced by rotating the dishpan. The temperature gradient within the fluid can be varied during rotation. As the

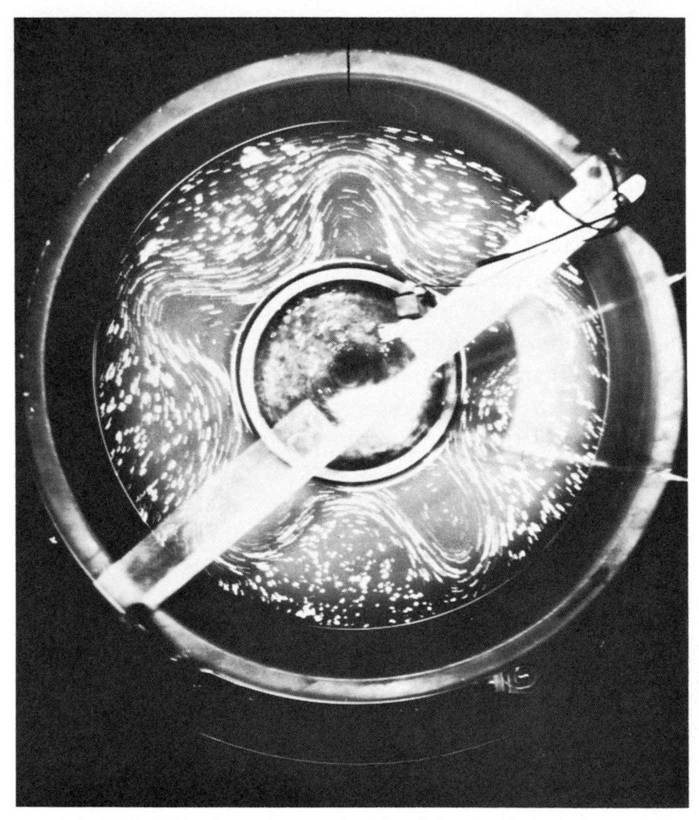

Fig. 12 • 3 Dishpan model showing the development of waves at appropriate temperature gradient and rotation rate. (Courtesy D. Fultz)

gradient is increased, the point of hydrodynamic instability is reached, whereupon the uniformly circulating fluid assumes a wave or eddy pattern much like that observable in the upper westerlies. Fig. 12 • 3 is a copy of a dishpan photograph illustrating the formation of waves at the appropriate temperature gradient and rotation speed. The pattern is revealed by the illuminated metallic powder in the moving fluid. The comparison between this experimental result and the chart of long-wave activity (Fig. 10 • 5) is very striking.

The full description and explanation of the general circulation of the atmosphere is by no means solved, but strong progress has been made toward the solution. To the factors of temperature gradient and rotation must be added the effect of latent heat of condensation. The attempt at a complete solution is now possible only through the use of high-speed computers in view of the tremendous amount of calculations necessary. Doubtless, when the general circulation is fully understood and its variations predictable, the forecasting of weather and climate will achieve considerably greater success.

EXERCISES

12·1 Describe the classical tricellular model of the general circulation.

12·2 Explain the importance of the upper-level wind observations in the theory of the general circulation of the atmosphere.

12·3 Briefly summarize the limitations of the classical convectional circulation model.

12·4 Explain the principle of angular momentum.

12·5 How is angular momentum conveyed from the earth to the atmosphere in the trade wind zone and the reverse in the westerlies zone?

12·6 What is the importance of this momentum transfer in the maintenance of the two wind systems?

12·7 Summarize the eddy theory of the general circulation, explaining the importance of the horizontal temperature gradient and the Coriolis force in this model.

CHAPTER 13

Air Masses and Fronts

Nature of Air Masses

Modern weather analysis involves the identification of air masses and their properties, movements, and changes. An air mass is a large, horizontally homogeneous or uniform body of air within the atmosphere as a whole. Its uniformity is principally one of *temperature* and *humidity.*

In size, air masses cover hundreds of thousands of square miles; vertically, they extend upward for thousands and tens of thousands of feet. There is no difficulty in conceiving of uniform ocean currents within the main ocean body. We can see them; ocean water is visible. The Gulf Stream is readily apparent by its movement, color, temperature, seaweed content, and so on.

Uniform bodies of air, or air masses, are not so obvious, but their presence is adequately shown by meteorological observations, particularly of their temperature and humidity. Although air masses are identified and their motion traced through instrumental rather than visual observation, their presence is often felt very noticeably by our senses. We are all aware of the oppressively hot, sticky, summer heat waves. We are also aware of the dramatic end of such a hot-weather spell, when, following a violent thunderstorm, a wave of cool, dry air is experienced for several days. A large, hot, humid air body responsible for the heat wave was simply replaced by a cool, dry air mass with its consequent relief for the heat sufferers.

The study of air-mass characteristics and behavior is known as *air-mass analysis.* A primary weather concern is to determine the conditions within the air mass, its direction of movement, and the changes in its properties as it moves. The resulting properties of this moving air mass are the weather conditions that are experienced along its line of motion.

Although air masses and their boundaries (fronts) are physically inseparable and are intimately related to the structure of cyclones, it is easier to consider them separately and then combine the information as in Chap. 14.

ORIGIN AND TYPES OF AIR MASSES. Air masses derive their original properties from the surface over which they form. The temperature and humidity characteristics of an air mass are determined directly by the nature of the surface beneath.

In considering the relatively large volume of air masses and the poor powers of heat conduction that air possesses, it is apparent that such uniform bodies will not form too rapidly. A large volume of air must remain stagnant or circulate for some time over a particular portion of the earth, gradually acquiring its distinguishing temperature and humidity characteristics.

Air masses develop more commonly in some regions than in others, the areas of formation being known as *source regions.* We may note, for example, that the common source regions for air masses affecting American weather are the Northern Pacific west of Canada; the northern interior of Canada; the North Atlantic east of Canada; the Pacific west of southern California; the desert areas of southwestern United States and northern Mexico; and the Gulf of Mexico and the Caribbean Sea.

It is noticed that the source regions tend to bound the belt of prevailing westerlies. One set of source regions exists along the northern boundary in the vicinity of the subpolar low-pressure circle, while the other set exists along and to the south of the horse latitudes. The basic difference between the air originating in the northern source regions and that in the southern is therefore one of temperature. A second difference is that of humidity.

Cold northern air masses are called *polar air masses,* while the warm-air bodies originating in low latitudes are called *tropical air masses.* Then, depending on whether they form over land or water, the air masses will be dry or humid, respectively. This leads to two subdivisions for the above air types. Dry polar air of continental origin is known as *polar continental air,* and when of oceanic origin, *polar maritime air.* Similarly, tropical air is known as *tropical continental* and *tropical maritime* air.

Arctic and *equatorial* air masses form in the far north and in equatorial regions, respectively. Table 13 · 1 summarizes the basic information concerning the six principal air masses.

Table 13 · 1 Classification of Air Masses

NAME OF MASS	PLACE OF ORIGIN	PROPERTIES	SYMBOL
Arctic	Polar regions	Low temperatures, low specific but high summer relative humidity, the coldest of the winter air masses	A
Polar continental	Subpolar continental areas	Low temperatures (increasing with southward movement), low humidity, remaining constant	cP
Polar maritime	Subpolar and arctic oceanic areas	Low temperatures, increasing with movement, higher humidity	mP
Tropical continental	Subtropical high-pressure land areas	High temperatures, low moisture content	cT
Tropical maritime	Southern borders of oceanic subtropical, high-pressure areas	Moderately high temperatures; high relative and specific humidity	mT
Equatorial	Equatorial and tropical seas	High temperature and humidity	E

The symbol w or k following air-mass designations indicates whether the air is warmer or colder than the surface over which the air is passing. Thus cPk is polar continental air, colder than the underlying land or water.

MOVEMENT OF AIR MASSES. Air must remain over a fixed location for air masses to form. Ultimately, the very factor that allows for their development, namely, the general air circulation, starts these masses in motion. Whole masses or large portions of them may move away from the source region. In the case of polar air, tongues of cold air will lap southward transgressing the areas previously beyond their limits. Tropical air set in motion will conform to the primary air motion and overlap to the northward.

Clearly, areas entirely within the permanent boundaries of either the tropical or the polar air masses will experience more or less uniform weather conditions. But areas that are outside the source regions, in *the middle latitudes or westerlies belt,* will undergo continual changes resulting both from the passage of warm and cold air masses and from the effects of the meeting of tropical and polar air.

As a result of their motion, the basic properties of air masses are often strongly modified, the nature of the modification depending on the surface conditions beneath. The actual amount of change within a given body of air depends on (1) the type and conditions of the underlying surface and (2) the velocity of the air in passing over this surface. Consequently, continental polar air (cP) may undergo a transition to maritime polar (mP) air, after protracted motion over the sea. Continental tropical air (cT) behaves similarly. Remember, however, that even maritime tropical air (mT) is warm and also dry at high levels in the subtropical high-pressure belts (horse latitudes).

PROPERTIES OF WARM AND COLD AIR MASSES. Conditions within moving air masses often depend very strongly on the temperature of the underlying surface. If a body of air is traversing a cold surface, it is usually termed a *warm air mass* and would have the symbol w attached. If the underlying surface is relatively warm, the air mass is considered a *cold air mass,* with the symbol k attached.

Warm air masses are usually of tropical origin, moving toward higher latitudes, but they may be warm marine air bodies moving inland over colder land, or warm continental air moving offshore over colder sea surfaces. The effect of the cooler surface beneath is to cool the air slowly from the surface upward. This uniform cooling of the air by a large, cool surface tends to produce stratified conditions within the air, with an absence of vertical air motion or turbulence. Consequently, any clouds that exist will be of the stratiform type. Any precipitation existing will be of the drizzle or light rain variety. The absence of turbulence produces poor visibility in warm air masses owing to the settling of dust and other foreign particles in the vertically calm air. Fog will be more common in such air as a result of the surface cooling.

Cold air masses may arise from polar air moving toward lower latitudes or from cold marine air moving over warmer land, or relatively cold continental air traversing warmer sea areas. In the passage over warmer surfaces, convection and turbulence develop rapidly in the cold air. Clouds of the cumuliform type tend to form, with precipitation, if present, being heavy or showery. Visibility is usually good to excellent in such air owing to the general stirring and overturning inherent in convectional turbulence.

CONSERVATIVE PROPERTIES OF AIR MASSES. Our knowledge of the existence and nature of air masses would be of less value if there were not some method of tracing their motion. To do this we must use some property of air masses that remains fairly constant regardless of the common modifications experienced through turbulence and other factors in the course of traveling over different surfaces. Temperature, relative humidity, pressure, cloudiness, and other weather features already studied vary too widely for the purpose of positively identifying air masses from day to day. There are, however, several characteristics of air masses that serve as valuable "tags" in keeping track of the flow of air: (1) the potential temperature, (2) the equivalent potential temperature, (3) the specific humidity and (4) the dew point. These properties were discussed in Chap. 5.

Fronts

In considering air masses it was observed that the weather properties within are relatively uniform. But, when air masses differing in temperature meet in the course of air movement, a sharp transition in weather conditions (temperature, humidity, pressure, wind, etc.) occurs across their

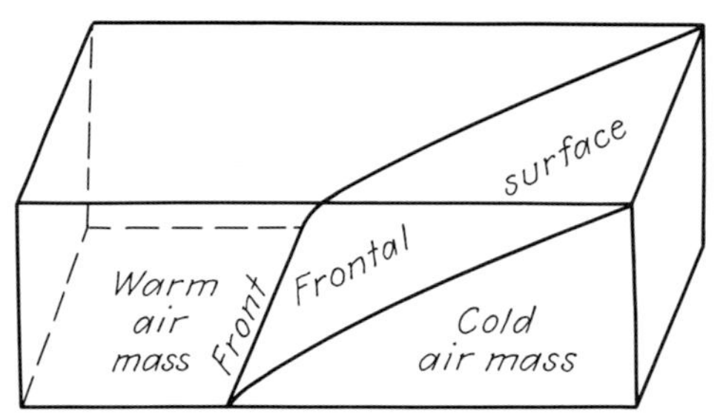

Fig. 13 · 1 Three-dimensional characteristics of a front.

boundaries. If one were traveling northward in a warm or tropical air mass, a slight but steady temperature decrease would be encountered. Then on crossing into the colder polar air to the north a sudden sharp drop in temperature would result. The uniform slow change in weather conditions gives way to an abrupt discontinuous change in leaving an air-mass boundary. This led to the term *line of discontinuity* being applied to the boundary of an air mass. The term *front* is synonymous with line of discontinuity and has pretty much replaced it. More specifically, fronts are the boundaries of, or separations between, air masses.

It should be remembered that air masses have a large vertical as well as horizontal extent. Hence the surface separating adjacent air masses vertically is known as the *frontal surface.* The ground front is therefore the line formed by the intersection of the frontal surface with the ground.

These three-dimensional aspects should always be considered and kept in mind when dealing with air masses and fronts. Most of the weather charts and maps that will be encountered, either in this text or elsewhere, will be maps showing the horizontal distribution of air masses, fronts, isobars, etc. The vertical extension of the sloping frontal surface should not be forgotten in viewing such charts.

General Frontal Characteristics

Although fronts differ as to type, as explained in the following section, they have many weather properties common to all. As pointed out earlier, when cold and warm air masses meet, the cold air wedges beneath the warmer air, which in turn rises over the sloping upper surface of the cold mass. Figure 13 · 2 shows a vertical cross section through adjacent warm and cold air masses and indicates, again, this condition.

The slope of the upper surface of the cold air is actually very gentle, varying from 1 : 100 to 1 : 500 with different air masses. By slope is meant

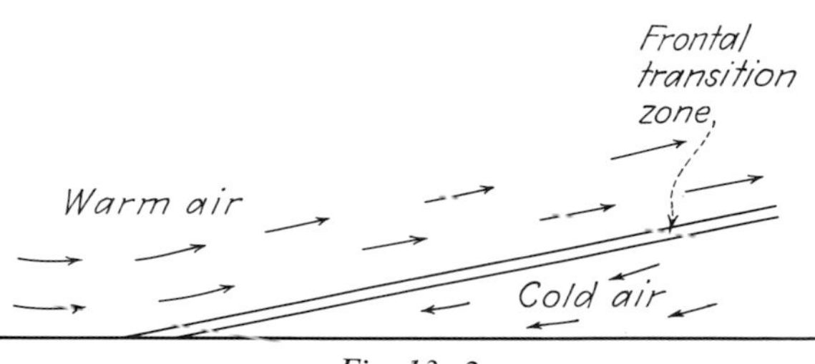

Fig. 13 · 2

the ratio of vertical rise to horizontal distance. Thus, a slope of 1 : 100 indicates a vertical change of 1 unit for each 100 horizontal units; e.g., a slope of 1 mile vertically over a horizontal ground distance of 100 miles. However, this slope is always greatly exaggerated in diagrams for explanatory purposes.

Although treated as such, the frontal surface is not actually a mathematical surface. In reality, a transition zone exists between the two different air masses. The frontal transition zone may vary from a few hundred to a few thousand feet, depending on the contrast in properties between the air masses. The greater the temperature and humidity contrast, the less is the mixing of the air bodies and the thinner is the transition zone. Owing to the gentle slope of the frontal surface, the transition area, even though of small thickness, will cover many miles when intersecting the horizontal ground surface.

1. TEMPERATURE. The temperature conditions across a front may vary through wide ranges and may take place very abruptly or more or less slowly. Air masses that have strong temperature contrasts will

exhibit very abrupt changes across the frontal zone, not only in temperature but in the dependent weather conditions as well. In addition, the frontal transition zone will be relatively thin when temperature conditions between the air masses differ markedly.

Further, it follows from this and our previous discussion of temperature that an inversion exists along a vertical line through the front. The temperature in the cold wedge of air will decrease with altitude until the front is reached. Then there will be a rise in temperature in the transition zone, the abruptness and amount depending on the temperature difference between the two masses. With continued increase in altitude the temperature will again fall in the warm mass.

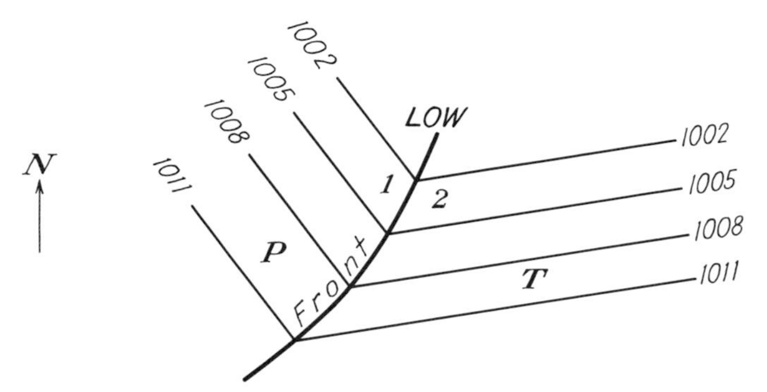

Fig. 13 · 3 The kink in isobars at a front.

2. PRESSURE. A pronounced difference in pressure gradient occurs between adjacent points on either side of the front. The nature of this pressure discontinuity depends on the basic temperature and resulting density difference between the air bodies. This pressure gradient discontinuity across fronts is clearly shown by the isobars. Within an air mass, isobars are always smoothly curving lines. But, when crossing air-mass boundaries, or fronts, they bend sharply in order to conform to this abrupt change in the pressure gradient.

Figure 13 · 3 shows a front separating a polar from a tropical air mass. The atmospheric pressure at the earth's surface is shown by isobars with the pressure being lowest at the northern end of the map. The pressure at point 1 in the cold polar air is higher than at the adjacent point 2 in the warm air. The pressure decreases north of point 1 in the polar air, whereas it becomes increasingly higher to the south. Thus, the 1,002-millibar isobar cannot continue directly across the front but must bend sharply to the north on encountering the polar air in order to remain on points with pressure readings of 1,002 millibars. Hence isobars always bend toward the lower pressure when crossing fronts, and the wedge formed by this bending points toward higher pressures. Note that the pressure increases in a direction perpendicular to the front, so that a front always lies in a trough of low pressure.

3. WIND. It will be recalled that wind, in response to the pressure gradient and the deflecting forces, blows with a slight angle to the isobars and crosses them from higher to lower pressure. It is clear, then, that to obey this rule an abrupt wind shift must take place at fronts. This

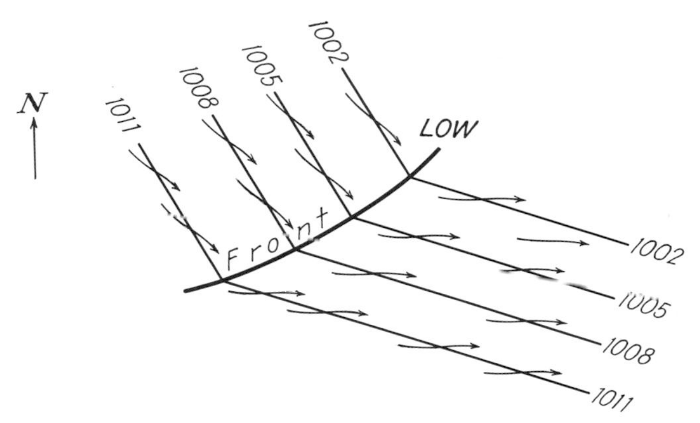

Fig. 13 · 4 The wind shift at a front.

is shown more graphically by adding wind arrows to the case considered above, as in Fig. 13 · 4.

We see in this case that the wind in the tropical air blows from the southwest while the wind immediately across the front, in the polar air, is northwesterly. Long before the true identity of air masses and fronts was recognized, these lines of abruptly shifting winds were observed. They were known simply as *wind-shift lines.*

4. CLOUDS AND PRECIPITATION. Fronts exhibit characteristic cloudiness and precipitation. These result from the adiabatic cooling of the warm air as it ascends the sloping frontal surface. The exact nature of the clouds and precipitation depends on the moisture content of the air and the slope of the front. Clouds of frontal origin thus extend for hundreds of miles because the warm air ascends along the entire extent of the frontal surface. The heavy low clouds formed in the rising and cooling warm air may yield precipitation. It should be

noted that the main cloud body forms in the air above the cold wedge, and the resulting precipitation originates in clouds in the warm air. Clouds in the cold air mass owe their development to local heating and are usually of the cumulus type except for nimbostratus formed beneath the rain-producing clouds. More specific cloud-front relationships are discussed in the following section.

FORMATION OF FRONTS. In many if not most cases, existing fronts form through the motion and meeting of air masses whose original properties are contrasting. Frequently, however, fronts may form where none existed or may have been suspected previously. The process of formation of a front is known as *frontogenesis*. Should the air motion in a particular area

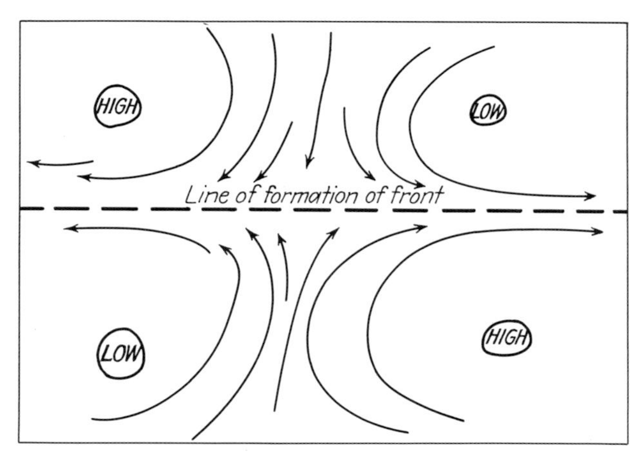

Fig. 13 · 5 Pressure and wind conditions conducive to frontogenesis.

be such, as a result of the pressure systems, as to cause air of different temperatures to be brought together, continuation of this process may cause a front to develop along the line of meeting of the different air bodies.

Figure 13 · 5 shows a pressure distribution with the resulting wind motion most conducive to the formation of a temperature discontinuity between warm air brought in from the south and cold air flowing in from the north. This is for the Northern Hemisphere.

Technically, this is known as a *deformation field*, for there is a contraction along a north-south axis and an expansion along an east-west axis as a result of the wind motion induced by the pressure arrangement. A careful study of pressure, wind, and temperature conditions often warns of future frontal development.

Frontolysis is the process in which a front dissolves as the contrasting conditions causing the discontinuity between the air masses disappear.

Types of Fronts

Depending on the motion of the air masses involved, several different types of fronts, each having particular properties, develop. These fronts will be treated individually: *warm fronts, cold fronts, stationary fronts,* and *occluded fronts.*

1. WARM FRONTS. A warm front is defined as a front along which warm air replaces cold air. Thus, if a cold and a warm air mass are adjacent, they will be separated by a front. Should the direction of motion of the air masses be such that the warm air progressively passes over ground surface previously covered by the cold air, the front becomes a warm front. On printed weather maps, the warm front is commonly

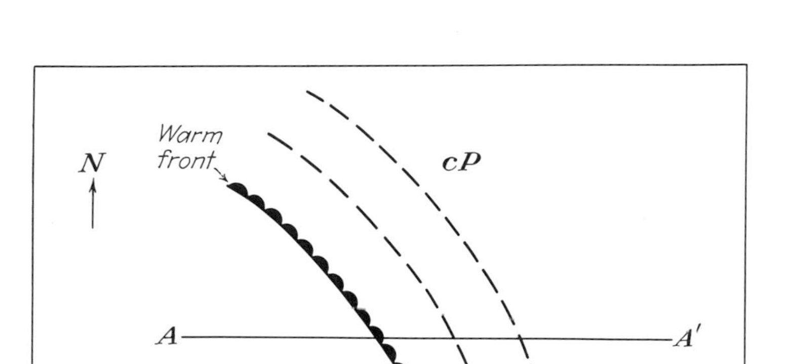

Fig. 13 · 6 Map view of a warm front. The broken lines show later positions of the front.

indicated by black semicircles drawn on the side of the front toward which it is moving and thus pointing in the direction of the cold air. On the working copy of a weather map it is drawn as a solid red line.

As shown in Fig. 13 · 6, we have a warm front separating continental polar from maritime tropical air. This front is moving in the direction of the cold air, later positions of the front being indicated by the dashed lines.

Stations east and northeast of this front will therefore experience a sequence of weather conditions. The weather prevailing in the cold air will undergo a more or less sudden transition as the front passes. These changes will follow the principles developed in the previous section regarding the temperature, pressure, wind, and cloud relationship to the fronts. Following the warm-front passage, the weather conditions of a tropical air mass will then prevail. A study of the weather map indicates just what conditions exist in the respective air masses.

The characteristic cloudiness and precipitation associated with a warm front will now be examined further. To do this we consider not only the horizontal air-mass distribution but the third-dimensional or vertical picture as well. In Fig. 13 · 7 we have a vertical cross section running west to east through the preceding warm front, along *AA′*. As shown on this vertical section, the warm front is perhaps best defined as the receding or trailing edge of a cold-air wedge.

The cold polar air sloping away from a warm front must always slope upward in the direction of motion of the front, here from west to east. Thus the warm tropical air ascends the frontal surface as a continuous sheet all along the warm front. When the rising air cools to the dew point, an extensive cloud sheet will form in the warm air

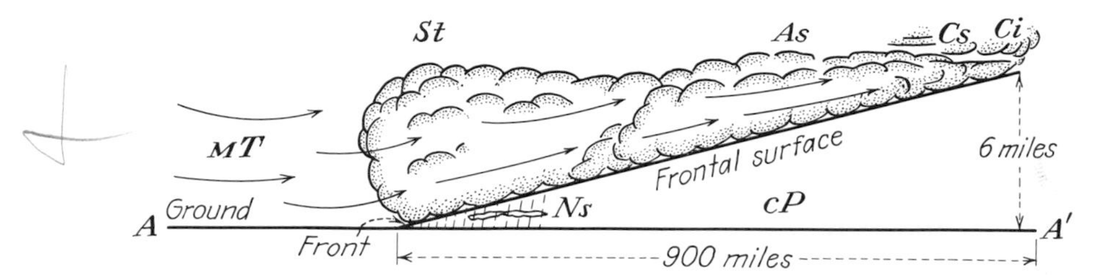

Fig. 13 · 7 Cross section of a warm front showing typical cloud deck formed in the rising warm air.

and will blanket the frontal surface up to a distance of 1,000 miles ahead of the front, extending along its entire length.

With a warm front having an average slope of 1:150, or 1 mile vertically for every 150 miles of horizontal distance, the high clouds forming in the warm air at heights around 6 miles will be 900 miles ahead of an approaching warm front. Consequently, as the warm front approaches, the cloud cover usually shows a typical sequence of cirrus, cirrostratus, altostratus, stratus, and associated nimbostratus. The relatively thick cloud sheet overlying the warm frontal surface near its tapering end usually yields steady precipitation which may extend over a long distance ahead of the front.

We see now that our earlier explanation of the formation of the above cloud system (Chap. 6) was nothing more than the consideration of a warm front.

As the warm front passes the observer, the heavy stratus-type clouds and any existing precipitation give place to clearing or partially clearing skies in the warm air.

2. COLD FRONTS. Cold fronts are defined as fronts along which cold air replaces warm air. Figure 13 · 8 shows a surface map with cold and warm air masses separated by a cold front. The black wedges are standard cold-front symbols and are always shown on the side toward which the front is moving, that is, they point into the warm air. The cold front is represented by a solid blue line on the working

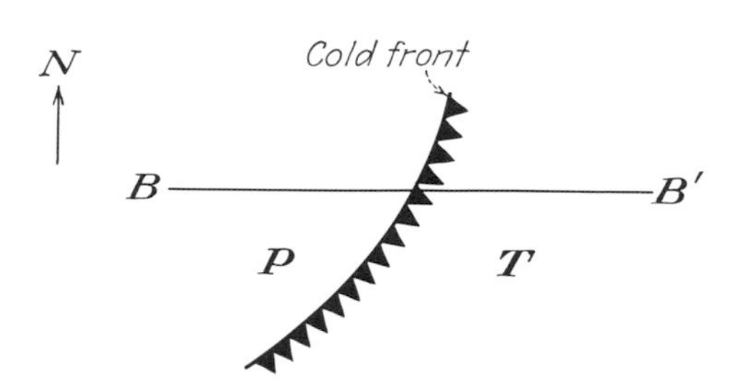

Fig. 13 · 8 Map view of a cold front.

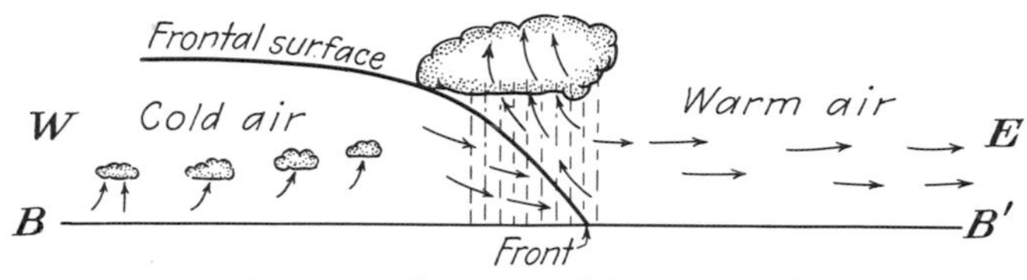

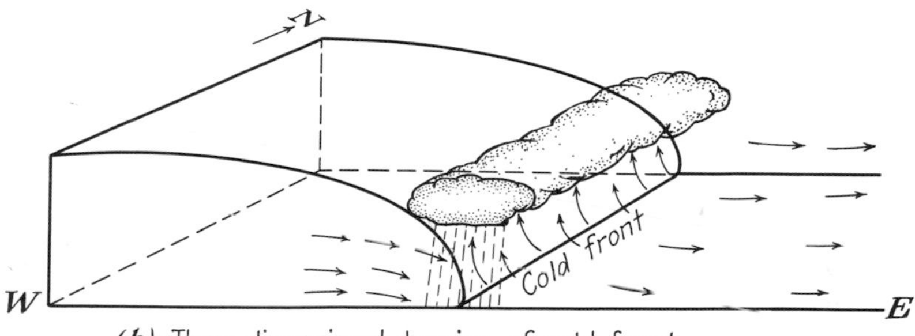

Fig. 13 · 9 (a) Cross-sectional view and (b) three-dimensional view of a cold front.

map. Here the air masses and front are moving from left to right, or west to east.

In this case the frontal surface will slope downward in the direction of motion as shown in the cross section along BB' through the cold front (Fig. 13 · 9). The cold front is therefore the advancing or leading edge of a steep cold wedge of air.

It is noticed that the cold frontal surface steepens considerably as it approaches the ground. This is a consequence of the motion of the cold air in the same direction as the slope of the front. The air motion near the ground is retarded from the effect of friction, while the free air at higher altitudes has a higher velocity than the lower. This tends to push the front forward at higher altitudes at a faster rate than the front at the ground, causing it to become much steeper than the warm front considered previously. The steepness of the front will depend on its velocity. With a low rate of motion, the slope of the cold front will become very gentle and approach that of a warm front. With higher velocities the lower part of the frontal surface begins to buckle and steepen.

In most cases the cold air advances faster than the warm air ahead. Consequently, the warm air immediately preceding the front will be forced violently upward along the steep cold-front surface. The rapid vertical uplift of the warm air results in rapid cooling in the air column. This yields clouds of the cumulus and cumulonimbus types, with frequent precipitation in the form of showers.

Depending on the steepness of the front and the temperature conditions in the warm air, the resulting clouds and showers develop as severe thunderstorms with extremely turbulent air conditions, which extend all along the front. The precipitation, although very heavy, will usually cover but a narrow zone along the cold front. For moderately sloping fronts the area of cloudiness and precipitation will be greater, but the intensity of any precipitation will be less.

We see that the storms associated with relief after a summer heat wave are of the cold-front type. Actually the thunderstorms have not brought about cooler weather. The colder air brought in the thundershowers, which are merely incidental to the drop in temperature as a cold front ends a heat wave. Local showers within a warm air mass yield no appreciable relief.

As the cool air traverses the surface previously warmed by the tropical air, local heating within the cold mass yields local rising-air columns, whose summits are capped by typical fair-weather cumulus clouds. The bases of these clouds, representing the height of the dew point of the rising air, are nearly on a level plane that stretches away into the distance.

3. STATIONARY FRONTS. Suppose that we consider two air masses, a cold and a warm, separated by a front. Will this front be a cold or warm front? The answer of course is that the identity of the front depends on its behavior. If the front is progressively displaced in the direction of the warm air, it will be a cold front; if in the direction of the cold air, it will be a warm front. But if the air masses are not in motion,

the front will be stationary and is called a *stationary front,* which is represented on weather charts by a combination of the warm- and cold-front symbols. No new cross section of the stationary front is shown, for it resembles in vertical section the picture of the warm front. The weather conditions are similar to those of a warm front.

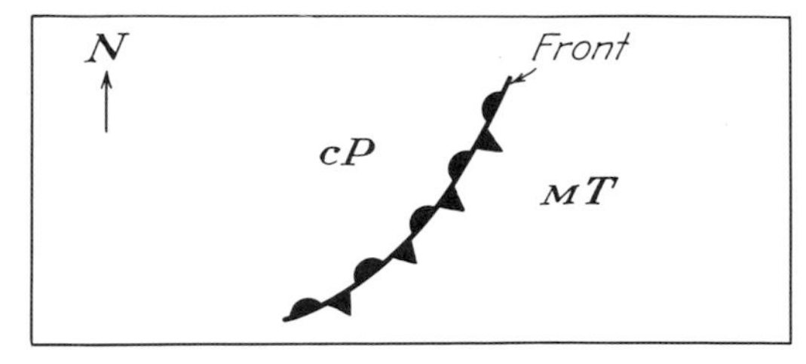

Fig. 13 · 10 Map view of a stationary front.

Should the cold air develop a movement toward the warm, the frontal surface, which will be sloping downward in the direction of motion, will buckle and steepen near the ground and develop cold-front characteristics. Or, if a cold front slows, it will assume characteristics of the stationary and warm fronts.

4. OCCLUDED FRONTS. Occluded fronts are those formed by the merging

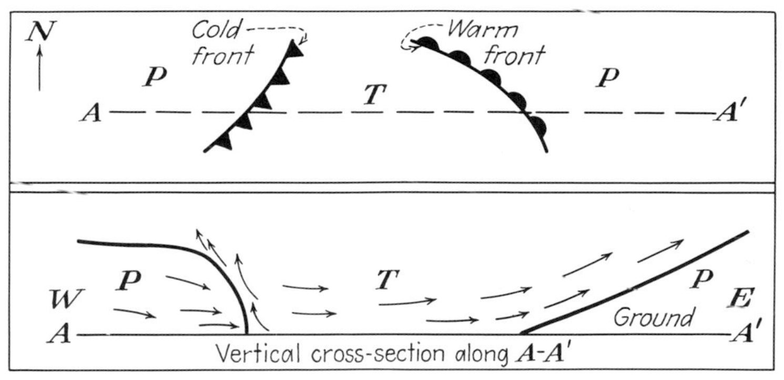

Fig. 13 · 11 Map and cross-sectional views of a cold and warm front.

of cold and warm fronts. If a cold front overtakes a warm, the result is an occluded front. It was stated above that the cold air with its bounding cold front moves faster than the warm air that may be ahead. If we have three air masses, a cold, a warm, and a cold in succession, moving from west to east, two fronts will exist (Fig. 13 · 11).

As a consequence of its higher velocity, the advancing cold front will overtake the warm front ahead. This results in the occluded front shown in Fig. 13 • 12, which is represented by a combination of warm- and cold-front symbols pointing in the direction of motion of the front. When drawn on a map, the occluded front is shown by a solid purple line.

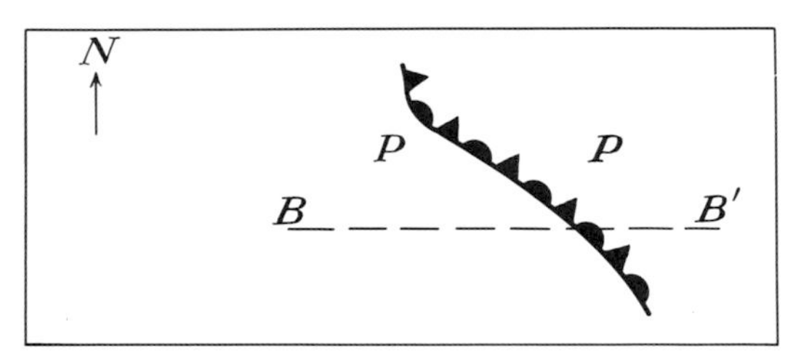

Fig. 13 • 12 Map view of an occluded front. The warm T air, being forced aloft, does not show at the surface.

But where is the warm air? It has been squeezed upward by the meeting of the cold- and the warm-front surfaces and is no longer present at the ground. To see it, we must examine the vertical air section across the front. Two possible cross sections may exist (Fig. 13 • 13). Exactly what has taken place to result in these cross sections is shown in Fig. 13 • 14*a–e*, which traces the movement of the fronts as the advancing cold overtakes the slower warm front.

The type of occlusion (warm- or cold-front) that results depends

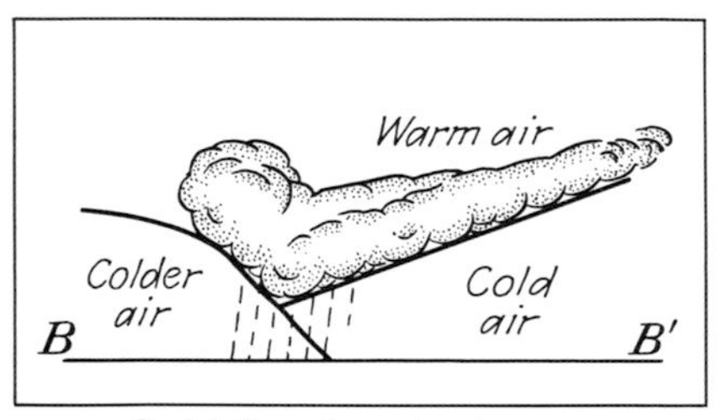

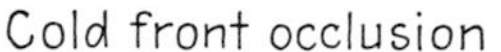

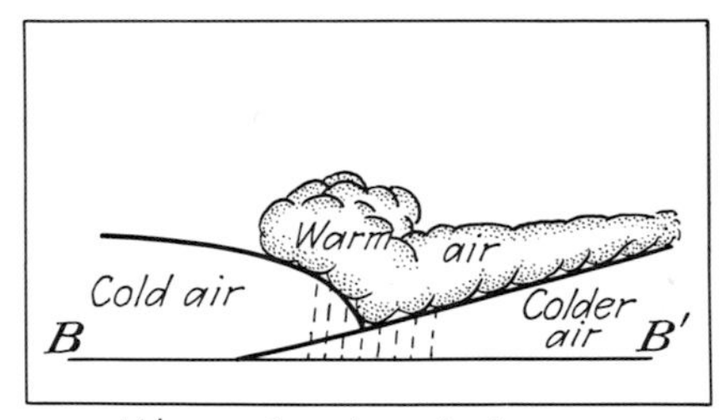

Fig. 13 • 13

on which of the polar air masses is the colder. If the air following the cold front is colder, the cold-front occlusion occurs; if the cold wedge preceding the warm front is the colder, the warm-front occlusion forms.

In either case, as the occluded front approaches, the cloud system with resulting precipitation will be very similar to that of the warm front, for the shape of the tapering wedge has not been altered prior to the front itself. As the front passes, the clouds and precipitation will be of the cold-front type.

From a knowledge of local weather conditions alone it is often im-

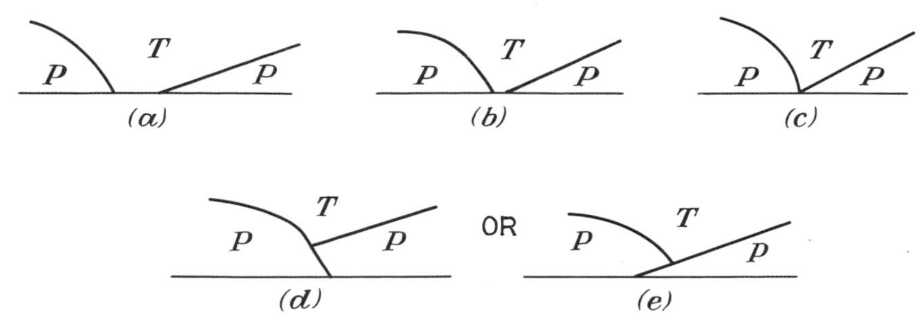

Fig. 13 · 14 Cross-sectional views showing a cold front overtaking a warm front to form a cold-front occlusion (d) or a warm-front occlusion (e).

possible to distinguish between the approach of a warm and an occluded front. The weather conditions at the zone of occlusion, where cold and warm fronts meet, are seriously affected. However, the conditions out ahead of the occlusion, overlying the warm frontal surface, are "unaware" of the changes associated with the occlusion and consequently show no indications of this process.

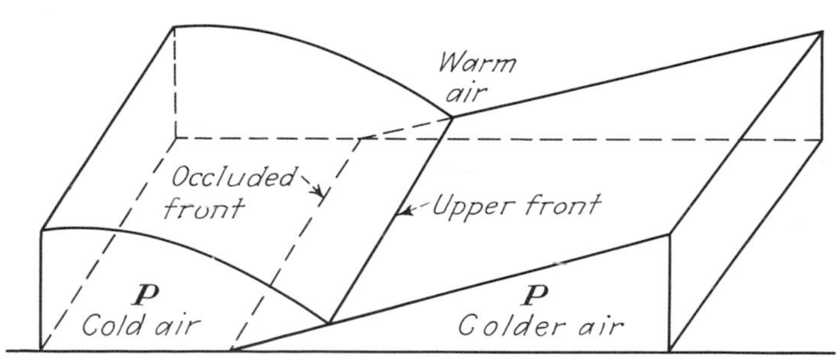

Fig. 13 · 15 Three-dimensional view of a warm-front occlusion.

UPPER FRONTS. Although they are of primary importance in aviation, our treatment of fronts should include a brief examination of upper fronts, for they represent a significant weather feature. As the name implies, upper fronts are those which exist in the upper air but whose effects on the weather may often be experienced at the ground. Most upper fronts occur in connection with, and as the direct result of, occlusions.

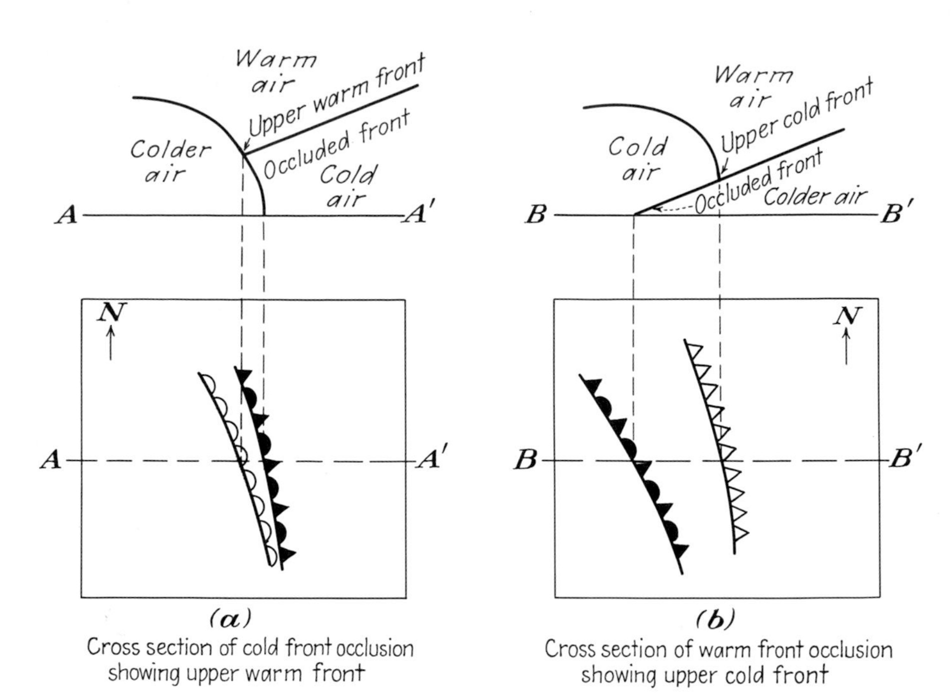

Fig. 13 · 16 Illustration of cold- and warm-front occlusions.

TYPE OF FRONT	ON WORKING MAP COPY	ON PRINTED MAPS
Warm front	Red line	
Cold front	Blue line	
Stationary front	Red and blue line	
Occluded front	Purple line	
Upper warm front	Broken red line	
Upper cold front	Broken blue line	
Formation of a front	F.G.	← Frontogenesis →
Dissipation of a front	F.L.	← Frontolysis →

Fig. 13 · 17 Summary of frontal symbols appearing on working and printed weather maps.

In the occlusion process, a cold front overtakes a warm one and, depending on the relative temperatures of the respective cold air masses involved, one of the fronts is forced up over the surface of the other. The front remaining at the ground we defined above as the occluded front. *The front that is forced aloft is the upper front.*

In the case of the warm-front occlusion, the cold front ascends the warm frontal surface and becomes an upper cold front. In the cold-front type of occlusion, the warm front being forced over the cold becomes an upper warm front. Figures 13 · 16*a* and *b* indicate this relationship. On working maps, upper cold fronts are shown by broken blue lines; upper warm fronts by broken red lines. On printed maps the symbols are the same as those for surface fronts, with the exception that they are not in black.

Upper fronts may not always be drawn on normal surface weather maps. Notice, however, that the upper warm front lies much closer to the occluded front than does the upper cold front. It is clear that owing to the relative steepness of the cold front, the rising warm front, in the occlusion process, has a large vertical component of motion. When the cold front rises over the warm, as shown in *b*, a large horizontal component of motion exists, owing to the much more gentle slope of the warm frontal surface.

World Air Masses and Fronts

We have now considered the general weather features of air masses and fronts. It might be well at this time to examine the important principal air masses and associated frontal zones of the world, since they are so important in determining the weather conditions experienced. Brief mention was made, above, of the main air-mass types. The generalized maps shown in Figs. 13 · 18 and 13 · 19 present the average world air-mass and frontal positions for the periods stated thereon.

ARCTIC AIR (A). Arctic air originates farther north than polar air, over the arctic north and the Greenland icecap. It is distinguished from the polar continental air by lower temperatures, although this distinction may often be absent, in which case no definite frontal zone occurs between these air masses. On occasions arctic fronts do exist in the far North Atlantic and Pacific, with polar maritime air existing to the south. Except for occasional outbreaks, the weather of the middle latitudes is rarely influenced by arctic air.

POLAR CONTINENTAL AIR (cP). Polar continental air is characteristically cold and dry. In the winter it originates over the cold, often frozen areas of central Asia and Canada. During the summer these cold-air bodies become more restricted as their sources necessarily retreat farther northward.

In the winter, the great polar air mass of Asiatic origin overflows to the

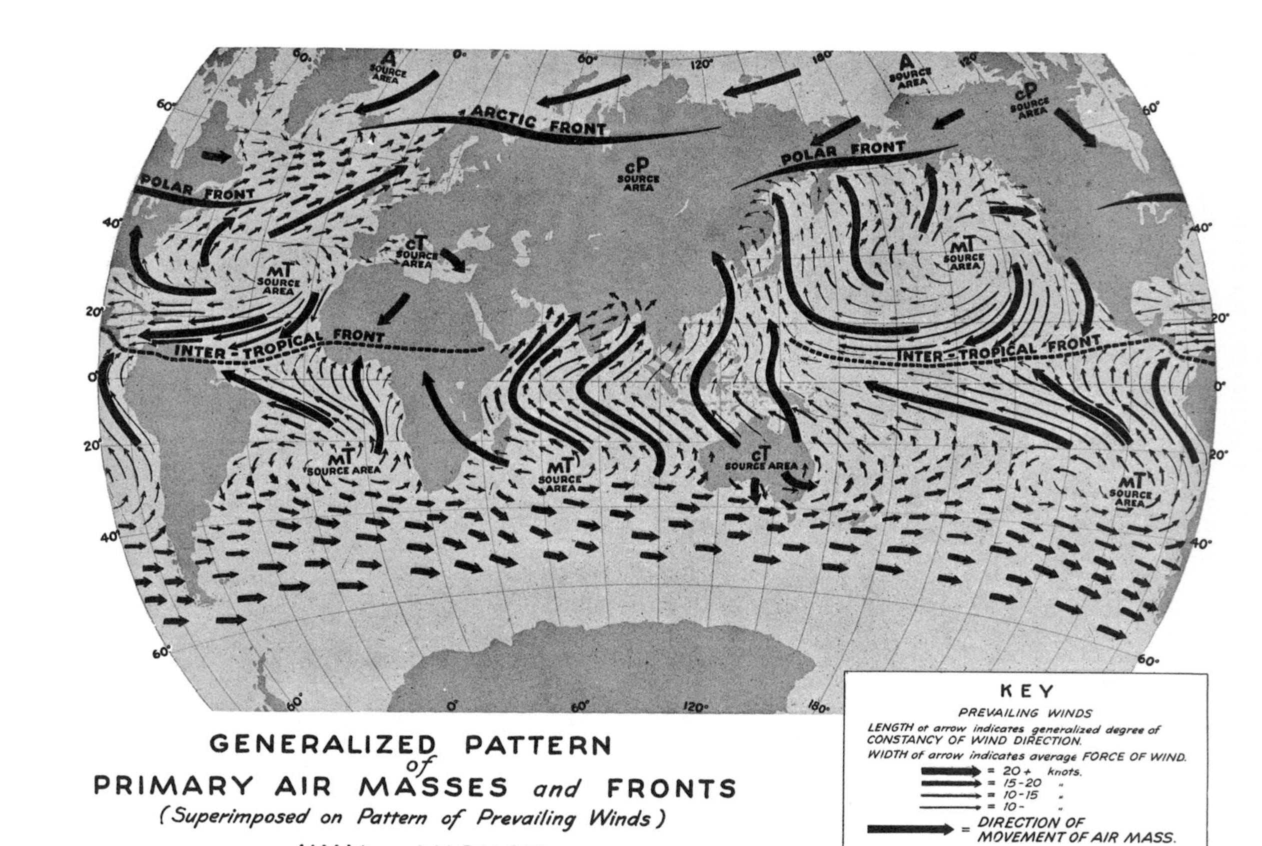

Fig. 13·18 (U.S. Navy)

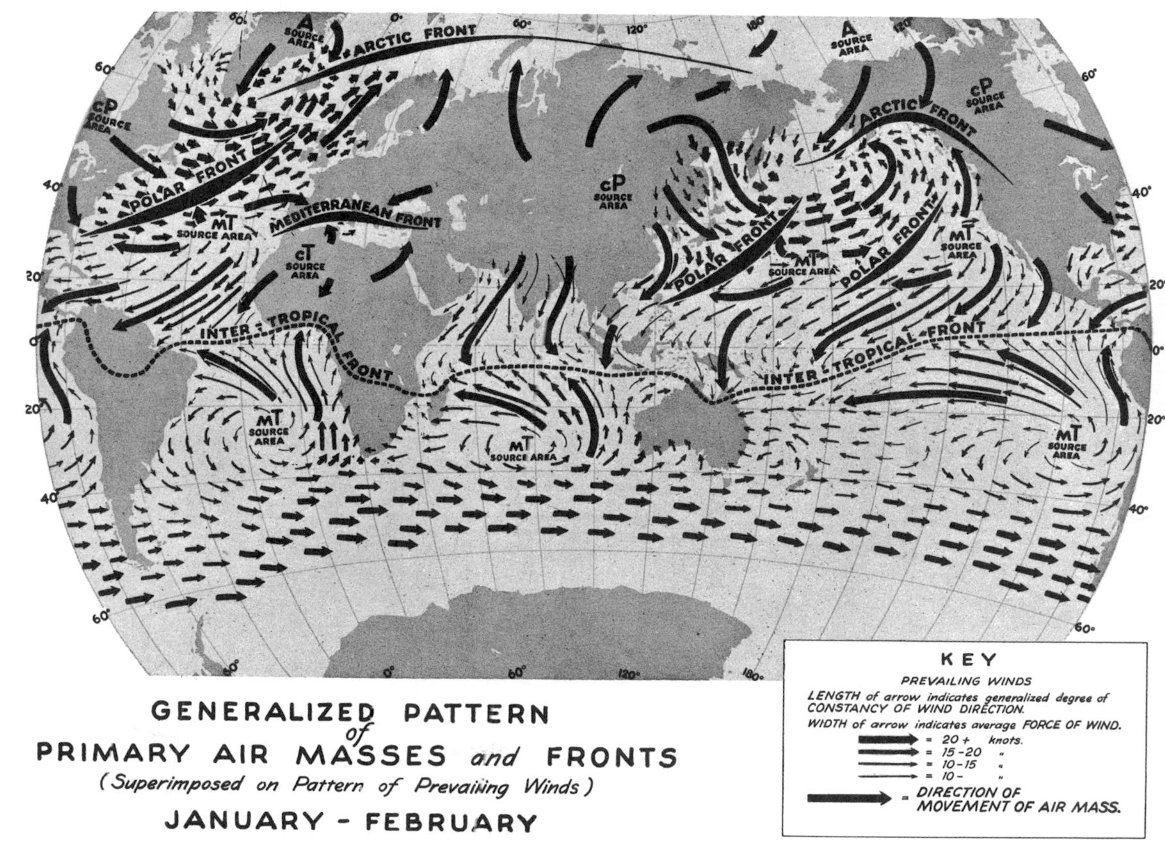

Fig. 13·19 (U.S. Navy)

south and east. The air moving southward travels with the northeast Indian monsoon, spreading cold air over the Indian Ocean and even reaching the South Pacific. The air traveling eastward continues into the North Pacific Ocean, where it plays a dominant role in shaping the weather not only of the Pacific, but of western North America as well.

When in the source region, polar continental air is very stable and clear, since the cold surface beneath prevents convection. The temperature in such air often increases with increase in altitude, yielding an inversion. This results essentially from the pronounced cooling of the surface air. After leaving the source region, the nature and amount of modification depend on the underlying surface conditions. Winter continental polar air suffers little change while crossing cold land surfaces. Upon reaching warmer land areas, local surface heating causes the formation of cumulus clouds which may become heavy and join to form stratocumulus.

When this air of winter origin advances over sea surfaces, arctic sea smoke or steam fog may form in the low levels of the air masses, if the water is warm. With the evaporation of sea water into it, the dry polar continental air mass undergoes modification to polar maritime air. Over cold water surfaces this air remains stable but tends to develop clouds of the stratus type, yielding drizzle and light rain. Fog may occur with or without the formation of the stratus sheet. Over warmer ocean waters, convectional-type clouds tend to form, with associated showers.

In the summer season, polar continental air is not nearly so prominent, remaining in higher middle latitudes for the most part. Modifications of the air in summer are very similar to conditions observed in winter air of the same origin.

POLAR MARITIME AIR (mP). Polar maritime air is for the most part polar continental air that has remained over the sea surface a sufficient length of time to absorb relatively large quantities of moisture. Since air masses ultimately move eastward, owing to rotational forces, the eastern sections of the oceans are characteristically overlain by polar maritime air. The western portion of the oceans may be influenced by polar continental air or polar continental air acquiring maritime characteristics. When moving over warmer surfaces, whether water or land, this air mass tends to yield cumulus-type clouds and associated showers, as a result of convection. When crossing colder surfaces, stratus clouds, fog, and often drizzle may result.

TROPICAL CONTINENTAL AIR (cT). In its source region, tropical continental air is warm and very dry. This air is limited in occurrence, originating over desert areas of North Africa, where it affects Mediterranean weather, and over the deserts of southwestern United States and northern Mexico. Continental tropical air, being dry and warm, has a high moisture capacity and a low relative humidity and will therefore absorb moisture rapidly when traversing water areas. It is thus modified rapidly to tropical maritime air.

TROPICAL MARITIME AIR (mT). Tropical maritime air originates in the subtropical high-pressure zones of the oceans (Azores and Pacific highs in the Northern Hemisphere). The weather here, as considered previously in connection with the horse latitudes, is mostly calm and clear. As maritime tropical air moves outward from its source, with either the westerlies or trade-wind circulation, its properties are modified. When this air moves with the westerlies, colder water surfaces are encountered. The uniform chilling of the air mass tends to produce fog and stratus clouds and occasionally light rain. With the retreat of the polar air masses in summer, the tropical maritime air extends farther northward than in winter. The northward surge of warm, humid air, upon meeting cold Arctic Ocean currents, is responsible for the prevalent summer fogs of the North Atlantic and North Pacific Oceans. In moving equatorward with the trades, maritime tropical air becomes warmer and more humid. For the most part, clear skies with scattered cumulus clouds prevail in this air in the poleward portion of the tradewind belt. The closer the approach to the doldrums, the greater is the tendency for the formation of convection-type clouds, with associated clouds and thunderstorms.

The maritime tropical air, moving eastward or westward from the source regions, encounters shore currents of varying temperatures and coastal zones of varying slopes. Consequently, California coastal fog develops when warm, moist, tropical air from the Pacific high moves across the colder California current (and the cold coast, in the winter). Upon ascending the steep slopes of the coastal ranges that border both California and Chile, heavy precipitation results. If cooler land masses are approached, fog and stratus-type clouds tend to form. If warm land exists to leeward, cumulus and cumulonimbus clouds with associated showers and thundershowers are common.

EQUATORIAL AIR (E). Equatorial air plays a prominent part in weather conditions over equatorial and tropical seas. When air becomes stagnant in equatorial areas, properties of high temperature and humidity are acquired. Cumulus clouds predominate in such air, with frequent thunderstorms.

When equatorial air crosses land areas, the effect of solar heating of the ground is to cause surface air temperatures to be highest in the afternoon. Consequently, towering cumulus and cumulonimbus clouds with resulting showers are most common in that part of the day. However, over ocean areas, as explained previously in considering thunderstorms, the lapse rate tends to increase during the night and finally provides the atmospheric instability necessary for thundershowers. Owing to the high absolute humidity of hot equatorial air, any precipitation therein is usually very heavy.

Equatorial air is encountered in great quantities in the vicinity of the doldrums in the Southwest Pacific. On occasions when equatorial air is carried to relatively high latitudes, it may be the source of dense fog, should cold currents be encountered.

ARCTIC FRONTS. Strong air-mass contrasts are occasionally produced when arctic air meets relatively warmer maritime air. Typical frontal weather disturbances then develop. Their effects are usually beyond the sphere of our normal activities, so that we shall omit further discussion of this situation.

POLAR FRONTS. Polar fronts are the boundaries of polar-air outbreaks. The polar front advances as the polar air advances, and retreats accordingly. Thus, in wintertime polar fronts advance with the polar air masses to much lower latitudes than in summer. An examination of the world air-mass maps shows clearly the northeast-southwest trend of the polar front during the winter. This becomes more nearly east-west during the summer. For the most part the transgressions and regressions of the polar fronts are characteristic of the middle latitudes. This accounts for the great variability of weather conditions in this zone, for invasions of widely differing air masses of either polar or tropical origin are continually recurring. Note these polar fronts well. We shall give them much greater attention in the following chapter, when their significance will become still more apparent. The locations and movements of polar fronts are controlled by the jet streams aloft.

INTERTROPICAL FRONTS. The exact nature and behavior of intertropical fronts are topics that require further research and explanation. These fronts form as a result of the meeting and convergence of the trade winds of both hemispheres in tropical regions. The position of these boundaries must shift seasonally and geographically in accordance with the migration of the doldrums. The intertropical front of the Atlantic is therefore always north of the equator.

The definite identification and evaluation of the intertropical fronts have been uncertain because so much of the equatorial region is in the oceans. The establishment of extended regions of storminess in the low latitudes indicates that some structural feature of the air, rather than simple convection, is responsible.

When the sun is on or near the equator (during the equinoctial periods), temperature conditions in the air masses on either side of the fronts are very nearly uniform. However, near the middle or end of summer or winter, temperature contrasts reach a maximum, for the temperature differences between the two hemispheres are then most pronounced.

PRESSURE-JUMP LINES. Cold fronts are often preceded by squall lines that may show more violent weather conditions than those associated with the front. Very recent research has shown that these squall lines are invariably associated with abrupt increases in pressure which are considered to result from the passage of a wave on an inversion surface aloft. The wave is believed to be generated by the pressure of a cold front on an inversion surface preceding the steeply sloping front. These pressure-jump lines have become important elements in modern observations and forecasts.

EXERCISES

13 · 1 Define the term "air mass."

13 · 2 What are the principal air-mass source regions and resulting air masses?

13 · 3 Explain the distinction between warm and cold air masses, and describe the general weather properties of each type.

13 · 4 Define front, frontal zone, and frontal surface.

13 · 5 Estimate the elevation of a warm front at a point 500 miles in advance of the front. What type of clouds and weather might be expected at this point?

13 · 6 Show how a front may form in a region where there was no prior contrast in air-mass properties.

13 · 7 What are the principal differences between the type and duration of weather phenomena of warm and cold fronts?

13 · 8 Describe the process of formation of occluded fronts, distinguishing between warm- and cold-front occlusions.

13 · 9 Describe and explain the stability changes that would be expected in (*a*) tropical air masses moving toward higher latitudes and (*b*) polar air masses moving toward lower latitudes.

13 · 10 Define the pressure-jump line.

CHAPTER 14

Development and Structure of Cyclones

The weather of the middle latitudes is associated not only with air masses and their motion, but also with cyclones and anticyclones, studied in a previous chapter. It will be recalled that these low- and high-pressure areas move approximately eastward in the belt of the prevailing westerlies, between horse latitudes and subpolar low. An approaching low brings with it a host of weather changes, mostly bad. We shall consequently be concerned here mainly with the study of the low-pressure area.

Considerably more information about the weather conditions in a low or cyclone is gained when the latter is considered in the light of modern weather knowledge of air masses and fronts. The application of this knowledge explains the origin, growth, and typical weather conditions of a cyclone. Such knowledge is requisite for the purpose of interpreting and forecasting the weather. Let us therefore proceed to the study of the cyclone in relation to air masses, noting its origin, development, and structure.

The Polar-front Theory or Wave Theory of Cyclones

The modern weather theory of cyclones, air masses, and fronts had its inception in Bergen, Norway. The great Norwegian meteorologist V. Bjerknes was one of the leaders of this famous Bergen school of meteorologists and is one of the principal founders of the newer concepts. Certain United States universities advanced these ideas still further. This was accomplished in two ways: (1) by sending United States scientists to Norway to study the new developments and (2) by actually inviting the

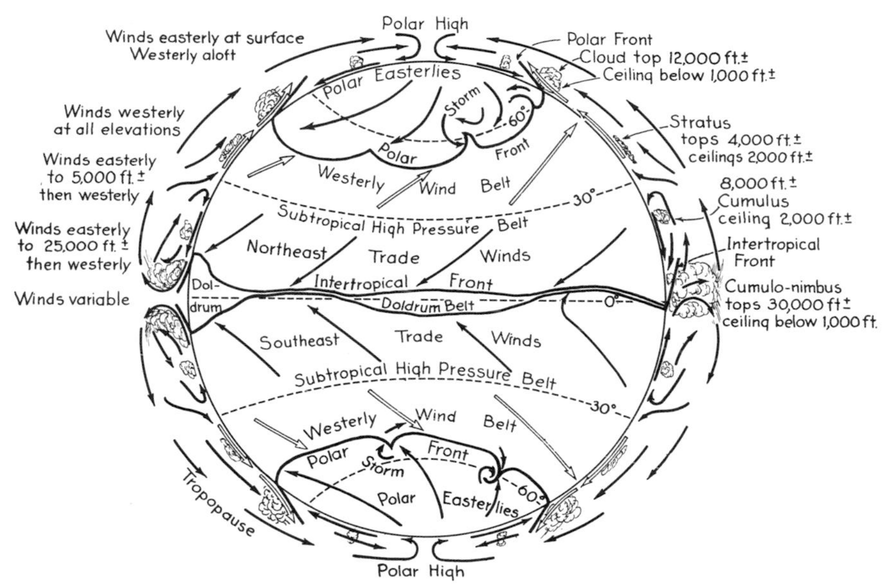

Fig. 14 · 1 Relation of the polar front to the wind and pressure belts. (U.S. Navy)

more eminent Norwegian specialists to come to the United States to engage in their research and teaching in American universities. The results have been most successful.

THE POLAR FRONT. We have seen that large, relatively uniform masses of cold air overlie the polar areas of the earth. The southern limit of this polar air is by no means a regular one. It is irregular and in constant, fluctuating motion advancing southward or receding to the north under the influence of the fluctuating jet stream aloft. Nor is this polar-air boundary a perfectly continuous one around the earth, since the cold polar air forms in more or less distinct source regions.

The natural name for the mobile southernmost limit or boundary of this polar air is the *polar front*—the front of the polar air. The polar front may have warm- or cold-front characteristics, depending on its direction of motion.

We may further consider the polar front to be the southern limit of the polar easterlies, or the line along which the cold polar winds from the north meet the warm prevailing westerlies that originate in much lower latitudes. This relationship of the polar front to the general atmospheric circulation is shown in Fig. 14 · 1.

The development of waves on the polar front is shown in a generalized way in this diagram. It was recognized that the waves, whose apexes point poleward, go through a fairly standard cycle of development from small, minor disturbances to those hundreds to a thousand miles across. The related wind and weather structure developing with the frontal wave constitutes the middle-latitude cyclone, to be considered in detail in the following section. Waves that grow in size to form these storms are called *unstable waves.* Others, which remain small and finally die out with little important weather effects, are known as *stable waves.*

During the growth of unstable waves, tremendous volumes of polar and tropical air are drawn far from their source regions and play a major part in the weather changes of the middle latitudes.

Development of Cyclones—Cyclogenesis

Most cyclones of the middle latitudes grow as unstable waves on the polar front. Other forms of cyclogenesis are usually restricted to particular localities, as in the case of the *thermal desert lows,* or involve relatively small areas.

Owing to marked temperature differences between tropical air and polar or arctic air, the air masses on opposite sides of the polar front also have strong differences in their energy quantities, as well as in their density structures. Whenever fluids of any kind are in contact, the surface between is a zone of great instability, or a zone in which growing disturbances are likely.

Prior to the initiation of an unstable cyclonic wave, the atmospheric circulation is characterized by an outbreak of polar (more rarely, arctic) air into the middle latitudes. The advancing edge of the polar air (polar front) becomes increasingly unstable as it encroaches on the region of the tropical air, until a small initial disturbance occurs that deforms the uniform slope and extent of the front. The cause of the disturbance may be one of many factors. The effects of irregularities in topography, such as mountains, of thermally induced eddies or irregularities in air flow, or of upper-level wind factors may act locally to retard the motion of the polar front. Remember that the polar front is a surface, so that the disturbance produces a wave not only along the ground position but also on the upward slope as shown in Fig. 14 · 2. The horizontal ground wave on the polar front is simply the pattern formed by the intersection of the trough depression in the frontal surface and the ground or horizontal.

The latitude of the initial disturbance varies seasonally because the polar

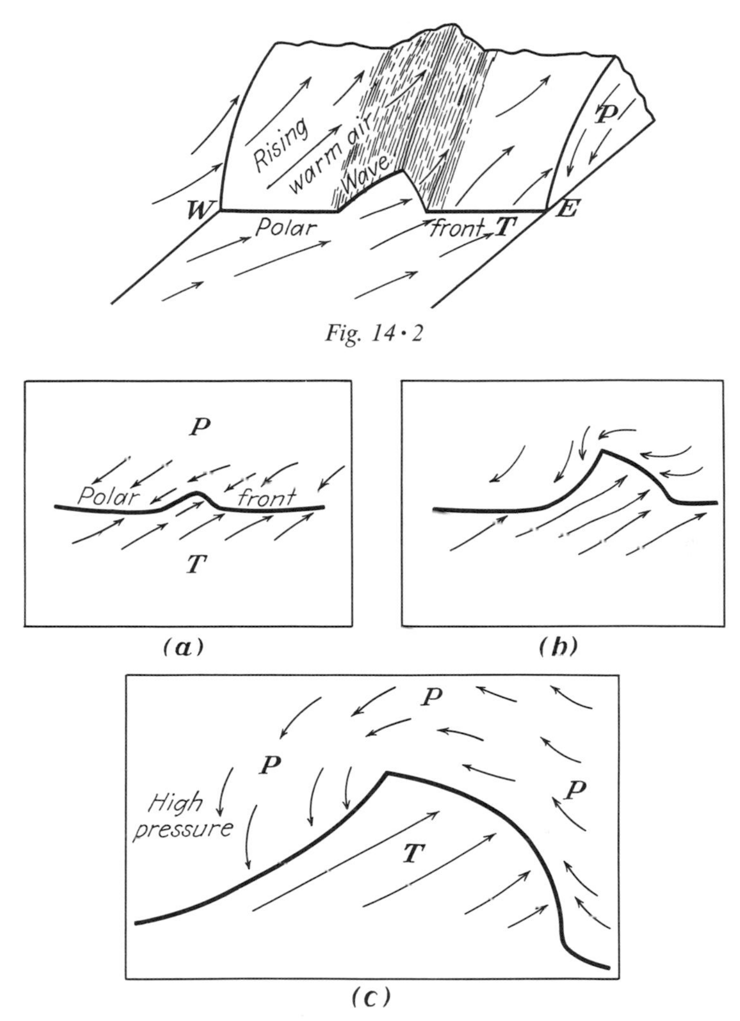

Fig. 14 · 2

Fig. 14 · 3 Development of cyclonic circulation with growth of polar-front wave.

fronts frequently reach latitudes around 30°, although the average winter position is farther north.

If conditions are appropriate for the generation of an unstable wave, the initial disturbance grows rapidly and travels eastward along the polar front. Figure 14 · 3 shows three stages in the growth of an unstable wave.

As the wave develops, the rather uniform linear motion of air on opposite sides of the polar front is disturbed into the flow pattern shown in Fig. 14 · 3. A consequence of this motion is the formation of a low-pressure center at the apex of the wave and a region of high pressure in the cold air to the west. The central low which begins to form at about the same time as the initial disturbance in (*a*) becomes more intense as the circulation about the apex increases. In moving along the polar front, the western portion of the wave, in advance of the cold air, moves faster than the eastern part, which leads the warm air, resulting in the asymmetry evident in Fig. 14 · 3.

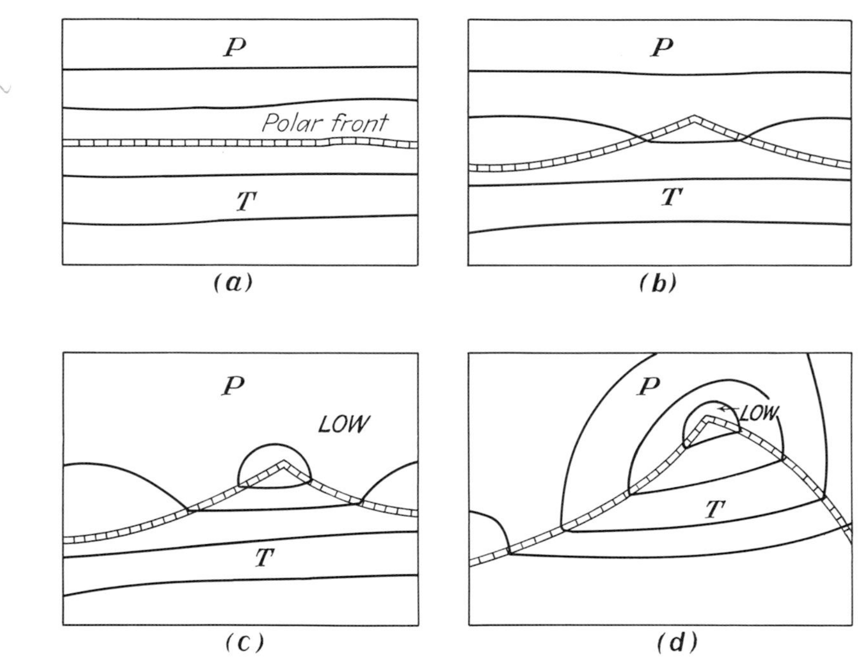

Fig. 14 · 4 Development of the isobar pattern about a low.

The changes in the pressure pattern and gradient as the wave grows are shown in Fig. 14 · 4, where the isobars become deformed from a pattern more or less parallel to the front to a roughly circular pattern showing discontinuities at the front. The time for the development shown in Figs. 14 · 3 and 14 · 4 is roughly 12 to 24 hours. When the well-formed cyclonic pattern with related storm conditions is reached, the stage is described as that of a mature cyclone, whose characteristics are summarized in the following section.

The Mature Cyclone

The ideal mature cyclone is shown in Fig. 14 · 5, which includes two vertical cross sections on lines north and south of the center. A view of the vertical structure is necessary in order to understand the origin and distribution of clouds within the low.

Some striking differences exist between this picture of the cyclone and

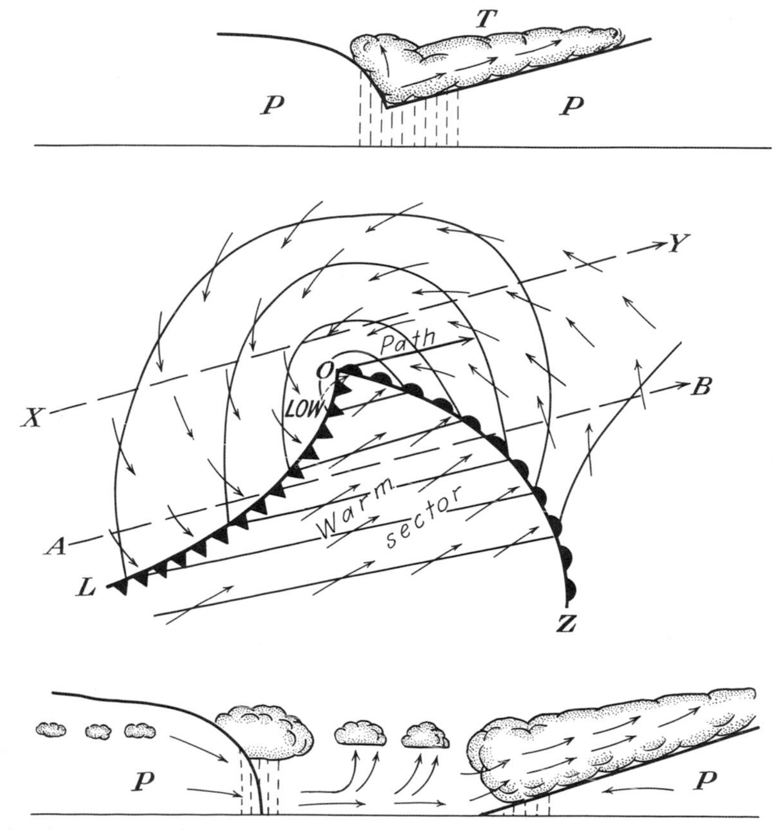

Fig. 14 · 5 A mature low-pressure area with cross sections showing cloud and precipitation conditions north and south of the center.

the more classical view given in Chap. 11. We no longer recognize the continuous circulation of a single air mass about the center. The mature cyclone is composed of at least two well-defined air bodies, a cold and a warm, with the former predominating at the surface level.

The warm tongue of air extending into polar air in the southern part of the low is known as the *warm sector.* As a result of this warm sector in the cyclone, the isobars about the mature low will show pronounced kinks when crossing either side of this warm zone. This differs also from the older picture showing smooth circular isobars in the low. Isobars in the warm sector of a well-developed cyclone are usually straight and parallel.

As a consequence of the two air masses present, with their distinct differences of temperature, pressure, and direction of movement, a pronounced wind shift occurs in crossing into the warm sector from the polar air on either side. No such abrupt wind shift was stressed in our earlier cyclonic model.

Since the cyclone with its attendant conditions moves easterly, it is clear that the line *OZ* along the original and now distorted polar front will be a line along which warm air will replace cold air. This is therefore a warm front and *has been a warm front ever since the wave first formed.* Similarly, the line *OL is and has been a cold front.*

It is interesting and instructive to examine a weather chart of the North Atlantic Ocean drawn prior to the time of routine air-mass and frontal analysis, as shown in Fig. 14 · 6. Examine the low in the center of the chart. Although not drawn, the positions of the warm and cold fronts can be easily located on the basis of the kinks in the isobars and the wind-shift zones. On the basis of this analysis, the locations of the polar and tropical air masses can also be estimated despite the absence of temperature reports.

The marine weather series for the Atlantic Ocean off eastern United States shown in Fig. 11 · 6 illustrates a typical pattern of cyclonic development in this region. Such storms often bring severe weather to east-coastal areas as well as to the waters of the western North Atlantic Ocean.

Refer to the map of Fig. 14 · 6. The intense low in central North Atlantic offers a good example. A pronounced kink in the isobars exists along a line extending from the center of the low approximately southward just east of the 60° meridian. To the west of this kink the winds are from the northwest; to the east, the winds are from the southwest. This wind-shift line is clearly the cold front of the cyclone. Extending northeastward from this cold front, the isobars are nearly straight and parallel, with winds prevailing from the southwest, marking the warm-sector zone. Another wind shift, this time from southwest to southeast, occurs in the forward or eastern section of the low. The isobars bend again along this wind-shift line, although the abruptness of the bend is not quite so pronounced. This represents the warm front. Similar conditions can be developed for other cyclones on this and the succeeding charts.

The vertical cross sections through the cyclone show the typical cloudiness and precipitation picture (Fig. 14 · 5). It is seen that basic areas of clouds and precipitation are those associated with the ascent of the warm

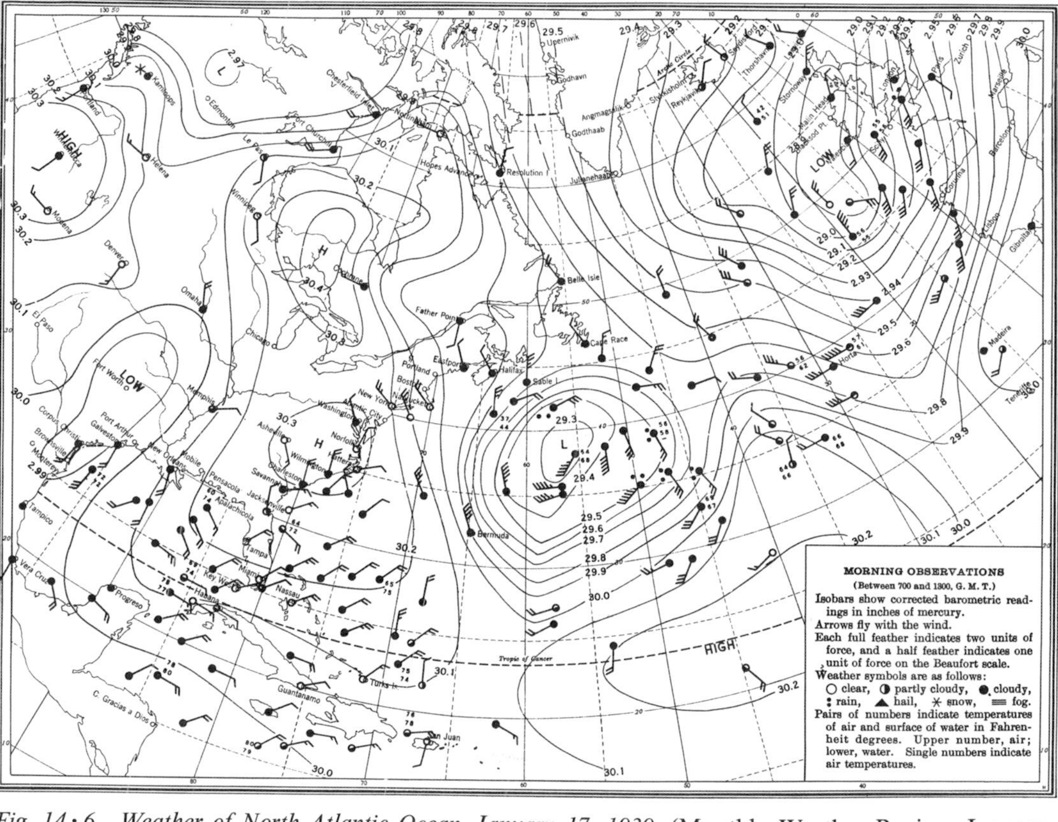

Fig. 14·6 Weather of North Atlantic Ocean, January 17, 1939. (Monthly Weather Review, January, 1939)

air over the warm and cold fronts. These conditions were given in detail in the previous chapter. We note that as a consequence of the uniform and extended cloud sheet lying above the gently sloping warm front, the precipitation covers a much larger area than that of the cold-front system.

Among the earliest local indications of an approaching low, or depression, are the cirrus clouds thickening into the cirrostratus that lie on the outer fringe of the advancing warm-front cloud deck. If at all possible, this should be related to pressure and wind conditions before any positive prediction is made. The precipitation falling through the cold air from the stratus clouds above the warm front frequently saturates the cold air and causes low clouds and fog to form. This greatly reduces both ceiling and visibility of the air preceding the warm front.

The warm sector is characterized by horizontally moving warm air. Vertically rising air is a local feature, resulting from overheating or topography. The clouds in the warm sector will therefore be of the cumulus type—either cumulus or altocumulus. Should the cumulus form in heavy bands, they become stratocumulus. Pronounced vertical movement of local air columns will form cumulonimbus clouds, yielding the local thunderstorms so frequent in the warm sector in spring and summer.

The cloudiness and precipitation at the cold front are very different from that at the warm front. Owing to the greater steepness of the cold front, the cloud band formed by the forced uplift of the warm air over the advancing front is much narrower in extent and in area covered. At the same time the more violent uprush of air over the steep cold front forms thick cumulus and cumulonimbus cloud types yielding precipitation of great intensity in the form of showers and thundershowers. Since this precipitation is usually not nearly so widespread as that of the warm front, its duration is thus much shorter. The cold-front portion of the cyclone was long known as the *line of clearing showers,* before the nature of the front itself was understood. The approach of the cold front in a low is often marked by a long, ominous band of rolling black clouds with associated severe winds and precipitation approaching from the west. After the passage of the cold front, the observer is situated in the relatively cool dry-air mass following the front. Local convection caused by the passage of the cool air over the previously warmed ground surface causes thin, flat, fair-weather cumulus to form, very typical of the fair, post–cold frontal weather.

We should note here that *secondary cold fronts* frequently form in a low following the passage of the first or primary cold front. The leading portion of the cold air following the warm sector is often so modified in passing over the previously warmed surface that it develops characteristics different from the main cold air mass. This gives rise to the secondary cold front, which separates the main mass from the modified mass and is also distinguished by a temperature, pressure, and wind discontinuity. Secondary-front weather conditions are often weaker than those of the primary front.

This front also extends outward from the low-pressure center but is farther to the west.

To avoid confusion in Fig. 14·5 the areas and types of cloudiness and precipitation are shown in Fig. 14·7, where only the fronts are drawn.

Along the path of the storm represented by the cross section XY in Figs.

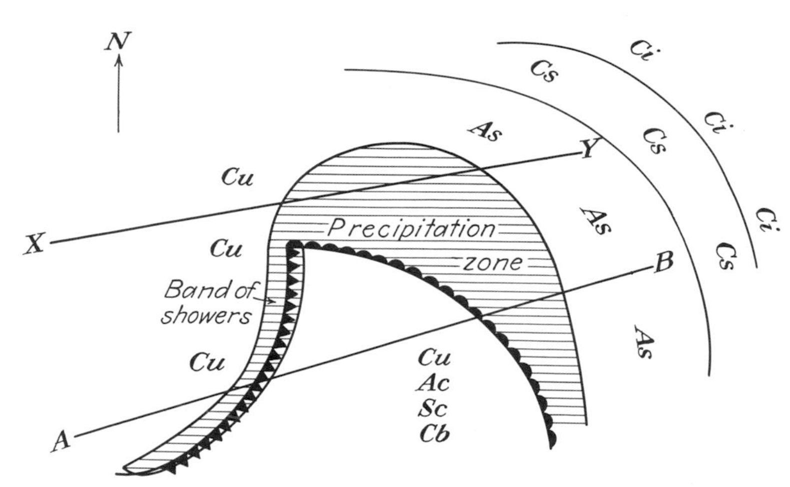

Fig. 14·7 Typical cloud precipitation zones of a mature low.

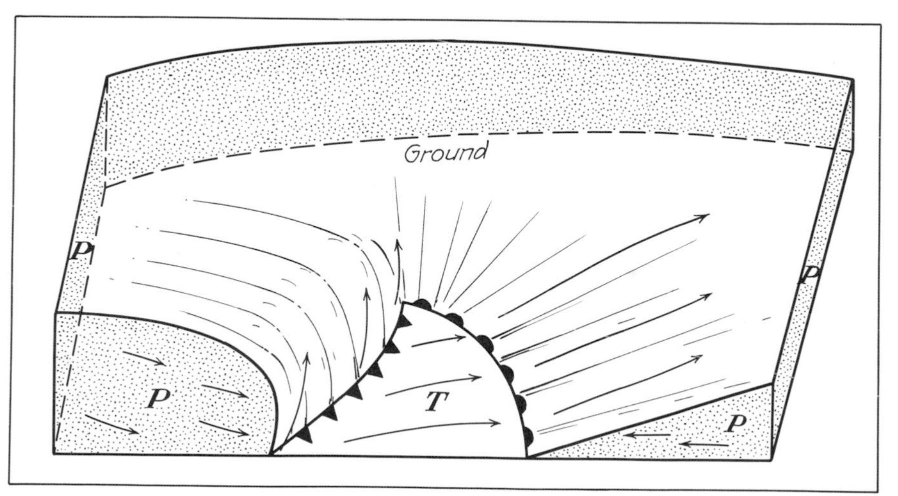

Fig. 14·8 Three-dimensional air mass and frontal structure of a mature cyclone.

14·5 and 14·7, it is noted that no front or warm sector is encountered north of the cyclone center as a result of the warm air rising over the fronts which here are above the ground.

The three-dimensional diagram (Fig. 14·8) is given in an attempt to clarify further the air-mass and frontal picture of the mature cyclone.

The Occluded Cyclone

We noted earlier that the cold-front portion of the cyclonic wave travels faster than the warm-front. Hence, continued development of the mature cyclone results in the cold front overtaking the warm, producing an occluded front as described in Chap. 13. The occlusion begins first near the apex of the wave, where the distance between the fronts is least, and gradually extends toward the more open part of the wave. Ultimately the warm sector is completely lifted above the surface, producing the fully occluded cyclone.

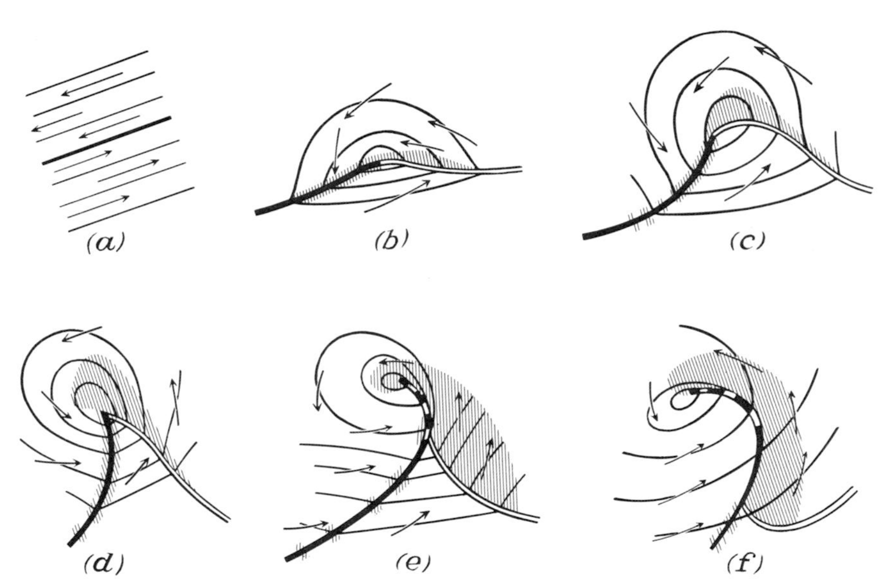

Fig. 14 · 9 Life history of a cyclone from inception to occlusion.

Depending on the density contrast between the cold air masses on both sides of the occluded front, the latter develops as either the warm- or cold-front occlusion. The cyclone usually reaches its greatest intensity during the occlusion process, with the maximum weather disturbance extending up to 100 miles north of the point of occlusion. Soon after complete occlusion, the two cold air masses mix across the front, causing dissolution of the occluded front and the weakening and disappearance of the low. The cloud and precipitation structure associated with the occluded front was given in Chap. 13. The life of a single-wave cyclone is usually about five to seven days. This life history is reviewed in Fig. 14 · 9.

A well-developed storm slightly younger than stage (*e*) of Fig. 14 · 9 is shown in the satellite telephoto picture of Fig. 14 · 10. The center of the

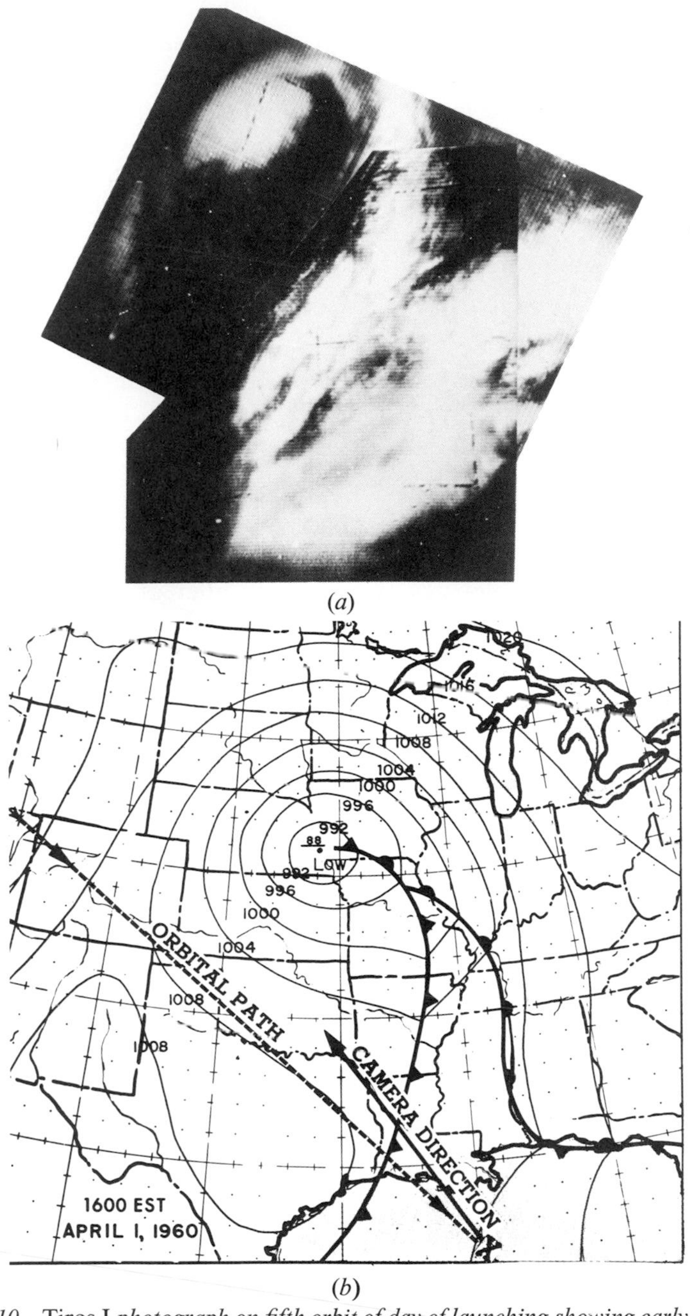

Fig. 14 · 10 Tiros I *photograph on fifth orbit of day of launching showing early occluded stage of a well-developed cyclone centered over eastern Nebraska. The cloud pattern in the photograph is correlated easily with the details shown in the surface weather map.* (U.S. Weather Bureau)

storm is revealed by the circular cloud mass in the upper part of the picture. The cloud band trailing south and southwest from the center outlines a prominent cold front; the large cloud mass in the foreground was located near the Gulf coast of Louisiana and Mississippi. The equivalent surface weather map which is shown below the photograph also indicates the path of the satellite, the camera direction, and the place (x) where the picture was taken.

Cyclonic Development in the Southern Hemisphere

The development of a cyclone in the Southern Hemisphere follows the same pattern as in the Northern. However, the polar air comes from the south and the tropical air from the north in this case. Hence the apex of the wave points southward rather than northward. See Figs. 14 · 20 to 14 · 22.

Paths and Movement of Cyclones

The directions and velocities of motion of different cyclones may vary widely, depending on surrounding pressure, temperature, air circulation, and upper-air conditions. Cyclones have an average rate of motion varying from 20 to 30 miles per hour, or from 480 to 720 miles per day. The lower velocities prevail during late spring, summer, and early autumn, when atmospheric circulation in general is rather sluggish. This velocity then increases, with the higher rates prevailing during the winter period. Cyclonic activity as a whole is far more active during the cold part of the year in either hemisphere, when the vigorous polar front encroaches toward tropical latitudes, causing the production of numerous and active cyclonic waves.

Cyclones, on the average, follow certain well-defined paths or tracks. The determination of speed and path of a particular cyclone is a problem to be considered under Weather Interpretation. However, we may note here an important observational fact: *Cyclones tend to move parallel to the isobars in their warm sectors.* Thus, the path of an individual cyclone with a definite warm sector can often be determined by noting the isobar pattern in this zone. Figure 14 · 11 illustrates the average world-cyclone tracks.

As a greater appreciation of the upper troposphere's circulation was gained, it became clear that the motion of surface lows and highs is predominantly controlled by the upper-air flow. Cyclones move parallel to the warm-sector isobars as just noted because these isobars reflect quite accurately the direction of movement of the upper westerlies, which actually guide the storm motion. The jet stream in particular exerts a strong steering effect on wave cyclones on the polar front, whose surface location is frequently just beneath the jet.

The synoptic charts for the same time at sea level and the 500-millibar (c. 18,000 feet) level in Figs. 14 · 12 and 14 · 13 provide a good comparison between the positions of sea-level fronts and cyclones and upper-level wind flow. Figure 14 · 12 shows two major outbreaks of polar air and related cyclones—one outbreak over the northeastern Pacific off northwestern North America and a second extending from Central America northeastward across the eastern United States and then into the North Atlantic. The airflow and isobaric pattern of Fig. 14 · 13 shows several long waves

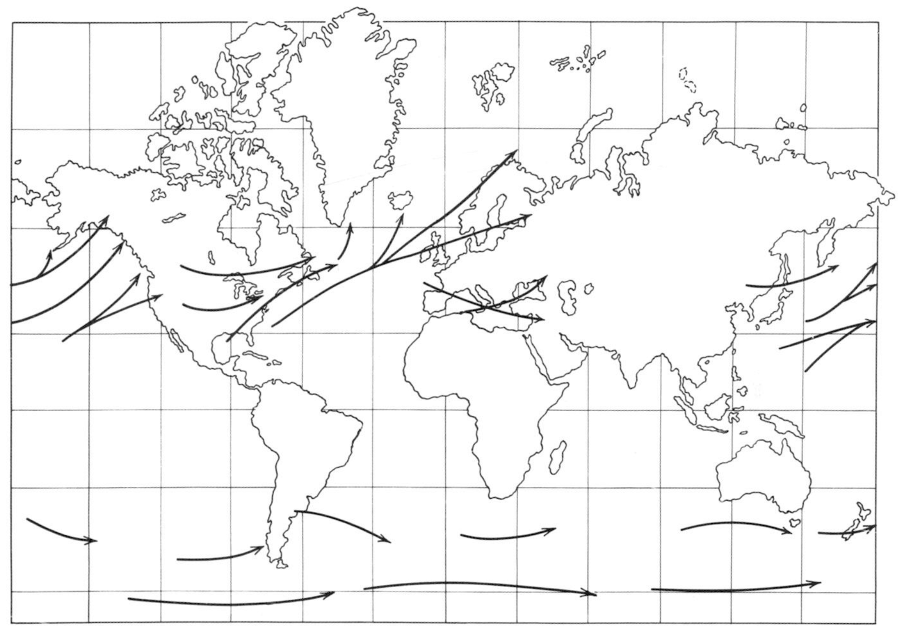

Fig. 14 · 11 Generalized world-cyclone paths.

(as discussed in Chap. 10) in the upper westerlies over the same area. As shown, the southern margin of the maximum wind pattern corresponds almost exactly with the wavelike patterns of the surface polar front.

Since the wave cyclones travel along the front, their paths must also correspond with that of the long-wave motion in the winds aloft.

Secondaries and Cyclone Families

SECONDARIES. Frequently, during the late maturing of the occluded stage of a cyclone, a small, new low develops on the fringes of the original depression. This usually occurs in either the southeastern or the southwestern quadrants, along the warm or the cold front, respectively. The new cyclone is known as a *secondary*. A secondary cyclone sometimes matures very rapidly, often at the expense of the primary, or as the latter becomes extinct. After occlusion, the secondary may follow along in the track of the primary cyclone, or it may move in a new path. Whatever the case, the weather conditions in the area affected by the development of

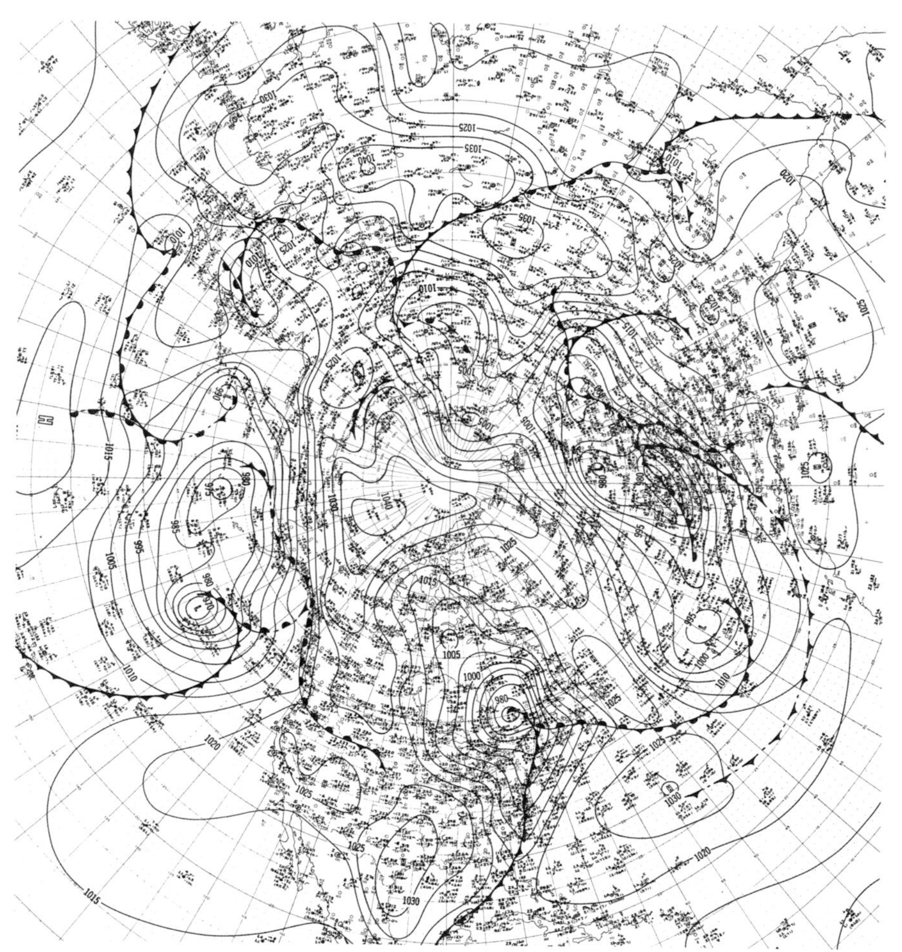

Fig. 14 · 12 Synoptic weather map at sea level for most of the Northern Hemisphere. Compare with the circulation at 500-millibar (c. 18,000 feet) level.

the new depression alter very rapidly. Care should be given to any tendency for the development of a secondary after a cyclone matures.

CYCLONE FAMILIES. So far, we have considered the development of a wave cyclone along a relatively small section of a polar front. However, with each outbreak of polar air into the middle latitudes, there usually develops not one but a series of waves along the polar front, each wave forming its own cyclonic disturbance. Each wave westward along the polar front is at a successively earlier stage in development. The polar-front boundary of the advancing cold air extends roughly as a northeast-southwest line. The earliest wave forms near the northeastern end of the

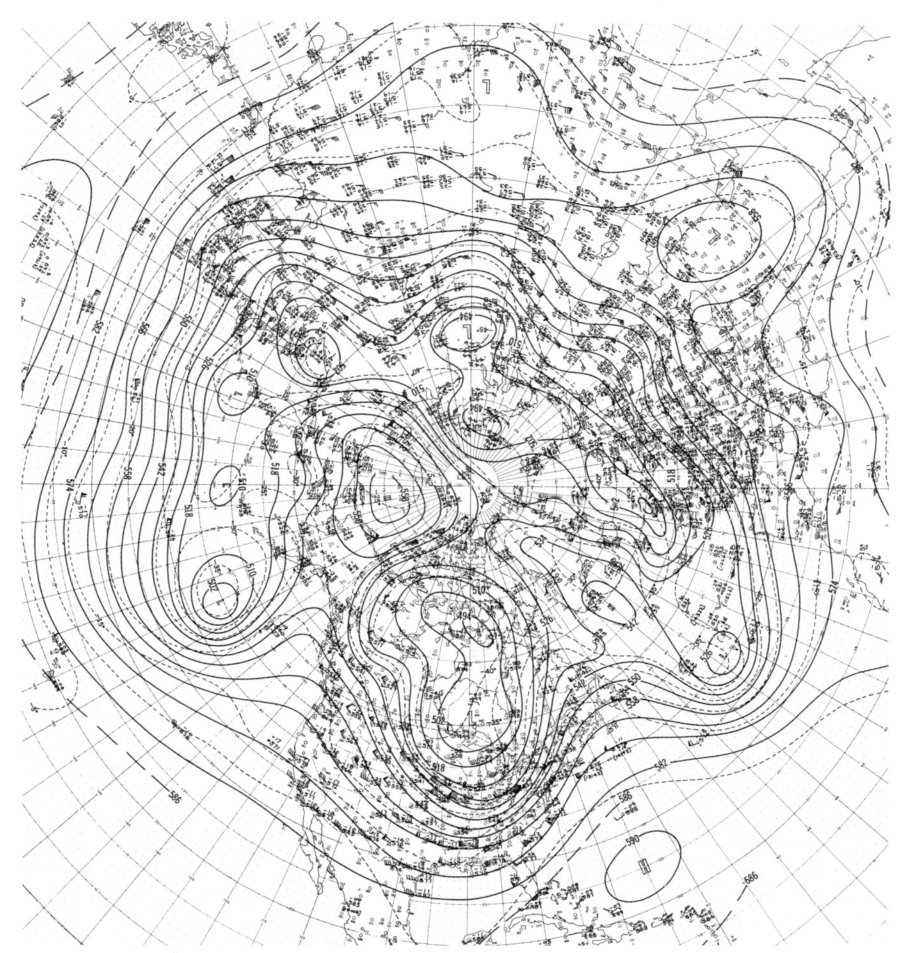

Fig. 14 • 13 Synoptic map at 500-millibar (c. 18,000 feet) level. Compare with Fig. 14 • 12.

original front and, as this wave develops into a true cyclone, the polar air to the west continues moving southward. As it does so, further reaction between the advancing polar air and the warmer prevailing westerlies leads to the development of newer waves. Anywhere from two or three to five cyclones may thus form along a particular polar front. These waves then move east or northeastward, so that the polar front has local characteristics of warm, cold, or occluded fronts. Figure 14 · 14 illustrates a family of wave disturbances. Following the passage of one cyclone family, another outbreak of polar air may occur, and with it a new system of disturbances.

The accompanying U.S. Navy weather maps illustrate three weather sequences for the North Atlantic, the North Pacific, and the South Pacific Oceans, respectively. These maps are printed in accordance with the standard weather symbols studied previously. Areas of moderate precipitation, such as those associated with warm or occluded fronts, are shown by closely spaced, slanting lines. Areas of heavy or showery precipitation along cold fronts are indicated by widely spaced lines crossing the cold fronts. These sequences all show very clearly the development of cyclone families, with the formation of young disturbances as those to the east become mature or occluded.

Fig. 14 · 14 A cyclone family along the polar front—Northern Hemisphere.

Figures 14 · 15 and 14 · 16 show the development of a cyclone family on a single polar frontal outbreak over the North Atlantic Ocean. In Fig. 14 · 15 the oldest wave, over the British Isles, is in the dissipation stage following occlusion. Along the front to the southwest are three other cyclones in successively younger stages of development. On the following day, the dissipating storm is no longer visible and the second, original storm of February 16 has advanced to the European coast, with marked decrease in size. The weather of the North Atlantic is now (February 17, Fig. 14 · 16) dominated by the cyclone which was second from the west a day earlier. It has just become fully occluded and is in its most intense stage. At the same time, the youngest storm of Fig. 14 · 15 is developing as a strong secondary, and to the southwest of this, a new wave can be seen well-enough established to yield precipitation.

Figures 14 · 17 and 14 · 18 show a similar development of a cyclone family in the North Pacific Ocean. A progressive increase in age and development of the storms is evident from southwest to northeast. And

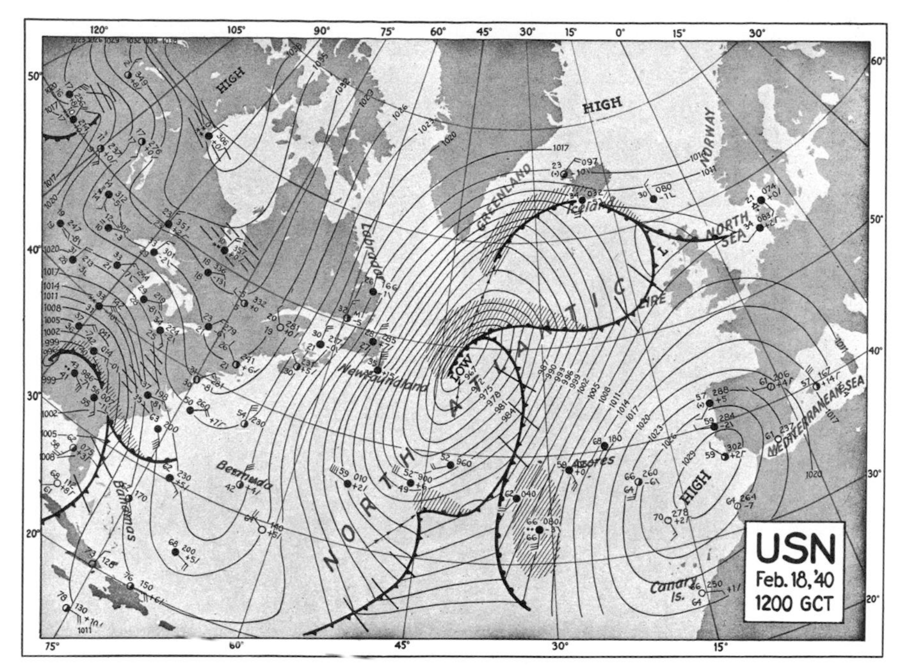

Fig. 14 · 15

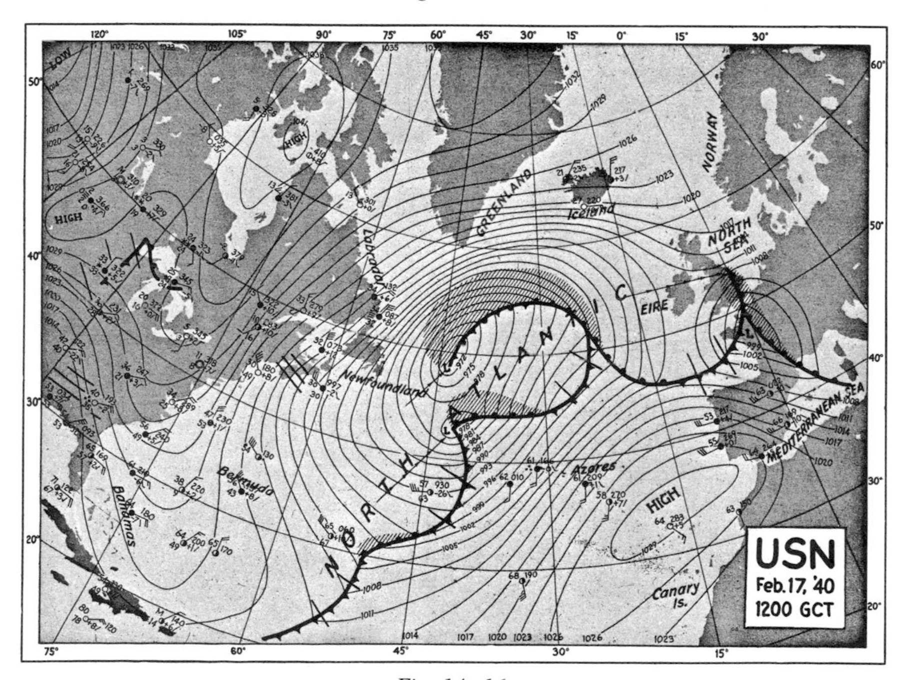

Fig. 14 · 16

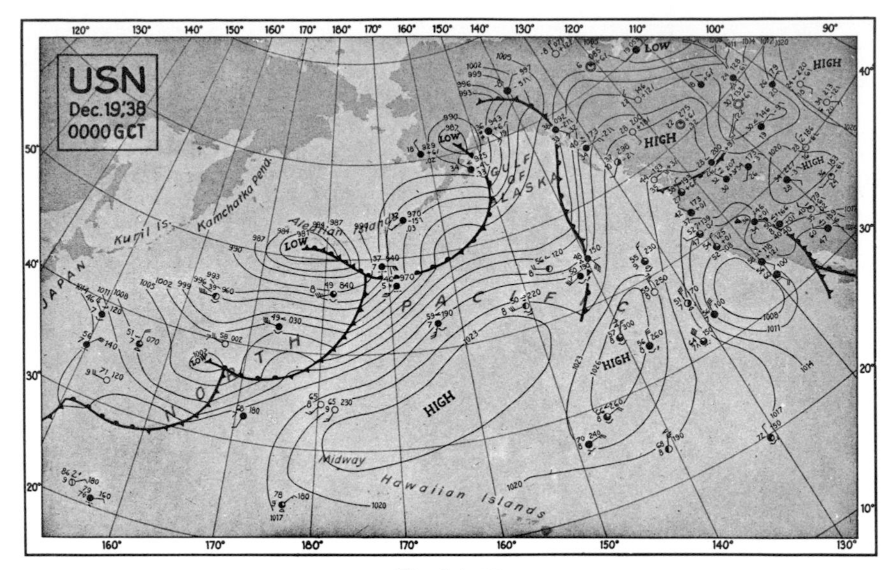

Fig. 14 • 17

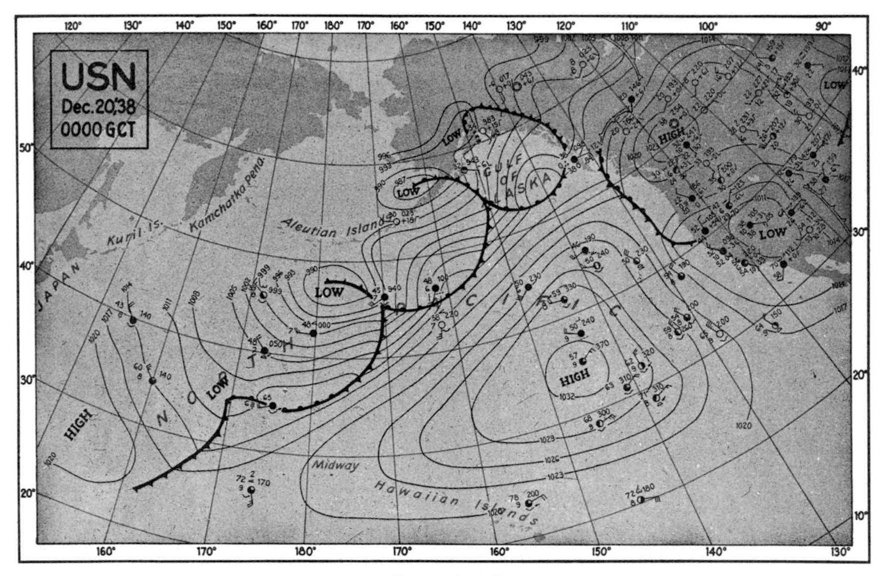

Fig. 14 • 18

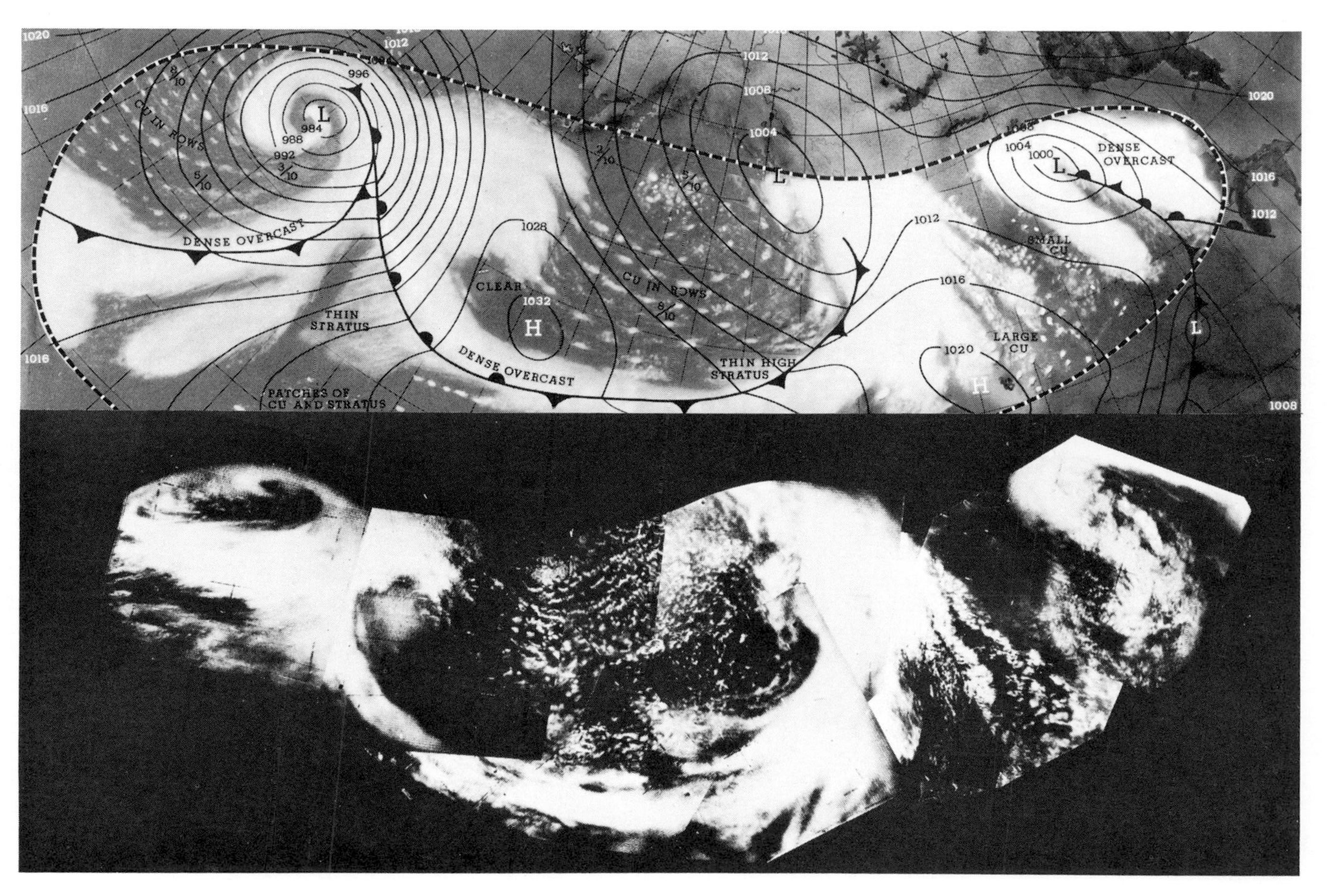

Fig. 14 · 19 Tiros *photograph of cyclone family in the North Pacific Ocean. The actual photograph in the lower portion of the figure is superimposed over the surface weather map in the upper portion.* (U.S. Weather Bureau)

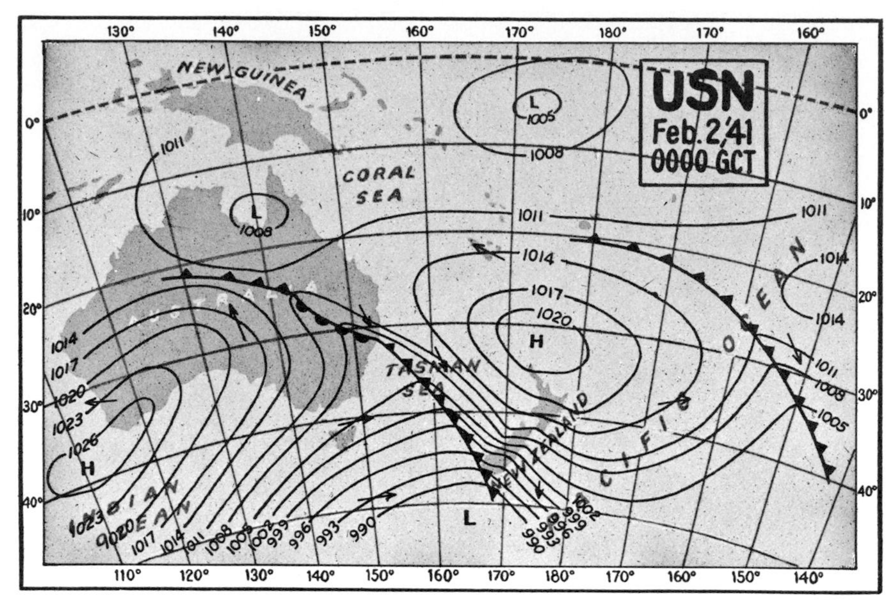

Fig. 14 · 20

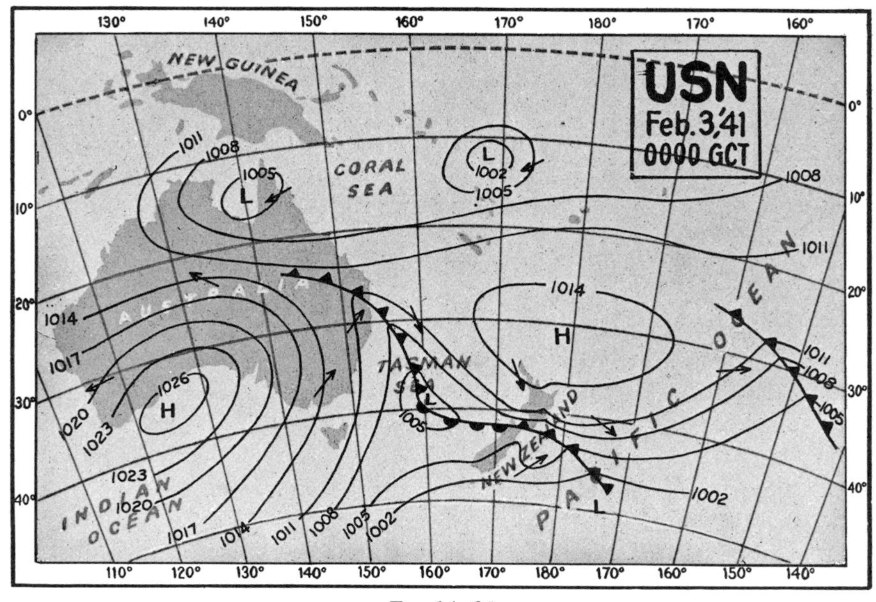

Fig. 14 · 21

again each storm can be seen maturing further during the 24-hour interval from Fig. 14 · 17 to Fig. 14 · 18.

Figures 14 · 20 to 14 · 22 show this pattern of cyclone development for the Southern Hemisphere in the vicinity of Australia and the southwestern Pacific Ocean. The development of two wave cyclones takes place in the two-day interval covered by these weather maps.

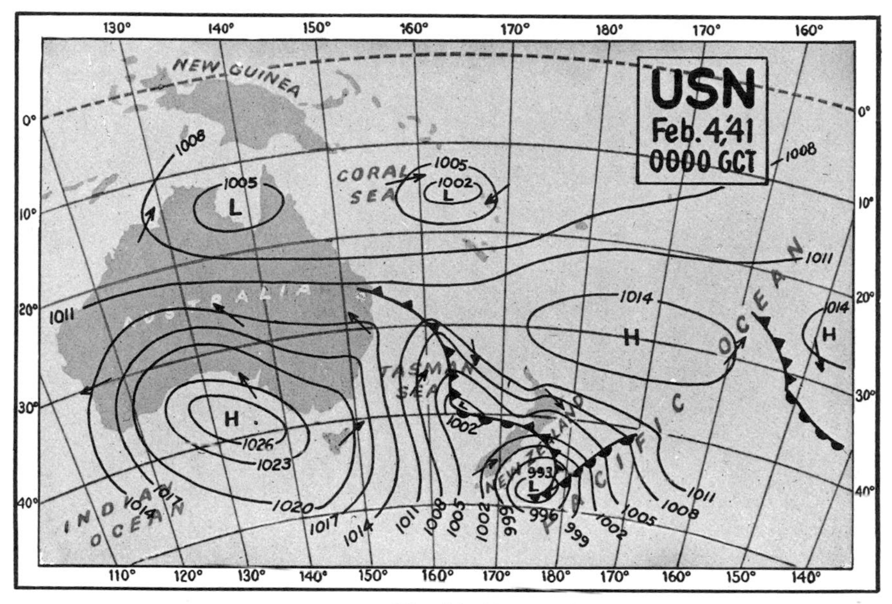

Fig. 14 · 22

In the lower part of Fig. 14 · 19 a *Tiros* photograph is shown which covers a large area of the North Pacific Ocean. The three organized cloud masses correspond with three storms in different stages of development. This pattern is superimposed over the surface weather map in the upper portion of the picture, which indicates the region of the photograph by the black and white line. Cloud types are also labeled on this upper illustration.

EXERCISES

14 · 1 Define the term "polar front," and describe its average annual range in latitude in the Northern Hemisphere.

14 · 2 How is the position of the polar front related to that of the jet stream?

14 · 3 Distinguish between stable and unstable waves on the polar front.

14 · 4 Briefly summarize the development of an extratropical cyclone as an unstable wave.

14 · 5 Describe the types and changes in weather expected as a mature cyclone passes with the center to the north of the observer in the Northern Hemisphere.

14 · 6 Do the same as in 14 · 5 when the cyclone center passes to the south of the observer.

14 · 7 Locate the warm and cold fronts in the Atlantic cyclone with its center south of Newfoundland in Fig. 14 · 6.

14 · 8 Project the expected path of the storm center based on the wind motion in the warm sector of the storm of Fig. 14 · 6 and the cyclone paths of Fig. 14 · 11.

14 · 9 With reference to Fig. 14 · 15, draw a line from 54°N, 8°E to 25°N, 55°W through the four cyclonic waves. Draw a vertical section along this line indicating frontal surfaces, air-mass types, and probable clouds and weather.

14 · 10 Draw a similar line through the four Pacific storms in Fig. 14 · 18 and show the air-mass, frontal, cloud, and weather structures in a vertical section along the line.

14 · 11 With reference to Figs. 14 · 12 and 14 · 13, describe the relationship between the surface high- and low-pressure systems and the circulation in the upper troposphere.

CHAPTER 15

Tropical Cyclones—Hurricanes

Tropical cyclones are the most violent storms experienced by the mariner. In West Indian waters these storms are known as *hurricanes;* in the East Indian and Japanese waters they are called *typhoons;* in the Indian Ocean they are called *cyclones;* off Australia, *willy-willies;* and off the Phillipines, *baguios.* Technically they are *tropical cyclones.* Owing to common American usage, we shall use the terms *hurricane* and *tropical cyclone* interchangeably.

Although tropical cyclones are not regular features of the low latitudes, storms resembling them but of lesser intensity are much less rare. By international agreement this group of related storms has been classified according to intensity as follows: *tropical depression*—winds up to 34 knots (39 miles per hour); *tropical storm*—winds of 35 to 63 knots (40 to 72 miles per hour); *hurricane* or *typhoon*—winds of 64 knots (73 miles per hour) or higher.

A somewhat different classification that includes four storm divisions is widely used in the United States, as follows: *tropical disturbance*—shows a slight surface circulation with one or no closed isobars; *tropical depression*—has one or more closed isobars with wind equal to or less than 27 knots (31 miles per hour); *tropical storm*—wind from 28 to 63 knots (32 to 72 miles per hour); *hurricane*—wind speed greater than 63 knots (72 miles per hour).

Pressure and Wind

Tropical cyclones are relatively small, intense, low-pressure areas having a more or less circular shape. These storms are characterized by the circulation of a single air mass—tropical marine in character without fronts in

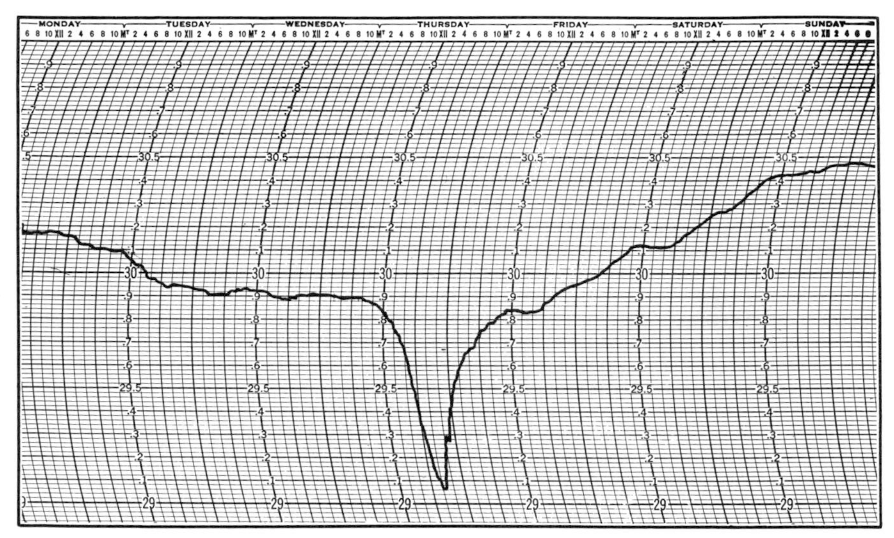

Fig. 15 • 1 Microbarograph trace showing passage of tropical cyclone at Kings Point, N.Y., September 14, 1944.

contrast to the two-fold air-mass structure of mid-latitude extratropical cyclones which has a typical polar-front wave. Hurricanes are also much smaller than extratropical cyclones. Their size is rarely much greater than 300 nautical miles in diameter, and the very intense part is often more restricted.

The pressure gradient within the small hurricane area is extremely steep. The fall of pressure from the periphery to the center of the storm commonly varies between 20 to 70 millibars. A record low pressure for Atlantic hurricanes was the 26.35 inches (892 millibars) reported for the storm of September 2, 1935 in the Florida Keys. The record low was reported as 26.185 inches (887 millibars) near Luzon on August 18, 1927. The gradient resulting from the severe drop is not radially uniform but increases in steepness toward the center.

The extremely high winds of the hurricane, which are directly related to this steep gradient, reach their maximum intensity near the center, owing to the increasing pressure gradient. This effect is shown schematically in Fig. 15 • 2. Note that as the wind spirals in toward the center it suffers increased deflection and finally travels in a circular path, never quite reaching the center. Thus, a central region about 10 miles in diameter

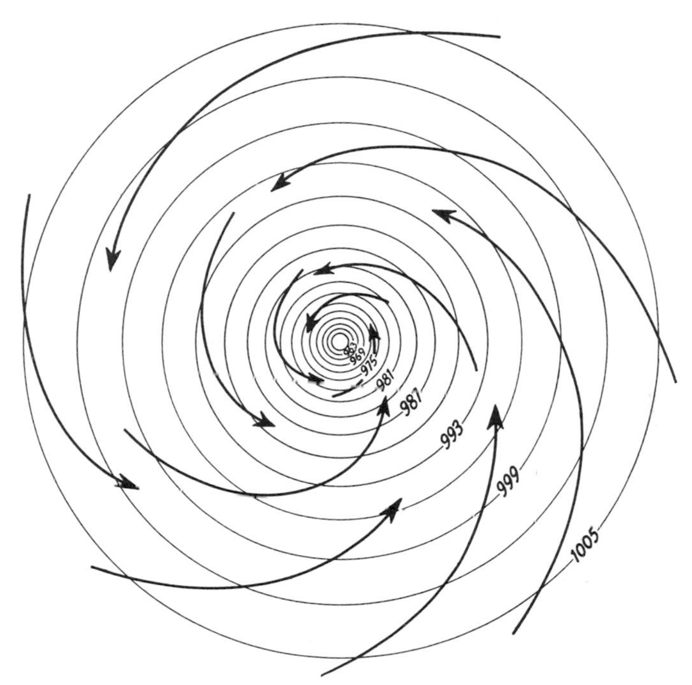

Fig. 15 • 2 The steepening pressure gradient and the associated wind system within a tropical cyclone (Northern Hemisphere).

remains relatively calm. On opposite sides of the calm center the winds blow in opposite directions. The abruptness with which the wind changes as the center crosses a station is shown very strikingly by the wind recording in Fig. 15 • 3. Central winds between 100 and 200 knots are not at all uncommon.

It is often said that the winds set in with greater violence as the center passes over a vessel or station. The truth is that as the center is approached, the winds build up to maximum violence more or less slowly, compared to the change from dead calm to maximum velocity as the center passes. At the same time, the destructive effect of a sudden high wind is greater than that of a wind building up slowly to the same velocity. Thus a line may snap under a sudden strain while it has withstood the same strain developed gradually. If the calm storm center passes across the vessel, preparations should always be made in anticipation of this sudden onslaught of the winds and their reversal. Many a vessel has keeled over and submerged following this sudden attack and reversal of the wind. The following

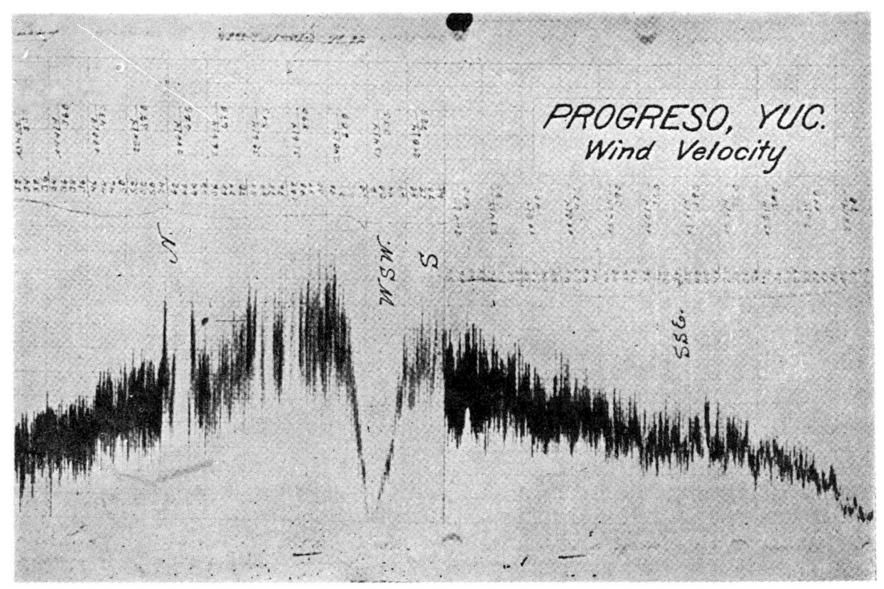

Fig. 15 · 3 Wind velocity recording showing intensity and gustiness of winds preceding and following the calm center of the Yucatan hurricane, August 26, 1938. (Monthly Weather Review, January, 1939)

quotation is from the report of the second officer on a ship passing through the hurricane of September 19, 1941, off western Mexico:

"The barometer fell so rapidly that it could be seen moving down—from 29.25 to 27.67. At 4:30 the ship passed through the center of the cyclone. The wind died down to almost 0 and the low clouds opened up so that high cirrus could be seen through a small opening. There was a peculiar yellow light and the sea became bright green in color. The extremely low atmospheric pressure caused discomfort in the ears. High confused swells broke aboard the ship with terrific force from all sides. In about 10 to 15 minutes the center passed and the wind came from the southwest, force 12 and over."

The Vertical Structure: Clouds and Precipitation

As the wind spirals into the low-pressure storm center, the air rises as well. This rising motion is most nearly vertical surrounding the storm center. Rising air always cools, and in the humid marine air this cooling results in condensation in the form of clouds. These clouds vary from thin cirrostratus in the outer portions of the storm through altostratus and nimbostratus to cumulonimbus, where the steep, almost vertically rising air is encountered around the storm center. The calm center, having no generally rising air and often descending air, will be relatively clear. Thus this calm center has a clear opening or shaft penetrating the surrounding heavy clouds and is called the *eye of the storm.* The surrounding dense, black, cumulonimbus mass which rings the calm center is often called the *bar* of the storm, since it appears as a dense cloud bank when viewed from a distance.

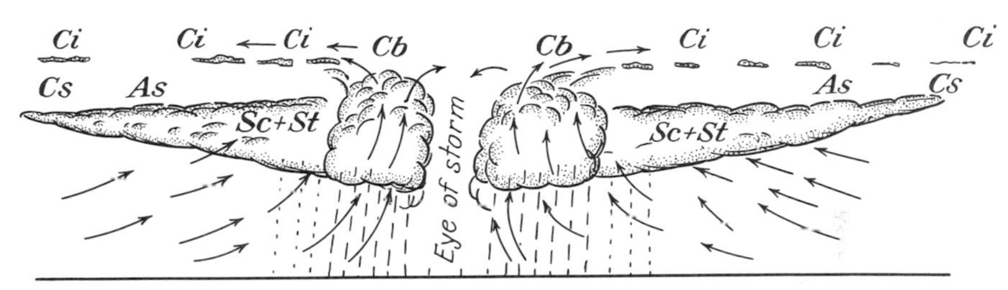

Fig. 15 · 4 Cross section of atmospheric conditions in a typical tropical cyclone.

The cross section in Fig. 15 · 4 indicates this air motion responsible for the cloud formations. It should be noted that the tops of the central cumulonimbus clouds produce cirrus at their peaks. These cirrus clouds blow outward from the storm center in all directions. Being masked by the storm clouds, the cirrus forms are not visible until the margin of the hurricane is reached. Rain falls in the inner central area of the storm, becomes very heavy in the ring surrounding the center, and is often attended by lightning and thunder in this area. The intensity of the rain in the region surrounding the eye is usually torrential in character. Measurements of 6 to 12 inches in 24 hours are not uncommon, and some extremes totaling several feet have also occurred. The effect is often like that of a sheet of water.

As regards the tremendous release of water, tropical cyclones play an important role in the general water budget of the atmosphere. Recall that the absolute moisture content of the atmosphere is highest in the summer as a consequence of the strong evaporation that occurs with high temperatures. Autumn hurricanes are then one of the important mechanisms by which much of the excess water is released during the season of declining air temperature, in addition to intense mid-latitude extratropical cyclones.

It must be realized in the above connection that a hurricane continuously entrains air as it travels its path. The total moisture content within a hurricane at any instant might amount to only about a couple of inches of water if it were entirely precipitated. Only by the continuous addition of heavily moisture-laden air can the storm release the heavy precipitation

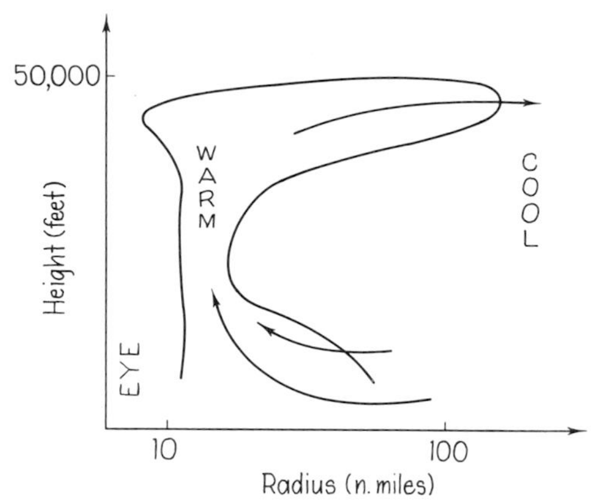

Fig. 15 • 5 Vertical radial cross section through a hurricane giving summary of chief features. (After H. Riehl in Science, Vol. 141, p. 1001, 1963)

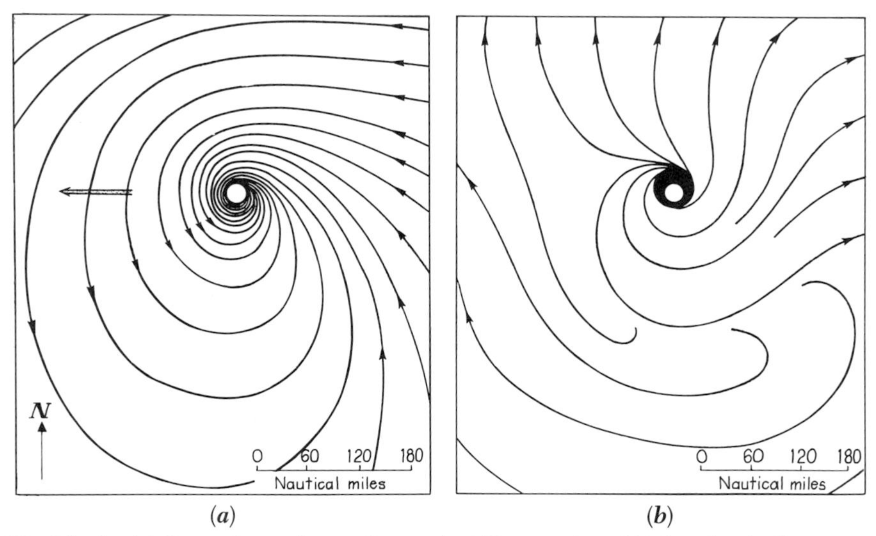

Fig. 15 • 6 (a) Streamlines of air inflow at the 300-meter (c. 1,000 feet) level of hurricane Donna southeast of Florida in September, 1960. (b) Streamlines showing outflow at a height of 13,700 meters (46,500 feet). (H. Riehl in Science, Vol. 141, p. 1001, 1963)

during its life. If the amount referred to earlier is summed over the entire region affected by the storm, the amount of precipitated water can be seen to be enormous. Examples of very heavy precipitation as recorded at particular stations are given in Table 15 · 1. If not given specifically in the last column, the recording interval was 24 hours.

Table 15 · 1 Some Heavy Rainfall Totals in Connection with Tropical Cyclones

AMOUNT (INCHES)	DATE	LOCATION
40.8	June 14, 1876	Cherrapunji, India—may not have been tropical storm. Averages 426 inches per year
21.4	June 15–16, 1886	Alexandria, Louisiana
26	Feb. 15–21, 1896	Pamplemousse, Mauritius
47	Feb. 15–21, 1896	Reunion, Mauritius
41	Feb. 15–21, 1896	La Marie, Tamarind Falls, and l'Etoile, Mauritius
63	Date unknown	Mt. Malloy, Queensland, Australia, in 3 days
96.5	Nov., 1909	Silver Hill, Jamaica, in 4 days
46	July, 1911	Baguio, Philippine Islands
88	July, 1911	Baguio, Philippine Islands, in 4 days
81.5	July 18–20, 1913	Funkike, Formosa, in 3 days
22.22	July 14–15, 1916	Altapass, North Carolina
23.11	Sept. 9–10, 1921	Taylor, Texas
29.60	Sept. 13–14, 1928	Adjuntas, Puerto Rico
19.76	Aug. 6–10, 1940	Crowley, Louisiana, 33.71 for 5-day period
31.66	Aug. 6–10, 1940	Abbeville, Louisiana, for 5-day period
29.65	Aug. 6–10, 1940	Lafayette, Louisiana, for 5-day period

A summary of the temperature relations, wind, and cloud patterns of a hurricane is shown in the radial vertical cross section in Fig. 15 · 5, which also gives average dimensional relations. The pattern of lower inflow and upper outflow of air in a hurricane has been shown very well from a study of hurricane Donna off Florida in September, 1960. Streamlines of air motion of the lower-level, counterclockwise, spiral inflow are shown in Fig. 15 · 6*a*, whereas the broad pattern of outflow streamlines aloft is essentially reversed in direction. The outflow generally occurs between 9,000 to 15,000 meters (30,000 to 50,000 feet). The broad cirrus cloud cover, which is usually masked except near the storm margins, forms in this outflow.

An interesting radar photograph of the above storm (Donna) taken from Key West, Florida, at roughly the time of the observations given in Fig. 15 · 6, is shown in Fig. 15 · 7. The low-level, counterclockwise, spiral inflow culminating in circular motion about the clear eye is pictured very effectively by the cloud organization in the photograph. The thinner clouds of the outflow pattern do not show here.

The eye of hurricane Esther (September 20, 1961) is shown in Fig. 15 · 8, which was taken from a U.S. Weather Bureau research plane at 20,000 feet. In this case the eye is broken by cloudy zones. The circular wind motion is shown by the peripheral bands; the cumuliform or convective central cloud mass is a characteristic "hub" cloud in many such storms.

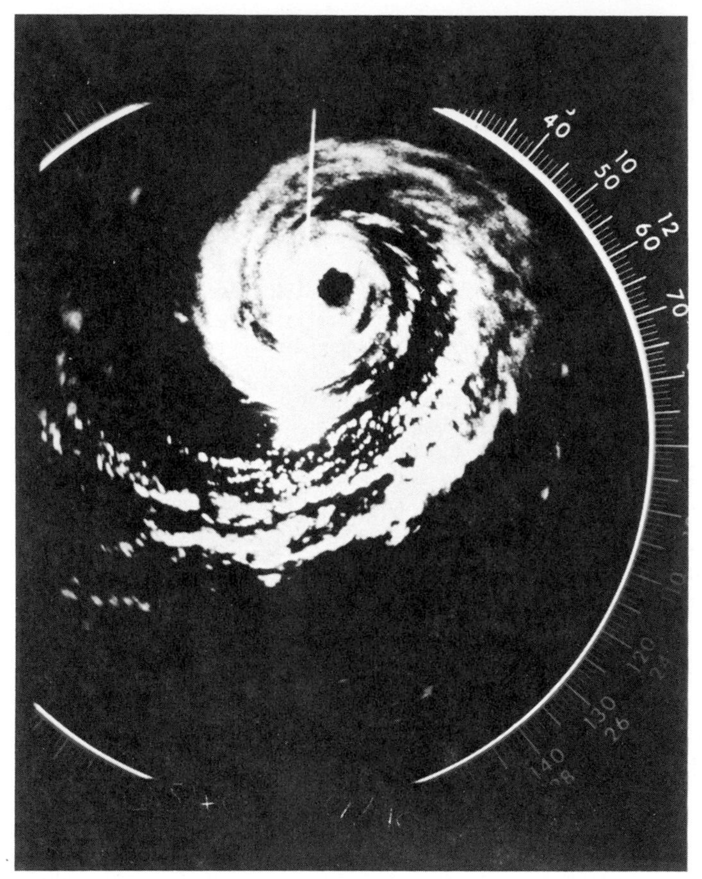

Fig. 15 · 7 Hurricane Donna at 1340 GMT (0840 EST), September 10, 1960 as photographed from the PPI scope of the WSR-57 Radar in Key West, Florida. (U.S. Weather Bureau)

Hurricane Sea and Swell

Since ocean waves are raised by the wind, it follows that waves generated by the violent winds of hurricanes should rank with the highest. The term *sea* refers to the ocean surface condition as it exists within the storm area where growing wind waves predominate. *Swell* refers to the waves that have outrun or left the storm. The height of wind waves depends on some combination of wind speed with the duration of a wind of steady direction or the *fetch* (distance that the wind blows in a straight line over the water).

Because a hurricane is small and because its winds follow strongly curved paths, the fetch is rarely very great. Also, the duration or time a given wind blows in the same direction is also small for most of the storm, except for that part of the storm in which the wind direction and direction

Fig. 15 · 8 Eye of hurricane Esther from 20,000 feet, taken by a U.S. Weather Bureau research plane on September 16, 1961.

of storm travel are about the same. The region of maximum wave generation can be determined from a study of either Fig. 15 · 2 or Fig. 15 · 6*a*, as follows. Imagine that the line of motion of a hurricane in the *Northern Hemisphere* divides the storm into two semicircles, which, as you look along this line of motion, are defined as *right* and *left* semicircles, respectively. Select an arbitrary line of motion in Fig. 15 · 2, or the direction given in Fig. 15 · 6*a*. Then the wind in the central and rear right quadrant of the right semicircle is parallel or closely parallel to the direction of travel. The waves growing in this region experience the longest duration and fetch of wind, thus developing the greatest heights and lengths within the storm.

In the Atlantic, they usually reach reported heights of 35 to 40 feet, although waves in excess of 45 feet have been reported for Pacific tropical cyclones. The waves which develop in the left rear quadrant travel opposite to the storm direction, experiencing the shortest duration and fetch of wind and therefore are of the smallest height in the storm. Also, as will be explained later, the wind speed in the right semicircle (in the Northern Hemisphere) exceeds that in the left side, contributing further to the greater wave height in this part of the hurricane.

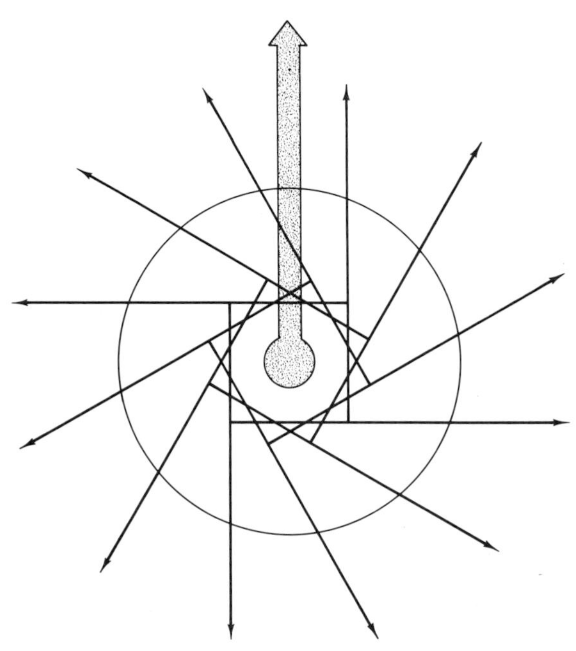

Fig. 15 · 9 Radiation of swell from a tropical cyclone. The open arrow shows the direction of storm travel.

At times waves generated in different parts of the storm may cross. In these cases the interference waves reach greater heights than either of the component interfering waves. Although heights up to 55 feet have been reported in the Atlantic, and greater than 60 in the Pacific, such waves are not common and develop only sporadically in a particularly violent storm when interference conditions are appropriate.

As the hurricane moves along its path, the calm center or eye travels across waves generated in the adjacent violent regions. Consequently a ship in the eye may encounter calms or light winds, but the seas will be very disturbed and confused as a result of the waves entering from various directions.

A further inspection of the diagrams illustrating the wind path in a tropical cyclone shows that the waves developing within the storm area must travel almost radially out from the center and therefore away from the storm. This is true even for those maximum-height waves generated in the right side which move parallel to the storm direction. Because waves move with a speed that is a large fraction of the wind speed, they travel much faster than the storm itself, and even if developed in the right side, they will soon travel through and then outrun the storm. While under the influence of the wind in the storm the waves are irregular and often confused-looking. But after leaving or outrunning the storm, the waves become more and more regular and long-crested as they slowly diminish in height, and are then known as *swell.* The arrows in Fig. 15 • 9 indicate the general way in which swells radiate after development in the different portions of the hurricane. Note that the maximum swell (generated in the right side) tends to move ahead of, but in the direction of, the storm and often is an early forerunner of a hurricane directly approaching an observer.

A comparison between Figs. 15 • 9 and 15 • 2 showing wind motion indicates that as waves move through the storm area, they must encounter winds that travel to the left of the wave direction, for the swell from upwind always deviates to the right of the existing wind. The amount of this deviation depends upon the speed and path of the storm.

The Storm Surge

By far the greatest damage and loss of life and property from hurricanes are due not to the wind or torrential rains but to the catastrophic flooding that usually accompanies a hurricane whose path crosses a coastal region, whether continent or island. This rise in sea level is known as the meteorological or storm surge.

In part, the surge results from the impounding of water against the coast by the wind. Then, waves tens of feet high, superposed on this raised sea level, inflict additional alternating onslaughts of water which create tremendous havoc. In addition to the slow buildup of the wind-driven seas and the impulsive effects of high wind waves, many hurricanes exhibit a further, rather abrupt surge of water that strikes the coast about the same time as the center does. This surge is thought to be an effect of resonance produced in the following manner. The low-pressure center of the hurricane causes the ocean to rise like an inverted barometer, the uplift depending on the pressure decrease. Under appropriate conditions of water depth, the central "mound" of water becomes amplified as it travels and strikes the coast as a wave of as much as 12 feet higher than normal sea level. If all of the above effects are superimposed on the normal diurnal high tide, particularly a spring tide, the results are all the more disastrous.

Paths and Regions of Occurrence of Tropical Cyclones

Although tropical cyclones often travel well into the middle latitudes, especially in the North Atlantic Ocean, they are essentially tropical and semitropical storms. Fig. 15 · 10 shows the regions and average paths of motion of most of the tropical cyclones of the world. Fully developed tropical cyclones have never been reported from the South Atlantic nor from the South Pacific east of about 140° west longitude. The path and motion of an individual storm may depart considerably from the mean paths shown in the diagram. An example of individual paths over a fairly

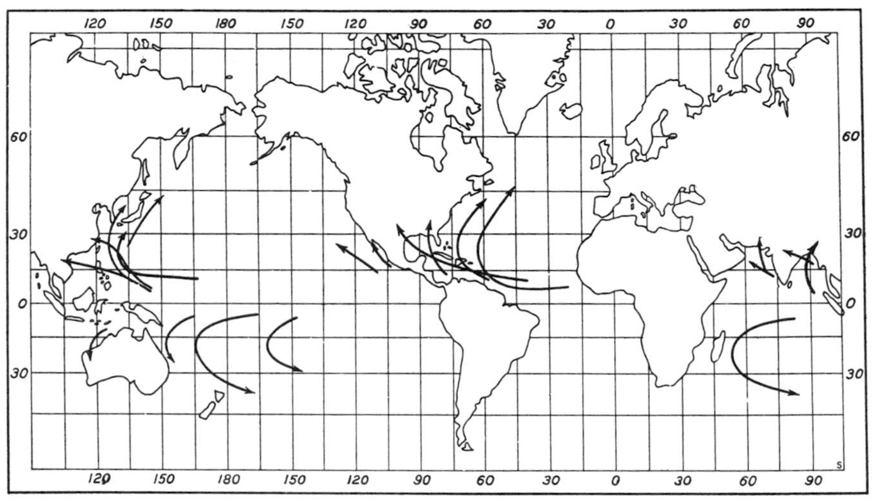

Fig. 15 · 10 The principal hurricane regions of the world and their average paths of motion.

long interval is shown for the North Atlantic Ocean in Fig. 15 · 11, which illustrates well the diversity of individual storms.

Despite this diversity, some important generalities can be given. As is evident in Fig. 15 · 10, hurricanes begin near, but never on, the equator. The region of origin is usually from 6 to 10° away from the equator. At first they move very slowly away from the equator but with the easterly tropical flow of air, developing a curving path west-northwest or west-southwest, in the Northern and Southern Hemispheres, respectively. Normally, this path recurves to the northeast or southeast when the storms encounter the westerly flow after crossing the horse latitudes, the total path thus assuming the parabolic shape shown in Fig. 15 · 10.

Prior to recurvature, hurricanes have a relatively low forward speed—about 10 to 12 knots on the average. This motion accelerates upon recurva-

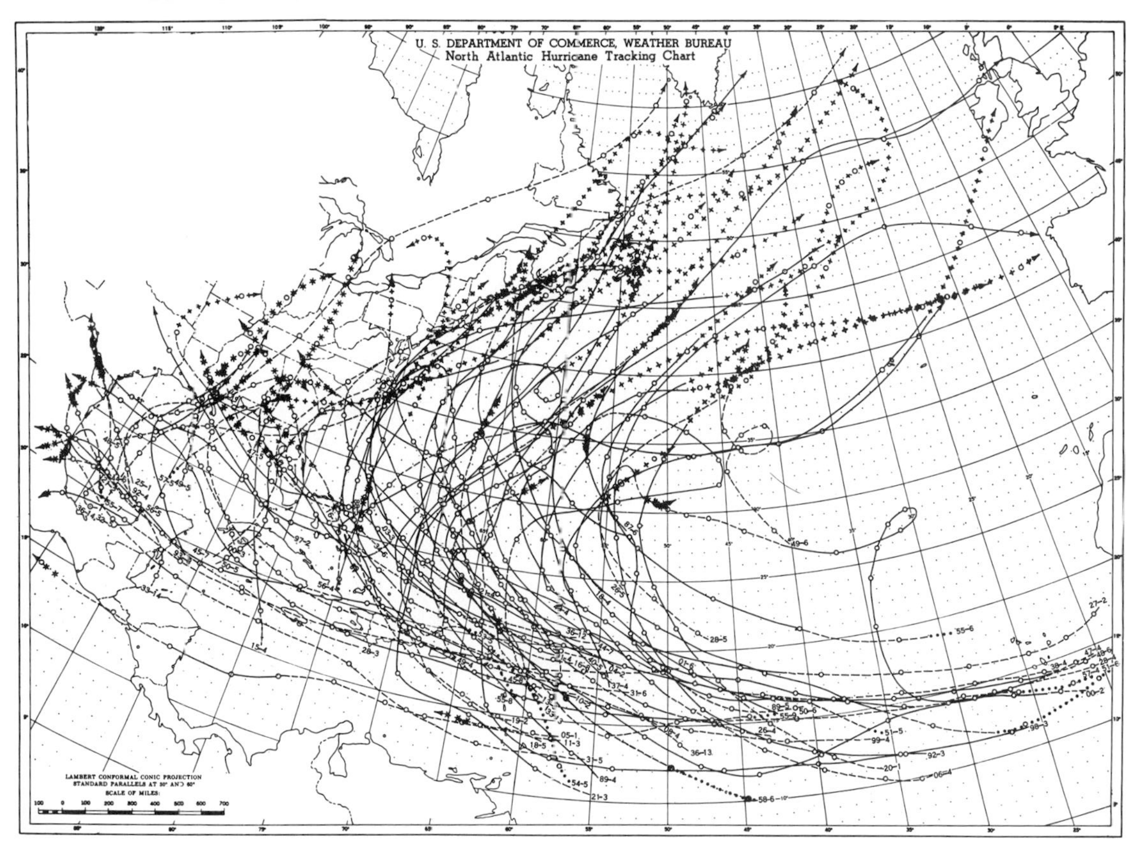

Fig. 15 • 11 Tracks of North Atlantic tropical cyclones and tropical storms between September 1 and 10 from 1886 to 1958. Dotted portion represents tropical depressions or the development stage; dashed portion, the tropical storm stage; solid line, the hurricane stage; crosses, extratropical cyclone stage; and stars, the dissipation stage. (U.S. Weather Bureau)

ture to an average of 20 to 30 knots although individual storms have reached 60 knots, such as those along the Atlantic Coast in 1938 and 1944. They ultimately peter out or merge with extratropical wave cyclones and lose their identity.

Although hurricanes tend to follow the tracks influenced by the trades

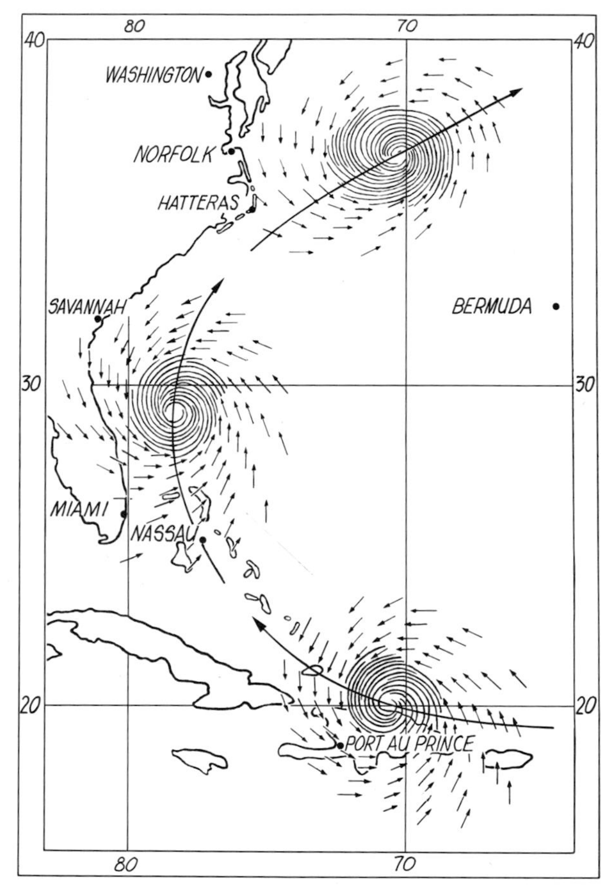

Fig. 15 · 12 Typical track and wind system of the West Indian hurricane.

and then the westerlies, the paths of individual storms are also specifically controlled by the circulation aloft, which often involves the presence of strong low and high surface pressure cells whose effects reach high elevations. A strong high-pressure region, particularly north of the horse latitudes, may block the normal recurvature of a hurricane and deflect the storm to the westward, following the circulation about the anticyclone. Many West Indian hurricanes are thus directed northward along the

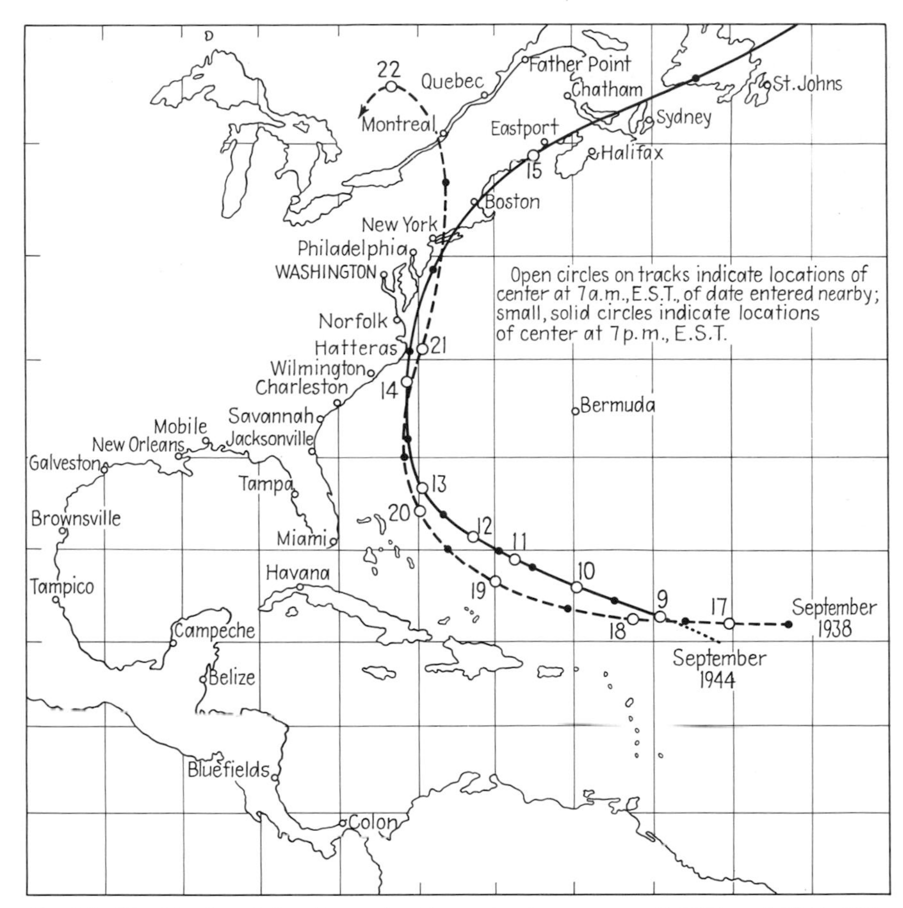

Fig. 15 • 13 Tracks of the destructive September hurricanes of 1938 and 1944. (U.S. Weather Bureau)

eastern coast of the United States, around the western flank of a well-developed Bermuda high instead of recurving out to sea. Figure 15 • 13 gives the tracks of two such storms that wrought great havoc along the coast of the United States.

Frequency of Occurrence

In regions such as the West Indies or the far western Pacific, few if any hurricanes go undetected, and the distinction between storms of true hurricane or tropical cyclone intensity and tropical storms of somewhat lesser intensity is usually well made. But in more remote parts of the oceans, particularly where little-traversed by merchant ships, this information is less accurate. This is especially true for the eastern North and South Pacific Ocean. As reconnaissance improves, the number of tropical cyclones and tropical storms usually increases for regions of known occurrence. For example, from 1947 to 1953 an average of five cyclones per year were reported to reach tropical storm intensity. This number doubled between 1954 and 1961, when aerial and shipping reconnaissance increased in the region. A far greater increase occurred when the *Tiros* weather satellite

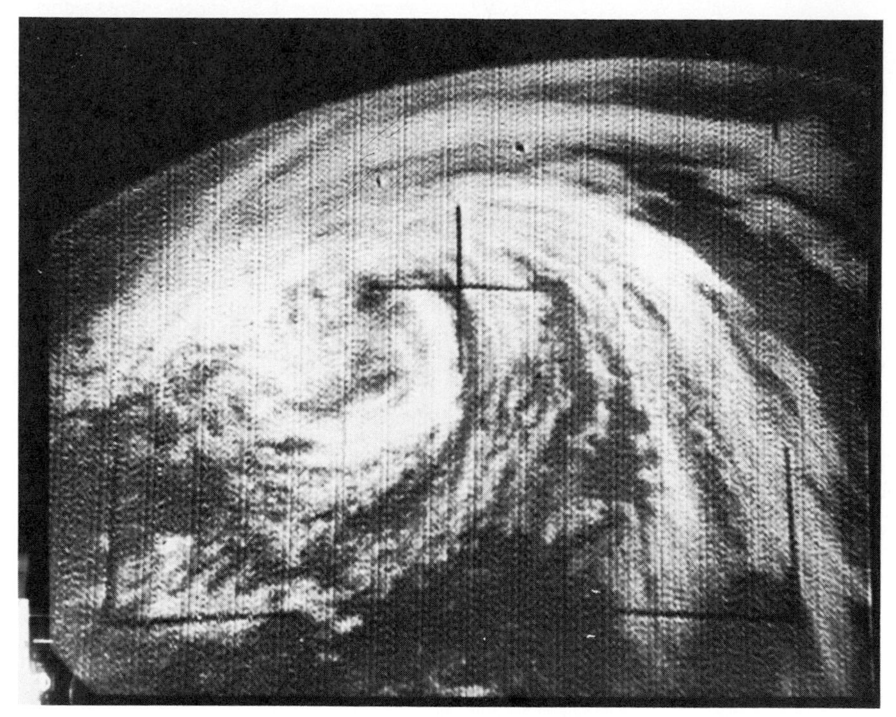

Fig. 15 · 14 Tropical cyclone Liza photographed by Tiros III *about 600 miles SSW of Los Angeles in coordinates 25° N, 120° W.* (U.S. Weather Bureau)

data became available. From August to September, 1962, only five tropical storms were reported by conventional means, out of the 22 detected by means of *Tiros* observations. The most recent estimate for the eastern North Pacific based on *Tiros* data is 30 tropical storms per year. Unfortunately, *Tiros* observations cannot now distinguish between tropical storm and hurricane intensities. Fig. 15 · 14 illustrates a tropical cyclone observed by *Tiros* III in the region under discussion.

Table 15 · 2 gives the average frequency per month of tropical cyclones

*Table 15 · 2 Monthly Frequency of Occurrence of Tropical Cyclones of Hurricane or Near-hurricane Intensities**

REGION	YEARS OF DATA	J	F	M	A	M	J	J	A	S	O	N	D
North Atlantic Ocean	68					0.1	0.4	0.5	1.5	2.6	1.9	0.5	
Eastern North Pacific* (off west coast of central America)	30					0.1	0.8	0.7	1.0	1.9	1.0	0.1	
North Pacific Ocean (west of 170° E)	36	0.4	0.2	0.3	0.4	0.7	1.0	3.2	4.2	4.6	3.2	1.7	1.2
North Indian Ocean (Bay of Bengal)	36	0.1		0.2	0.2	0.5	0.6	0.8	0.6	0.7	0.9	1.0	0.4
North Indian Ocean (Arabian Sea)	23	0.1			0.1	0.2	0.3	0.1		0.1	0.2	0.3	0.1
South Indian Ocean (west of 90° E)	70	1.3	1.7	1.2	0.6	0.2							0.1
South Indian Ocean (off N.W. Australia)		0.3	0.2	0.2	0.1								0.1
South Pacific Ocean (east of 160° E)	105	0.7	0.4	0.6	0.2							0.1	0.3

* To be regarded as absolute minimum values as they do not include data from *Tiros* observations.

that reached hurricane or near-hurricane intensity in the regions illustrated in Fig. 15 · 10. Note that the data for the eastern North Pacific taken from recently published data of Dunn and Miller of the National Hurricane Center in Miami can no longer be regarded as complete or accurate, but indicate absolute minimum values.

The western North Pacific Ocean is certainly the region with the highest frequency of hurricanes, followed by the western North Atlantic and the Bay of Bengal. It is not at all uncommon for two storms to be present simultaneously, and occasionally three hurricanes occur in the same region. During the month of September in both 1950 and 1951, three simultaneous hurricanes were present in the western North Atlantic, including the Caribbean–Gulf of Mexico area.

Hurricane Origin and Maintenance

A hurricane can be considered a rather simple heat engine driven by the difference in temperature between the center and the margins. The central column must be warmer than the surrounding area at each level in order to maintain the strong convection upon which the existence of the storm depends.

The energy necessary to continue the violent motions characterizing the storm comes from the continuous supply of latent heat of condensation brought in by moisture-laden winds. In this thermally driven heat engine tremendous amounts of energy are converted from heat to the mechanical energy of motion in a rather short time, as is apparent in the fury of the winds. According to a published estimate, the power equivalent of the energy liberated within a hurricane in one day is about ten thousand times the daily power consumption in the entire United States. A different form of the calculation indicates that the daily energy release within a hurricane is equal to about one hundred thousand bombs of megaton strength.

When hurricanes enter continents, they usually weaken very quickly and soon dissipate unless they recurve out to sea, as often happens with storms crossing the southeastern coast of the United States. The weakening over land emphasizes the importance of ocean-derived moisture in the maintenance of hurricane energy.

Exactly how this concentration of energy occurs is not yet fully understood, but certain basic relations have been established from the great bulk of observational material now available.

Hurricanes all form in the doldrums, or the region of the intertropical convergence (frontal) zone, described in Chap. 13. They can only develop in this region when at some distance from the equator, as noted earlier, because the curving wind motion requires an adequate Coriolis force to be initiated; this force is too small at and very close to the equator to initiate cyclonic circulation.

It is also known that hurricanes do not originate over water colder than about 27°C (80°F). This relatively high surface temperature seems to be required for the heating of the lower air to form the high lapse rate necessary to begin and maintain the convection of a tropical cyclone.

Hurricanes also appear to originate in *easterly waves,* which are troughs of low pressure embedded in and moving with the trade-wind air stream. Such waves are also recognized from the discontinuity in normal trade-wind motion. Ahead of or to the west of the wave, the air subsides from higher levels, becoming dry and clear, and spreads westward. Behind the wave, the air shows a pattern of convergence and uplift producing cumulus and cumulonimbus clouds with or without showers (see Fig. 15 · 15). The inclined surface which can be imagined through the center of the wave is a surface of pressure and wind discontinuity within the same air mass. It is not a frontal surface in the sense of separating dissimilar air masses. Most easterly waves are stable, but occasionally, for uncertain reasons, an easterly wave becomes unstable and develops a hurricane vortex.

Despite the frequent occurrence of easterly waves and other conditions favorable to the origin of hurricanes, they remain relatively rare features of the tropics. The exact triggering mechanism is yet to be ascertained.

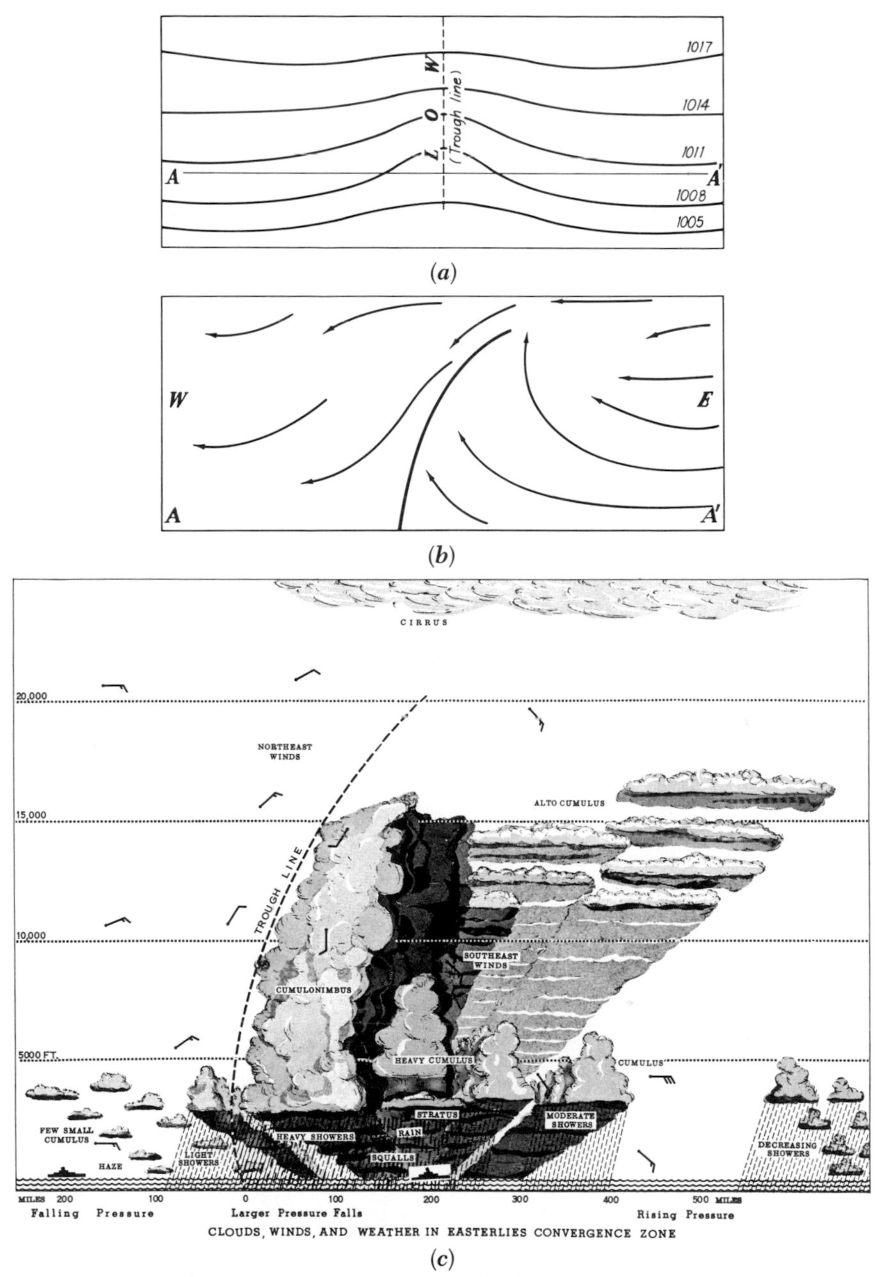

(*c*)

Fig. 15 · 15 Surface (a) and vertical section (b) of an easterly wave in the tropics. The wave moves from east to west. (c) Typical distribution of clouds, wind, and precipitation in an easterly wave in vertical cross section. Horizontal and vertical scales are given. (U.S. Weather Bureau)

Dangerous and Navigable Semicircles

Although all parts of a hurricane are dangerous to mariners, the danger is greater in the right than in the left semicircle in the Northern Hemisphere. In general, the wind is stronger in the right semicircle because the observed wind in this region is the sum of the cyclonic wind and the planetary wind stream parallel to the track. As drawn in Fig. 15 · 16, the wind in some part of the right semicircle in the Northern Hemisphere and the left semicircle in the Southern Hemisphere is parallel to the direction of the main stream in which the storm is embedded. And nearly all of

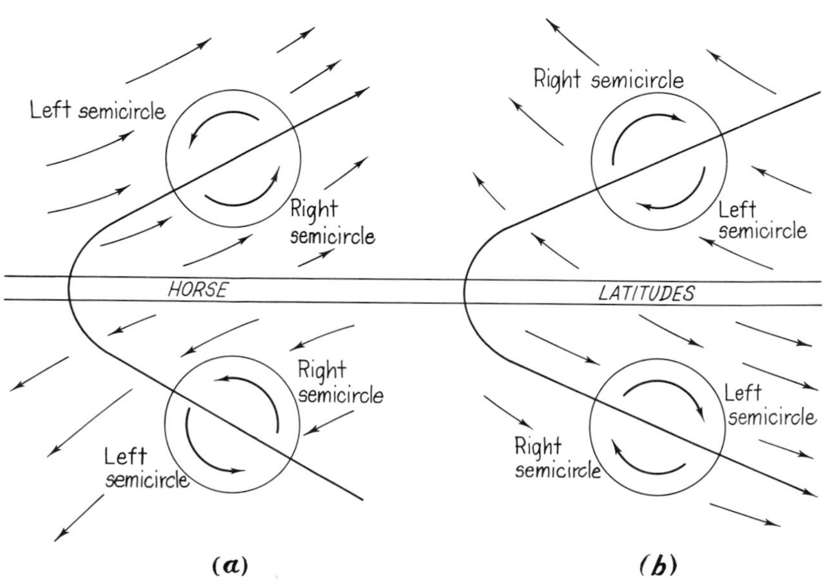

Fig. 15 · 16 Explanation of the dangerous and navigable semicircles in the Northern (a) and Southern Hemispheres (b).

the winds in these semicircles are parallel to at least some component of the main stream. In general the wind in the right and left semicircles in the Northern and Southern Hemispheres, respectively, is increased by the appropriate component of the main air stream. In the opposite semicircles, the net wind is less than the cyclonic wind from the subtraction of the main air stream which always has a greater or lesser component opposite to the cyclonic wind. The wind-speed difference between the two semicircles of a particular storm may be quite large.

Owing to both the higher wind speed and the greater duration and fetch of the wind in the right (left) semicircle, as noted earlier, this half of the storm is also of greater hazard to the mariner.

It is also evident from Fig. 15 • 16 that a vessel in the right (left) semicircle may tend to drift directly around into the storm path, owing to the wind circulation. This is especially true for a vessel whose ability to make way has been diminished.

A final reason for the designation as dangerous applies to vessels in the right (left) semicircle prior to, but in the latitude of, recurvature. The imminent danger exists that recurvature may occur, placing the vessel directly in the line of the storm center and in the path of maximum violence.

Some general rules worth repeating here have been issued by the Oceanographic Office of the U.S. Navy Department. These are for vessels maneuvering within the storm and are designed for the purpose of avoiding the center and also leaving the storm area. The following rules, which can be developed from the information already given in this chapter, apply to the Northern Hemisphere:

1. *Right or dangerous semicircle.* Steamers: Bring the wind on the starboard bow, make as much way as possible, and if obliged to heave to, do so head to sea. Sailing vessels: Keep close-hauled on the starboard tack, make as much way as possible, and if obliged to heave to, do so on the starboard tack.
2. *Left or navigable semicircle.* Steam and sailing vessels: Bring the wind on the starboard quarter, note the course, and hold it. If obliged to heave to, steamers may do so stern to sea; sailing vessels on the port tack.
3. *On the storm track in front of center.* Steam and sailing vessels: Bring the wind two points on the starboard quarter, note the course, and hold it, and run for the left semicircle, and when in that semicircle maneuver as above.
4. *On the storm track in rear of center.* Avoid the center by the best practicable route, having due regard for the tendency of cyclones to recurve to the northward and eastward.

Locating the Storm

TECHNICAL INVESTIGATIONS. Modern methods of technical exploration have greatly minimized the danger to the mariner of unexpected encounters with hurricanes. Once discovered and tracked, the storm information is relayed rapidly to all ships and island and coastal stations in the vicinity of a tropical cyclone.

Weather satellites such as the *Tiros* series have provided probably the greatest single advance in the early detection of tropical storms—often before they reach hurricane intensity. The first such storm to be detected by these means was a typhoon discovered by *Tiros* I in the South Pacific

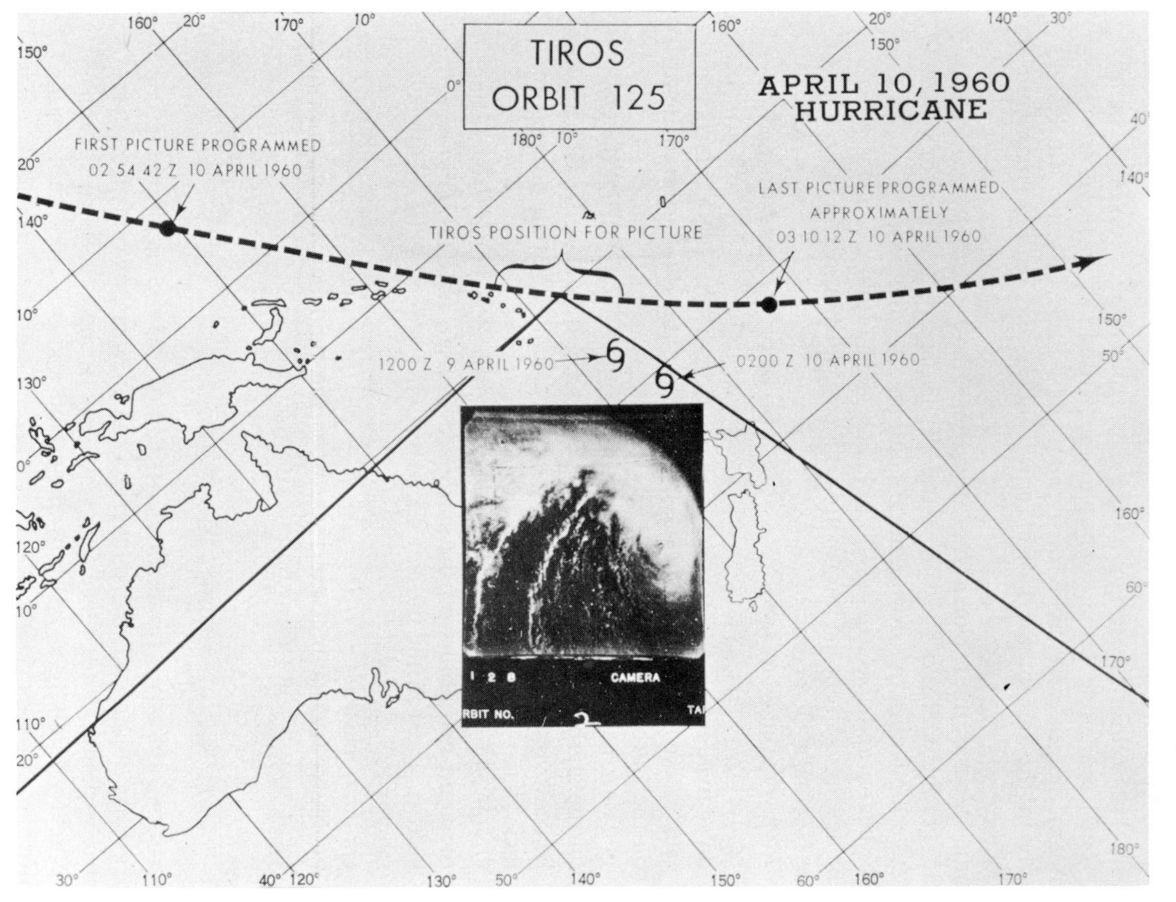

Fig. 15 · 17 The first picture of a hurricane taken by a satellite. The photograph shows a typhoon discovered by Tiros I *on April 10, 1960 during orbit number 125, near Australia. The clockwise spiral structure and the cloudless eye (right center) show clearly.* (U.S. Weather Bureau)

off Australia. A *Tiros* photograph showing the typical Southern Hemisphere *clockwise* cloud spiral is illustrated in Fig. 15 · 17, which also shows a map of the storm path and the satellite orbit over the region. Discoveries such as this certainly lessen one of the great perils of the seas.

Routine air reconnaissance of suspected storms has provided another valuable means of early detection and tracking of tropical cyclones and lesser but still intense tropical storms. Air investigation is of course limited to the regions where such flights are logistically possible and are therefore of greater value in the tropical Atlantic-West Indies region than in the much broader, less accessible, tropical Pacific.

Once detected and within reach of modern long-range radar instru-

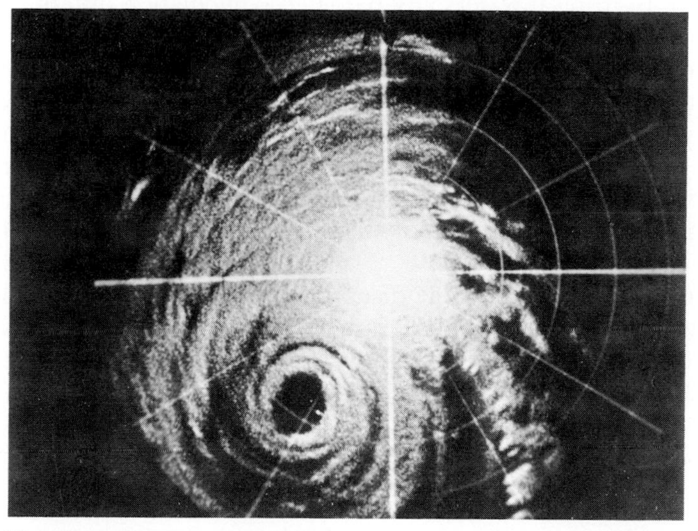

Fig. 15 · 18 Cape Hatteras radar photo of hurricane Helene, September 27, 1958, 1805 EST. Station is at the center of the grid; range markers are at 20 nautical mile intervals. The diameter of the clear eye (southeast quadrant) is about 25 nautical miles. (U.S. Weather Bureau)

ments, a continuous surveillance is maintained, which gives very good determinations of storm paths and changing internal structure. This is of particular value to both ship and air navigation. An example of the details provided by this means is shown in Fig. 15 · 18, which shows a photograph of hurricane Helene taken from the Cape Hatteras radarscope on September 27, 1958. The radar station is at the center of the grid, accounting for the anomalously bright circular region in the center which is not related to any storm condition. The characteristic spiral, banded, cloud structure culminating in a ringlike cloud around the clear center is very well shown. The small but somewhat brighter or more "solid" radar echoes in the northeastern (upper right) quadrant represent characteristic thunderstorm images. Tornadoes are known to develop within such thunderstorms in the forward regions of a hurricane.

DIRECT LOCAL OBSERVATIONS. Despite the availability of technical information, it is nevertheless necessary for a navigator to understand and relate direct observations to the location and tracking of a hurricane. Basically, this consists of integrating the pertinent information described so far in an attempt to avoid the storm when more positive technical communications are either unavailable or not sufficiently accurate or timely.

Pressure Characteristics. With the continued approach of the storm to within a few hundred miles, unusual barometric conditions are frequently observed. In tropical waters any barometric fluctuation differing from the normal diurnal variation with peaks at 10 A.M. and 10 P.M. and troughs at 4 P.M. and 4 A.M., especially if coincident with the above indications, serves as a further herald of the hurricane. As the edge of the storm area itself is actually encountered, the barometer may rise abruptly and then fall, or simply be erratic, or it may fall much faster than normal. If a hurricane is definitely approaching, the barometer will, after a possible initial unsteadiness, fall more and more rapidly as the eversteepening pressure gradient moves in.

An approximate idea of the distance to the storm center can be gained by noting this increasing hourly rate of fall. Table 15 · 3 furnishes a rough guide to this distance by showing the average relationship between rate of fall and distance to center.

Table 15 · 3 Relation of Pressure Change to Tropical Cyclone Distance

HOURLY RATE OF FALL, IN.	DISTANCE TO CENTER, MI.
0.02 0.06	250–150
0.06–0.08	150–100
0.08–0.12	100–80
0.12–0.16	80–50

Wind Characteristics. With the approach of a hurricane, the wind usually shifts from the prevailing direction and increases in velocity. This change is a noticeable one in view of the normal steadiness of the trades that prevail in the tropics and subtropics. The wind increases in squalls or gusts, becoming more and more violent closer to the center.

In the absence of reports from other ships or stations, the observer must rely upon his own observations to locate the center. From the law of Buys-Ballot, if one faces the wind, the low-pressure or storm center is 8 to 10 points on the right. One observation will yield only a line of position, somewhere along which the storm center lies. The force of the wind and the rate of fall of the pressure indicate the approximate distance of the storm center *provided it is approaching directly.*

Under hurricane conditions the true *wind* direction can often best be determined by the motion of the *lower* clouds. Recall that the upper wind blows more freely and consequently is parallel to the isobars. Thus by facing into the low-cloud movement one may judge the storm center to be directly (8 points or 90°) on the right in the Northern Hemisphere and to have a similar bearing on the left in the Southern Hemisphere.

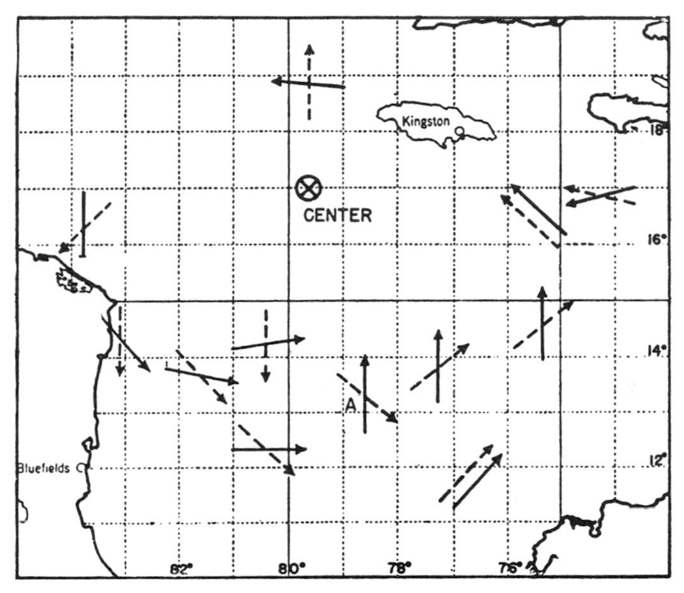

Fig. 15 · 19 Direction of wind (solid arrows) and direction of swell (dashed arrows) in hurricane whose center is shown in the diagram.

The more numerous the reports, the more accurately can the storm center be located. Figure 15 · 19 indicates an example of this situation. Occasionally, the exact position of the center of the tropical cyclone is secured when some vessel happens to be within it.

The observer can usually locate his position with reference to the storm center and the dangerous and navigable semicircles by a knowledge of the wind direction and the nature of the wind *shifts.* It is clear that if the wind direction remains constant, with the force increasing steadily, he must be on the storm track *ahead* of the center. With a steady wind direction but with a diminishing velocity, he is *behind* the center.

When in any position other than on the track of the center, *shifting* of the wind will occur. In the Northern Hemisphere, when the wind sets in from points southerly to easterly, the ship is most likely in the right or dangerous semicircle. Northerly and westerly winds place the observer in the left semicircle. This results from the normal counterclockwise whirl of the winds about the hurricane or any low-pressure center in the Northern Hemisphere.

East to northeast wind indicates the forward right quadrant of the storm, one of the most dangerous areas. The reader can construct a similar

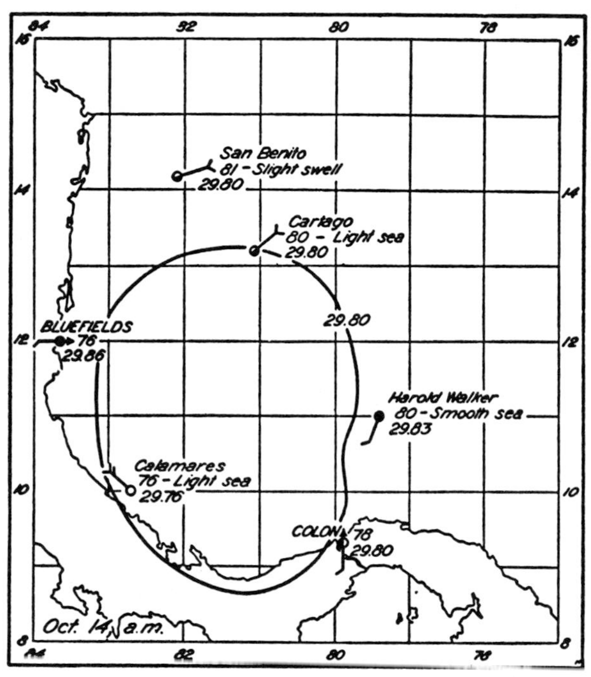

Fig. 15 · 20 Pressure and wind conditions at 7 A.M., *October 14, 1926, showing gentle cyclonic wind circulation of an incipient disturbance. The amount of filling of the small circles at the heads of the arrows indicates the state of the sky.* (U.S. Weather Bureau)

picture for a tropical cyclone in the Southern Hemisphere, remembering the clockwise circulation prevailing there about a low-pressure center.

Clouds. We noted previously that the cirrus clouds spread out far in advance of the storm itself and are often, when in combination with an abnormal swell, accurate warnings of an existing tropical cyclone. If the storm is approaching, the cirrus often take the form of long bands radiating from the position of the storm center. As the storm approaches, the cirrus give way to increasing cirrostratus. These form a typical film or veil over the sky and cause halos about the sun or moon. Under these conditions, the sky is frequently a brilliant red at sunrise or sunset.

Swell. Swell or ground swell, referred to previously, consists of long, low, undulating waves that have outrun the winds producing them. The stronger the wind, the higher and longer are the resulting waves. Consequently, in a hurricane zone, any swell differing in length and direction from the swell characteristic of that locality is a possible warning. The

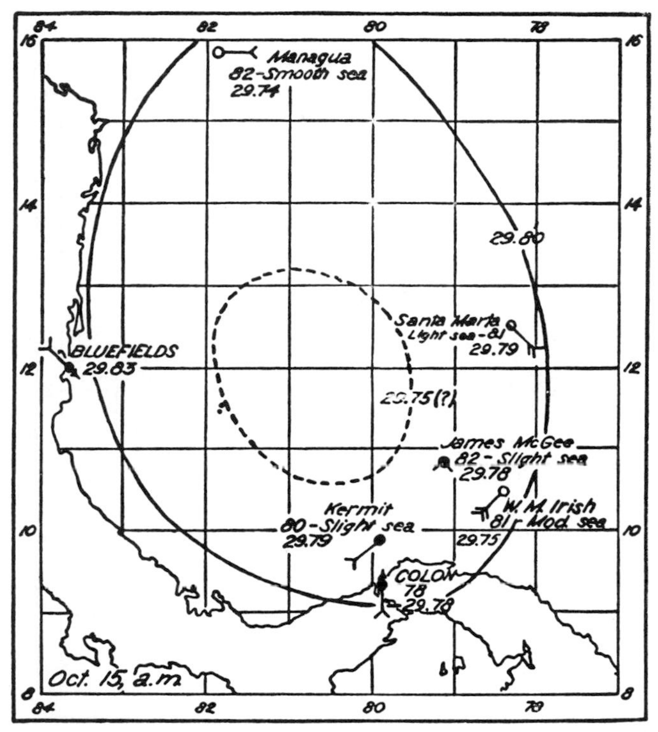

Fig. 15 · 21 Pressure and wind conditions at 7 A.M., *October 15, 1926. Lower pressures and slightly more vigorous wind circulation are evident. Conditions of the sea surface are indicated here and on the following figures. Dotted line indicates probable central area of storm.* (U.S. Weather Bureau)

storm itself has been in the direction from which the swell moves. For example, swell with a period of 10 to 15 seconds, when encountered in the Caribbean or the Gulf of Mexico, is usually a reliable sign of a tropical cyclone in that area. From the previous discussion of the nature and distribution of swell in a tropical cyclone, it is evident that swell of increasing height coming from a constant direction indicates that the ship or observer is on the path ahead of the center. If constant direction but decreasing height is observed, the observer is on the path but to the rear of the storm. If to either side of the storm, the swell will shift direction as well as height and period. The relationship between wind and swell, the latter moving to the right of the wind, is also shown in Fig. 15 · 19.

A summary of significant changes in observations of pressure, wind, and swell associated with the development of a hurricane in the southern part of the Gulf of Mexico is given in Figs. 15 · 20 to 15 · 23.

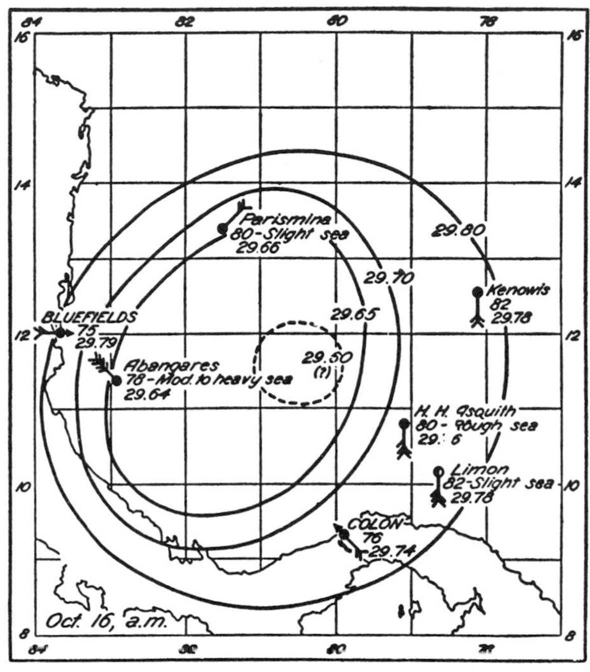

Fig. 15 · 22 Conditions at 7 A.M., October 16, 1926. Marked increase in winds with further decrease in pressure seen. (U.S. Weather Bureau)

Hurricane Warnings

The information about existing hurricanes obtained from weather satellites, radar, and airplane reconnaissance is transmitted immediately over special teletype circuits to all U.S. Weather Bureau offices in the regions that are or may be involved. The most detailed reports are of course those from direct airplane observations. In addition, merchant ships of all countries that contain qualified observers, plus Coast Guard weather ships, radio complete observations every six hours. When pertaining to hurricanes that may affect coastal areas, the information is also transmitted via the special teletype circuits. In case of coastal storms, observations are also obtainable from lighthouses and light ships, Coast Guard installations, and defense and private installations concerned with weather.

Ships at sea in the critical areas receive information about hurricane

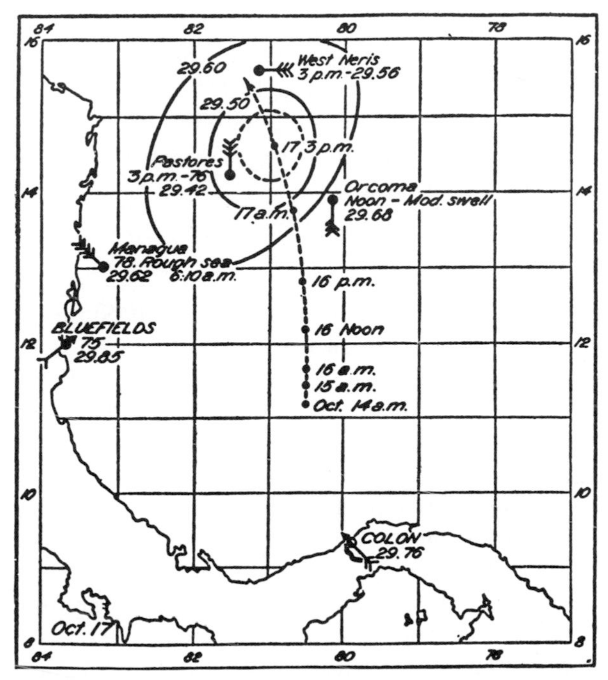

Fig. 15 · 23 Conditions on October 17, 1926, showing, in addition, the track of the storm center with dots indicating its position at the times given. A rapid increase in intensity followed with the storm passing over Havana and Bermuda, exhibiting winds of full hurricane force. (U.S. Weather Bureau)

conditions over regular radio channels. The public receives a variety of warnings that are not directly meaningful. The types of warnings related to hurricanes issued by the U.S. Weather Bureau are described below.

1. *Hurricane watch*—issued to alert the public that the hurricane may endanger a locality or area in some way (tides or winds) within 36 hours, although indications are not yet sufficiently definite to justify a hurricane warning. *The issuance of a hurricane watch is not a hurricane warning.*

2. *Gale warning*—issued to warn the public and maritime interests that within 24 hours winds of Beaufort force 8 (39 miles per hour) or higher are expected either on the coast or over waters adjacent to the coast, or that there is a possibility of the wind's attaining a velocity sufficiently high and sustained for a period long enough to interfere with safe operation of seagoing vessels. This form of warning is frequently the *first warning* issued in connection with a hurricane, although a hurricane watch may have been issued earlier for the same portion of the coast. A gale warning may be issued when only the fringe effects of the hurricane are expected to be felt in the area.
3. *Hurricane warning*—issued to warn the public that within the next 24 hours winds of Beaufort force 10 (55 miles per hour) or higher are expected and either there is a possibility of increase to full hurricane intensity or the accompanying waves or tides or other conditions justify emergency action.

EXERCISES

15 · 1 Compare the international and the United States divisions and descriptions of the types of tropical storms.

15 · 2 Summarize the nature and changes of the following elements within a tropical cyclone: pressure, wind, clouds, and precipitation.

15 · 3 Why is the calm center or eye of the storm so dangerous to mariners?

15 · 4 (*a*) On a diagram of a hurricane including the path of motion, indicate the relative water-wave heights in each of the four quadrants (formed by a line normal to the path and through the storm center). (*b*) Explain why the forward right quadrant generates the greatest waves.

15 · 5 Distinguish between hurricane waves and hurricane swell.

15 · 6 How do observations of hurricane swell indicate the probable motion of the storm relative to the observer?

15 · 7 Describe the relationship between wind and wave direction with a hurricane.

15 · 8 Define and explain the significance of the storm surge.

15 · 9 Give the average number of hurricanes expected per year in each of the regions of occurrence.

15 · 10 Explain the significance of the water budget of a tropical cyclone in maintaining the energy of the storm.

15 · 11 (*a*) In general, how does the speed of a hurricane in the mid-latitudes compare with the speed in the low latitudes? (*b*) Determine the speeds of the storms of September, 1938 and 1944 for each 12 hours of motion. Show the results in tabular form.

15 · 12 Why cannot hurricanes develop closer than several degrees from the equator?

15 · 13 Describe easterly waves and explain their relationship to hurricane origin.

15 · 14 Describe the location of, and explain the reasons for, the existence of the dangerous semicircle.

15 · 15 Construct a hypothetical chart of a hurricane, showing positions of different vessels in the fringe region of the storm on the track ahead of, and to the rear of, the center and in each of the quadrants. Summarize for each ship the expected observations and changes of pressure, winds, and clouds, precipitation, and swell.

CHAPTER 16

Weather Analysis and Interpretation

The complete process of developing a forecast can be broken down into four distinct stages:

1. Taking and transmitting weather observations
2. Plotting the weather map
3. Analyzing the weather map
4. Interpreting, developing prognostic charts, and forecasting

Although it is not the purpose of a general introductory book like this to provide adequate training for a professional weather forecaster, it is the purpose to give a background in the procedure that will be useful for those who may not get further training but who need some experience in weather interpretation. It is also intended to give the more general reader a grasp of the techniques involved.

Taking and Transmitting Weather Observations

For the most part, the taking of observations has been described in connection with the various weather elements described and discussed earlier in this book. In addition to the information provided here, the U.S. Weather Bureau has published a number of small booklets designed to give specific aid to actual weather observers in both the taking and transmitting of weather data. All of these publications are listed in the bibliography references in the Appendix. The transmission of weather observation data to central stations is necessary in order to obtain a large area synthesis of the conditions of the atmosphere at any time. Remember that weather is related to global conditions; hence the need to see and interpret the atmosphere on a large area basis in order to understand and forecast weather. Remember also that the globe is about 70 percent ocean covered, so that weather transmission from both fixed land and more mobile sea stations is involved.

For purposes of efficiency and uniformity in the transmission of weather data, the International Meteorological Congress adopted the International Weather Code. Ships and stations of all nationalities can freely interchange weather messages. Each code table giving the numerical values to be used in preparing and decoding reports is designated by a standard international code letter or symbol. The coded messages for marine and land stations usually contain observation groups common to both and some that will differ. The latter, for example, may include code numbers for observations unique to the sea, such as state of the sea, waves, swell, air-sea temperature differences, and so on. In addition, a land-station report begins with a three-digit number identifying the station, whereas a marine report uses codes locating the ship by the appropriate octant of the globe and then by latitude and longitude. Day of the week and time (GCT) are also included in the beginning of a marine message but omitted from regular land reports.

The complete coded message is usually transmitted in five-digit groups for convenience in decoding. The transmission is by radio from marine sources and by teletype for land sources. The times of transmission are 00, 06, and 12, and 18 hours GCT for all stations, thus assuring simultaneous transmission of information. Any land station connected to the teletype transmission system can obtain the observations from all other stations and thus do individual plotting and forecasting of weather. After marine messages are all received, they are rebroadcast for the use of marine stations at which weather maps may be drawn. The marine observations are also retransmitted over particular teletype circuits for use by coastal stations concerned with marine weather. The stations, frequencies, and times of the rebroadcast of marine observations are provided in U.S. Weather Bureau documents in addition to the complete manuals of weather codes (synoptic codes) and observation and transmission manuals (see Appendix).

An example of a coded message showing the observations normally plotted on printed maps of the U.S. Weather Bureau is given below, first in symbolic form and then in numerical code.

iii	Nddff	VVwwW	PPPTT	$N_hC_LhC_MC_H$	T_dT_dapp	$7RRR_ts$
405	83220	12716	24715	67292	14228	74542

The explanation of the symbolic and sample coded message is given in Table 16 · 1. If this were a marine message, the first two five-digit groups would symbolically be

YQLaLaLa LoLoLoGG

Table 16 · 1 Explanation of Symbols and Map Entries

SYMBOLS IN ORDER AS THEY APPEAR IN THE MESSAGE	EXPLANATION OF SYMBOLS AND DECODE OF EXAMPLE ABOVE	REMARKS ON CODING AND PLOTTING
iii	Station number 405 = Washington	Usually printed on manuscript maps below station circle. Omitted on Daily Weather Map in favor of printed station names.
N	Total amount of cloud 8 = completely covered	Observed in tenths of cloud cover and coded in Oktas (eighths) according to code table N (Fig. 16 · 2*a*). Plotted in symbols shown in same table.
dd	True direction from which wind is blowing 32 = 320° = NW	Coded in tens of degrees and plotted as the shaft of an arrow extending from the station circle toward the direction from which the wind is blowing.
ff	Wind speed in knots 20 = 20 knots	Coded in knots (nautical miles per hour) and plotted as feathers and half-feathers representing 10 and 5 knots, respectively, on the shaft of the wind direction arrow.
VV	Visibility in miles and fractions 12 = 12/16 or 3/4 miles	Decoded and plotted in miles and fractions up to 3⅛ miles. Visibilities above 3⅛ miles but less than 10 miles are plotted to the nearest whole mile. Values higher than 10 miles are omitted from the map.
ww	Present weather 71 = continuous slight snow	Coded in figures taken from the "ww" table (Fig. 16 · 2) and plotted in the corresponding symbols same block. Entries for code figures 00, 01, 02, and 03 are omitted from this map.
W	Past weather 6 = rain	Coded in figures taken from the "W" table (Fig. 16 · 2*d*) and plotted in the corresponding symbols same block. No entry made for code figures 0, 1, or 2.
ppp	Barometric Pressure (in millibars) reduced to sea-level 247 = 1024.7 mb	Coded and plotted in tens, units, and tenths of millibars. The initial 9 or 10 and the decimal point are omitted.

Table 16 · 1 Explanation of Symbols and Map Entries (continued)

SYMBOLS IN ORDER AS THEY APPEAR IN THE MESSAGE	EXPLANATION OF SYMBOLS AND DECODE OF EXAMPLE ABOVE	REMARKS ON CODING AND PLOTTING
TT	Current air temperature 15 = 15°C	Coded in °C and plotted in actual value in whole degrees F.
N_h	Fraction of sky covered by low or middle cloud 6 = 7 or 8 tenths	Observed and coded in tenths of cloud cover. Plotted on map as code figure in message. Table N, Fig. 16 · 2*a*.
C_L	Cloud type 7 = Fractostratus and/ or Fractocumulus of bad weather (scud)	Predominating clouds of types in C_L table are coded from that table and plotted in corresponding symbols.
h	Height of base of cloud 2 = 300 to 599 ft	Observed in feet and coded and plotted as code figures according to code table in Fig. 16 · 2*d*.
C_M	Cloud type 9 = Altocumulus of chaotic sky	See C_M table in Fig. 16 · 2*a*.
C_H	Cloud type 2 = Dense cirrus in patches	See C_H table in Fig. 16 · 2*a*.
T_dT_d	Temperature of dewpoint 14 = 14°C	Coded in °C and plotted in actual value in whole degrees F.
a	Characteristics of barograph trace 2 = rising steadily or unsteadily	Coded according to table in Fig. 16 · 2*a* and plotted in corresponding symbols.
pp	Pressure change in 3 hours preceding observation 28 = 2.8 mb	Coded and plotted in units and tenths of millibars.
7	Indicator figure	Not plotted.
RR	Amount of precipitation 45 = 0.45 in.	Coded and plotted in inches to the nearest hundredth of an inch.
R_t	Time of precipitation began or ended 4 = 3 to 4 hours ago	Coded and plotted in figures from table in Fig. 16 · 2*d*.
s	Depth of snow on ground	Not plotted. Coded in inches.

Note: Although temperature (TT) and dew point (T_dT_d) are now reported in °C, aviation reports and weather maps continue to use °F.

in which Y is the day of the week, numbered 1 to 7; Q, the octant of the globe (1 to 8); LaLaLa, the latitude (in degrees and tenths); LoLoLo, the longitude (in degrees and tenths); and GG, Greenwich Time (00, 06, 12, 18).

In addition to the standard synoptic reports every 6 hours by means of the international numerical code, airports in the United States transmit hourly and special observations over the national weather teletype system. In this transmission a different code is used, in which observations are represented by appropriate descriptive symbols and word abbreviations.

Plotting the Weather Map

After decoding, the weather observations for each station must be plotted at the appropriate location in a systematic manner following the international station model illustrated in Fig. 16 · 1. In actual practice,

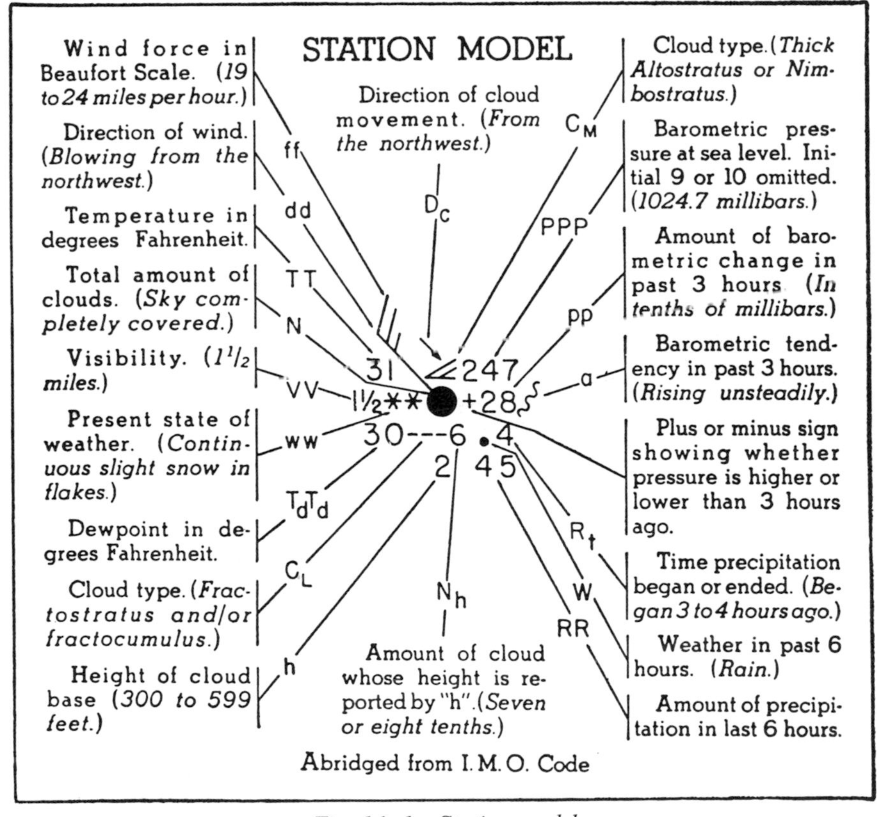

Fig. 16 · 1 Station model.

C_L Clouds of type C	C_M Clouds of type C	C_H Cloud of type C	W Past Weather	N Total amount all clouds	a Barometer characteristics
0 No Sc, St, Cu, or Cb clouds.	0 No Ac, As or Ns clouds.	0 No Ci, Cc, or Cs clouds.	0 Clear or few clouds.	0 No clouds.	0 Rising then falling. Now higher than 3 hours ago.
1 Cu with little vertical development and seemingly flattened.	1 Thin As (entire cloud layer semitransparent).	1 Filaments of Ci, scattered and not increasing.	1 Partly cloudy (scattered) or variable sky.	1 Less than one-tenth or one-tenth.	1 Rising, then steady; or rising, then rising more slowly. Now higher than, as, 3 hours ago.
2 Cu of considerable development, generally towering, with or without other Cu or Sc; bases all at same level.	2 Thick As, or Ns.	2 Dense Ci in patches or twisted sheaves, usually not increasing.	2 Cloudy (broken or overcast).	2 Two- or three-tenths.	2 Rising steadily, or unsteady. Now higher than, 3 hours ago.
3 Cb with tops lacking clear-cut outlines, but distinctly not cirriform or anvil-shaped; with or without Cu, Sc or St.	3 Thin Ac; cloud elements not changing much and at a single level.	3 Ci, often anvil-shaped, derived from or associated with Cb.	3 Sandstorm, or duststorm, or drifting or blowing snow.	3 Four-tenths.	3 Falling or steady, then rising; or rising, then rising more quickly. Now higher than, 3 hours ago.
4 So formed by spreading out of Cu; Cu often present also.	4 Thin Ac in patches, cloud elements and/or occurring at more than one level.	4 Ci, often hook-shaped, gradually spreading over the sky and usually thickening as a whole.	4 Fog, or smoke, or thick dust haze.	4 Five-tenths.	4 Steady. Same as 3 hours ago.§
5 Sc not formed by spreading out of Cu.	5 Thin Ac in bands or in a layer gradually spreading over sky and usually thickening as a whole.	5 Ci and Cs, often in converging bands, or Cs alone; the continuous layer not reaching 45° altitude.	5 Drizzle.	5 Six-tenths.	5 Falling, then rising. Same or lower than 3 hours ago.
6 St or Fs or both, but not Fs of bad weather.	6 Ac formed by the spreading out of Cu.	6 Ci and Cs, often in converging bands, or Cs alone; the continuous layer exceeding 45° altitude.	6 Rain.	6 Seven- or eight-tenths.	6 Falling, then steady; or falling, then falling more slowly. Now lower than 3 hours ago.
7 Fs and/or Fc of bad weather (scud) usually under As and Ns.	7 Double-layered Ac or a thick layer of Ac, not increasing; or As and Ac both present at same or different levels.	7 Cs covering the entire sky.	7 Snow, or rain and snow mixed, or ice pellets (sleet).	7 Nine-tenths or overcast with openings.	7 Falling steadily, or unsteady. Now lower than 3 hours ago.
8 Cu and Sc (not formed by spreading out of Cu) with bases at different levels.	8 Ac in the form of Cu-shaped tufts or Ac with turrets.	8 Cs not increasing and not covering entire sky; Ci and Cc may be present.	8 Shower(s).	8 Completely overcast.	8 Steady or rising, then falling; or falling, then falling more quickly. Now lower than 3 hours ago.
9 Cb having a clearly fibrous (cirriform) top, often anvil-shaped, with or without Cu, Sc, St, or scud.	9 Ac of a chaotic sky, usually at different levels; patches of dense Ci are usually present also.	9 Cc alone or Cc with some Ci or Cs, but the Cc being the main cirriform cloud present.	9 Thunderstorm, with or without precipitation.	9 Sky obscured.	

§ The symbol is not plotted for "ww" when "00" is reported. When "01, 02, or 03" is reported for "ww," the symbol is plotted on the station circle. Symbols are not plotted for "a" when "3 or 8" is reported.

Fig. 16 • 2a Explanation of symbols used on station model.

WW
Present weather

00 Cloud development NOT observed or NOT observable during past hour.§	01 Clouds generally dissolving or becoming less developed during past hour.§	02 State of sky on the whole unchanged during past hour.§	03 Clouds generally forming or developing during past hour.§	04 Visibility reduced by smoke.	05 Dry haze.	06 Widespread dust in suspension in the air, NOT raised by wind, at time of observation.	07 Dust or sand raised by wind, at time of ob.	08 Well developed dust devil(s) within past hr.	09 Duststorm or sandstorm within sight of or at station during past hour.
10 Light fog.	11 Patches of shallow fog at station, NOT deeper than 6 feet on land.	12 More or less continuous shallow fog at station, NOT deeper than 6 feet on land.	13 Lightning visible, no thunder heard.	14 Precipitation within sight, but NOT reaching the ground at station.	15 Precipitation within sight, reaching the ground, but distant from station.	16 Precipitation within sight, reaching the ground, near to but NOT at station.	17 Thunder heard, but no precipitation at the station.	18 Squall(s) within sight during past hour.	19 Funnel cloud(s) within sight during past hr.
20 Drizzle (NOT freezing and NOT falling as showers) during past hour, but NOT at time of ob.	21 Rain (NOT freezing and NOT falling as showers during past hr., but NOT at time of ob.	22 Snow (NOT falling as showers) during past hr., but NOT at time of ob.	23 Rain and snow (NOT falling as showers) during past hour, but NOT at time of observation.	24 Freezing drizzle or freezing rain (NOT falling as showers) during past hour, but NOT at time of observation.	25 Showers of rain during past hour, but NOT at time of observation.	26 Showers of snow, or of rain and snow, during past hour, but NOT at time of observation.	27 Showers of hail, or of hail and rain, during past hour, but NOT at time of observation.	28 Fog during past hour, but NOT at time of ob.	29 Thunderstorm (with or without precipitation) during past hour, but NOT at time of ob.
30 Slight or moderate duststorm or sandstorm, has decreased during past hour.	31 Slight or moderate duststorm or sandstorm, no appreciable change during past hour.	32 Slight or moderate duststorm or sandstorm, has increased during past hour.	33 Severe duststorm or sandstorm, has decreased during past hr.	34 Severe duststorm or sandstorm, no appreciable change during past hour.	35 Severe duststorm or sandstorm, has increased during past hr.	36 Slight or moderate drifting snow, generally low.	37 Heavy drifting snow, generally low.	38 Slight or moderate drifting snow, generally high.	39 Heavy drifting snow, generally high.
40 Fog at distance at time of ob., but NOT at station during past hour.	41 Fog in patches.	42 Fog, sky discernible, has become thinner during past hour.	43 Fog, sky NOT discernible, has become thinner during past hour.	44 Fog, sky discernible, no appreciable change during past hour.	45 Fog, sky NOT discernible, no appreciable change during past hr.	46 Fog, sky discernible, has begun or become thicker during past hr.	47 Fog, sky NOT discernible, has begun or become thicker during past hour.	48 Fog, depositing rime, sky discernible.	49 Fog, depositing rime, sky NOT discernible.

§ The symbol is not plotted for "ww" when "00" is reported. When "01, 02, or 03" is reported for "ww," the symbol is plotted on the station circle. Symbols are not plotted for "a" when "3 or 8" is reported.

Fig. 16·2b Explanation of symbols used on station model (continued).

W W
Present weather

50	Intermittent drizzle (NOT freezing) slight at time of observation.
51	Continuous drizzle (NOT freezing) slight at time of observation.
52	Intermittent drizzle (NOT freezing) moderate at time of ob.
53	Intermittent drizzle (NOT freezing), moderate at time of ob.
54	Intermittent drizzle (NOT freezing), thick at time of observation.
55	Continuous drizzle (NOT freezing), thick at time of observation.
56	Slight freezing drizzle.
57	Moderate or thick freezing drizzle.
58	Drizzle and rain slight.
59	Drizzle and rain, moderate or heavy.
60	Intermittent rain (NOT freezing), slight at time of observation.
61	Continuous rain (NOT freezing), slight at time of observation.
62	Intermittent rain (NOT freezing), moderate at time of ob.
63	Continuous rain (NOT freezing), moderate at time of observation.
64	Intermittent rain (NOT freezing), heavy at time of observation.
65	Continuous rain (NOT freezing), heavy at time of observation.
66	Slight freezing rain.
67	Moderate or heavy freezing rain.
68	Rain or drizzle and snow, slight.
69	Rain or drizzle and snow, mod. or heavy.
70	Intermittent fall of snow flakes, slight at time of observation.
71	Continuous fall of snowflakes, slight at time of observation.
72	Intermittent fall of snow flakes, moderate at time of observation.
73	Continuous fall of snowflakes, moderate at time of observation.
74	Intermittent fall of snow flakes, heavy at time of observation.
75	Continuous fall of snowflakes, heavy at time of observation.
76	Ice needles (with or without fog).
77	Granular snow (with or without fog).
78	Isolated starlike snow crystals (with or without fog).
79	Ice pellets (sleet, U.S. definition).
80	Slight rain shower(s).
81	Moderate or heavy rain shower(s).
82	Violent rain shower(s).
83	Slight shower(s) of rain and snow mixed.
84	Moderate or heavy shower(s) of rain and snow imxed.
85	Slight snow shower(s).
86	Moderate or heavy snow shower(s).
87	Slight shower(s) of soft or small hail with or without rain or rain and snow mixed.
88	Moderate or heavy shower(s) of soft or small hail with or without rain or rain and snow mixed.
89	Slight shower(s) of hail††, with or without rain or rain and snow mixed, not associated with thunder.
90	Moderate or heavy shower(s) of hail††, with or without rain or rain and snow mixed, not associated with thunder.
91	Slight rain at time of ob.; thunderstorm during past hour, but NOT at time of observation.
92	Moderate or heavy rain at time of ob.; thunderstorm during past hour, but NOT at time of observation.
93	Slight snow or rain and snow mixed or hail† at time of ob.; thunderstorm during past hour, but not at time of ob.
94	Mod. or heavy snow, or rain and snow mixed or hail† at time of ob.; thunderstorm during past hour, but NOT at time of observation.
95	Slight or mod. thunderstorm without hail†, but with rain and/or snow at time of ob.
96	Slight or mod. thunderstorm, with hail† at time of observation.
97	Heavy thunderstorm, without hail†, but with rain and/or snow at time of observation.
98	Thunderstorm combined with duststorm or sandstorm at time of ob.
99	Heavy thunderstorm with hail† at time of ob.

† Refers to "hail" only. †† Refers to "soft hail," "small hail," and "hail."

Fig. 16·2c Explanation of symbols used on station model (continued).

Code Number	W	PAST WEATHER
0		Clear or few clouds (Not Plotted)
1		Partly cloudy (scattered) or variable sky (Not Plotted)
2		Cloudy (broken) or overcast (Not Plotted)
3	[symbol]	Sandstorm or dust-storm, or drifting or blowing snow
4	≡	Fog, or smoke, or thick dust haze
5	,	Drizzle
6	•	Rain
7	*	Snow, or rain and snow mixed. or ice pellets (sleet)
8	▽	Shower(s)
9	[symbol]	Thunderstorm, with or without precipitation

h	HEIGHT IN FEET (Round)	HEIGHT IN METERS (Approximate)
0	0 - 149	0 - 49
1	150 - 299	50 - 99
2	300 - 599	100 - 199
3	600 - 999	200 - 299
4	1,000 - 1,999	300 - 599
5	2,000 - 3,499	600 - 999
6	3,500 - 4,999	1,000 - 1,499
7	5,000 - 6,499	1,500 - 1,999
8	6,500 - 7,999	2,000 - 2,499
9	At or above 8,000 or no clouds	At or above 2,500 or no clouds

R_t	TIME OF PRECIPITATION
0	No precipitation
1	Less than 1 hour ago
2	1 to 2 hours ago
3	2 to 3 hours ago
4	3 to 4 hours ago
5	4 to 5 hours ago
6	5 to 6 hours ago
7	6 to 12 hours ago
8	More than 12 hours ago
9	Unknown

Fig. 16 · 2d Explanation of symbols used on station model (continued).

only weather maps in first-class forecasting centers approach the completeness of this model. Printed maps and maps used for plotting usually have an appropriately numbered circle corresponding to each reporting land station, and observations are plotted about this location in the appropriate position regardless of the number of observations shown. The symbols used for different observations are shown in Fig. 16 · 2*a* to *d*. On marine maps, fixed station circles are only given for Coast Guard weather ship positions owing to the shifting positions of merchant vessels submitting reports. The plotted station model corresponding to the coded message given above and in the explanation in Table 16 · 1 is shown in Fig. 16 · 3. It must be emphasized again that the weather pattern affecting a locality is an integral part of the much larger hemispheric weather pattern. Because these weather structures generally move from west to east, particularly above the tropics, it is necessary to plot a map over a rather large area, especially to the west of a station.

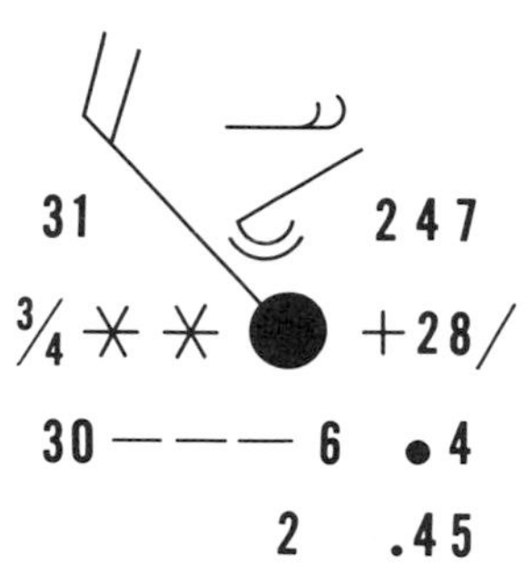

Fig. 16 · 3 Plotted station model corresponding to the coded weather message on page 322 and the explanation in Table 16 · 1.

Weather Map Analysis

The analyzed weather map is often said to present, both directly and indirectly, more information than any other chart in existence. Such a map, after being plotted, is a maze of apparently disorganized facts. Potentially, it has an important story to tell. The process of weather map analysis consists of organizing the information on it into a logical meteorological picture.

We can divide the procedure into three basic steps:

1. Frontal analysis
2. Isobaric or pressure analysis
3. Analysis of air-mass conditions

In accordance with the scope of this book, special attention will be given to the first two steps. We shall mention only briefly the problems of analysis and interpretation of air-mass conditions. The reason for this is twofold: (1) A detailed study of air masses requires more observational data than are usually available for the marine or lay forecaster. (2) The problem involves a technical study of atmospheric conditions beyond our present scope. Each of these three steps can of course be refined or subdivided further. For a more complete treatment of the analysis and interpretation of air masses as applied to weather conditions, the interested reader is referred to the many advanced texts covering this subject, some of which are recommended in the Bibliography.

In all weather analysis it is of prime importance to consider the weather map or maps previous to the one at hand. Obviously, conditions on the new map must tally with those of the previous ones. This *continuity* also simplifies analysis, since the analyst thus obtains a knowledge of how conditions ought to appear on the new map. If no previous maps are available, the first one may be more difficult to prepare. Only the analysis of surface maps will be considered.

Frontal Analysis

Fronts, being the boundaries of air masses, clearly have a pronounced influence on weather changes. The distinguishing of the existing fronts on a weather map not only indicates the air-mass boundaries, but also aids greatly in the identification of the position of the different air masses themselves. In Chap. 13 on Air Masses and Fronts, a study was made of the more obvious weather differences on opposite sides of a front. *It follows then, if we can determine the position of a line on either side of which more or less marked weather differences exist, that the line is probably a front.*

The important criteria employed in the location of fronts may be enumerated as follows:

1. Temperature differences
2. Wind shifts
3. Humidity differences
4. Clouds and precipitation conditions
5. Pressure and isobaric conditions

As many of these factors as possible should be utilized in frontal analysis.

1. TEMPERATURE CONDITIONS. We have noted that temperature conditions are the essential feature of air masses. The determination of the temperature difference or discontinuity between air masses is very useful in establishing the location of a front. However, the surface air temperatures are often strongly affected by local heating and cooling, in addition to topography and other factors. The true temperature discontinuity is best shown above the ground, away from the influence of surface conditions. Marine air temperatures are also more indicative of the true discontinuity than those over land.

   Further, the air bodies separated by the front may have very nearly the same temperatures. In such cases, this factor may be unreliable; in any case, it should be supplemented by other criteria.
2. WIND SHIFTS. The wind shift or wind discontinuity on either side of a front is often one of the best criteria in locating it. There is nearly always some degree of shifting of the wind associated with fronts.

This usually shows quite clearly on the synoptic weather chart and enables the proper placing of the front.

On occasion, abrupt wind shifts may occur without an attending front. A pronounced wedge or ridge of high pressure will cause such a shift along the axis of the ridge, in order for the wind to conform to the isobars. A further empirical rule is of great value in distinguishing such cases. Consideration of the relationship between winds and fronts developed in the past two chapters shows that in the Northern Hemisphere, *if one places his back to the wind, any associated front will be on the left.* Hence, as a result of the clockwise wind circulation about a high, this rule places the front in a direction opposite to the axis of the high-pressure ridge, where the wind shift occurs.

It is apparent that the direction of the wind is of greater significance than its velocity, for this purpose. However, there may also exist a marked difference in velocity on opposite sides of the front, particularly in the case of an advancing cold, blustering, polar air mass. This property is also of value in frontal analysis.

3. HUMIDITY DIFFERENCES. Dew-point observations are usually the only direct humidity observations shown on the synoptic chart. A careful examination of dew-point differences along the position of a suspected front often aids materially in analysis. At times these differences may be the only definite indication of the presence of a front. Remember that the dew point of a given air mass remains quite constant, barring gain or loss of water vapor.
4. CLOUDS AND PRECIPITATION CONDITIONS. The relationship between areas of cloudiness and precipitation is often well defined when these conditions are of frontal origin. The association of the types and areas of clouds, with the fronts responsible, has been explained in the chapters on fronts and cyclone development. This knowledge can be used only if these factors are included in the report. They are nearly always available on land-station reports but may be omitted from ship reports. If cloud and precipitation data are recorded on the weather map, a study of their type and distribution is often an excellent guide to the type and location of existing fronts. Thus a narrow band of heavy cumulus or cumulonimbus clouds, especially if associated with storms, may indicate a cold front. A warm front has its own sequence of clouds and precipitation, as studied earlier. This information, together with wind observations, may clearly mark warm- or cold-front positions.
5. PRESSURE AND ISOBARIC CONDITIONS. Normally, any existing fronts are located before the drawing of the isobars. However, an examination of pressure conditions is still of value since some pressure change characterizes nearly all fronts. The pressure tendency, given in the

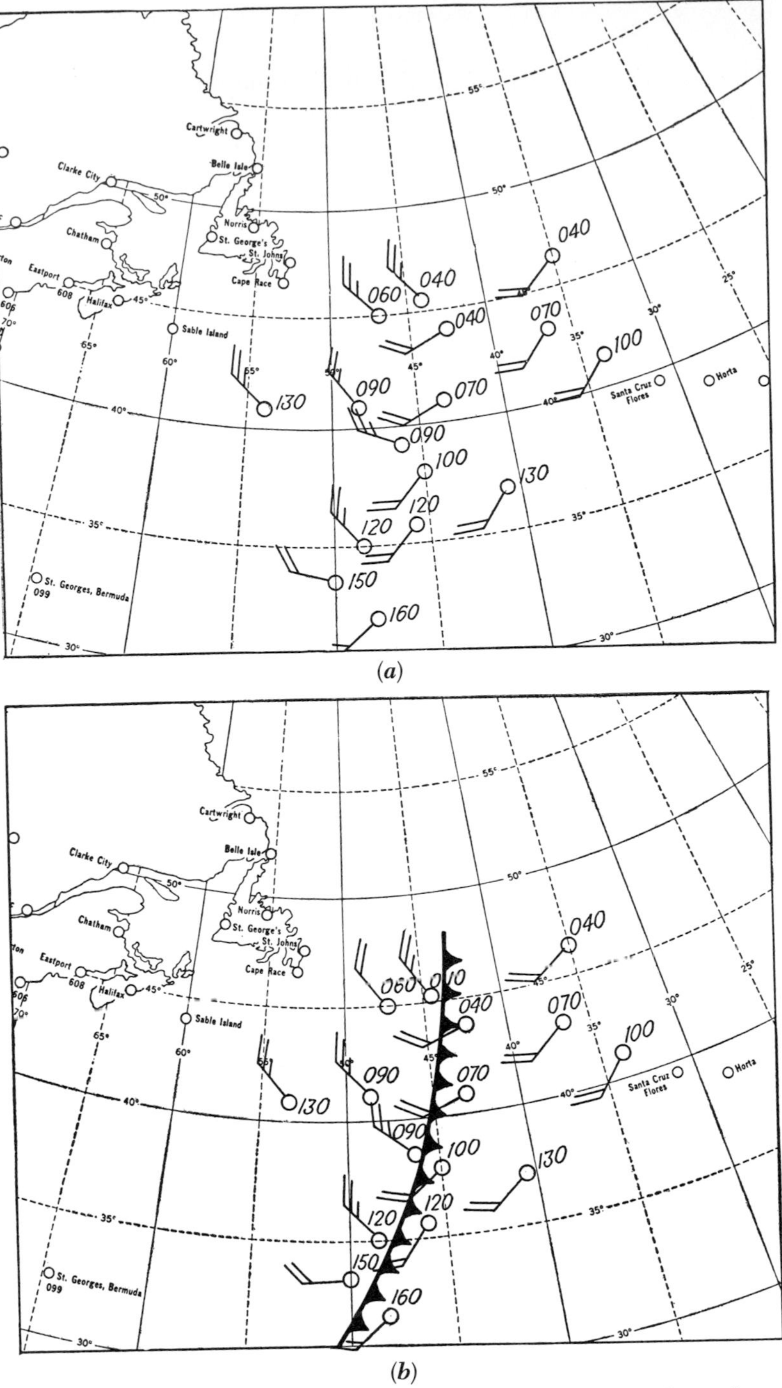

Fig. 16 • 4 (a) Reports of pressure and wind plotted. (b) Frontal analysis of conditions in (a).

supplemental section of a weather report, is very useful. Thus, the position of a front may be determined by the fact that a negative tendency (falling pressure) exists on one side, while a positive tendency (rising pressure) exists on the other. Fronts lie in pressure troughs!

Then there is the angular bend in the isobars where they cross a front. Regardless of how much an isobar bends, it always does so sharply. After the position of the front is located roughly, the drawing of the isobars in the vicinity of it will locate it more accurately.

ILLUSTRATIONS OF SIMPLE FRONTAL ANALYSIS. At times, conditions may be such that a front will stand out like the proverbial sore thumb. At other times, the determination of existing fronts may require a much closer examination of the plotted weather map. The more knowledge and experience the analyst can apply, the more easily and accurately will the fronts be located.

Figure 16 · 4*a* shows ship reports of pressure and wind plotted on a small portion of the marine radio weather map. A rapid survey indicates that two distinct wind streams are present: one from the northwest and one from the southwest. It seems clear that a front probably lies between the two air masses distinguished by the different wind directions. An examination of the pressure reports shows that the isobars would make a very sharp bend at the position of the suspected front. Figure 16 · 4*b* illustrates the result of this survey, with the front being drawn so as to have northwest winds on one side and southwest on the other. Note that if the observer places his back to the wind, the front is to the left in accordance with the rule. Note also that if isobars were drawn, those to the right of the front would have a northeast-southwest trend, whereas those to the left would have a northwest or north-northwest to southeast trend, marking a definite sharp angle at the front.

In consideration of the direction of the two air masses present, from the cold northwest and warm southwest, respectively, the front is considered to be cold, despite the absence of temperature reports. Since both air masses are moving generally eastward, the front will clearly move in the same direction, or toward the warmer (tropical) air. From our knowledge of the cyclone structure we can conceive of the air mass from the southwest as being part of the warm sector of a cyclone.

Figure 16 · 5*a* illustrates an example of reports where wind, temperature, and state of the sky are given. A study of these reports indicates the presence of a front running approximately northwest-southeast, through the island of Horta. One group of reports shows temperatures in the low sixties, with clear to partly cloudy conditions and southwest winds. The other group shows southerly winds with temperatures in the fifties and overcast conditions. This leads us to place the front in the position

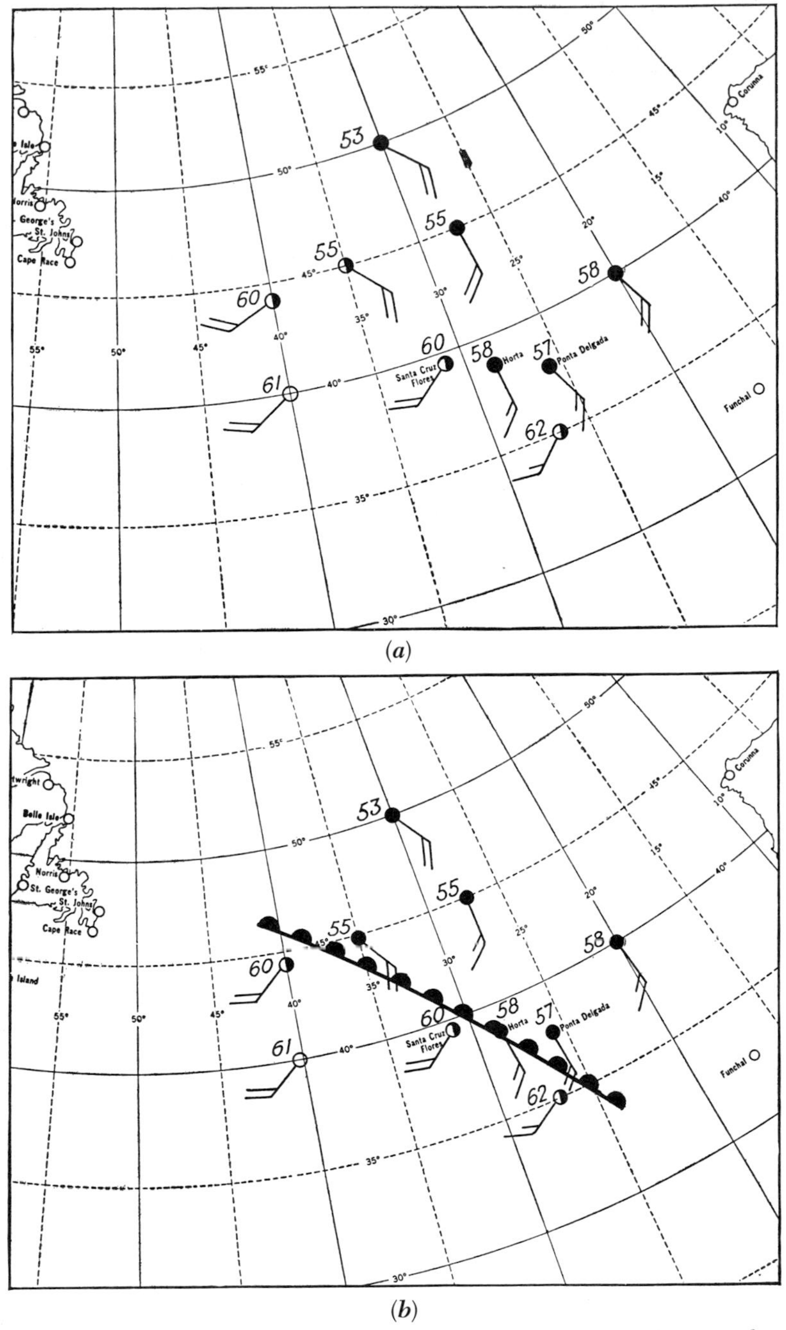

Fig. 16 • 5 (a) Reports of wind, temperature, and state of sky given. (b) Frontal analysis of conditions in (a).

shown in Fig. 16 · 5*b*. A consideration of temperatures and directions of motion leads to the assumption that it is a warm front, which might well be the eastern boundary of a cyclone warm sector.

Figure 16 · 6*a* is slightly more complicated than the others. Here we have reports on wind, temperature, clouds, and present weather conditions. Three sets of wind systems prevail: a southeasterly set, a southwesterly set, and a west-to-northwesterly set. The ships with northwesterly winds on the 30th and 35th meridians show thundershowers and intermittent heavy rains, respectively. The temperatures at these ships are lower than those to the east. Three other ships, to the west, also show low temperatures but have fair-weather cumulus clouds. Apparently we have here a cold air mass, with the line of showers marking the cold-front boundary.

South of latitude 50°, the cold northwest air is separated from the southeast air by a sector of warm air from the southwest. North of latitude 50°, the cold air masses are adjacent to each other, with heavy rains or showers prevailing in both. The rain in the southeast air, south of 50°, is moderate, and out ahead of the rain altostratus clouds prevail. Here we have indications of a warm front. From a consideration of wind direction, clouds, and precipitation it seems that a front exists between the southwest and southeast air and extends south-southeast from 50°N 27°W. Northwestward from this point, the warm front joins the previously determined cold front, yielding an occluded front separating two cold air masses.

Figure 16 · 6*b* shows the completed frontal analysis. Note that an east-west line, north of 50°, crosses but one front or windshift line. South of 50°, both cold and warm fronts are encountered, with associated wind and temperature changes. If no fronts can be located on a plotted weather map, it is likely that only one air mass exists in the area covered by the observations. In such a case, the weather is determined by conditions within the air mass and will show no effect of frontal disturbances.

Isobaric Analysis

Isobaric analysis is necessary (1) to indicate the distribution of high- and low-pressure areas, with their relation to existing air masses and fronts and (2) to show the nature of the pressure gradient in different areas, which aids greatly in determining the future movement of weather conditions. The difficulty in drawing isobars lies mainly in the inadequate number of observations available. Consequently, their positions must usually be estimated. Isobars are usually drawn only for every 3 or 4 millibars, depending on the number of observations available.

Another source of difficulty lies in the errors in plotted pressure observations. These errors may arise from a number of causes, such as instrumental, observational, transmissional, plotting, or coding. To simplify the

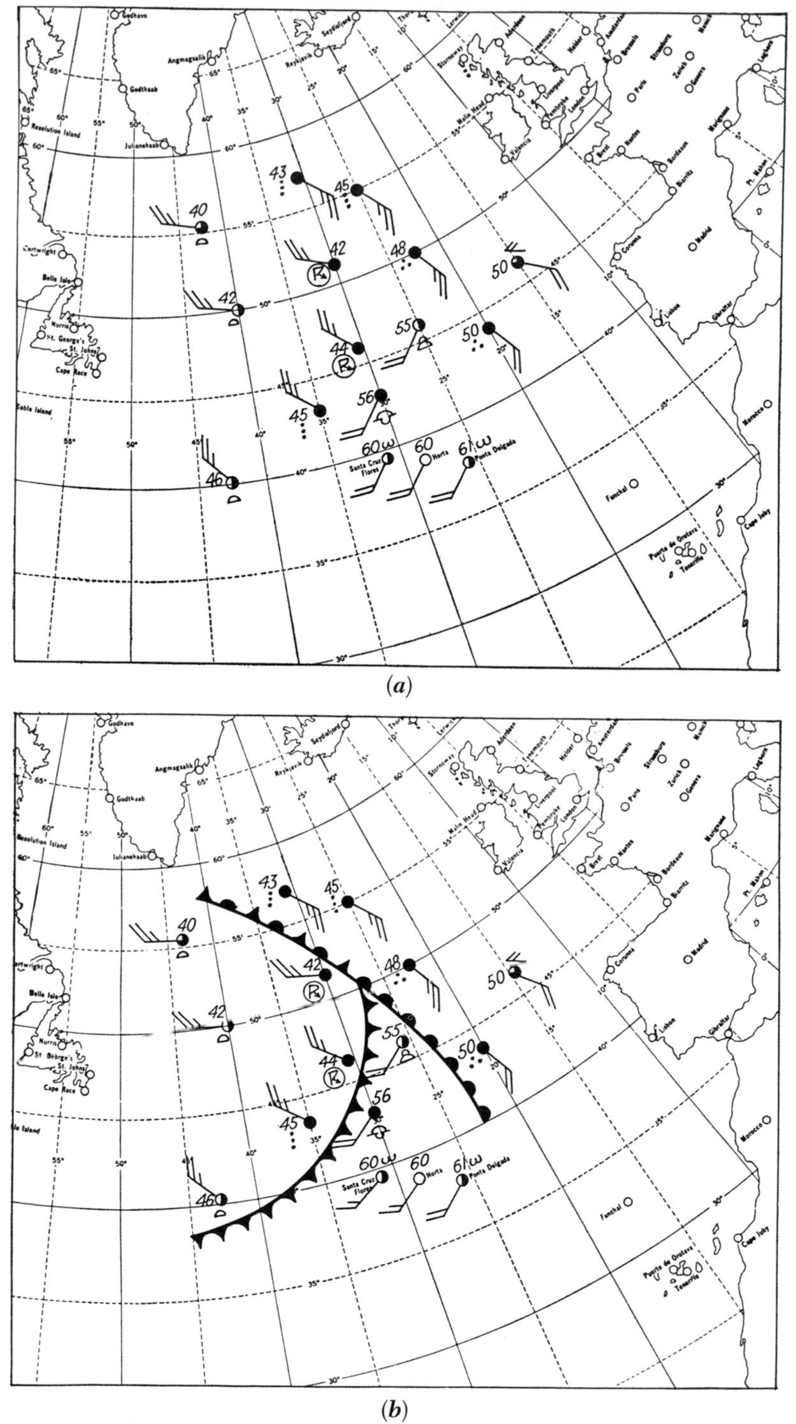

Fig. 16 • 6 (a) Reports of wind, temperature, clouds, and present weather conditions shown. (b) Frontal analysis of conditions in (a).

process of drawing isobars, a number of empirical rules have to be developed. These rules and their explanations follow.

Before studying these rules, recall the generalities drawn from our consideration of isobars in Chap. 8 on Atmospheric Pressure.

Isobars are always closed curved lines. No loose or dangling ends can exist on a map. An isobar may not close up within the confines of a map, but it is terminated at the map border. Obviously, isobars may never cross each other or meet at a right angle, with one coming to a dead end. All this is equally true of topographic contour lines for much the same reason.

RULES FOR DRAWING ISOBARS.

1. *Smooth isobars with simple patterns are more usual than irregular isobars or complicated patterns.* Minor irregularities that are not common to other isobars are probably due to errors, such as those listed

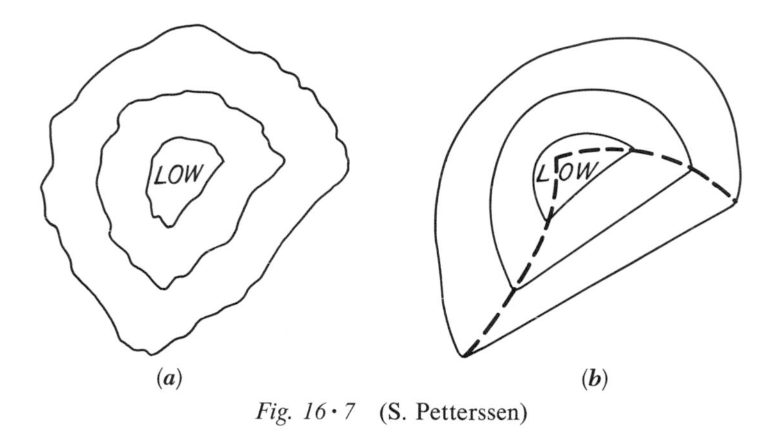

Fig. 16 · 7 (S. Petterssen)

above. The isobars should then be smoothed or *faired* to eliminate minor wiggles and bends.

If pressure observations on a weather map are followed faithfully, results shown in Fig. 16 · 7*a* might develop. The smoothed isobars are shown in Fig. 16 · 7*b*.

If systematic irregularities exist, as indicated by several isobars, the condition shown is probably real. Note the sharp bend of the isobars in the eastern and southwestern sections of the low. This is common to the whole pressure pattern. Our knowledge of fronts leads to the supposition that these kinks represent the warm and cold fronts of a cyclone.

At points *A* and *B* in Fig. 16 · 8, we note two apparent anomalies in the isobars. These may be the result of an error, or they may indicate the beginning of a secondary low-pressure center. Rule 2 aids us in this case.

2. *The direction of the isobars should conform to the proper relationship with wind directions; the spacing of the isobars should agree with the pressure gradient and dependent wind velocity.* As a result of deflection, recall that the wind blows along the isobars with but a slight inward drift, from higher to lower pressures. At sea, with normal wind velocities, the acute angle between wind direction and the isobars is about 15°. Thus, the isobars should be drawn *down wind,* keeping this approximate relationship. The wind should not cross the isobars at, or nearly at, right angles.

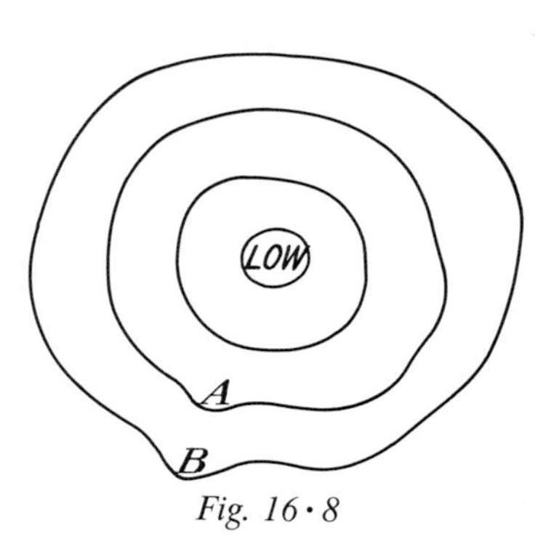

Fig. 16 · 8

Assume that, in accordance with the pressure readings, the isobars are drawn. In Fig. 16 · 9*a* the wind reports do not agree with the isobars at points *D* and *E.* Either the wind or the pressure reports are in error here, for the wind crosses the isobars nearly perpendicularly. Wind reports are usually more dependable than pressure reports at sea. Thus, in accordance with rule 1, we fair the isobars as shown by the broken line, noting that wind directions and isobars now agree.

In Fig. 16 · 9*b* we see that, despite the irregularity, the isobars *do*

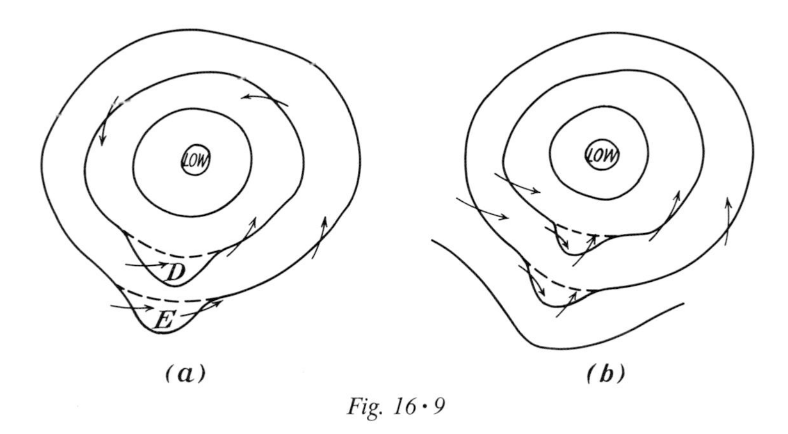

Fig. 16 · 9

conform to the wind direction, with the wind having slight drift across them from higher to lower pressures. If we smooth the isobars as shown by the broken line, the wind either crosses them at a large angle or blows from lower to higher pressure. Such a situation may represent the beginning of a secondary low, which must then be watched closely on succeeding maps for further development, since a major cyclonic disturbance often develops from original pressure and wind irregularities of this type.

Where the pressure gradient is steep, with relatively high wind velocities, the isobars should be closer together than on other parts of the map where lower velocities prevail. Care should be taken in drawing the map to see that the isobars conform to this principle.

3. *All pressure observations on one side of an isobar must be higher in value than the isobar, and all readings on the other side must be lower.* In accordance with our knowledge of isobars, it is clear that they may not have higher and lower pressure readings on the same side. Thus, in Fig. 16 · 10*a*, the 1,017 isobar cannot stand alone. An examination

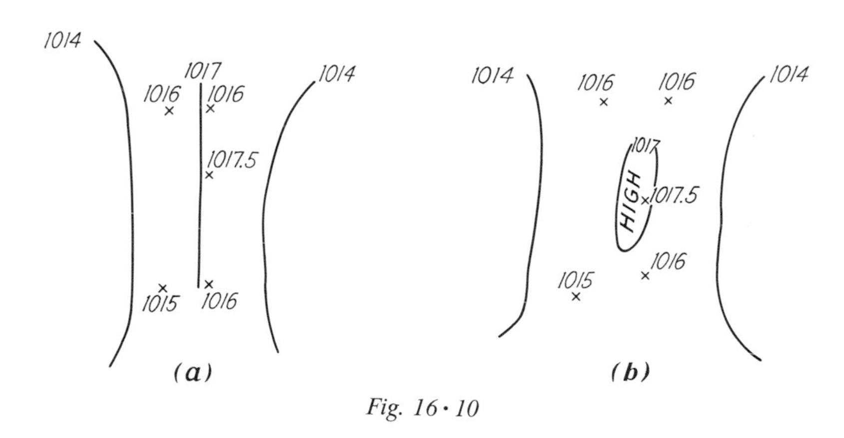

Fig. 16 · 10

of the surrounding observations shows that it actually encloses a small ridge or dome of high pressure, as in Fig. 16 · 10*b*. Now the area within the 1,017 isobar is all higher in pressure, while the outer area is lower.

4. *Isobars crossing fronts should be drawn with sharp bends, the kinks pointing to higher pressures; this brings out the pressure gradient discontinuity.* According to Petterssen's technique, the isobars on either side of a front should be continued across it as a straight line. A line drawn connecting the points of intersection of the isobars thus serves as a further accurate method of locating the front. This is illustrated in Fig. 16 · 11.

Figure 16 · 12*a* illustrates isobars drawn without careful attention to the wind direction along the axis of the low. The application of rules 2 (wind direction) and 4 gives the more correct picture shown in

Fig. 16 · 12*b*, with the front in the position *xy*. Pressure observations are here given in inches and tenths.

5. *Isobars should be drawn first where the pattern is apparently simplest and most easily seen.* Accordingly, the analysis is best started where the number of reports is the densest. On marine maps, the drawing of isobars should start at coastal regions where the reports may be more numerous than over the sea and where they will probably be more accurate. The isobars should then be extended seaward. They should also be drawn first in one air mass and then in another, starting if possible with the pressure centers. After analysis is complete within each air mass, the isobars should be continued and linked across the fronts.

Fig. 16 · 11 (Modified from S. Petterssen)

6. *Isobars, when drawn, should show a logical agreement with the preceding map.* Since weather conditions move, and often change as they move, the isobars on a new map should agree with the expected movements and changes of the pressure areas on the previous map. If the pressure centers show an erratic path of movement, it is probable that the

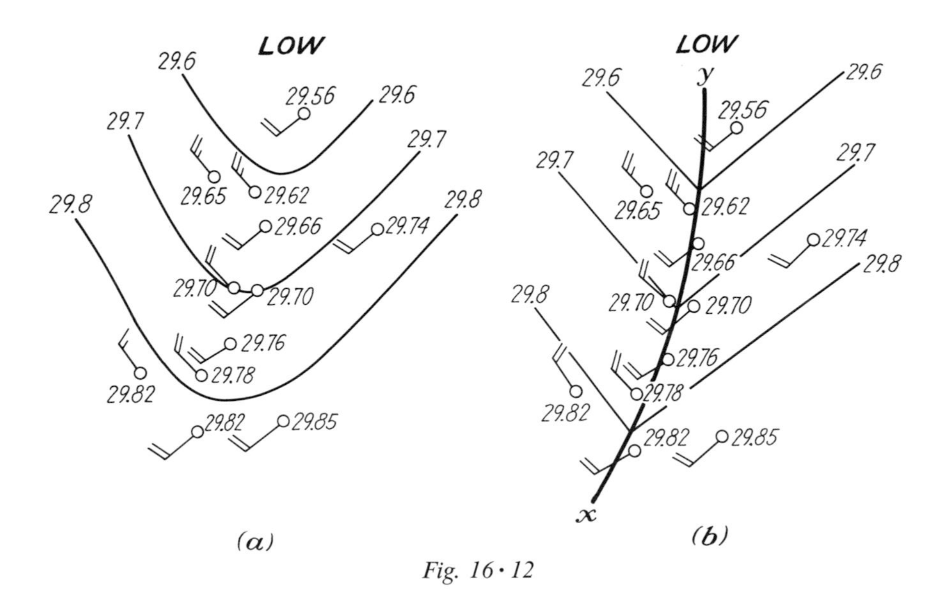

Fig. 16 · 12

new map is wrongly analyzed, provided the earlier map is correct. This may not always be the case. The earlier map may be shown to have been incorrect on the basis of the new one.

A very interesting and unusual, though by no means unique, example of the application of frontal and isobaric analysis to the problem of navigation at sea was given in a letter by Chief Officer Duggan, of the Canadian M. V. Ary Larsen, to the Hydrographic Office. Part of this follows:

"We left Belgium bound for Hampton Roads in ballast. The weather was very bad, and for three days after leaving St. Catherines we saw nothing by which we could fix our position. Our leeway was purely guesswork, and at times we estimated about five points. Our reckoning was bound to be faulty, as we had nothing to check by, and the course the ship was steering was, to say the least, wild. On the fourth day the second officer and the Captain had fixed a position by dead reckoning, and I, in the meantime, had done Rugby's weather, which showed that a very deep depression was to the west of Ireland (it took no map to show that we were in it) with a very marked cold front. I plotted our barometer reading in our assumed position, but it seemed to be a great deal out. However, I ignored it, and reckoned the possible time the cold front would meet us. I argued that either our position was wrong or that the barometer was sadly in error. Further, I argued that if we were on the isobar our barometer gave, the cold front should pass over us in about an hour, whereas if we were in the D.R. position, it would not meet us for about two hours. In approximately one hour there was a heavy downpour of rain, the wind increased to about force 10 and shifted violently and suddenly to the northwest. Later that afternoon the sky cleared and we were able to get a good fix, which found the D.R. position some 60 miles in error, and proving that the ship was on the isobar indicated by the barometer. I am not going to say that the position of the ship can be accurately fixed by means of a weather map and isobars, but it could, with a good barometer, be a good check against such errors as large as these, and which can happen in the winter time in ships such as these which carry very little ballast, and which are at the mercy of the elements, literally and absolutely."

Analysis of Air Masses

After the fronts and isobars have been drawn, attention is then given to an examination of the air masses present. Many of the recorded observations are employed in this process. The data of particular importance include temperature, movement, dew point, clouds, and present weather conditions.

1. By means of the study of the above data and reference to preceding charts, the source of origin, and hence the type, of air mass may be determined.

2. The existing characteristics of the air mass are determined from a consideration of the cloud and present weather conditions. Clouds of the cumulus or vertical development type, and associated showers, indicate convection and unstable conditions within the mass. Fog and stratus clouds, with or without light rain or drizzle, indicate stable air conditions, with a lack of pronounced vertical movement.
3. Modifications of the air mass may be estimated by noting the conditions of the underlying surfaces, over and toward which the air mass is moving. The temperature of the lower layer of air may be noticeably affected by the surface temperature beneath and cause surface air to become either warmer or colder. If the air becomes colder, as a result of a colder surface beneath, the lapse rate will tend to decrease, increasing the stability of the air. Heating of the surface air, while the upper-air temperature remains fairly constant, will increase the lapse rate and thus increase the tendency toward instability.

   Continental air moving out across sea areas tends to develop a higher humidity through increasing the water content by evaporation.
4. As explained earlier, the air-mass boundaries are indicated by the discontinuities in weather conditions prevailing at the fronts. A band of clouds of the cirrostratus-altostratus sequence is indicative of an approaching front, although it may be beyond the confines of the particular map.

   Once again it is emphasized that air-mass conditions on a map must exist in logical sequence with conditions exhibited on previous maps. We have considered here mainly a descriptive or qualitative treatment of air masses.

Weather Forecasting

Weather forecasting requires considerably more skill, training, and experience than can be acquired from the general introductory treatment given in this book. Modern procedures involve the use of upper-level as well as surface charts and prognostic charts which show the expected weather pattern for various levels at selected intervals in the future. Further, numerical forecasting, to be described later, requires the use of high-speed computers in the solution of meteorological equations well beyond the scope of this book.

Despite the complexity of professional forecasting, it is nevertheless possible to understand and apply some useful rules in making semiprofessional forecasts from surface-weather synoptic weather patterns. It is also possible to make reasonable local forecasts based only on single observations of significant weather elements.

We have already considered a wealth of facts and principles applicable to weather forecasting. For example, it can now be predicted with certainty that the temperature will normally increase toward midafternoon and decrease toward early morning. Our study of the inverse temperature and relative humidity relationship leads to an anticipation of decreasing relative humidity as the temperature rises, and an increase toward evening and early morning when temperatures fall. This routine may of course be upset by the passage of a front, introducing a new air mass with new properties.

We have noted the variations of wind direction to be expected, depending on the paths of lows or highs past a ship or station. We have considered also the changes in cloud forms and states of the sky as different types of fronts come and go, together with the nature of the precipitation attending these fronts. We have studied the basic features of air masses and the principal air masses that affect weather conditions, in addition to some of the changes suffered by these air masses in transit. A combination of all these conditions has been considered in connection with the weather structure of the middle-latitude cyclone. Further, we have learned the weather patterns and associated features of the violent tropical cyclones that beset the seas in certain areas and at certain seasons of the year. The guiding, general wind circulation of the atmosphere, on which all the above conditions are superposed, has also been examined. These weather factors have not only been studied in their relation to each other, but the physical causes for their variations have been noted as well.

FORECASTING FROM SURFACE WEATHER MAPS. Recall that weather patterns move generally eastward in the middle latitudes. However, a particular weather system may move either north or south of a due-east line. Also, conditions within the system may change as it moves. As a general procedure, forecasting involves an estimation of (1) the movement and changes in highs, lows, and other pressure configurations; (2) the movement and changes of fronts and associated frontal weather conditions; and (3) the changes in air-mass characteristics shown on the previous map.

Although modern forecasting requires the careful analysis of the upper-level wind field which guides the synoptic features seen on surface-level maps, it is with these latter that we are concerned here. Remember that the forecasting process requires a thorough examination of previous weather maps, as well as of the current one. From this examination, an attempt is made to project these conditions into the future and determine the probable appearance of succeeding weather maps, thereby showing the weather conditions 12 to 36 hours later. Thus, forecasting is the estimation of future weather conditions from a study of past and present conditions. The forecast must therefore be in logical conformance with these conditions. Based on the long history of observations and synoptic weather studies of the U.S. Weather Bureau, a number of useful general rules have been developed for use with surface weather maps. It should be realized, however, that each situation must be interpreted within the weather pattern at the time and that actual professional forecasting is based more on upper tropospheric than on surface patterns.

GENERAL FORECASTING RULES. It should be realized that the following are general rules that may not apply in individual cases but are of value in the absence of more complete weather data and forecasts. Also, professional forecasting relies more on the evaluation of upper tropospheric than on surface patterns.

Rules Related to Lows

1. An occluded low will move very slowly or remain almost stationary until it dissipates.
2. In general, lows tend to follow the record of previous days, that is, their historical sequence; they will also tend to move toward regions of greatest rainfall during the last 24 hours.
3. Secondary depressions tend to move in the direction of the wind circulation around the primary low, moving faster in winter than in summer.
4. Lows moving southeastward (northeastward in the Southern Hemisphere) usually increase in intensity on recurving to the northward (southward in Southern Hemisphere), but their movements will be slow while recurving.
5. Lows with isobars closely crowded on the west and northwest (west and southwest in the Southern Hemisphere) generally move slowly to the east or southeast (east or northeast in the Southern Hemisphere).
6. Rainfall and strong winds will continue for a longer period than usual in western and northern sections of the low (western and southern sections in the Southern Hemisphere).
7. A young low usually moves in the direction of the isobars in the warm sector.
8. Until a low occludes, its rate of movement increases.
9. After a low has occluded, its speed decreases until the maximum deepening of its center occurs.

Rules Relating to Highs

1. Ridges of high pressure between lows tend to move in the same direction and with the same speed as the lows.
2. Highs that separate a series of lows tend to move southward (northward in the Southern Hemisphere).
3. In general, highs move with the same speed and in the same direction as the frontal cyclone (low).
4. Small (closed) highs usually move faster than large ones.

5. Warm highs tend to move slowly or become stationary.

Rules Applicable to Pressure Centers

1. When the warm sector narrows, the disturbance deepens until occlusion takes place.
2. When fresh cold air intrudes in a filling depression, it will start to deepen again.
3. In winter a low will usually deepen when passing over the ocean from land; in summer, deepening usually takes place as the low passes from ocean to land.
4. A fully occluded depression rarely deepens, but it may fill up slowly or persist for several days.
5. Old depressions may fill up rapidly when new deepening lows move into their circulation.
6. Large depressions, when completely occluded, move very slowly and sometimes on an irregular course.
7. Depressions tend to move around large, warm, well-established highs in the direction of the air flow around the boundaries of the highs.

Rules Applicable to Fronts

1. Fronts tend to dissipate (frontolysis) in anticyclonic regions.
2. When cold air lies to the left of the general air flow, the sharpness of the front increases.
3. Fronts with warmer air in the cold air mass to the northward (southward in the Southern Hemisphere) are short-lived.
4. A fast-moving cold front soon dissipates; warm fronts usually move more slowly than cold fronts.
5. When the pressure increases slowly after passage of a cold front, continued poor weather is indicated; a rapid rise in pressure and drop in temperature following a cold-front passage indicate clearing weather.
6. A nonfrontal depression tends to move in the same direction as the strongest winds circulating around it, that is, in the direction of the isobars where they are nearest together.
7. The more a secondary depression deepens, the more it approaches the center of the primary depression. Eventually it will absorb the old primary and become the primary depression itself.

Rules Relating to Weather along Fronts

1. Frontal precipitation normally will be more intensive the sharper the frontal convergence of winds as indicated by the angle of isobars and the resultant shift of the wind.
2. The prefrontal rainfall zone on a warm front will be narrow if the axis of the high-pressure ridge ahead is relatively near the warm front.

3. An extended area of prefrontal precipitation occurs when there is a strong pressure gradient within the warm sector and no marked ridge of high pressure ahead.
4. In subtropical latitudes, weather activity of cold fronts is more pronounced than that of warm fronts; in polar latitudes, the greater weather activity accompanies warm fronts and occlusions of the warm-front type.
5. A slow-moving cold front normally has a broader zone of precipitation than a rapidly moving cold front. In this connection, there may not be a continuous zone of precipitation along a rapidly moving cold front but some prefrontal showers and squalls instead.
6. A continuous precipitation zone may be expected along and behind a slowly moving cold front but no squalls and showers.

FORECASTING FROM LOCAL INDICATIONS. It is only in recent years that the mariner has had the aid of radio bulletins to provide synopses and forecasts of weather conditions at sea. As a result many, if not most, mariners have developed a so-called "weather sense." Actually, they are simply drawing conclusions after noting certain observational conditions with their results during a period of years at sea. We have now studied the scientific basis for many of these conclusions.

From the general knowledge of weather conditions and the factors underlying their changes, one can, by observation, make local predictions for short periods, the accuracy of which will increase with practice. Fairly accurate forecasts may be developed for 12 hours in advance of the observations. At times, the forecast may be safely extended for 24 hours, but this is definitely the limit in the middle latitudes.

The process of forecasting from local observations consists essentially of relating one's observations to the general pattern of weather. Thus, by knowing this relation of observed conditions to the moving weather pattern, a particular set of observations indicates what part of this pattern is in the observer's vicinity and therefore what part is likely to approach with its associated weather.

At the very outset of this book, we stressed the fact that clouds and wind direction were the most important local features to be observed for the purpose of weather determination. The behavior of the barometer may further clarify the picture. It might be well to consider a few examples.

Observations of the wind direction enable the placing of the direction of the low-pressure center. When the wind sets in from northeast or southeast, a low is approaching. Observations of the barometer may confirm this. The relation of wind shifts and approaching lows was completely developed in Chap. 11. An even earlier indication than either the wind or the barometer is the type of cloud observed. Thus, the appearance of cirrus and cirrostratus clouds in the western sky may presage the approach of a distant warm or occluded front. If these clouds move eastward and slowly thicken and lower, together with a fall in the barometer and a shift of the winds to an eastern quadrant, the approach of the warm or occluded front seems fairly certain.

Whether the low-pressure center, from which the front extends, will pass to the north or south of the observer is determined from the wind direction, as explained earlier. If the wind direction indicates a passage to the north, weather changes may be anticipated in accordance with the analysis of cyclone structure south of the center, as given in Chap. 14. It is of course very difficult, if not impossible, to judge whether or not the approaching front is occluded. If it should be, no warm-sector conditions will pass.

Should the wind indicate the passage of the low to the south, again no warm sector will be observed. Rather there will be a continuous overcast with moderate precipitation, ending in heavier precipitation or showers, and with backing winds followed by clearing and invariably colder weather.

In general, in the middle latitudes of the Northern Hemisphere, a shift of the wind to the northwest, with or without typical cold-front showers, is the common associate of clearing weather. Fair-weather cumulus often prevails, serving as a further indication that the storm area has moved away to the east.

Some common weather proverbs may be mentioned in connection with forecasting from local observations:

1. When the dew is on the grass,
   Rain will not come to pass.
2. A morning fog that obscures the sun's ray
   Indicates the coming of a clear day.
3. Mackerel sky, twelve hours dry.
4. A veering wind means weather fair,
   A backing wind, foul weather's near.
5. Rainbow at night, sailor's delight,
   Rainbow in the morning, sailor take warning.

Proverbs 1 and 2, of course, refer to conditions that require clear nights, and hence some time lapse is necessary for any bad weather that may follow. Proverb 3 refers to cirrocumulus which is often associated with the outer fringe of the stratus cloud deck preceding the warm front. As a rule, precipitation will not fall for another 12 hours, until the front draws much closer. The fourth proverb may often, but not necessarily, be true. The reader can explain this one for himself. The fifth is slightly more involved. Rainbows are common features of showers or thundershowers and are always seen in the opposite part of the sky from the sun. A rainbow at night, with the sun setting in the west, must be in the eastern sky, indicating that the storm has already passed. A rainbow in the morning must, in the same way, exist in the western sky, thereby heralding the approach of the storm.

Numerous other weather adages and examples of local forecasting could be cited. On the whole, however, such forecasting simply requires that the observer explain and interpret to himself the conditions he may see, drawing on his knowledge for this purpose.

Numerical Weather Forecasting

For the most part, the weather rules described above represent a qualitative and often subjective attempt to estimate quantitative values involved in moving weather systems. The true motions involved are governed by complex hydrodynamic equations. In the past, the solution of these equations involved the substitution of reliable observational data and an amount of time much longer than that for the forecast to be realized. Now we have better data and high-speed computers which permit solutions to be obtained in a matter of minutes.

At present, weather information can be transmitted directly from a teletype circuit to the computer, which checks and stores the data until the computation is begun. This application is especially valuable in the preparation of forecast (prognostic) maps of upper-level winds and other conditions. Forecasts up to 72 hours in advance are made in minutes. It has been found that the machine-produced forecasts are at least equal, and often superior, to maps developed by qualified weather forecasters using the more qualitative and subjective methods. The U.S. Weather Bureau now produces upper-level wind charts on a routine basis by the machine method.

However, the problems involved in developing operational forecasts of all of the weather elements are so complex that these are still carried out by experienced forecasters using the numerical weather charts as guides, particularly in determining the control of upper-level circulation on surface weather systems.

Extended Weather Forecasts

Extended or long-range forecasts attempt to give a somewhat generalized weather picture for intervals up to 30 days. At present, both 5- and 30-day forecasts are issued, the latter type being revised after 15 days, if necessary. Although the procedure and theory involved are beyond our present scope, an outline of the technique can be given in order that some understanding of the nature of the forecasts can be gained. Extended forecasts give a rather statistical statement of certain weather elements—commonly temperature and precipitation, rather than the detailed picture of the weather elements common to the more traditional short-range forecast.

The latter involves the instantaneous synoptic weather pattern at the ground and upper levels for a confined part of the hemisphere—mostly to the west of the station as described previously. Clearly, the weather over a prolonged interval involves not only the adjacent areas, but the state of the atmosphere over the entire hemisphere. It also includes the effects not only of instantaneous synoptic conditions, but also of the changes that occur with time as weather patterns move. This in turn involves aspects of the atmospheric circulation over the entire hemisphere. The extended range procedure therefore attempts to determine trends and changes in the general circulation to which other weather changes can be connected.

In order to "smooth" very short-period changes that may be unimportant in the extended outlook, the method uses mean or average values of the general circulation—usually pressure values for upper levels in particular. Means over a five-day interval have been found very useful in the work so far. In the forecast procedure, a mean-pressure chart is usually drawn for an upper level (700-millibar level) first, and from this a mean sea-level pressure chart is constructed. Continuity with observed conditions during the preceding (5-day) interval is quite critical in this work.

Based on the mean picture of the expected circulation, prognostic maps of temperature and precipitation are prepared with reference to normal conditions for the region. The forecasts of temperature are made for specific areas in terms of the departure from normal using the descriptions *near normal, above or below normal* and *much above or much below normal.* These terms are defined on the basis of local frequency of occurrence of the values being forecast; that is, normal and above and below normal values each occur one-quarter of the time, and much above and much below values occur one-eighth of the time. Precipitation in the 30-day forecast is given as *near normal, above normal, below normal.*

Figure 16 · 13 shows examples of prognostic charts prepared in accordance with the procedure sketched above. From the mean sea-level chart (upper right) prepared first, the mean sea-level chart (upper left) was developed. The analysis of these charts in terms of mean weather led to the 5-day mean prognostic charts of temperature departure and precipitation totals, lower left and right, respectively.

It is hoped that continued basic research on the general circulation, and refinement of the procedures involved, will give improved accuracy to extended forecasts.

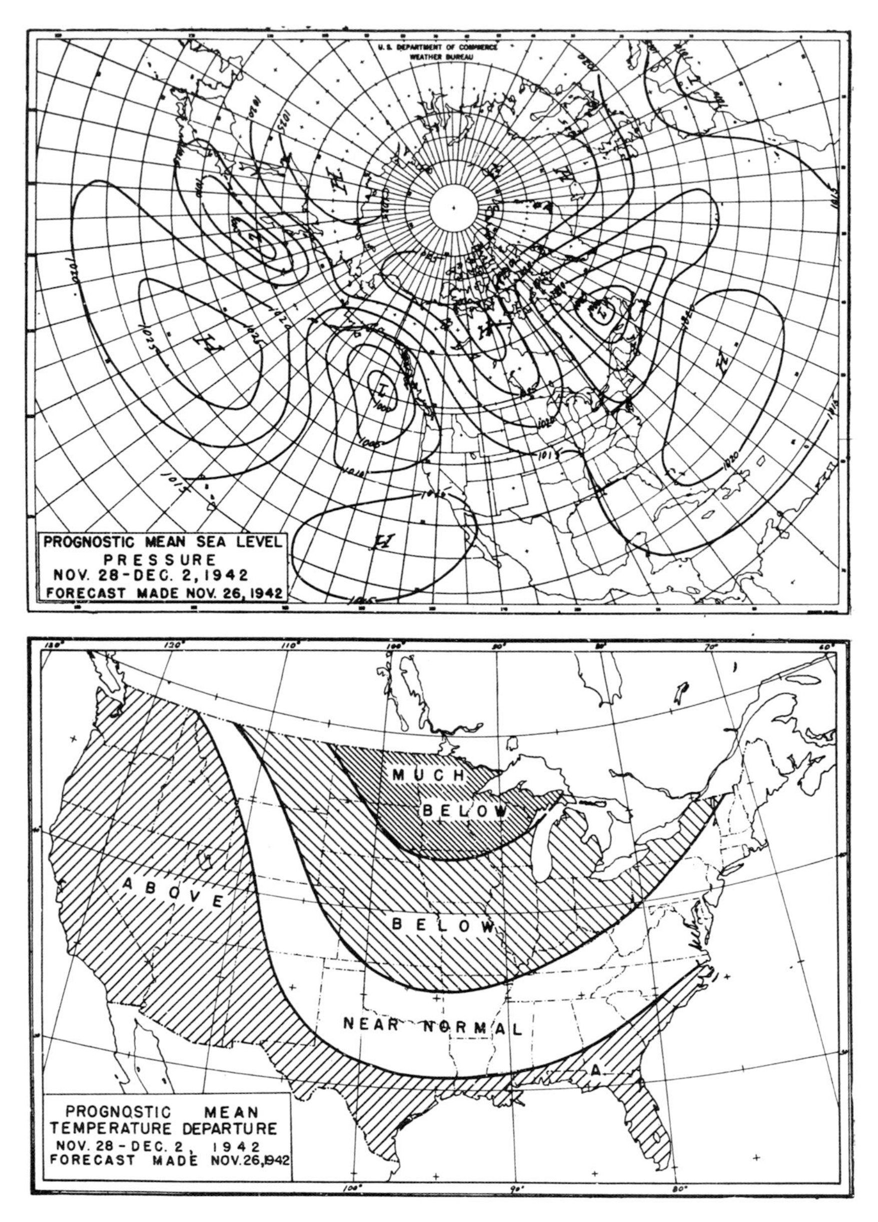

Fig. 16·13 Prognostic charts for the period November 28 to December 2, 1942 used

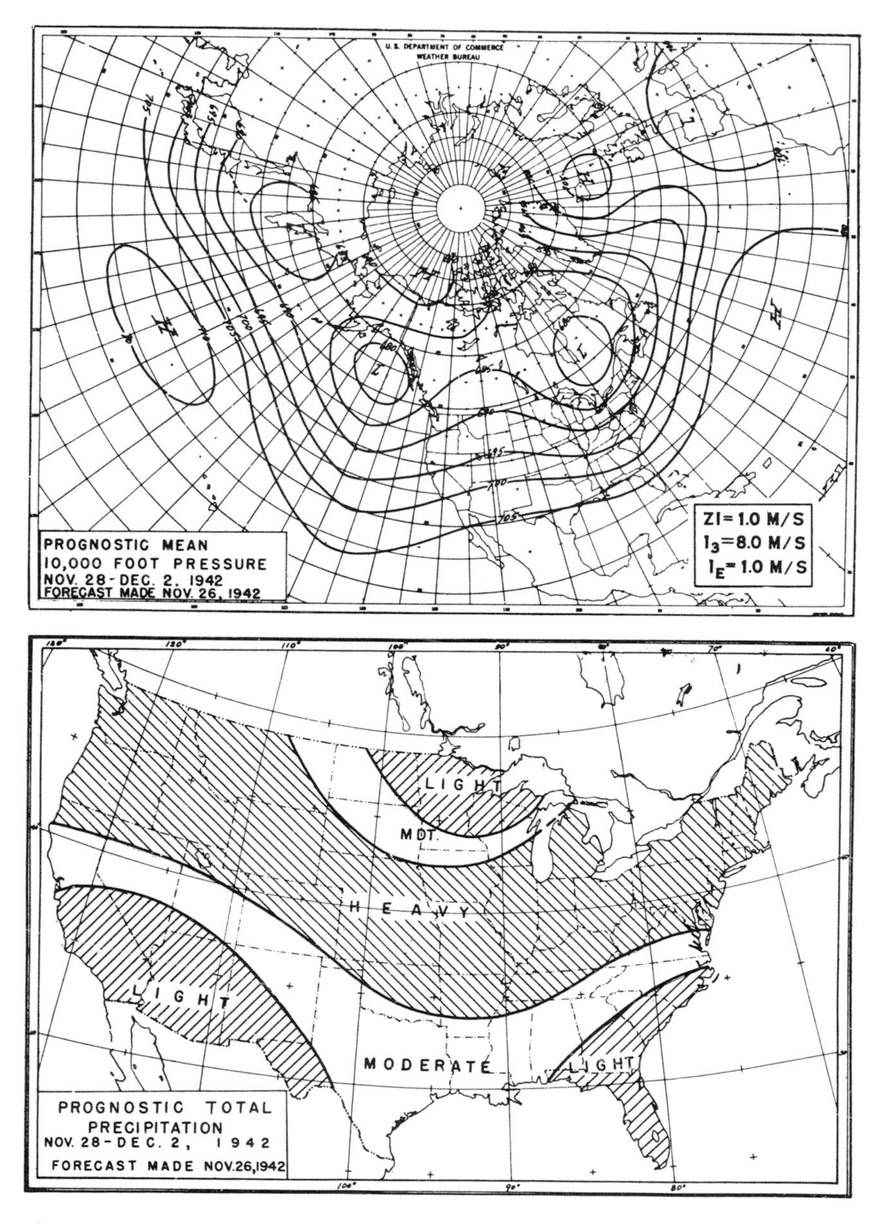

in five-day forecasts. (J. Namias, U.S. Weather Bureau)

EXERCISES

16 • 1 By reference to Table 16 • 1 and the tables of codes and symbols in Fig. 16 • *a* to *d*, complete the coded message for the following observations:

Station Bermuda: 016
Total amount of cloud cover: five-tenths
Wind direction: from due south
Wind speed in knots: 25
Visibility: 7⅛ miles
Present weather: continuous slight rain
Past weather: rain
Barometric pressure in millibars: 1,008.4
Current air temperature: 58°F (Convert to °C)
Fraction of sky covered by low or middle clouds: entirely overcast
Low-cloud type: fractostratus and fractocumulus
Height of cloud base: 1,400 feet
Middle-cloud type: thick altostratus
High clouds: none
Dew point: 55°F (Give in °C)
Character of barograph trace: falling, then steady
Pressure tendency: −1.8 millibars
Indicator figure: 7
Amount of precipitation: 0.18 inch
Time precipitation began: 8 hours ago
Depth of snow: none

16 • 2 Plot the above information on a station model.

16 • 3 Describe the criteria of chief importance in the location of fronts in the course of weather map analysis.

16 • 4 Explain the reason for the rule in forecasting that cyclone centers usually move parallel to the isobars in the warm sector of the low.

16 • 5 Justify the statement that in the middle latitudes of the Northern Hemisphere a shift to southeast winds plus thickening cirrostratus clouds usually indicate the approach of bad weather.

16 • 6 Compare and explain the differences in precipitation associated with warm, cold, and occluded fronts.

16 • 7 What will be the principal differences in weather experienced during the passage of a low in the Northern Hemisphere by observers to the north and south of the center?

16 • 8 Summarize the steps in the preparation of extended forecasts.

CHAPTER 17

Weather at Sea

Throughout this book, marine applications of meteorology have been introduced whenever appropriate. However, in addition to the maritime references already given, a large amount of special information and procedures is available for the professional mariner and the deep-water smaller-boat navigators.

Weather Bulletins and Data Broadcasts

The meteorological agencies of many maritime nations make arrangements for plain-language radio broadcasts giving current weather summaries as well as marine forecasts twice and in some cases four times daily. These broadcasts, which include special storm warnings when appropriate, apply to a large part—often the particular quarter—of the ocean adjacent to the country's coastline. The United States, having coastlines on two major oceans, transmits forecasts from appropriate coastal stations for each of the oceans involved.

In addition to the verbal weather synopsis and forecasts, coded weather observations appropriate to the area in question are rebroadcast after collection, either at the end of a regular marine shipping broadcast or according to a separate schedule. The times and station frequencies for all of these weather transmissions are readily available from the meteorological services in any major coastal city.

After received aboard ship, the coded weather data can be plotted for each ship or station in accordance with the station model described in the preceding chapter. The message can be decoded by means of the complete code tables available in various forms from the U.S.—or other—Weather Bureaus. The plotted data are then analyzed to give a complete marine weather map that may be of particular use to the mariner. Detailed examples of the decoding, plotting, and analysis of marine data can be found in U.S. Weather Bureau Circular R, *Preparation and Use of Weather Maps at Sea.*

Meteorological agencies of some countries (the U.S. Weather Bureau, for example) also transmit especially valuable bulletins in the form of code messages that describe actual analyzed weather conditions, which greatly simplifies the drawing of weather maps at sea. The information included in these messages describes the types, characteristics, central pressures, locations, courses, and movements of low- and high-pressure systems and gives positions of points for use in drawing fronts and isobars. Numerical values of the isobars as well as the types of fronts are also included. The weather data encoded for these broadcasts are taken from maps which are plotted and analyzed at central forecasting offices of the particular agency.

In addition to broadcasting all of the above aids for the preparation of marine maps, the U.S. Weather Bureau also transmits maps completely drawn and analyzed at the joint Weather Bureau–Air Force–Navy analysis center. These maps can be recorded on land or sea for the particular area desired on commercial radio-facsimile recorders. The entire recording operation is completely automatic. As weather maps of different regions (sections) are broadcast according to a regular program available from the U.S. Weather Bureau, the recording time can be easily determined in advance.

Use of Radar

Although examples of radar photographs of storms and clouds have been given previously, the only formal reference to the meteorological use of radar has been given in connection with rawins in the section on methods of observing winds. In addition to this valuable use, a more direct and less technical application is available for any ship equipped with a radar system employing a PPI (Plan Position Indicator) scan.

Radar gives the observer up-to-the-minute information on clouds, precipitation, and storm movement as well as providing a means for detecting otherwise unknown storms. It is not possible for synoptic meteorological procedures to give the specific and detailed information possible by this means.

The ability to "see" weather depends on the radar returns or echoes from meteorological sources. These sources consist of light, moderate, and heavy rainfall, hail, sleet, and snow, the latter not being a good reflector at all times. The intensity of the echo is a function of the size of the droplet, the frequency of the system, and the total amount of water particles per unit volume of atmosphere. Returns do not occur from fair-weather cumulus or very light rain or drizzle.

The weather features capable of good radar detection and identification are: thunderstorms, active convective clouds, well-defined fronts, hurricanes and tropical storms, tornadoes and waterspouts, and occasionally dense stratus-type clouds.

THUNDERSTORMS. A thunderstorm echo is usually one of the most easily detected meteorological radar signals. An individual thunderstorm appears as an isolated, bright, dense central area with an indistinct boundary. The cumulonimbus structure of the storm can be identified from the large vertical extent of the cloud echo when elevation angles from top to bottom can be obtained. As several square miles are covered by single storms, the size can be checked from the range and angle grid on the scope. Thunderstorms of a simple thermal convective nature within air masses are usually scattered rather randomly compared to the organized storm patterns related to fronts.

COLD FRONTS AND SQUALL LINES. An active cold front is commonly characterized by a band of well-developed thunderstorms and is identified on the radarscope by the linear pattern of strong cloud echoes. The clouds are very bright, with a rounded and solid appearance. A qualitative estimate of the structure and activity of the front can be made from the intensity of the cloud echoes, the spacing between the bright areas, the area covered by the individual clouds, the vertical extent of the clouds, and the velocity of the cloud line as measured on the radarscope. Weak cold fronts are usually much less well defined and may be missed entirely by radar if convective activity is low.

Squall lines are usually narrower than frontal zones but are otherwise much the same in appearance and in convective activity. They commonly precede cold fronts, so that their proper identification from radar and other sources may provide a means of forecasting the cold front to follow.

WARM FRONTS. Cloudiness and precipitation related to warm fronts cover a very wide area. Radar echoes from the characteristic stratus-type clouds which may yield precipitation are usually hazy and extend over much of the range. Regions of differing intensity of rainfall are marked by radar returns of variable brightness. The outer boundaries of the precipitation areas are much more diffuse than occurs for thunderstorms. When unstable warm fronts exist, scattered thunderstorms may develop which will be marked by typical solid, bright echoes embedded in the overall hazy return.

HURRICANES AND TROPICAL STORMS. Some discussion of hurricane radar observations with related examples has been given previously in the chapter on hurricanes. In summary, a hurricane appears as a very large mass of cloud echoes all extremely intense, and especially so when thunderstorms are present. When the center of the storm is within the range of the PPI, the concentric or tightly spiraled cloud organization is quite characteristic. If the center is beyond range, its location can be well approximated from the nature of the curvature of the cloud bands.

In the application of radar to weather, continued examination of the features described will provide very valuable experience for the observer. Its use will greatly supplement all forms of weather information available from remote sources and will permit much more efficient use of this information.

Ship Routing

The application of climatological knowledge of marine winds, waves, and ocean currents, together with the meteorological information of existing and forecast winds and waves, has been shown to have a tremendous advantage in the routing of ships. Use of such routes gives increased safety to ship, passengers, and cargo when security is the requirement. From the standpoint of economy, large savings can be realized through very significant savings in fuel or in time of crossing or both.

In general, *heavy* weather requires reduced speed, or, to maintain speed, an increased fuel consumption, as well as increased hazard. The development of *optimum ship routing* followed from earlier meteorologic and oceanographic procedures involved in amphibious operations during World War II. At present, an *optimum track* is determined for a particular type of vessel from long-range forecasts of winds, waves, and currents and modified on the basis of current weather. Tracks can be prepared on the basis of least time, maximum safety and comfort, minimum fuel consumption, or a combination of these goals. When the former is the primary consideration, the plan is called a *least time track,* which is usually favorable for the other factors as well.

The U.S. Navy and the U.S. Military Sea Transportation Service (MSTS) now practice this procedure routinely as do some commercially operated vessels. The U.S. Navy Oceanographic Office (formerly Hydrographic Office) supplied the MSTS with 1,000 routes during a two-year interval with an average reduction in travel time of 14 hours plus other advantages. Savings of two million dollars a year have been estimated from the time reductions.

Ship routing may be based on the development of an individual track for each crossing or upon the use of climatological information. The former is only as accurate as the ability to forecast wind and wave conditions well in advance. The latter, using average conditions, is based more upon a probability that certain conditions will be encountered.

INDIVIDUAL ROUTING. The U.S. Oceanographic Office has developed the following procedure, used with the success referred to previously. In estimating the expected wind and wave conditions, the primary tool is the

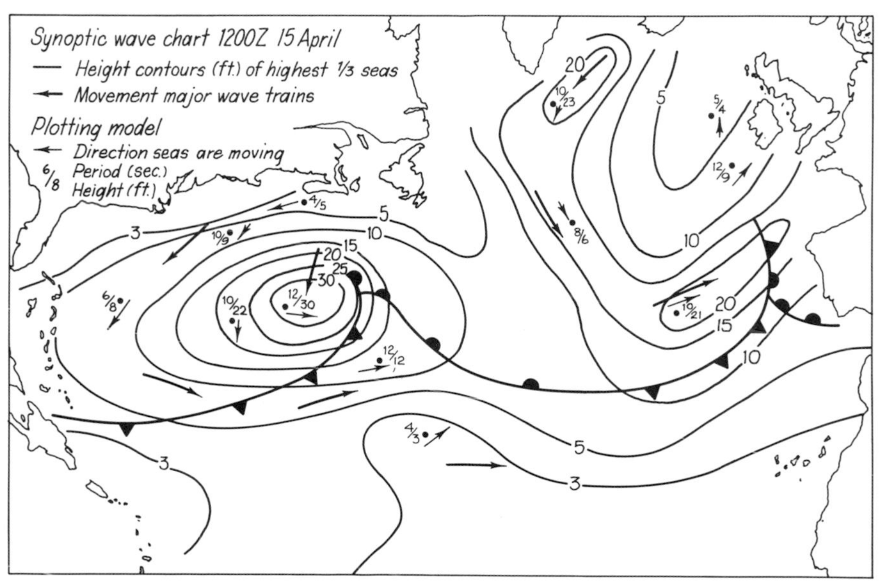

Fig. 17 · 1 Synoptic wave chart of the North Atlantic Ocean used in ship route construction. (G. Hanssen and R. James, Journal of the Institute of Navigation)

5-day sea-level prognostic weather map series issued several times a week by the Extended Forecast Section of the U.S. Weather Bureau. A further basic data guide is the synoptic wave chart which is constructed from marine radio reports of wave heights, period, and direction of travel. This information, together with contours of equal wave height, is plotted for the ocean area involved. Figure 17 · 1 is an example of such a map together with meteorological fronts for reference. (Note that the continuous lines represent equal wave height, not isobars.)

Of greater aid is the prognostic wave chart, which shows expected wave characteristics forecast on the basis of wind estimates obtained from the prognostic sea-level pressure maps referred to above. The prognostic wave chart corresponding to the time of the synoptic wave chart in Fig. 17 • 1 is shown in Fig. 17 • 2, where the continuous lines are again wave-height contours, not isobars. Although winds also directly affect the speed of the vessel, depending on whether they are head, beam, or following winds, the effect of the waves is more serious and more direct. Also, in allowing for wave conditions, an automatic optimum allowance for wind is essentially built into the procedure.

The meteorological conclusions must be interpreted in terms of ship size and type as they affect performance under different wave conditions.

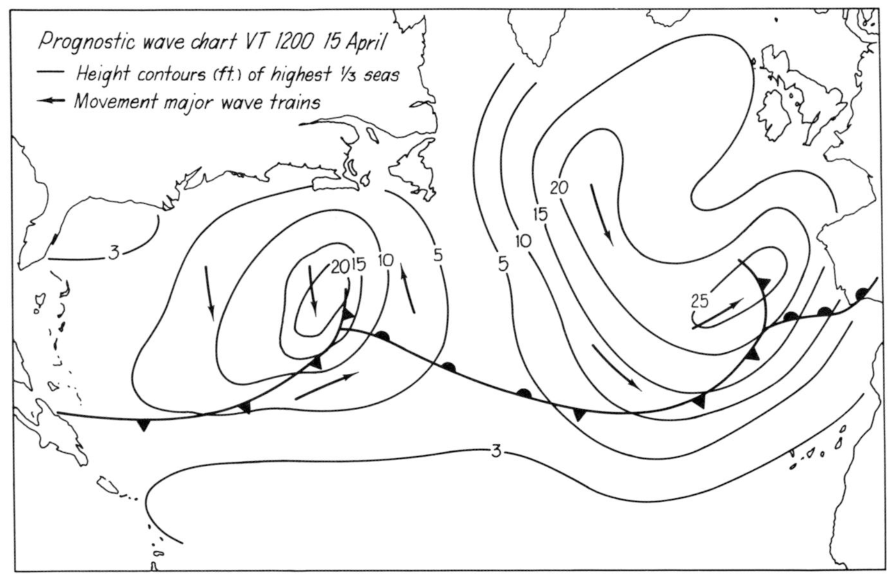

Fig. 17 • 2 Prognostic wave chart used in ship route construction. (G. Hanssen and R. James, Journal of the Institute of Navigation)

In this evaluation, ship performance curves, as shown in Fig. 17 • 3, are utilized. From these curves, developed by R. W. James, the reduction in speed of standard ship types can be determined for head, following, and beam seas.

All of these procedures, including daily modifications, are used in constructing a number of daily tracks plotted on both sides of the great-circle route—the shortest navigable track, as in Fig. 17 • 4. In this figure, A is the point of departure and B the destination. The lines S_1, S_2, S_3, etc., connect the ends of the daily tracks constructed for each day. After 5 days, the shortest distance between point B and the line S_5 indicates the point to

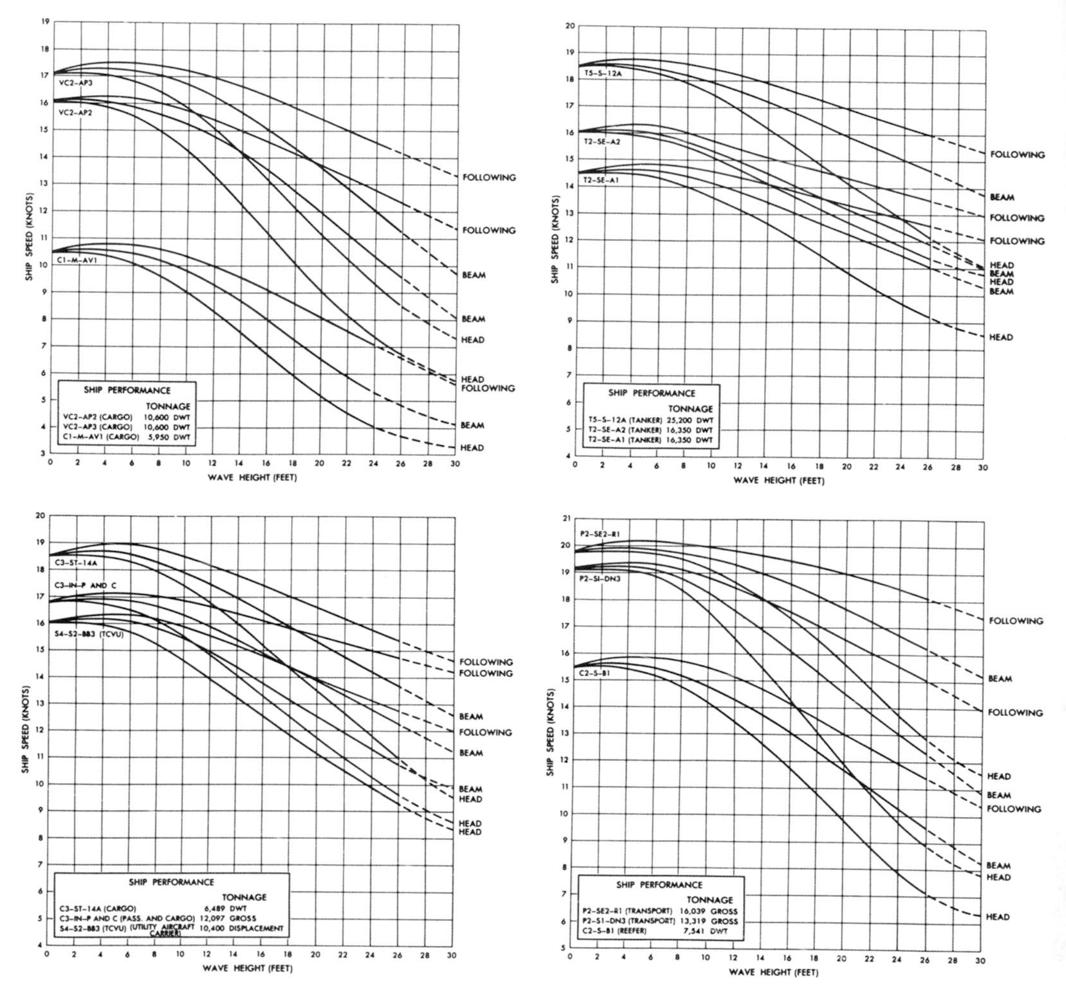

Fig. 17 • 3 Examples of ship performance curves for standard types of commercial vessels giving normal running speed for waves from different directions. Dashed lines indicate estimated data curves revised June, 1959. (R. W. James, U.S. Oceanographic Office Special Publication 1)

which the least time track from A should be drawn. It may be expected that this procedure will be modified from time to time as more experience and knowledge are gained.

CLIMATOLOGICAL ROUTES. In the absence of individual route construction, considerable advantage can be gained from the use of tracks constructed on the basis of average or climatological sea conditions—particularly ocean waves. It now seems well established that ship-speed reduction required by heavy seas will be encountered more often and for longer duration when traveling standard great-circle lanes than when traveling routes based on climatic data. A study by Holcombe, MacDougall, and Perlroth, of the U.S. Naval Oceanographic Office, of a large amount of

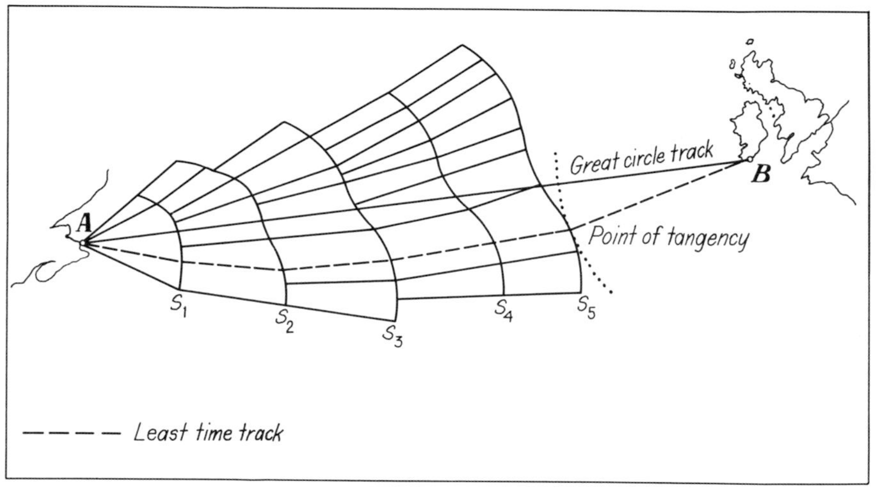

Fig. 17 · 4 Construction of least time track. (G. Hanssen and R. James, Journal of the Institute of Navigation)

available ship's logs and climatic data has shown that the great circle route represents the fastest course only 13 percent of the time for eastbound ships and 2 percent of the time for westbound ships in the Atlantic Ocean.

Examples of how wave climatology can provide more efficient and more comfortable Atlantic crossings are shown in Fig. 17 · 5. In the charts here, the percentage of time that reduction in ship speed is expected for various headings during January is shown by the curved lines. The lines C and C' are the traditional rhumb line–great-circle tracks determined originally by the southern limit to which icebergs may be expected. The lines X and X' indicate alternative routes, based on these charts, which would save time and fuel and give increased security to passengers, crew, and cargo.

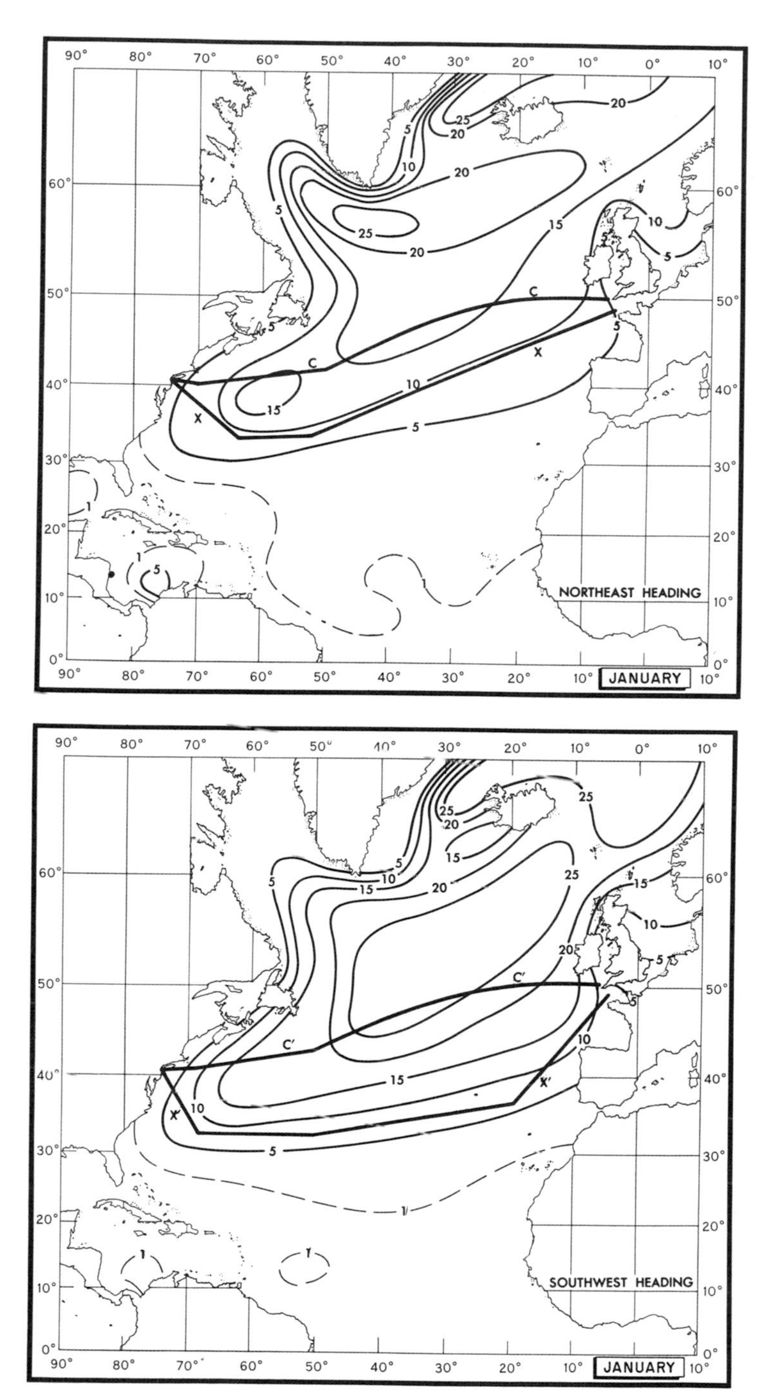

Fig. 17 • 5 Charts showing the percentage of time that reduced ship speed is required for different headings in January. (Continued on next page.) (E. Joseph and J. Kipper, Jr., U.S. Naval Oceanographic Office Publication TR-148)

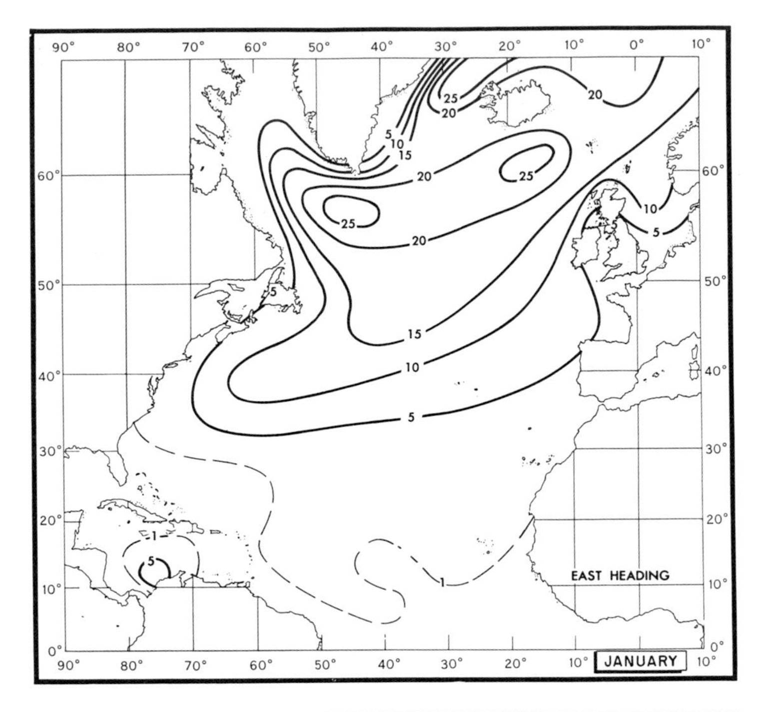

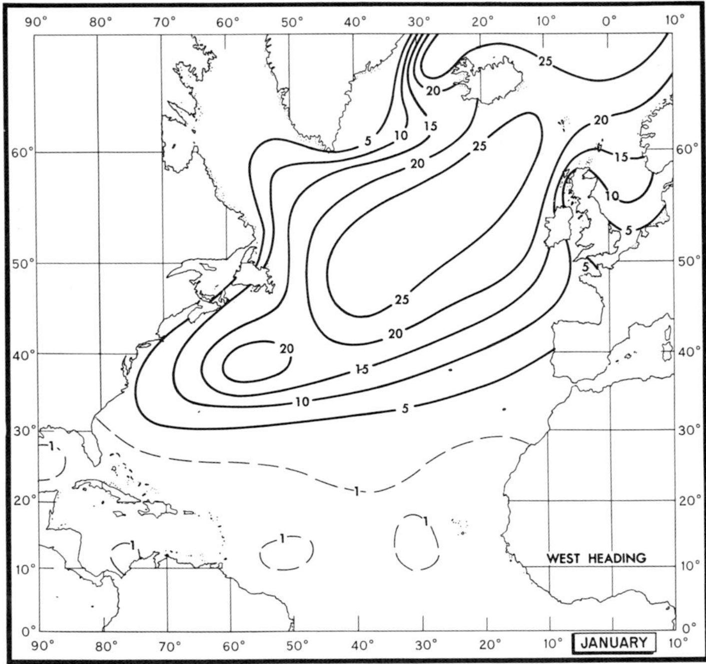

Fig. 17 · 5 (continued)

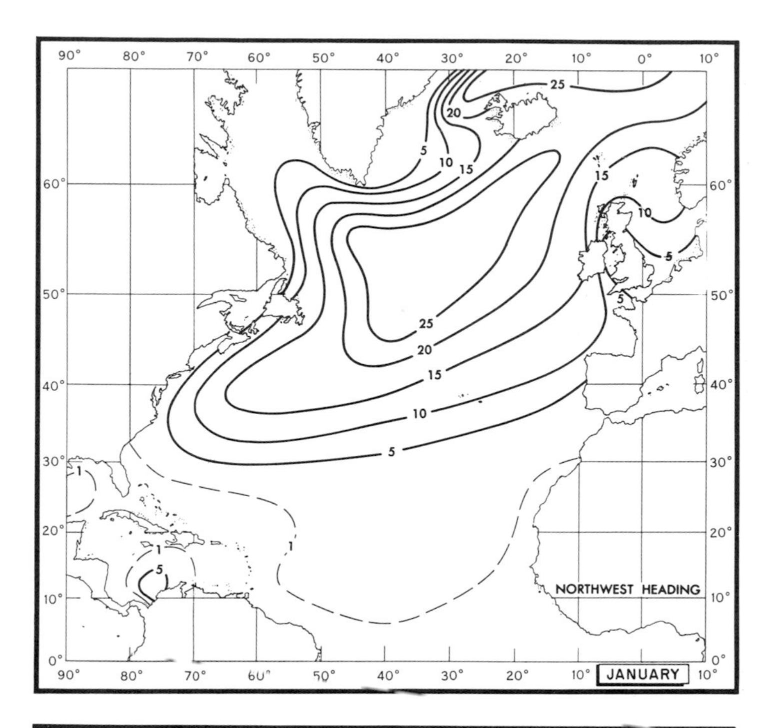

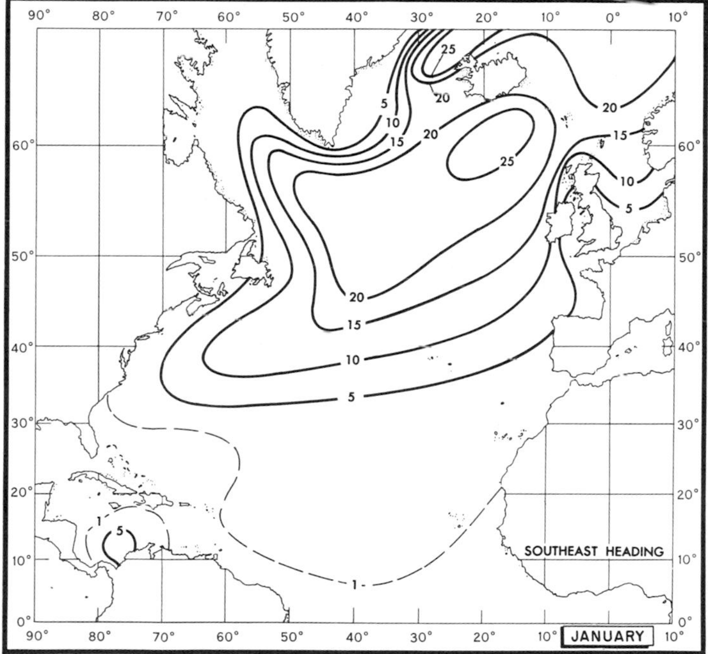

Fig. 17 · 5 (continued)

Similar charts for every month of the year can be found in the U.S. Naval Oceanographic Office Report TR-148, *Wave Climatology as an Aid to Ship Routing in the North Atlantic Ocean,* by E. J. Joseph and J. M. Kipper, Jr.

Trip Analyses

By correlating a log of carefully made observations of wind and waves with the log of actual distance covered versus the number of shaft miles covered, a very good appreciation can be gained of the effect of weather on a particular ship's voyage. An actual example of such an analysis from the records of the captain of the *U.S.S. Beckham* on a voyage from San Pedro to Samar has been given by L. Allen of the Office of Naval Research. The significant information is summarized in Fig. 17 · 6.

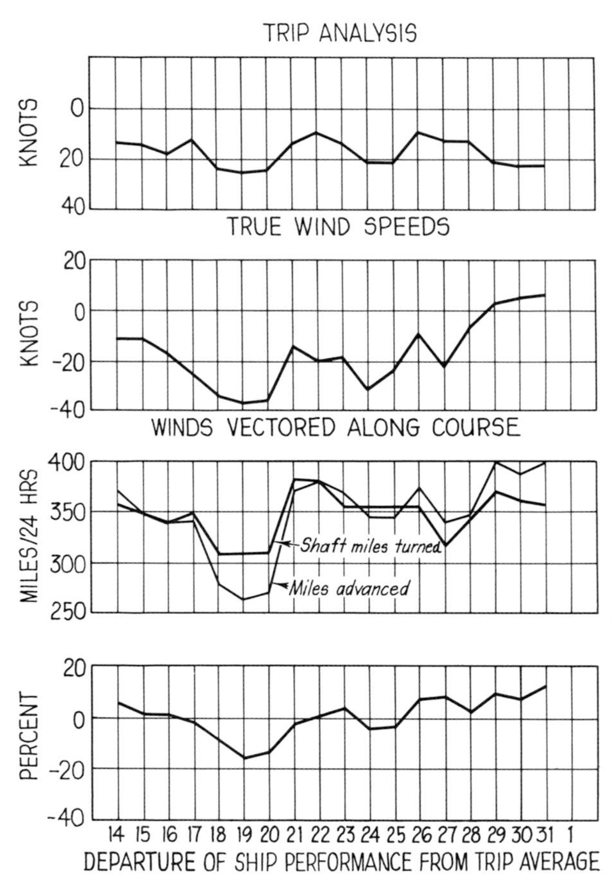

Fig. 17 · 6 Trip analysis of U.S.S. Beckham on voyage from San Pedro to Samar, December, 1945. (L. Allen and American Meteorological Society)

In this diagram the first (upper) graph gives the true wind speed averaged for each day from hourly observations for the interval December 14 to January 1, 1945. The second graph shows the component of wind speed parallel to the ship's course; positive values refer to stern winds and negative values to head winds. The third graph compares actual miles advanced in 24 hours (thin line) with the *average* number of miles expected in 24 hours on the basis of the number of shaft rotations (heavy line). The last graph gives the percent difference between the ship's performance and the trip average and is determined from the difference between the two curves in the third graph. In a qualitative way, it can be seen readily from a comparison of these curves that the ship's performance was generally below average during the first half of the voyage and exceeded the average performance during most of the latter half. The best performance occurred on January 1, when a 23-knot wind resolvable into an 8-knot stern wind resulted in a 40-mile excess over the trip average. Certainly any individual or climatically selected route that would give the vessel such benefits on a routine basis would contribute to greater efficiency and economy.

Average Weather for the North Atlantic Ocean

Enough meteorologic data have now been gained for both the North Atlantic and North Pacific Oceans to permit quite reliable summaries of average monthly marine weather to be developed. Although such summaries may not pertain to a specific part of the ocean at a specific time, they certainly indicate to a mariner or passenger what kind of weather and seas may be *expected* on a particular voyage across these oceans. The monthly summaries which follow are taken from the *Mariners Weather Log* of the U.S. Weather Bureau.

JANUARY—This month is generally characterized by rough weather over the middle and northern latitudes of the North Atlantic. LOWS frequently become deep and extensive, and associated winds often reach gale and sometimes hurricane force. The Icelandic LOW, average central pressure 995 mb. (29.38 in.), centered over the western Denmark Strait, has both its lowest pressure and greatest areal extent of the year. The Azores HIGH with a central pressure of about 1025 mb. (30.27 in.) covers a belt from the Strait of Gibraltar west-southwestward toward the Florida coast.

North of 40°N. the prevailing winds are westerly over most of the ocean. The average wind speeds are predominantly force 4 to 6, except south and east of the southern tip of Greenland where they are force 5 to 7, and in the Labrador Sea and Davis Strait where they reach force 6 to 8. Between 25° and 40°N. wind directions are mostly variable, except northerly near the European coast and a trend toward southwesterly along the American coast. Within this belt wind speeds force 3 to 5 predominate. From 5° to 25°N. the northeast trades persist with great regularity. More than 70 percent of the time, wind speeds range from force 3 to 5.

Gales, force 8 and higher, occur in 10 percent of the observations north of 35°N. over the western part of the ocean and north of 40°N. over the eastern part. The highest frequency, 30 percent, is found between 55° and 60°N. south and southeast of Greenland. Over the lower latitudes, gales are rare. On occasion during strong outbreaks of cold polar air southward across the Gulf of Mexico, gales described as "Northers" are experienced.

During the winter months of December, January, and February LOWS form most frequently in a band 150 to 350 mi. wide extending from the North Carolina coast northeastward to about the latitude of Cape Cod. This is part of a larger area of cyclogenesis that extends from the Gulf coast of the United States northeastward to 50°N., 30°W. Some LOWS also originate over the warm waters of the western Gulf of Mexico and pass northeastward across the Gulf States before intensifying near Cape Hatteras. From the Carolina coast the LOWS usually move northeastward toward Iceland. Over northern latitudes the characteristic Y-shaped track appears south of Greenland. LOWS from the Great Lakes region and the Gulf of St. Lawrence migrate northeastward toward the Denmark Strait or northward into the Davis Strait.

Visibility less than 5 mi. is noted in about 10 percent of the observations north of the 40th parallel. Along the northern shipping lanes the frequency is about 20 percent over the western half and near 15 percent over the eastern half. Entering the English Channel the frequency increases to more than 20 percent, and in the North Sea it exceeds 30 percent. The Labrador Sea and western portions of the Norwegian Sea also have a high incidence with more than 30 percent chance of encountering visibilities less than 5 mi.

Average air temperatures over the northern shipping lanes range from about 4°C. (39°F.) near New York to near 10°C. (50°F.) at the entrance to the Channel. Near the Nova Scotia and Newfoundland coasts average temperatures are slightly below freezing.

FEBRUARY—Usually weather conditions over the North Atlantic during the month of February are a continuation of the storminess characteristic of January, and there are years when February weather is the most severe of the winter months. The average pressure distribution remains quite similar to that of January. The Icelandic LOW fills to 1000 mb. (29.53 in.), and the central pressure of the Azores HIGH drops slightly to 1023 mb. (30.21 in.). This reduction in the average north-south pressure gradient is caused both by LOWS being slightly less intense on the average during February, and by the more frequent appearance toward the advent of spring of a blocking HIGH at higher latitudes. This forces the storm tracks farther south than normal and gives rise to periods of easterly winds over the northern shipping lanes.

The prevailing wind pattern shows only slight change from January. Above 40°N. there is very close agreement; however, between 30° and 40°N. west of about 60°W. and along the coast of the United States, the prevailing direction changes from southwesterly to westerly and northwesterly. The northeast trades remain well established between 5° and 25°N. The frequency of gales decreases slightly over the high latitudes near Greenland, but remains nearly unchanged south of the 50th parallel.

A study of wave heights for February shows that seas 12 ft. and higher can be expected 10 percent of the time north of a line from Cape Hatteras to Cape Finisterre. The highest frequencies, greater than 30 percent, are found over a triangular area between 57°N., 43°W.; 45°N., 40°W.; and 60°N., 15°W.

The extent and frequency of low visibilities (less than 5 mi.) are practically the same as in January. February is the coldest month on the North Atlantic. Over American coastal waters the average air temperature ranges from −2°C. (28°F.) south of Newfoundland to 25°C. (77°F.) over the Caribbean Sea. Over the eastern Atlantic the range is from about 2°C. (36°F.) along the coast of Norway to 15°C. (59°F.) near the coast of Morocco and 20°C. (68°F.) at Dakar.

MARCH—March is a transition month in which the weather retains many of the wintry aspects of January and February and at the same time begins to exhibit some features which are typical of spring. Weather conditions are generally a continuation of winter weather during the first part of March. They then taper off, approaching springlike characteristics near the close of the month. However, wide variations in climatic averages may be expected, and this pattern is not always the rule.

Winds from westerly quadrants generally prevail over the major part of the North Atlantic north of 30°N. However, near the coasts of Morocco and the Iberian Peninsula northerly winds are predominant. South of 30°N. the northeast trades are the dominant winds over most of the ocean with few exceptions. East of the Florida coast to about 68°W. wind directions are variable, and in the Gulf of Mexico there is a strong tendency for east and southeast winds. For the month as a whole wind forces between latitudes 40° and 60°N. average from force 4 to 6; between 20° and 40°N. force 3 or 4; and south of 20°N. only one observation in 4 or 5 is above force 4.

Gales, force 8 or higher, during the latter half of March, tend to decrease in strength and frequency. On the average, gale force winds have been noted in 10 percent of the ship observations north of a line extending roughly between Cape Hatteras and the Bay of Biscay. The maximum frequency of gale occurrence, 20 percent, may be expected from south of Cape Farewell to about 55°N. between 40° and 50°W. Only one tropical cyclone, a hurricane in the Lesser Antilles in 1908, has been reported in the North Atlantic area in the past 77 years.

Conditions of low visibility (less than 5 mi.), on the average, occur one time in 10 north of about 35°N., the frequency increasing to the north, especially in the western North Atlantic and along the North American coast. The highest percentages, 20 to 30 percent, occur north of 60°N. and south of Greenland west of 40°W. over the Grand Banks to about 42°N. The frequency of low visibility along the North American coast diminishes southward to the 10 percent level in the vicinity of Cape Hatteras.

In the spring months March, April and May, cyclogenesis is active over an area from the middle Atlantic coast of North America east-northeastward to the 10 degree square 40°–50°N., 30°–40°W. Within this area cyclogenesis is extremely active from the Virginia–North Carolina coast northeastward to 38°N., 65°W. The movement of the centers of LOWS between 40° and 60°N. in the western part of the ocean is generally on a northeastward course, while elsewhere north of 30°N. they generally move eastward. Storm centers usually move at a speed of 20 to 25 kt. when toward the east or northeast, but movement in other directions seldom exceeds 20 kt. The intensity of LOWS, that is, the steepness of the pressure gradient, is greatest between 40° and 50°N. from 30° to 50°W., and least between the Madeira Islands and the Iberian Peninsula.

APRIL—During the month of April weather conditions over the middle and northern latitudes are generally much more settled compared to the preceding month. Thus intervals of favorable weather are more frequent and usually of longer duration. There is a notable reduction in the frequency and intensity of winter-type LOWS. This fact is definitely established by the northward recession of the southern boundary of gales, with the 10 percent isoline moving from 40°N. in March to 52°N. in April.

The prevailing winds between latitudes 40° and 60°N. are generally westerly, but on the approaches to the coasts of North America and the Continent become variable. Within this large belt, 60 to nearly 70 percent of the observations report wind forces varying from 4 to 6. From approximately 40°N. southward to the northern boundary of the trades, the prevailing winds, while mainly westerly in the preceding month, become variable except near the Moroccan coast and Iberian Peninsula where northerly winds continue to dominate. Generally in the area between the westerlies and the trades, wind forces of 3 to 5 persist about 65 to 70 percent of the time. The trades are more firmly entrenched in April as compared to March. They usually prevail from 5° to 25°N. and in the eastern North Atlantic extend slightly north of 30°N. Sixty to 70 percent of the time their force is 3 to 4.

The area subject to gales compared to March decreases greatly in the middle and northern latitudes. The southern limit of the zone with gale observations 10 to less than 20 percent is near 52°N. This gale area extends northward to about 62°N. between 55° and 15°W. No tropical cyclones were reported during any April in the North Atlantic in the past 77 years. This is the only month of the year in which no tropical cyclone activity has been reported.

During the spring months average wave heights exceeding 12 ft. are found more than 10 percent of the time inside the triangular area between 43°N., 40°W.; 63°N., 58°W.; and 61°N., 03°W.

Occurrences of low visibility increase over the western part of the North Atlantic, especially west of 40° W. The greatest change from March takes place over the Grand Banks and the waters south and east of Newfoundland where over 30 percent of the observations show visibilities of less than 5 mi.

MAY—Weather over the North Atlantic in May is generally moderate. LOWS are less frequent and less severe than in April with gales south of 35° to 40° N. relatively infrequent. Over much of the southern part of the ocean the northeast trades prevail with increasing steadiness. In the Gulf of Mexico and the Caribbean Sea summer weather is established and the season for tropical storms is still ahead.

Visibility less than 5 miles, an indication of fog frequency, is seldom observed south of 30° N., but from the Virginia Capes northward along the coast and over the Grand Banks increases to 3 or 4 observations in 10.

JUNE—The weather is for the most part pleasant over the North Atlantic during June. Some LOWS occur but usually travel rather far north. They are generally mild and infrequently give rise to gales. LOWS of tropical origin average only 1 in 2 years, and of these less than half reach hurricane force. They generally form in the western Caribbean and move over the Gulf of Mexico and the Atlantic Coastal States. Fog is more frequent in June than in any other month over the northern and western portions of the ocean. Over the Grand Banks fog occurs on an average of 20 days during the month.

The wind regime of June is characteristic of summer and is largely controlled by the Azores HIGH centered about 30° N. Southward from the center of this HIGH to the vicinity of the equator winds are predominantly east and northeast. The strength of these winds, like their direction, is rather steady: 50% to 60% are force 3 or 4. Rarely are gales encountered here. Near the center of the HIGH winds are variable in direction and gentle in speed: from 60% to 80% are force 3 or less. Northward from about 35° N. to about 55° N. winds are variable in direction but with a preference for south, southwest, or west. Here, as south of the HIGH, 50% to 60% of winds are force 3 or 4. Winds of force 8 or higher occur in less than 5% of the observations. In the Davis Strait area they may reach over 10%.

June is probably the foggiest month of the year over the North Atlantic. Fogginess increases in general toward the north and toward the west. The Grand Banks east of Newfoundland are the foggiest area. Here 40% or more of observations report visibility less than 5 mi., fog undoubtedly being the principal element causing the restriction. These fogs are usually caused by south and southwest winds bringing warm moist air over the cold ocean surface.

JULY—From June to July a marked northward shift of cyclonic activity normally occurs over the North Atlantic. Consequently weather conditions are relatively settled during July with the buildup of the Azores HIGH to a seasonal maximum and a displacement of the primary storm tracks north of 45° N.

The Icelandic LOW has a normal pressure of about 1007 mb. centered over the Davis Strait. The Azores HIGH is centered near 35° N., 35° W. with pressures near 1026 mb.

The prevailing winds over the middle and northern routes are from southerly and westerly quadrants, and wind speeds average 10 kt. or less 40 to 60 percent of the time. Northerly winds are common near the entrance to the Mediterranean, while over the Mediterranean Sea northwesterly winds are steady. The northeast trades blow steadily across the ocean between 30° and 10° N., and in the Gulf of Mexico winds from between south and east are most frequent. Severe storms rarely occur, and while the frequency of gales is at a minimum for the year, past records indicate occasional gales in the waters immediately south of Greenland eastward to Iceland and over the northern Norwegian Sea.

Principal areas of cyclogenesis during the summer months are along the North American coast from the Carolinas to north of Newfoundland, in the Denmark Strait and north of Scotland. The primary cyclone tracks lead from the Hudson Bay region northeastward through the Davis Strait, from the Grand Banks or the Gulf of St. Lawrence toward Iceland, and from north of Scotland across central Scandinavia to Arctic European Russia.

Tropical cyclone activity in the North Atlantic hurricane belt is still limited during July. Over a period of 77 years, 1886–1962, 44 tropical cyclones have been logged, 25 of which reached hurricane force. July hurricanes usually originate in the Gulf of Mexico or in the area just east of the Lesser Antilles. Those originating in the Gulf of Mexico generally move northward striking the Gulf Coast. Those forming east of the Lesser Antilles usually move westward across the Caribbean Sea toward Yucatan or curve northward into the Gulf of Mexico, while a few follow tracks taking them northward off the east coast of the United States.

Wave heights of 12 ft. or greater are encountered more than 10 percent of the time in a small area located south of Greenland near 58°N., 46°W.

Visibility of less than 5 mi. occurs frequently over the North Atlantic along the coastal regions from Cape Hatteras to Newfoundland with a maximum occurrence of more than 40 percent over the Grand Banks area. Similar frequencies are found in Arctic waters along the west coast of southern Greenland. July is the foggiest month over the western North Atlantic.

The air temperatures have increased considerably from June, and the 59°F. (15°C.) isotherm follows roughly a line from Cape Sable to central Ireland with a southward dip across the Grand Banks.

AUGUST—The favorable weather, characteristic of July, generally continues into August with the temperature pattern reaching its maximum northward displacement and making this month the warmest of the year. A few storms, generally weak in force, occur north of 40°N. Those attaining severe intensity usually are of tropical origin.

The average pressure distribution in August shows the Icelandic LOW at 1009 mb. extending across the northern latitudes from the Davis Strait to the western Norwegian Sea. The subtropical HIGH is centered near 35°N., 37°W. with a pressure of 1025 mb.

The prevailing wind direction in most areas conforms closely with the preceding month. Wind directions north of 40°N. are slightly more variable than in July. The westerly components are generally the dominant ones. Between 40° and 30°N. the prevailing direction is from north and northeast in the eastern part of the ocean and from south and southwest in the western portion. The etesians (northwesterlies) blow across the Mediterranean Sea and are quite steady over the eastern half. The northeast trades lie principally within the belt between 30° to 15°N. extending southward to near 10°N. over the western half of the ocean. Near the approaches to the American coast at these latitudes, the trades become more easterly and even southeasterly in some areas.

Winds of gale force, except in tropical cyclones, are very infrequent south of 45°N. Between 50° and 60°N., 5 to 10 percent of ship observations report winds of force 8 or higher in the western portion of the ocean. Gales are also encountered over 5 percent of the time in a small area west of Ireland.

August is one of the principal months in the North Atlantic hurricane season, ranking a close 3d behind September and October in tropical cyclone development and 2d behind September in the number of these storms which attain hurricane force. From 1886 to 1962, 137 tropical cyclones occurred in August, and 101 of these reached hurricane intensity. A maximum of 7 cyclones occurred in August 1933, and in 14 of these 77 years no storms were reported in August. The level of tropical cyclone activity increases as the month advances with the likelihood of storm occurrence being about twice as great in the last 10 days as during the first part of the month. The spawning area of tropical cyclones is much larger in August than during the preceding month. Some tropical cyclones originate in the vicinity of the Cape Verde Islands, gather strength as they are carried across the lower latitudes of the North Atlantic by the prevailing easterlies and then enter the Caribbean, Gulf of Mexico, United States, or western Atlantic as fully developed hurricanes. Characteristic of this activity is the split mean storm track around the West Indies with one branch passing to the north and one to the south of the islands.

Wave heights 12 ft. and greater are encountered more than 5 percent of the time over most of the ocean north of the 50th parallel. Frequencies greater than 10 percent are found in the area south of Greenland, in the Denmark Strait and just west-northwest of Ireland.

Fog is both less frequent and less extensive in August than in June or July. Visibilities less than 5 mi. occur about 30 percent of the time in the Grand Banks area. Southwestward along the coastal region occurrences decrease to about 20 percent in the Nova Scotia-New England area, and diminish to the 5 percent level at points off the Virginia coast.

August is the warmest month of the year over the North Atlantic. The 59°F. (15°C.) isotherm extends roughly from Cape Race to the coast of Scotland. South of 38°N. in the western part of the ocean normal temperatures exceed 77°F. (25°C.). Dew point temperatures exceed 68°F. (20°C.) south of 40°N. in the western part of the ocean and approach 77°F. (25°C.) over areas of the northern Caribbean and southern Gulf of Mexico.

SEPTEMBER—With the approach of autumn, subdued weather conditions that generally characterize the summer over the higher latitudes are gradually replaced by moderate intrusions of colder air masses and by increased cyclonic activity. Tropical cyclone frequency reaches a peak during this month and occasionally severe extratropical LOWS may be encountered. The Icelandic LOW generally deepens slightly and the Azores HIGH is a little weaker than in August. The average air temperatures decline only slightly from the August values particularly in the middle and higher latitudes; however, near the close of the month much cooler weather may be experienced above 40°N.

From 40°–60°N., the prevailing winds are westerly, generally of force 3–5. From 30°–40°N., the most frequent wind directions over the eastern Atlantic are from the northerly quadrants, while over midocean and off the United States coast the winds are variable. From 10°–30°N., the northeast trades are predominant. They are modified somewhat in the western part of the ocean, blowing more from the east and southeast, due to the configuration of the Azores HIGH. The southeast trades prevail south of 10°N., except near the South American coast where a considerable percentage of east and northeast winds are recorded.

The frequency of gales, force 8 or more, generally increases—particularly along the northern routes. The highest percentage of gales, 10 percent, occurs north of 55°N., between longitudes 40° and 50°W.; and near 60°N., south of Iceland. Along the New York to Channel routes the occurrence of gales varies from 1–3 percent near the coasts to more than 5 percent in midocean. South of 40°N., gales are unlikely to be encountered except in storms of tropical origin.

The hurricane season reaches its maximum in September. From 2–4 tropical cyclones may be expected, and in the past, 60 percent of these storms reached hurricane intensity. During September 1949 as many as 7 tropical cyclones were reported, while there were none in 1930. This illustrates that the average is not definitely a dependable criterion. The entire West Indian region is subject to these storms. Many enter the Gulf of Mexico while a greater number recurve off the American coast. Some, however, strike the eastern coast of the United States. Many continue to advance with considerable energy into northern waters and if encountered by shipping may be of considerable danger.

The July 1959 issue of the *Mariners Weather Log* contains a comprehensive article showing the monthly frequencies and movements of North Atlantic tropical cyclones by 5° squares for the period 1886–1957.

During the autumn months wave heights greater than 12 ft. are generally encountered from 45°–60°N., between 12° and 55°W., 20 percent of the time. Maximum frequencies of 30 percent are encountered in an area 200–600 mi. south of the western coast of Iceland, and in another small area about 400 mi. south of the southern tip of Greenland.

Occasions with visibility less than 5 mi. become less frequent than during the summer months. Over the Grand Banks and southeast of Greenland low visibilities continue to plague shipping on an average of more than 20 percent of the time.

OCTOBER—During October LOWS in the middle and higher latitudes increase in frequency and energy. The main belt of westerlies recedes southward, and the temperature contrast off the Atlantic coast, from the Carolinas northeastward to Newfoundland, is conducive to cyclogenesis. Occasionally these storms are of great intensity, causing gales of hurricane or near hurricane force. This is not the rule, however; in many cases a week may pass with no more than a single gale.

During the month a noticeable reduction in the average air temperature takes place north of 35°N., generally ranging from 4°–10°F., with the largest falls occurring from Newfoundland southwestward to Cape Hatteras. Along the Channel routes, from 40°–60°N., temperature averages vary from 48°–60°F. (9°–16°C.).

The mean sea level pressure distribution for the month shows that the Azores HIGH has diminished in extent since September, but its central pressure remains at 1020 mb. (30.12 in.). The LOW over Baffin Bay and Davis Strait has deepened to 1005 mb. (29.68 in.) and the LOW over Iceland also now has a closed isobar of 1005 mb.

Westerly winds generally prevail north of 40°N. However, from 30°–40°N., near the European coast northerly components predominate, while off the United States coast the winds are variable. The northeast trades are well established from 10°–30°N. while the southeast trades are experienced south of 10°N.

The frequency and the area subject to gales increase markedly during October. Winds of force 8 or higher may be expected oceanwide as far south as 40°N. and somewhat farther south off the American coast. The highest percentage, 20, occurs near 57°N., south of Greenland. Percentages from 10–19 occur from Davis Strait over a roughly triangular area to near 45°N., 40°W., thence to the areas around Iceland and also over most of the Norwegian Sea. An analysis of 6 locations north of 40°N. indicates gales generally last less than a day but recur in 2–4 days.

The West Indian hurricane season continues into October, but after the middle of the month the probability of occurrence of a tropical cyclone diminishes rapidly. One or 2 usually occur before the 20th of the month. Proportionately fewer tropical cyclones in October reach hurricane strength than in August or September. Most of the October storms form in the western Caribbean; however, some are still spawned in the Lesser Antilles.

Instances of low visibilities (less than 5 mi.) are rather infrequent; the average over the northern routes being 1 day in a week or 10 days.

NOVEMBER—In November several weather features which are characteristic of the winter months make their appearance. With an intensification of the Icelandic LOW, both in depth and area, the major primary cyclonic track shifts from the east to the west side of Iceland, and a center of maximum cyclone frequency occurs in the Denmark Strait, where it remains during the next five months. Additional cyclonic paths which appear in November and intensify during the winter months are those in the Gulf of Mexico and in the southern portion of Davis Strait. While normally November is considered a transitional period between fall and winter weather over northern waters, this month occasionally develops into the severest of the winter season. Most rough weather is confined to north of 45°N. except along the American coast where there is an increase in wintertype LOWS. In the Great Lakes region two primary storm tracks, composed of Alberta and Colorado LOWS, converge into an area of maximum cyclone frequency which contributes in a large measure to rough weather at the close of the Lakes navigation season.

Westerly winds, force 4–6, prevail over most of the ocean between 40° and 60°N. Wind directions are variable from 30° to 40°N. except along both the American and European coasts where a noticeable northerly component tends to prevail. The northeast trades, average force 3–4, extend from 30°N. southward to 5°N.

Gale frequencies over northern and middle latitudes increase materially after October. Winds of force 8 or higher may be expected 10–19 percent of the time north of 40°N. and southward to about 35°N. off the American coast. The highest percentage, 20, occurs in the Davis Strait north of 63°N.; south of Greenland in the vicinity of 55°N., 45°W.; and over most of the Norwegian Sea.

Tropical cyclones are infrequent in November. Usually one storm may be expected in three years. In the 77-year period 1886–1962, 28 tropical cyclones occurred and 12 of these reached hurricane strength.

Visibility less than 5 mi. is more frequent in November than October, particularly north of 40°N. in the western part of the ocean. Low visibilities persist on the average 20 percent of the time over the Grand Banks and northward to the southern tip of Greenland and over most of the Davis Strait and Labrador Sea. The highest percentage, 30, occurs over much of the Greenland Sea north of 65°N. and near the coast of Greenland in the Davis Strait.

DECEMBER—December is generally one of the stormiest months of the cold season over the North Atlantic, particularly north of 35°N. Deep and extensive LOWS traverse the middle and northern shipping lanes, producing strong winds and high seas. Usually extended periods of rain, sleet, or snow attend these storms.

A comparison with the normal pressure distribution of the preceding month shows that in December the Azores HIGH has increased to 1022.5 mb. (30.19 in.) with its center just northwest of the Canary Islands. The Icelandic LOW has deepened to 997.5 mb. (29.46 in.) and is located southwest of Iceland. As a result of the intensification of these major centers of action, the pressure gradient has increased sharply in December over the middle latitudes.

Westerly and southwesterly winds prevail over most of the ocean north of 35°N. Near the American coast, at these latitudes, northwesterly winds are frequent. Between 30° and 35°N. winds are variable except the approaches to the European coast where north and northeast directions prevail. The northeast trades, average force 3–4, persist between 5° and 30°N. except in the western ocean near the American coast. The usual wind forces are 4–5 in latitudes 30° to 40°N. and 5–6 in the higher latitudes, including the steamer routes to northern Europe.

In the winter months (Dec.–Jan.–Feb.) there are three principal regions of cyclogenesis in the North Atlantic. The most prominent and extensive area is along and off the North American Continent from northern Florida to the Gulf of St. Lawrence. This area extends eastward to about 60°W. between latitudes 35° and 40°N. A second irregular area, situated over the Grand Banks, extends southeastward to about 40°N., 50°W., then northeastward to near 55°N., 35°W. The third area exists over the waters to the north, west, and southwest of Iceland.

The movement of LOWS during the winter months is more rapid than in the other three seasons. Between 35° and 40°N. and west of 30°W., 65 to more than 85 percent of the LOWS move in an easterly or northeasterly direction at speeds varying from 25–29 kt. Beyond 30°W. direction of movement of storms is more variable with 30–40 percent following an easterly course at speeds of about 25 kt. in midocean, diminishing to 12 kt. near the continent. From 40° to 50°N. the movement of low pressure systems is about equally divided, more than 80 percent of the time, between easterly and northerly directions at average speeds ranging from 20–32 kt. The higher average speed of movement occurs in the area of cyclogenesis east and southeast of Newfoundland. North of 50°N. about half of the storms move in a northeasterly direction, and easterly and northerly directions rank second and third, respectively. Most LOWS between 40° and 50°N. travel at average speeds of 20–30 kt. and north of the 60th parallel between 14 and 28 kt.

Areas which have their highest cyclone frequency of the year during December include the Great Lakes and the waters near Spitsbergen. Furthermore, Iceland has more days with low centers in December than in any other month. Finally, the frequency of cyclogenesis in the Gulf of Mexico reaches its annual maximum in December.

The occurrences of gales are more frequent over northern and middle latitudes than in November. Winds of force 8 or higher occur 5 percent or more of the time north of about 35°N. In the western North Atlantic the area of gales extends southward paralleling the continental coastline to about 30°N. This reflects the intensification of LOWS as they move northeastward off the central and southern North American coast. The principal area of gales, however, is north of 45°N. from 30° to 45°W. Northers sometimes occur during December in the Gulf of Mexico and the western Caribbean Sea.

December is well outside the normal hurricane season of the North Atlantic. During the 76-year period 1886–1961, four tropical cyclones were charted and two reached hurricane strength during the month.

During the winter months (Dec.–Jan.–Feb.) wave heights greater than 12 ft. are generally encountered 30 percent of the time over a large area between 50° and 60°N. extending westward from the British Isles to near 50°W. A maximum frequency of 60 percent is found in a small area near 56°N., 20°W.

North of latitudes 38° to 39°N. occurrences of low visibilities (less than 5 mi.) increase from 10 percent of the recorded observations to 20 percent south of Iceland and more than 30 percent over the Greenland and Norwegian Seas. In the western third of the ocean the 20–30 percent limits dip southward to the Gulf of St. Lawrence and the Grand Banks area.

The air temperature along the Great Circle route from the East Coast ports to the Channel ranges from an average of 30°F. (4°C.) near Cape Sable and Cape Race to about 50°F. (10°C.) near the entrance to the English Channel.

Average Weather for the North Pacific Ocean

JANUARY—The most severe weather of the year generally occurs in January over the middle and northern latitudes of the North Pacific. The circulation over the ocean is mainly controlled by the major centers of action—the Aleutian LOW, the subtropical HIGH, and the Siberian HIGH which are all near their peak seasonal development. The Aleutian LOW with a central pressure of 1000 mb. (29.53 in.) and located over the western Aleutians, and the Pacific subtropical HIGH at 1022 mb. (30.18 in.) near 30°N., 135°W. govern the flow over the major part of the ocean, while the wind regimes near the Asiatic coast are controlled by the flow between the Aleutian LOW and the Siberian HIGH. The central pressure of this "mean" HIGH is 1035 mb. (30.56 in.) and its position is just south of Lake Baikal.

Westerly winds prevail over most of the ocean north of 40°N. and above 30°N. west of 160°W. They are most steady west of 180°, while in the Gulf of Alaska and southward to near 40°N. their regularity is interrupted during the frequent passage of LOWS developing as secondaries and moving along tracks having their origin farther south. The average force of winds over the belt of the "westerlies" is force 4 to 6. The northeast trades extend northward to near 30°N. in the eastern part of the ocean and to about 25°N. west of 170°W. South of 20°N. these winds are very steady. The wind speeds in the trades generally range from force 3 to 5. In Asiatic waters the northeast monsoon is steady over the South China Sea with northerly and northwesterly winds dominant over more northern coastal waters.

The frequency of gales is between 10 and 20 percent over most areas south of the Aleutians and north of a line from southern Honshu to Vancouver Island. A maximum incidence of slightly above 20 percent is found southeast of Kanchatka and in the central Gulf of Alaska near 50°N., 145°W. "Tehuantepecers" are encountered on occasion by vessels in the Gulf of Tehuantepec south of southern Mexico. These violent squally winds from the north or north-northeast occur when strong northers from the Gulf of Mexico funnel across the isthmus into the Pacific. They may be felt up to 100 mi. out at sea.

Principal areas of cyclogenesis during the winter months are found over southern portions of the East China Sea near the Ryukyu Islands and to the east of Honshu. The migratory LOWS move mostly northeastward from here across the western Aleutians into the Bering Sea or east-northeastward toward the Gulf of Alaska. Other primary tracks approach the Gulf of Alaska and the British Columbian coast from the southwest. A secondary track from the northwest toward the British Columbian coast is also followed at times.

Tropical cyclones are infrequent in January. On the average one can be expected every two or three years. Most of these storms develop between 6° and 10°N. and west of 140°E. and move toward the central Philippines and South China Sea.

Visibility less than 5 mi. occurs in about 20 percent of the observations over most of the ocean north of 40°N. Restricted visibility is seldom encountered south of 25°N. except over the waters around Taiwan and over the eastern approaches to the Philippines where it may occur as much as 10 percent of the time.

Average air temperatures over the northern shipping lanes from Japan to West Coast ports drop from 8°C. (46°F.) near Yokohama, Nagoya and Kobe to a low of 2°C. (36°F.) south of the central Aleutians and then increase to 10°C. (50°F.) near San Francisco. Average readings near the Hawaiian Islands are about 24°C. (75°F.).

FEBRUARY—February weather in general is equally as rough as that of the preceding month over the middle and higher latitudes of the North Pacific. A slight deepening of 1 to 2 mb. to about 998 mb. (29.47 in.) of the Aleutian LOW and a similar small increase in the pressure of the subtropical HIGH to 1023 mb. (30.21 in.) occur with no change in the mean position of these centers. This in turn signifies a slight strengthening of the westerlies over the middle latitudes from the January values. The high degree of similarity of the normal climatic values for the two months does not preclude, however, a mild February following a severe January or a reversal of this circumstance.

The average strength and general pattern of the westerlies, the northeast trades, and the northeast monsoon in general conform closely to the mean January distribution. The areal distribution and percentage frequency of gales are also practically identical to the preceding month with a few exceptions. The 20 percent area increases a little in the central Gulf of Alaska, and the other area of maximum incidence of near 20 percent southeast of Kamchatka moves farther southward to around the 40th parallel.

Tropical cyclone activity is at the annual minimum during February. On the average, one typhoon can be expected every 5 to 10 years. As in the other winter months, the principal genesis region is over the latitudes of the Philippines and mostly to the east of these islands. There is no record of any tropical cyclones in the areas off the Mexican west coast in recent years during February.

Seas of 12 ft. or more are generally encountered 10 to 20 percent of the time between latitudes 30° and 52°N. from 140°W. to 140°E. A small area with a similar frequency is located over the waters around Taiwan where the northeast monsoon blows strong and steady. The area with the highest frequency, between 20 and 30 percent, lies off the eastern approaches to Japan between 31° and 49°N. from 150° to 170°E.

Areas of limited visibilities (less than 5 mi.) are almost identical with the January patterns. However, due to additional cyclonic activity with associated increased incidence of precipitation over the waters east of Japan, percentages are increased slightly over this region. Some deterioration of visibility also occurs in Alaskan and Aleutian coastal waters.

Average air temperatures remain nearly unchanged from the January values at their minimum for the year. A slight tightening of the south to north temperature gradient has occurred due to minor increases of a degree Celsius or two over the low latitudes and similar decreases over northern regions.

MARCH—This month over the North Pacific is normally considered one of the transitional months between winter and spring. Compared to the North Atlantic ameliorative changes are somewhat delayed by the vast expanse of the ocean and the lingering winter climate over Siberia. Stormy weather is about as frequent as in the preceding month along the northern routes, especially from the western Aleutians southwestward to the vicinity of Japan. In the central and western North Pacific gales may be expected as far south as 30°N. In this area, north of 35°N. and west of 175°W., 10 to more than 20 percent of ship reports contain winds of force 8 or higher. In the eastern part of the ocean, east of 175°W., there is a large reduction in gale frequencies compared to February and occurrences are generally confined to latitudes north of 35°N. Percentage frequencies of gales in the central Gulf of Alaska, 10 to 20 in the preceding month, drop to 5 to 10 percent during March.

From the 40th parallel northward, winds from a westerly direction are most frequent. In 55 to more than 60 percent of the observations, the usual wind force is 4 to 6. However, near the North American coast the most frequent wind speeds are force 3 to 5. North and northwest winds are most prevalent in Japanese waters and the Yellow Sea where more than 50 percent of all winds vary from force 4 to 6. During March the northeast monsoon continues to prevail along the China coast south of Shanghai and over Philippine waters. From 25° to 40°N. wind directions are variable and the force is from 3 to 5 more than 60 percent of the time. The northeast trades are the dominant winds from 25°N. to the Equator and extend northward to about 30°N. in the eastern part of the ocean. The usual wind speeds, force 3 to 5, persist more than 70 percent of the time over the ocean area under the influence of the trades.

The southern limit of 5 percent frequency of low visibility (less than 5 mi.) extends from 10°N. east of the Philippines east-northeastward across the ocean to the coast of Lower California near 30°N. The area of highest incidence of low visibility, 20 percent or more, is encountered over the central and eastern ocean roughly between 40° and 50°N. from 140°W. to 180°.

Typhoons are infrequent during March with a probability of 1 in 3 years.

In the spring months, March, April and May, centers of low pressure in the North Pacific form principally over an area extending from Taiwan northeastward to about 45°N. Cyclogenesis is especially active over the ocean area east of Honshu, extending southwestward to the Ryukyu Islands. The greatest frequency of cyclogenesis in the Northern Hemisphere takes place in the area off the Ryukyus in March. For the 3-month period the great majority of LOWS in the Pacific north of 30°N. move northeastward or eastward. Centers west of 180° and north of 30°N. move northeastward and eastward at about 25 to 30 kt., and decrease to about 15 to 20 kt. near the North American continent. North Pacific LOWS at this time have their greatest intensity north of 40°N. and west of 160°W.

APRIL—The weather over the North Pacific generally shows a marked improvement over that of any month since October. Compared to the winter months, periods of storminess are fewer, but severe extratropical LOWS are still occasionally encountered.

While the mean position of storm tracks during April closely approximates that of the preceding month, there is a northward displacement near Japan and the Kamchatka Peninsula. This generally results in fewer gales and better weather over trans-Pacific routes. Two areas of high frequencies of gales, 10 to almost 20 percent, persist as a residual of the winter months in the middle and northern latitudes: one in the Gulf of Alaska south of Kodiak Island to about 53°N. and eastward to near 140°W.; the other, east of Honshu from about 35°N., 150°E. northeastward to about 45°–50°N. and 180°.

Over a large portion of the North Pacific, mainly between 35° and 55°N., the wind speeds are of force 4 to 6 in 55 to 65 percent of the observations, except near the North American coast and off the central Japanese coast, where winds of force 3 to 5 are generally encountered. The prevailing winds are from the westerly quadrants. Between 30° and 35°N. winds are variable with forces 3 to 5 recorded in 60 to 75 percent of the observations. In these latitudes, southeast of Japan, winds have a large easterly component, backing to northeast and north over the East China Sea. The northeast trades prevail over most of the ocean south of 30°N., except westward of midocean where their northern limit slopes to about 25°N. South of Shanghai over the China Sea to the Philippines the northeast monsoon continues with less strength and steadiness than in the colder months.

The probability of encountering weather limiting visibility to less than 5 mi. is similar to that of March.

Typhoon expectancy in Far Eastern waters, based on past records, is 1 in 2 or 3 years.

MAY—May's weather on the North Pacific is usually pleasant. There may be an occasional storm in the northern portion but the likelihood decreases as the month advances. Visibility less than 5 miles is present in 15% to 20% of the observations over the large areas between 30°N. and 50°N., particularly west of 180°. South of 30°N. there are few interruptions to the northeast and east trade winds except near the Philippine Islands and the Asiatic mainland. Over half of these trade winds are either force 3 or 4.

In tropical waters of the Far East typhoons begin to increase in frequency but their average is less than one a month. The probability of a vessel encountering one of these storms is, therefore, small.

JUNE—In June summer weather becomes well established to the North Pacific. The semi-permanent HIGH spreads over most of the ocean. The circulation in Asiatic waters, called the southwest monsoon, becomes well established this month. Over the entire ocean winds are light and gentle: from 40% to 80% being force 3 or less. Typhoons average only one in June in the western North Pacific. Occasionally a severe tropical storm forms in Mexican waters. Fog decreases on the southern and increases on the northern Asiatic coast. It increases greatly over the northern part of the ocean, and somewhat on the American coast.

JULY—The steady and rather settled summer weather conditions that commence in June over the North Pacific become widespread and firmly established during July. The Aleutian LOW has disappeared from the normal pressure chart and the subtropical HIGH with a pressure of 1027 mb. has moved northward to near 37°N.

Due to the strong development and northward position of the subtropical HIGH, the northeast trades extend over a large portion of the ocean. They prevail between 30° and 10°N. and in the eastern ocean extend northward to about 35°N. The southwest monsoon is well established in Asiatic waters blowing most steadily over the South China Sea. The westerlies of the middle latitudes, due to the absence of the Aleutian LOW, are less steady than during the colder months. Large southerly components are found over the western two-thirds of the ocean at these latitudes, while northerly components are the rule closer to the American coast.

Occurrences of winds of gale force are rare during July except in those areas affected by tropical cyclones. Over the northern shipping lanes, the usual wind speed is 10 kt. or less.

Cyclogenesis during the summer months occurs in Asiatic waters from Taiwan northward to Sakhalin with the greatest frequency just east of Honshu and Hokkaido. Another area is found near 47°N. from 160° to 175°W. The primary storm tracks lead from these two areas northeastward to the Bering Sea and the Gulf of Alaska, respectively.

The number of typhoons which occur in the western North Pacific averages about 3 or 4 during July. These storms originate mostly in the ocean areas east of the Philippines. In the early stages they generally move west-northwestward. After development some may continue westward across the Philippine Islands into the South China Sea, others may curve northwestward toward Taiwan, the coast of China or Japan. Those reaching the higher latitudes generally recurve toward the northeast under the influence of the upper westerlies.

Another area of tropical cyclone activity is over the waters off the west coast of Mexico. About two storms can be expected in July with one of these reaching hurricane force. These storms are usually short-lived but can be dangerous to both shipping and coastal areas.

Wave heights 12 ft. and greater can be expected 10 percent of the time in a small area near 50°N., 170°W. south of the Aleutians.

The occurrences of low visibilities over northern waters are most frequent during July compared to other months of the year. The visibility drops below 1 mi. in 30 to 40 percent of all observations near the Kuril Islands and southern Kamchatka. The 10 percent frequency line encloses most of the ocean north of 40°N. and west of 150°W.

Air temperatures over the North Pacific continue their rise toward the August maximum. The 59°F. (15°C.) isotherm stretches from central Hokkaido eastward along the 43d parallel to near the Oregon coast then southward to the Los Angeles area. The greatest temperature gradients are found off the coast between the Tokyo-Yokohama area and northern Hokkaido (from 76° to 56°F.), and between the southern tip of Baja California and the Los Angeles area (from 80° to 60°F.).

AUGUST—The mild weather of July continues into August over the North Pacific. The two months are very similar except in August fogginess decreases, typhoons are more numerous, and it is the warmest month of the year.

The Aleutian LOW still has not made its appearance on the monthly normal pressure chart. The subtropical HIGH is centered near 38°N., 148°W. with a central pressure of 1026 mb.

The prevailing wind pattern closely coincides with July conditions except in the waters south and east of Japan where winds are quite variable. The usual speed of the wind is 10 kt. or less. The occurrences of winds, force 8 or higher, are relatively few in areas not subject to tropical cyclones.

The number of extratropical cyclones is slightly higher in August than in the preceding month. Most of these storms which form off the coast of Japan move northeastward into the Bering Sea.

The frequency of tropical cyclones in the western North Pacific reaches a peak in August and September with similar frequencies for both months. About five tropical cyclones can be expected in August with most of these storms reaching typhoon intensity. Typhoons in August are displaced farther to the north, and they usually miss the central and southern Philippines. Some move directly toward Japan and Taiwan. Others may pass over Japan after recurvature in the Yellow Sea or the Sea of Japan.

Over the tropical waters west of Mexico usually two tropical cyclones occur. The average duration of these storms is three days and about half attain hurricane intensity.

During August wave heights of 12 ft. and more are generally encountered 5 percent of the time along the western half of the northern routes to Japan. A maximum occurrence of 10 percent is found in a small area near 51°N., 170°W.

Although visibility conditions improve in August, fog is still frequently encountered over the northern shipping lanes. The greatest frequencies are found over the northwestern portion with the 10 percent frequency line of visibilities less than 1 mi. encompassing most of the ocean north of 40°N. and west of 150°W.

Air temperatures reach their annual maximum over the North Pacific. Changes from July are the largest over the western part of the ocean while near the American coast increases are only small. The 59°F. (15°C.) isotherm crosses the ocean between the southern Kuril Islands and Vancouver Island. Near the central Aleutians air temperatures are near 50°F. (10°C.), while in the areas south of 25°N. in the western half of the ocean temperatures reach 82°–84°F. (28°–29°C.). Dew point temperatures in this latter area are near 77°F. (25°C.).

SEPTEMBER—Weather conditions over the North Pacific generally continue favorable in the early part of September, but as the month advances early wintertype storms appear over the northern portions. Over the northern shipping routes some rough weather may be expected, particularly over the Asiatic portion. The western parts of high latitude routes are subject not only to LOWS of extratropical origin, but also to tropical cyclones that move into the area. Gales are extremely rare in low latitudes, except in the region of tropical cyclone activity. Winds of 34 kt. or more may be expected about 5 percent of the time in the shipping lanes north of 38°N. and west of 140°W.

The formation of the Aleutian LOW and the Asiatic HIGH during September greatly modifies the prevailing wind patterns which existed in July and August. In September the northeast trades prevail over most of the ocean from 10°–30°N., with their most northward extension, near 35°N., between 130° and 160°W. In midocean, while somewhat weaker and unsteady, they extend southward almost to the equator. East of Honshu southerly winds still predominate. However, along the Asiatic coast a gradual change to the winter monsoonal winds commences. In the Sea of Japan the most frequent winds are from the north and the northeast. In the East China Sea the southwest summer monsoon changes to the northeast monsoon by the close of the month. Westerly winds prevail along the northern shipping routes north of 40°N., although, in the western part of the ocean, there is considerable variability in direction due to the movement of low pressure systems.

The expected frequency of Far Eastern typhoons, usually 3–4, maintains its seasonal maximum during September. These storms originate generally north of 8°N. and between 150°E. and the Asiatic Continent and move westward. Some travel across the Philippines and the South

China Sea to the mainland, while others recurving near 130°E., between 20° and 30°N., pass over or near the Japanese Islands. The tropical cyclones off the Mexican coast region are most frequent in September, with an average of about 2 this month. While the majority of these storms travel up the Mexican coast, some going inland, occasionally they develop near the Revilla Gigedo Islands or farther west, between Mexico and the Hawaiian Islands.

The occurrence of fog is less frequent in September than in the summer months. Along the North American coast fogs occur quite frequently from Vancouver southward to Cape San Lucas at the tip of Baja California. The percentage of low visibilities (less than 5 mi.) varies from 10–20 over the northern routes and attains a maximum value of 25 percent along the California coast.

OCTOBER—This month marks a transition from summer and early fall weather to wintry conditions over the northern parts of the North Pacific. The Aleutian LOW deepens during the month to an average central pressure below 1003 mb. (29.62 in.) and overspreads the eastern Aleutian Islands and the Gulf of Alaska. The strength of the Pacific HIGH diminishes somewhat to the northwest of the Hawaiian Islands while the Asiatic HIGH intensifies with a southeastward extension over the Yellow Sea.

During October a noticeable change in the average air temperature occurs over most of the ocean. In the western Pacific the average temperature ranges from 44°F. (7°C.) near the southern end of Kamchatka to 80°F. (27°C.) south of Taiwan. In American waters the range is between 36°F. (2°C.) in the northeast Bering Sea and 80°F. (27°C.) at the entrance to the Gulf of California. Along the Great Circle routes between San Francisco and Yokohama the temperature ranges from about 48°F. (9°C.)., in northern midocean to about 66°F. (19°C.) near Tokyo Bay.

North of 40°N., westerly winds are predominant, except on approaching the American coast where there is a gradual change to northerly winds. From 30°–40°N., the westerlies diminish and winds are more variable in direction. Near the American coast, the Japanese Islands, and approaches to the Asiatic mainland, winds from northerly directions prevail. The northeast trades largely persist from 10°–30°N., but extend northward to about 35°N., in the eastern part of the ocean and southward to near 7°N., in midocean. In Asiatic waters the northeast monsoon of the colder months is well established.

The likelihood of encountering gale winds of force 8 or higher increases appreciably, compared to September, particularly in the middle and higher latitudes. The area from 48°–57°N., extending from Kamchatka to the central portion of the Gulf of Alaska, has the highest incidence of gales. In this area more than 10 percent of the wind observations attain force 8 or higher. Southward there is a gradual decrease in frequency to 5 percent at 30°N. South of 30°N., gales are scattered and infrequent, except in the Far East, where in the waters near Taiwan and southwest of Japan, typhoon occurrences increase the chances of gale winds.

Two to 4 typhoons may be expected to develop in the tropical waters of the Far East. After the middle of October the probability of occurrence of these storms diminishes rapidly. Most of these tropical cyclones form at low latitudes, from 7°–20°N., and eastward from the Philippine Islands to about 150°E. Their early movement is generally westward or northwestward, and most storms recurve northward east of the Philippines and sweep up along the east coast of Japan. A lesser number move westward across the Philippine Islands into the South China Sea, and a still smaller number reach the mainland.

In the tropical waters of the eastern North Pacific the occurrence of tropical cyclones is less frequent than in September. About 50 percent of the tropical LOWS reach hurricane intensity along some part of their path. These storms have less regular tracks than during the previous month, with a large percentage striking the Mexican coast.

Visibilities of less than 5 mi. are infrequent over most of the North Pacific, but may be encountered along the northern California and Oregon coast and in the Gulf of Alaska.

NOVEMBER—November may be a stormy month north of 35°N. There have been years in which November has been the stormiest month of the winter season, but usually the weather is not too severe and there may be quiet periods of several days. South of 35°N. the probability of severe storms diminishes rapidly.

During November the Aleutian LOW is well established over northern latitudes with a central pressure of 1000 mb. (29.53 in.). The Pacific HIGH has not changed much since October in respect to position and intensity. Off the China coast pressure is higher than during October due to the buildup of the Asiatic winter HIGH.

The main storm tracks are north of 45°N. in the central and eastern North Pacific but in the western portion extend southward to about 30°N. off the southeastern Japanese coast. In the eastern Pacific three new paths lead to an area of maximum cyclone frequency off the coast of British Columbia. One track originates in the Gulf of Alaska, an area where there are more LOWS per 5 degree square than any other part of the Northern Hemisphere, not only in November but in 5 of the following 6 months. Except for an additional track over the Sea of Japan, the primary tracks are quite similar during November and December.

Over the ocean between 35° and 55°N. westerly winds predominate. On the average the force is 4–6 in more than 50 percent of the observations. Southerly winds are prevalent from southeastern Alaska to the latitude of northern California. North and northwest winds are most frequent southward to the central Mexican coast. The northeast trades persist along most of the central American coast. In Japanese waters northerly and westerly winds are most frequent. Off the China coast south of about 30°N. the northeast monsoon of winter is well established and is prominent over the South China Sea and Philippine waters. During this month the northeast trades, force 3–5, are most steady between 10° and 24°N.

The frequency of winds of force 8 north of 40°N. increases by 5–10 percent during November as compared to the previous month. South of 30°N. gales are infrequent except in the Taiwan Strait where they occur on the average in about 10 percent of the observations.

The average number of typhoons in the Far East is about two in November. The region of most frequent formation is between the Philippines and the Caroline Islands. Tropical cyclones are infrequent this month in Mexican west coast waters, but when they do occur, they can be quite dangerous.

During November wave heights greater than 12 ft. are generally encountered 10 percent of the time over most of the ocean between 40° and 55°N. Maximum frequencies of 20 percent are found in an area from 30° to 45°N. between 143° and 154°W.; in a 300-mi.-wide band south of the Aleutian Islands between 158° and 172°E.; in portions of the Sea of Okhotsk; and in a small area southwest of Taiwan.

Periods of low visibility (less than 5 mi.) are more frequent in November than in October over middle and northern latitudes. The greatest frequencies of low visibility, 20–25 percent, occur over two ocean areas, roughly from 42° to 45°N. between 165° and 175°E. and from 41° to 50°N. between 143° and 172°W.

The average air temperatures along the Great Circle route between San Francisco and Yokohama range from about 41°F. (5°C.) in the northwestern ocean to about 59°F. (15°C.) near the entrance to Tokyo Bay.

DECEMBER—December is usually a stormy month over northern North Pacific waters. Rough weather is encountered in the northern and middle latitudes. Occasionally gales occur as far south as the tropics, caused by a deep LOW that may be centered much farther north.

A comparison with the normal pressure distribution of the preceding month shows that the Aleutian LOW has increased in extent during December and dominates the pressure pattern north of 40°N. In the lower latitudes east of 160°E. the belt of high pressure is smaller than in the preceding months while in the Far East there is an intensification of high pressure over eastern China and southern Japanese waters. The number of primary storm tracks is more numerous this month than in November especially west of 170°E., indicating that more wintry weather, on the average, may be expected over the western half of the ocean.

The prevailing wind directions along the northern and mid-latitude routes are mainly from westerly quadrants. Westerly winds continue to predominate southward to about 35°N. in the middle and eastern sectors and as far south as 30°N. west of 160°E. The northeast trades are steadiest between 10° and 25°N. The wind conditions in American coastal waters are similar to those which prevail in November. In Asiatic coastal waters prevailing winds are largely from north to west over the Sea of Japan and along the east and southeast coasts of Japan. Northerly winds prevail over the Yellow Sea. From Shanghai southward to Borneo and throughout Philippine waters the northeast winter monsoon predominates, merging with the northeast trades at the more eastern longitudes.

A larger area of the North Pacific is subject to gales during December than in the preceding month. North of about 40°N. and westward to 178°E., 10 percent of the observations contain winds of force 8 or higher. West of 178°E. the 10 percent limit extends southward to about 33°N. and westward to near the Japanese coast. The greatest frequencies, 20 to about 25 percent, occur in areas from the southern tip of the Kamchatka Peninsula southward and southeastward to about 36°N. Excluding tropical cyclones, gales in southerly waters are sometimes noted in latitudes of the northern Philippine Islands and Taiwan due to strong development of the northeast monsoon. Also, at times, gale force northerly winds occur in the Gulf of Tehuantepec off the south coast of Mexico.

The average number of typhoons during December is between one and two. The most likely area of formation is in the neighborhood of the Caroline Islands. In the waters off the Mexican coast tropical cyclones are rare.

Occurrences of low visibilities are most frequent in the eastern part of the North Pacific roughly from 39° to 49°N. between 135° and 164°W. In this area, 20–25 percent of the observations report visibilities of less than 5 mi.

EXERCISES

17·1 Summarize briefly the meteorological services available to mariners.

17·2 What are the chief radar effects of the types of weather patterns that yield precipitation?

17·3 Outline the steps involved in planning a least time track.

17·4 Distinguish between climatological routes and least time tracks.

CHAPTER 18

Optical Features of the Atmosphere

The optical phenomena of the atmosphere are undoubtedly among the most weird and awe-inspiring spectacles in all nature. So strange are some of these features that observers have often ascribed them to tricks of their own fancy and imagination, rather than to plausible physical causes. Nowhere is there a more appropriate vantage point for the viewing of most of these atmospheric displays than on the open sea. Here the unobstructed horizon allows for the observation of nearly the whole of the celestial dome at one time and there is no artificial illumination of the skies to interfere with nocturnal optical displays.

In studying these features of the air, we shall consider them in accordance with the physical processes responsible for their existence. The optical effects resulting from these processes can be separated into (1) refraction features, (2) diffraction features, (3) diffusion and scattering features, (4) perspective features, and (5) electrically induced features.

Refraction

The features directly attributed to refraction are more numerous than those of the other groups. Refraction is the process in which rays of light are bent from a straight line in passing through a medium of varying density or from one medium to another of different density. Many common examples of refraction exist: e.g., a pencil or rod partly immersed in water appears to bend sharply at the water line and assume a distorted shape.

Light rays are refracted since they travel with different speeds in media of different density.

PRINCIPLES OF REFRACTION. An elementary examination of the principles of refraction will aid materially in understanding refraction features in the atmosphere. In Fig. 18 · 1, we note that the ray AO is bent abruptly upon emerging from the denser medium (water) into the rarer medium (air), and follows the path OB. An observer with his eye at B appears to see the source of the light in the direction OA' rather than in the true direction, as a consequence of refraction.

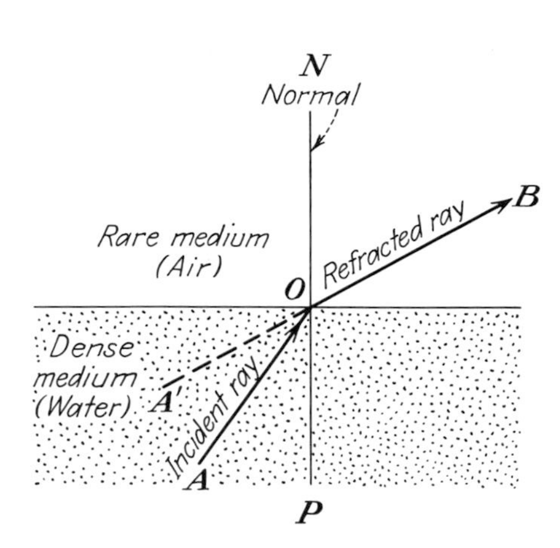

Fig. 18 · 1 Refraction and nomenclature involved.

The line NOP, perpendicular to the surface between the adjacent layers or media, is called the *normal.* The ray AO striking this surface is called the *incident ray,* and OB, the *refracted ray.* Further, the angle between the incident ray and the normal AOP is known as the *angle of incidence,* while NOB is known as the angle of *refraction.*

In the figure the incident ray is in the dense medium. What then is the path of the ray when the source of light is in the rare medium? If the light source were at B, then BO would be the incident ray, and OA, the refracted ray, would be just the reverse of the first case. An observer at A would then see point B displaced in the direction made by extending line AO into the rare medium.

Without further explanation, three facts are evident from this discussion:

1. A ray of light perpendicular to the surface between two media never suffers refraction. Thus a ray parallel to line NOP, above, undergoes no bending or refraction.
2. When a ray of light emerges from a dense to a rare medium, it is bent away from the normal. Thus, for the incident ray AO the angle of refraction NOB exceeds the incident angle AOP.
3. When a ray of light enters a denser medium, the ray is refracted

toward the normal. If *BO* is considered to be the incident ray, it is then bent in the direction *AO*, or toward the normal. Now *AOP* is the angle of refraction.

There is one other important fact which can be developed from principle 2. As the angle of incidence of a ray moving from the denser to the rarer medium increases more and more, there finally comes a situation where the refracted ray is bent so far from the normal as not to enter the rare medium at all but to travel parallel to the surface between the media. The angle of incidence at which this occurs is known as the *critical angle* and must naturally depend on the particular substance or medium involved. If the critical angle is exceeded by the incident ray, the refracted ray is bent entirely back into the dense medium, the phenomenon being known as *total reflection.* The critical angle and total reflection are illustrated in Fig. 18 · 2. The critical angle can now be defined as the angle of incidence beyond which total reflection occurs.

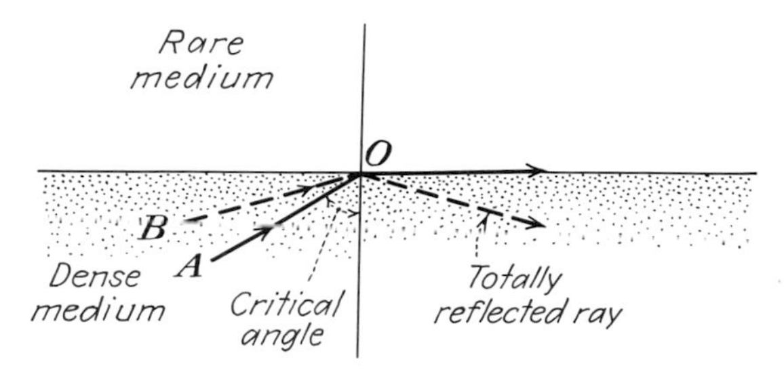

Fig. 18 · 2 Critical angle and total reflection. Ray BO, *exceeding the critical angle, is totally reflected.*

DISPERSION. Dispersion can be considered as a kind of differential refraction. We know that visible white light actually consists of light of many colors which when blended produces white light. When white light is passed through a medium of varying thickness, such as a glass prism, the components of the light are refracted differently, in accordance with their respective wavelengths. The resulting color band is known as the *spectrum,* consisting of red, orange, yellow, green, blue, and violet regions which grade into each other. These are the spectral colors commonly observed in rainbows. Dispersion is thus the process whereby white light is separated into its component colors. It is important to note that the wavelength of blue light is about half that of red, and the blue-violet end of the spectrum is refracted more than the red, which suffers the least bending, when white light is separated by dispersion. These last are important factors in causing certain of the prominent optical phenomena now to be studied.

Refraction Features

The behavior of light in the atmosphere shows the air to be extremely variable in density. For the most part this density difference is a direct result of differences of temperature and humidity within separate air layers. This nonhomogeneous character of the atmosphere is characterized both by a few well-defined and relatively thick layers and by numerous thin, tapering layers which give the air a streaky property. Many common refraction features are a result of this condition.

ALTITUDE EFFECT. Refraction does not always require two distinct layers of differing density. A simple variation of density within a single medium will also cause a bending of transmitted light rays. The atmosphere is an excellent example of such a medium. There is a steady increase in density of the air from its outer margins to the earth's surface. Consequently light from all celestial objects is refracted in passing through the atmosphere, except when at the zenith, for in this latter case the rays enter normal to the atmosphere. The effect of this refraction is to displace the apparent position of all celestial objects and give them a greater altitude above the horizon. The nearer the horizon, the greater is the refraction, since the thickness of air traversed by the light increases. The refraction or altitude correction is familiar to all navigators who have taken sights on astronomical bodies. The effect of refraction is such as to cause the sun to appear above the horizon when it has just set below.

TWINKLING, SHIMMERING, AND SHADOW BANDS. These are relatively simple and closely related effects resulting from streakiness in the atmosphere. The light from most astronomical bodies (except the more luminous ones) and from distant terrestrial sources undergoes a pronounced flickering known as *twinkling*. Small density irregularities in the atmosphere cause numerous deflections to the relatively thin beam of light from these distant objects and result in this familiar effect.

Shimmering is an equally familiar sight. Objects that are observed across relatively warm surfaces appear to dance, or shimmer, suffering distortion. Some common examples of this are found in the behavior of light passing over hot radiators, chimney stacks, dark roadways, etc. In all such cases, the atmospheric density differences resulting from convection over the warm or hot medium cause irregular refraction which produces the shimmering.

Shadow bands are alternate, narrow, light and dark bands that flit rapidly over the ground immediately preceding and following the instant of totality of a solar eclipse. This effect is also directly ascribed to distortion of the restricted sunbeam by numerous thin zones of different atmospheric density, resulting for the most part from temperature inequalities. These features are rare only because total eclipses are rare and not because of the lack of the necessary atmospheric conditions.

OPTICAL OR REFRACTION HAZE. Frequently, visibility becomes very restricted even though the air is perfectly free of both dust and water droplets. A definite white haze obstructs the vision, resembling very closely a thin fog or mist, despite the actual clarity of the air. *Tongues* and layers of air of different densities may be superimposed on each other or intermingled generally, producing a pronounced *optical heterogeneity* of the air. Irregular refraction is then so great that little light can travel any distance without suffering marked distortion, yielding the resulting poor visibility. Coastlines are frequently shrouded in this haze, being invisible from a short distance offshore as a result of marked temperature inequalities prevailing along shore boundaries.

HALOS. For ages past, the huge rings or halos about the sun or moon have been taken to have portents and meanings of all sorts. Now we know them to be optical results of the refraction of sunlight by *ice crystals,* in high, thin, cirrus or cirrostratus clouds. Halos of two different radii have actually been observed: one, the halo of 22°, and the other, the halo of 46°. The angles of refraction within the ice spicules determine the type of halo that forms, although the halo of 22° radius is by far the more commonly observed. Mild dispersion usually occurs with this refraction, and the red light, being bent the least, appears on the inner portion of the curve, with the other spectral colors following outward. The green and blue are usually too weak to be seen.

For the production of halos, the ice crystals in the cloud must have a very heterogeneous arrangement. In this case the refracted light forms a general circle completely around the sun. However, if a particular uniform arrangement of ice crystals exists, the refraction occurs in one plane only, having the same altitude as the sun. Instead of a circle, only two bright spots on either side of the sun result. These images have nearly the same radius as the halo would have if the crystals were situated in all positions. They are known as *parhelia,* or *sun dogs,* there being a parhelion of both 22° and 46°. Very rarely, the parhelia of 22° become extended in a direction concave to the sun and are then named *arcs of Lowitz.*

Although other associated phenomena are observed on rather infrequent occasions, we omit description of them. For reference purposes many of them are reproduced in Fig. 18 · 3, together with the features just mentioned.

RAINBOWS. The formation of a rainbow involves simple refraction, total reflection, and dispersion of light. Rainbows are visible on occasions when the sun is shining and the air contains water spray or raindrops. This condition is frequent during or immediately following local showers. The bow

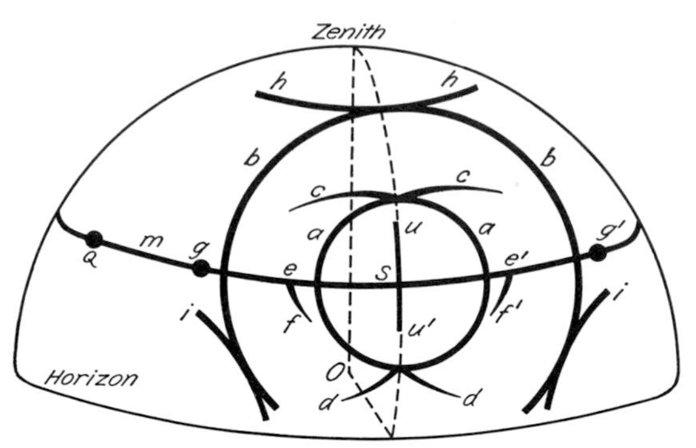

Fig. 18 · 3 Perspective view of the sky showing: sun (S); ordinary halo of 22° (a); *great halo of 46°* (b); *upper tangent arc of the halo of 22°* (c); *lower tangent arc of the halo of 22°* (d); *ordinary parhelia of 22°* (e, e′); *arcs of Lowitz* (f, f′); *parhelia of 46°* (g, g′); *circumzenithal arc* (h); *infralateral tangent arcs of the halo of 46°* (i); *the parhelic circle* (m); *paranthelion of 90°* (q); *plane of the horizon; and the observer* (o).

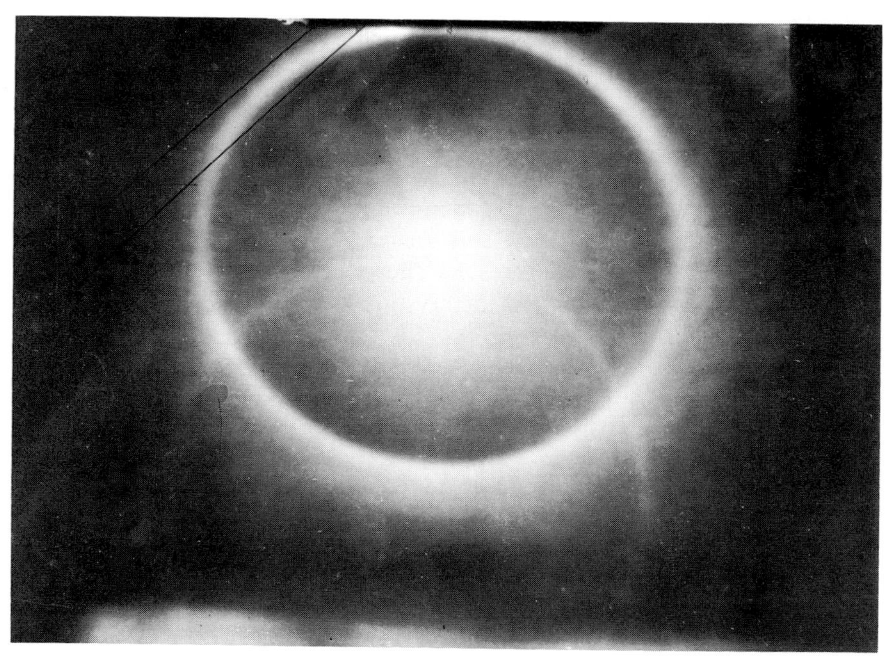

Fig. 18 · 4 Halo of 22° around sun with parhelic circle through the sun. (U.S. Weather Bureau)

is always observed in that portion of the sky opposite to the sun. The sun, the observer's eye, and the center of the rainbow arc are always on a straight line; thus, a rainbow formed at sunrise or sunset can appear as a complete semicircle on the horizon opposite to the sun.

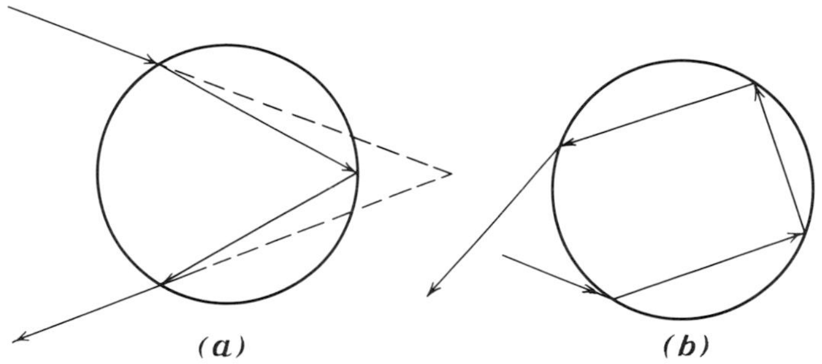

Fig. 18 · 5 Refraction and total reflection of light within raindrops resulting in primary rainbow (a) and secondary rainbow (b).

In the formation of the rainbow, light entering the denser waterdrop is refracted toward the rear of it. Some light strikes the rear surface at such an angle as to be totally reflected and then passes out of the front portion.

Fig. 18 · 6 Rainbow with secondary bow to the right. (Clarke, U.S. Weather Bureau)

This process, repeated in identical manner for myriads of drops, produces the primary rainbow, which has a radius of nearly 42°. The color bands forming the rainbow differ from those of the halo in that the red is on the outer, and the blue on the inner, edge.

If two internal reflections take place in the raindrop, a secondary bow of slightly larger radius occurs, with the color bands reversed. Tertiary bows are also observed.

LOOMING AND TOWERING. Looming and towering are phenomena in which objects actually below the horizon are brought to appear above the horizon (looming), and visible objects are stretched or elongated upward, with an apparent increase in their height (towering). These effects occur frequently in high middle-latitude or arctic waters, when a marked increase in the density of the air exists near the surface (the density at the surface may be normal, but a marked decrease in density with altitude would have

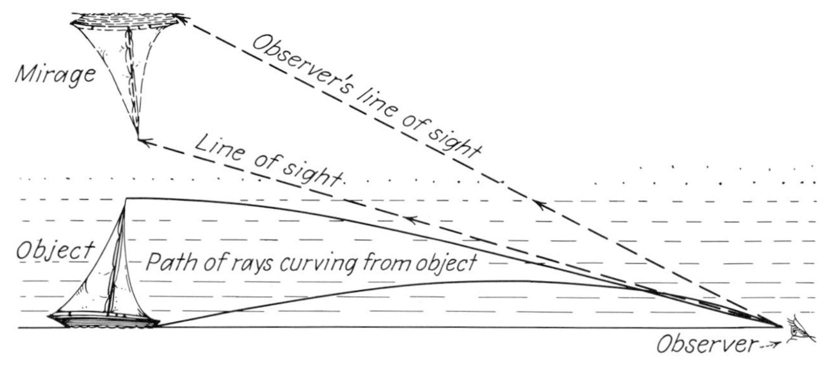

Fig. 18 · 7 The formation of superior mirage from refraction in a layer of air having a rapidly increasing density near the surface.

the same effect). In this case, the altitude of terrestrial objects is increased in much the same fashion as that of astronomical bodies, described above.

Owing to this effect, lights actually well below the horizon are often visible to the mariner. With close approach, the refracted rays may no longer be intercepted and the light disappears. When the actual light is seen, the effect of towering may greatly increase its apparent height. Hulls, masts, and stacks of ships are similarly affected.

SUPERIOR MIRAGE. The superior mirage requires much the same conditions as looming, with again a rapid decrease in density of the air with increase in altitude. By virtue of the strong refraction resulting, the image of a ship or of any object below the horizon may appear in the sky above the horizon in an inverted position. Such conditions may develop when a pronounced inversion occurs some distance above a relatively cold surface.

At times, the object itself may appear, together with its inverted counterpart directly above, almost as though a reflecting mirror surface were present. Occasionally, the density *stratification* of the air is such as to produce still a second mirage above the first, but this time in a normal upright position. Thus, ships have been seen near the horizon with an inverted image floating above them and a second upright image above that. Owing to the appearance of the mirage above the actual object, the name *superior* is given to this type.

INFERIOR MIRAGE. The deceptive inferior mirage is rarely seen over the oceans but is common over highly heated land surfaces, particularly deserts. The mirage derives its name from the fact that it always appears below the actual object, inverted in position. The mirage is normally manifested by the inverted image of distant objects which themselves are often beyond the range of vision and by the inversion of the distant low blue sky, giving the appearance of a lake, or water surface. The presence of

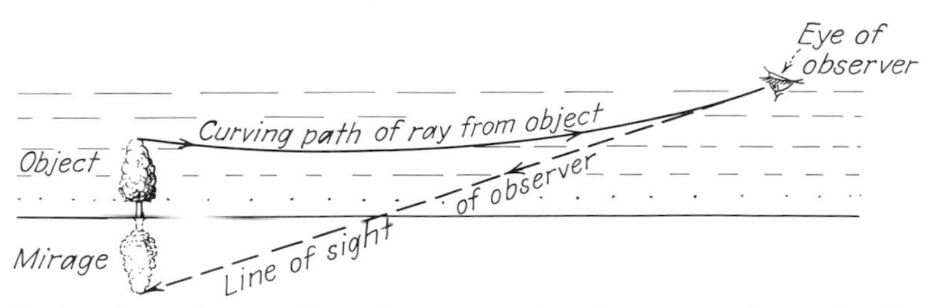

Fig. 18 · 8 The production of an inferior mirage by refraction in a layer of air having a density decreasing rapidly near the surface.

the distant inverted image, shimmering in the sky mirage, creates the appearance of reflection from a water surface, so misleading to desert travelers.

This mirage is the result of a thin rarefied layer of air (a few feet thick, at most), caused by the heating of the land surface. The observer's eye must be above this zone, in the denser overlying air. Light rays from the sky or distant objects, instead of being bent downward in passing through the rarefied layer, as in the case of looming or superior mirage, are refracted upward to the observer's eye. The observer's line of sight, being a straight line, projects this image back toward the earth in the direction from which the refracted rays came, and creates the inferior mirage.

FATA MORGANA. Some very remarkable, complicated, multiple mirages have been observed at times in the Straits of Messina, but more frequently in higher latitudes along the coasts of the Atlantic and the Pacific. These mirages combine the effects of inferior and superior mirages, looming, towering and the reverse, and general distortion, being produced by atmospheric zones of varying densities and thicknesses. The results of these complications transform the appearance of the coastline features into strange and often grotesque sights. Shore cliffs and buildings may be extended as tremendous swelling towers and castles or depressed and flattened into other strange shapes. Observers likened this phenomenon to the legendary castles of Morgan, le Fay. The term *Fata Morgana* has since been applied to any complicated mirage system.

In the same way, the legend of the Flying Dutchman is kept alive by the almost ghostly appearance and disappearance of entire ships. Mirages have had a profound part in the supernatural influence of nature on the unenlightened.

THE GREEN FLASH. The green flash is another strange and beautiful example of refraction that requires an unobstructed horizon and relatively clear atmosphere, two requirements again achieved at sea. On rare occasions, the last bead of the setting sun (or rising sun) and the surrounding sky flash out in a deep, more or less brilliant green or blue green, which lasts for but a short time. We have noted above that the spectral colors of white light are refracted differently—violet the most and red the least. Thus, conditions may just occur, when only a small rim of the sun is visible on the horizon, so that the red, orange, and yellow portions of sunlight are not refracted sufficiently to reach the eye, while the green light is so refracted. This results in the appearance of the green flash.

Diffraction Features

When rays of light encounter obstacles of very small size, the rays are deflected from their normal path, usually in a straight line. This is the process of diffraction. The various wavelengths composing white light are diffracted somewhat differently, causing dispersion, or the separation of white light into its spectral colors. *Coronas* are the most common examples of atmospheric diffraction. They are rims of small radius and of subdued color, commonly observed surrounding and sometimes apparently touching the solar disk, caused by the diffraction of sunlight or moonlight through thin cloud veils. When discernible, the colors of shorter wavelength, blue and green, appear on the inner portion of the corona, with red and orange forming the outer ring.

Diffusion and Scattering Features

Were it not for the diffusion or scattering of sunlight by the air, some of the greatest beauties of the atmosphere would be lost. Much compensation would exist however. Daylight, which is taken so much for granted, requires the presence of an atmosphere. The effect of the air is to spread much of the entering sunlight fairly evenly over the sky, so that, except in the immediate vicinity of "Old Sol," the sky is more or less uniformly illuminated. This diffusion of sunlight actually gives greater illumination to the earth's surface as a whole than if the rays were to shine in a direct beam.

Without the diffusion of sunlight the sky would appear black, with stars, planets, and moon always visible, in addition to the brilliant orb of the sun, whenever above the horizon. Such a condition seems paradoxical, yet it is only the brightness of the daytime sky (from diffusion) that prevents the observation of the other astronomic objects, for they are certainly there. In the absence of a diffusing atmosphere, the sun would shine forth brilliantly from the surrounding blackness of the sky. Intense contrasts of light would exist, with harsh shadows and glaring bright areas. Such a situation would of course be a great boon to the astronomer by greatly increasing his period of observation. Balloon observers ascending to heights above 14 miles, with most of the air below, have encountered such conditions.

The phenomenon of scattering explains, too, the existing blue of the sky, the brilliant colors of sunset, and the conditions of twilight. The blue of the sky has been a perplexing problem, long evading satisfactory explanation. It has finally been shown to be a feature resulting from scattering.

Scattering is the process whereby light, traveling in a straight path through the atmosphere, is disturbed by small particles of dust and air—small in comparison to the wavelength of light. The effect of this disturbance is to give a random distribution to the light striking the particles and is much more pronounced for the shorter wavelengths—blue light. Thus, most of the scattered light is blue, causing the color of the clear sky to be blue. The more intense the scattering, the deeper is the blue color.

The striking crimson and scarlet hues often accompanying the rising and setting of the sun are similarly explained. We have noted earlier that the thickness of the atmosphere traversed by sunlight increases markedly with decrease of altitude. The effect of this additional amount of intervening dust and air particles may cause so pronounced a scattering as to permit the direct passage of only the longest or red wavelengths in the sunlight. The results of this require no further description.

Twilight is the name applied to that interval after sunset during which the sky is still illuminated. Were it not for the presence of the atmosphere, darkness would set in almost immediately after sunset. As it is, scattering and refraction "preserve" the sunlight for varying periods after the sun itself is no longer visible. The interval between sunset and the time when it is too dark for normal outdoor activities is defined as *civil twilight*. Specifically, civil twilight is the period from sunset until the center of the sun has descended 6° below the horizon. *Astronomical twilight* is the period from sunset until the sun's center is 18° below the horizon. It is quite dark by that time. Clearly, the duration of the twilight period must vary with the path of the sun. When this path is very oblique to the horizon, as in summer, the period will be longer than when the path is more perpendicular. The *American Nautical Almanac* publishes tables giving the duration of twilight at different latitudes and different dates.

Perspective Features

Perspective effects are in part the results of physical conditions and in part the results of optical illusions. We have already examined some of the perspective phenomena associated with clouds. They included the darkening of the clouds and the narrowing of the sky interval between, as the observer's view approaches the horizon. In these cases, a direct physical explanation accounted for the observed features.

The expansion in size of the sun or moon when near the horizon is a familiar feature, explained on the basis of perspective. The distance of the sky apparently increases with approach to the horizon. Thus, although the distance of the sky appears to change, the actual distance of the sun or the moon, and their actual angular diameter, remain the same regardless of celestial position. Consequently, in order to compensate for their seemingly greater distance when on the horizon, the size of the sun or the moon increases. The size increase is therefore illusory.

In the same way, any angular arc of the sky itself appears to be magnified with decrease in celestial altitude. This causes an apparent increase in diurnal motion of astronomical objects near the horizon, in order that they may traverse this apparently greater sky interval in the same time.

Another very common perspective feature that is explainable by optical illusion is the phenomenon known as the *crepuscular rays*. Popularly, these rays are referred to by the phrase "the sun draws water." The sun is so distant from the earth that its light rays approach along parallel lines. In cloudy weather, when openings or rifts appear in the clouds, the path of the sunlight appears as a number of illuminated beams that seem to converge upward toward the opening. The beam itself is merely sunlight reflected by foreign matter in the air, along the path of the beam. The same effect is shown by the illuminated path of a bright searchlight at night or by the projection beam in a motion-picture theater. These sunbeams, diverging from the gaps between clouds, are the crepuscular rays. The convergence at the cloud level is a common perspective illusion in which parallel lines seem to meet in the distance. A familiar analogy is the approach of railroad tracks toward each other as the eye looks down the tracks.

Electrically Induced Features

There are few more strange or remarkable sights than the *aurora borealis*. When visible in the skies of the Northern Hemisphere, this phenomenon is often termed *aurora polaris,* and in southern skies, *aurora australis*. They are ghostly displays of light in the forms of streamers, rays, arcs, bands, curtains, draperies, sheets, or patches that seem to shimmer and flit across the sky. Auroras are most common in higher latitudes, centering around the magnetic poles. The shifting auroral patterns are most often a greenish white, though pronounced reds, yellows, and greens are very often observed.

Fig. 18 · 9 Crepuscular rays formed by a stratocumulus cloud masking the sun.

Auroras are associated with vast magnetic storms on the sun, which appear as sunspots and are nearly always attended by magnetic disruptions on the earth that seriously impair the normal behavior of communication devices. As such, auroras are explained as resulting from solar discharges of electrified particles emitted by the magnetic storm areas, some of which approach the earth and tend to be attracted to the magnetic poles. The effect of this electrical bombardment of the upper rarefied portions of the atmosphere is to cause excitation of the grass therein, with a consequent emission of radiation. We may compare this phenomenon to that of the familiar colored commercial sign displays. Here, relatively high-voltage electric discharges are passed through tubes containing particular gases under low pressure. Light of characteristic color is then emitted, depending on the gas employed.

It will be recalled that one method of estimating the height of the atmosphere is the determination of the elevation of auroral displays.

Fig. 18 · 10 Aurora borealis as seen from Ogunquit, Maine, Aug. 12, 1919; painting by Howard Russell Butler. (Courtesy of American Museum of Natural History)

EXERCISES

18 · 1 Explain the meaning of the critical angle and total reflection.

18 · 2 Explain the origin of the inferior and superior mirages in terms of total reflection.

18 · 3 What is the fundamental optical difference between rainbows and halos?

18 · 4 Why do distant, small objects and lights, particularly when their altitude is low, appear to shimmer or twinkle?

18 · 5 How does the aurora borealis (northern lights) form? Why is this phenomenon most common in high latitudes?

18 · 6 Distinguish between refraction and diffraction.

CHAPTER 19

The Oceans

A pronounced reciprocal relationship exists between the atmosphere and the oceans. The movements of the air are responsible for most of the wave action and much of the surface current motion in the seas and, together with pressure variations, greatly affect the level of the oceans, along coasts especially. The oceans in turn, covering nearly 71 percent of the earth's surface, have a most direct and important influence on the heating of the air, which, as we have noted earlier, is the cause of nearly all the other atmospheric variations. The moisture of the air is derived almost completely from the great expanses of ocean waters. So important are these effects of the oceans and the air on one another that no treatment of the conditions of the atmosphere from a marine viewpoint would be complete without at least a brief examination of the physical properties of the sea.

The scientific study of the oceans is limited to a period of less than a century. Hence, the science of *oceanography,* as this study is called, is one of the youngest of the natural sciences. With the publication of *The Physical Geography of the Sea,* by Lieut. Matthew Maury of the U.S. Navy, in 1855, oceanography emerged as a definite field of study. To Maury goes the credit for developing one of the first world wind and current charts. This was accomplished through his intensive study of ships' logs and records.

The voyage of the British warship *Challenger,* sponsored by the British Royal Society, was of further importance in advancing this new branch of knowledge. The *Challenger* cruised the oceans for 3½ years, engaged only in compiling oceanographic data, with the results filling 50 volumes. A number of other institutes and individuals have since carried on these original investigations and added greatly to the accumulation of our ocean knowledge. Nevertheless, to this day, owing to the vastness of the oceans and the basins they occupy, there is a tremendous amount of research still awaiting the oceanographer.

General Features of the Oceans

DIMENSIONS AND DISTRIBUTION. The surface area of the earth is very close to 200,000,000 square miles. Of this, nearly 71 percent is comprised of the waters of the oceans and their isolated parts (seas). Clearly the oceans are the dominant feature of the earth's surface. The continental masses divide these features of the earth into more or less distinct units. There is a great inconsistency in the definitions and nomenclature applied to the description of these water bodies. We follow here the terminology adopted by Sverdrup, Johnson, and Fleming.

The interconnected Atlantic, Pacific, and Indian Oceans surrounding Antarctica are often referred to by oceanographers collectively as the *Southern Ocean.* Many partly land-enclosed or more or less isolated sections of the primary oceans occur, being known commonly as *seas.* The distribution of the land areas is such that the ratio of land in the Northern Hemisphere to that in the Southern is greater than 2 to 1. Table 19 · 1 shows land-water relationships for each 10° latitude zone.

The average depth of the oceans is about 2.4 miles. This contrasts strongly with the average height of the land masses which is only 0.5 mile. Thus, the average ocean depth exceeds the average continental altitude by almost five times. The extreme conditions are not nearly so contrasting although again the marine basin has the advantage. Mount Everest, the highest peak, towers to a height of 29,100 feet. Modern sonic sounding methods have indicated the greatest known depths to lie in the Japanese trench and in the Philippine trench farther south. In both of these deeps, the ocean descends to a depth of 34,000 to 35,000 feet. Table 19 · 2 summarizes the areas, volumes, and mean depths of the oceans.

THE OCEAN BASINS. The ocean basins are the tremendous receptacles that contain the ocean waters. The continental limits form the sides of these basins. Although local differences are common, their chief features

*Table 19 • 1 Land–Water Distribution for Each 10 Degrees of Latitude**

ZONE LATITUDE	NORTHERN HEMISPHERE				SOUTHERN HEMISPHERE			
	LAND, MILLIONS OF SQ MI.	SEA, MILLIONS OF SQ MI.	LAND, PER-CENT	SEA, PER-CENT	LAND, MILLIONS OF SQ MI.	SEA, MILLIONS OF SQ MI.	LAND, PER-CENT	SEA, PER-CENT
0–10°	3.9	13.1	23	77	4.0	13.0	24	76
10–20°	4.3	12.2	26	74	3.6	12.9	22	78
20–30°	5.8	9.7	38	62	3.6	11.9	23	77
30–40°	6.0	8.0	43	57	1.6	12.5	11	89
40–50°	6.4	5.8	52	48	0.4	11.8	3	97
50–60°	5.7	4.2	57	43	0.1	9.8	1	99
60–70°	5.2	2.2	71	29	0.7	6.6	10	90
70–80°	1.3	3.1	29	71	3.3	1.2	73	27
80–90°	0.2	1.4	10	90	1.5	—	100	
Total	38.8	59.7	39	61	18.8	79.7	19	81

* After E. Kossinna, 1921.

Table 19 • 2 Dimensions of the Oceans

OCEAN	AREA, MILLIONS OF SQ MI.	VOLUME, MILLIONS OF CU MI.	MEAN DEPTH, FT
Atlantic	32.15	64.72	13,112
Pacific	64.45	141.51	14,302
Indian	28.65	58.21	13,236
Total	125.25	264.44	13,750
Atlantic—including adjacent seas	41.53	70.94	11,129
Pacific—including adjacent seas	70.07	144.74	13,454
Indian—including adjacent seas	29.24	58.38	13,016
Total	140.84	274.06	12,533

are essentially the same, and it is these features with which we are now primarily concerned.

In general there exists a gradual slope, known as the *continental shelf,* from the continental shore out to a depth of 500 to 600 feet. The width of this shelf may vary considerably from almost nil to more than 200 miles. Along rugged mountainous coasts, this zone may be very narrow. Along coasts exhibiting well-developed coastal plain areas, the shelf is simply the gently sloping seaward extension of the plain. The continental shelf is composed of sediments carried out from the land. Since the greatest depth achieved by storm waves is 100 fathoms, this limit marks the end of the depositional extension of the coastal plain.

Beyond the 100-fathom line there is a more or less abrupt transition in slope, known as the *continental slope.* Although the angle of this surface from the horizontal averages only a few degrees, it may locally be so precipitous that deeply towed fishing nets can become snagged as the fishing vessel crosses into the shallower waters over the continental shelf.

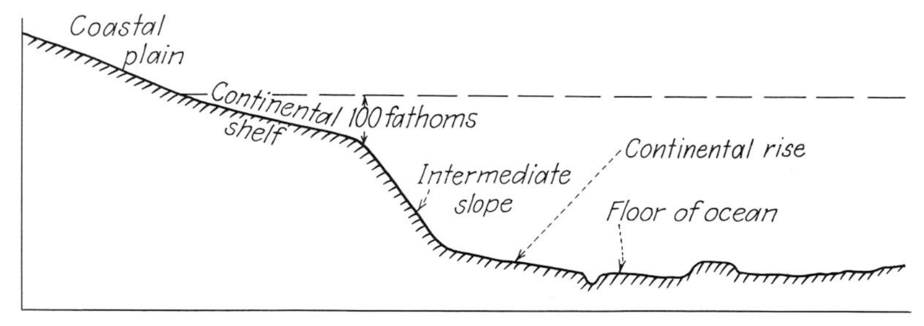

Fig. 19 · 1 Generalized profile of the ocean basins.

The continental slope continues as the gentler *continental rise* until the true floor of the oceans, or *abyssal region,* is reached. The ocean floor is now known, from extensive echo sounding, to be covered with plains, plateaus, volcanoes, mountain ranges, *deeps,* and other irregularities once thought to be limited to the continents. A summary of depths in the Atlantic Ocean is shown in Fig. 19 · 2, which has been constructed from continuous ocean-bottom profile records.

Static and Dynamic Properties of Seawater

In physical oceanography, the oceans are usually examined from two viewpoints: the static properties and the dynamic features. The static properties involve such factors as temperature, composition and salinity, density, and ice formation. The first three of these in particular are closely interrelated so that the density of seawater depends on both temperature and composition. The dynamic aspects of oceanography involve the study of the movements of seawater, particularly waves, currents, and tides. It will be seen later that currents in turn are partly related to the static property of density.

Temperature of the Oceans

The ocean temperatures are of direct importance in the effect of the sea upon the air. The most prominent feature of marine temperatures is their pronounced uniformity, compared to those on land. The reasons for the conservative thermal properties of the sea were discussed earlier, in Chap. 2. Since the overlying air temperatures depend closely on the sur-

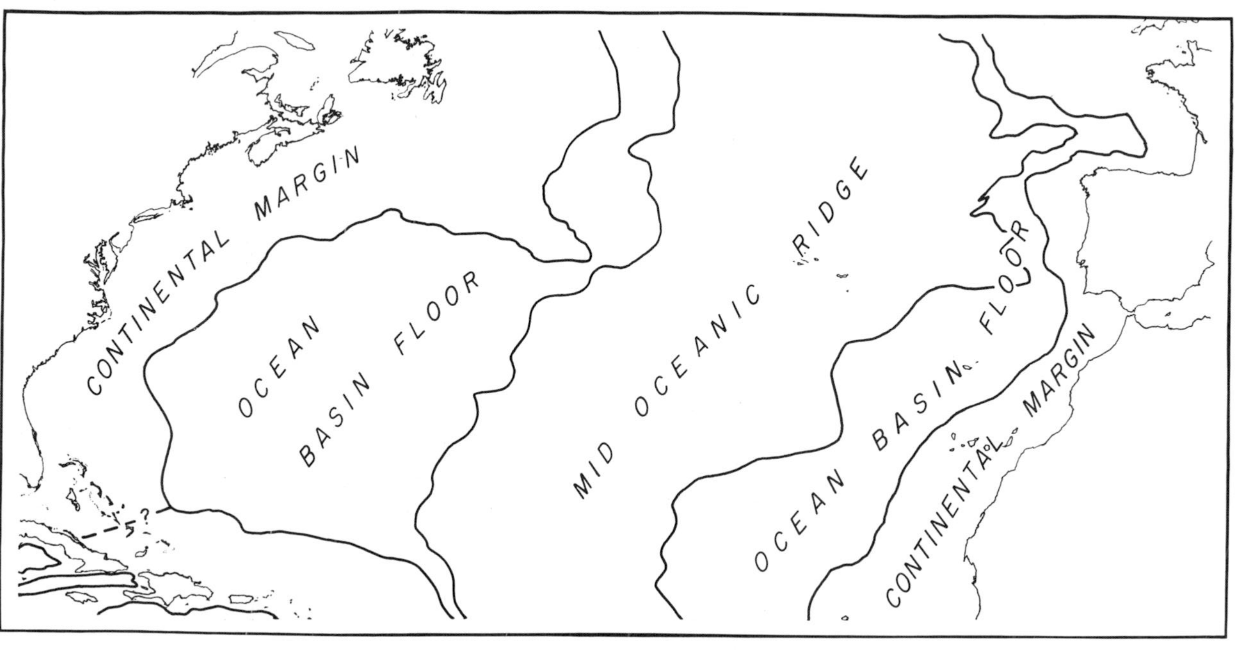

Fig. 19·2 Depth chart of North Atlantic Ocean together with a generalized profile from North America to Europe. (Courtesy of B. Heezen, M. Tharp, and M. Ewing)

face temperatures beneath, it follows that marine air should be very uniform in temperature. That such is the case is a matter of common knowledge and observation.

ANNUAL AND DIURNAL SURFACE TEMPERATURE VARIATIONS. The periodic temperature variation of the surface is twofold: annual and diurnal. The annual temperature variation of the oceans is by no means constant from place to place. Owing to the temperature lag considered earlier, the temperature range is never very large, rarely exceeding a maximum average annual range of 20°F, although the mean annual range is about 10°F. The greatest average surface temperature range is in the middle latitudes, occurring between 40 and 50° in the Northern Hemisphere and between 30 and 40° in the Southern Hemisphere. The mean annual variation in the tropics is considerably less than that in the middle latitudes, and the range in the Northern Hemisphere exceeds that of the Southern Hemisphere. The predominance of land north of the equator thus shows its influence on marine temperature. The warm winds blowing offshore in the summer tend to increase slightly the temperature maximum for the oceans and are similarly responsible for lowering the minimum somewhat, when cold winds blow seaward from the much colder continents. Thus the maximum average range in the northern middle latitudes reaches 16°F, while the maximum southern middle-latitude range reaches only 10°F. In the tropics in either hemisphere the range is from 1 to 5°.

The diurnal surface temperature variation is very small. This is to be expected after a consideration of the temperature lag considered earlier. This range averages only 0.35°F. It is greater for days with clear skies than for those with an overcast. Clear days permit a greater influx of insolation during the day and a greater amount of radiation at night than do overcast conditions. The range is also greater for calm days since wind causes wave action which produces mixing of the surface water with the water at a slight depth, thereby distributing the surface heat and lowering the temperature range.

HORIZONTAL VARIATIONS. A definite temperature gradient exists in both hemispheres from the equator to the poles. Observations of surface temperatures show decrease poleward from the equator at the approximate rate of 0.5°F for every degree of latitude. The temperature of the equatorial oceans averages about 80°F, while that of the polar seas approximates 30°F. It is significant to note that the peak marine-surface temperatures occur north of the equator. This again is a reflection of the predominance of land north of the equator. Figures 19·3 and 19·4 summarize the horizontal temperature distribution.

DEPTH VARIATIONS. Since the sun is the ocean's source of heat, it follows that the temperature will vary inversely with depth. The temperature usually decreases very rapidly from the surface downward. In lower

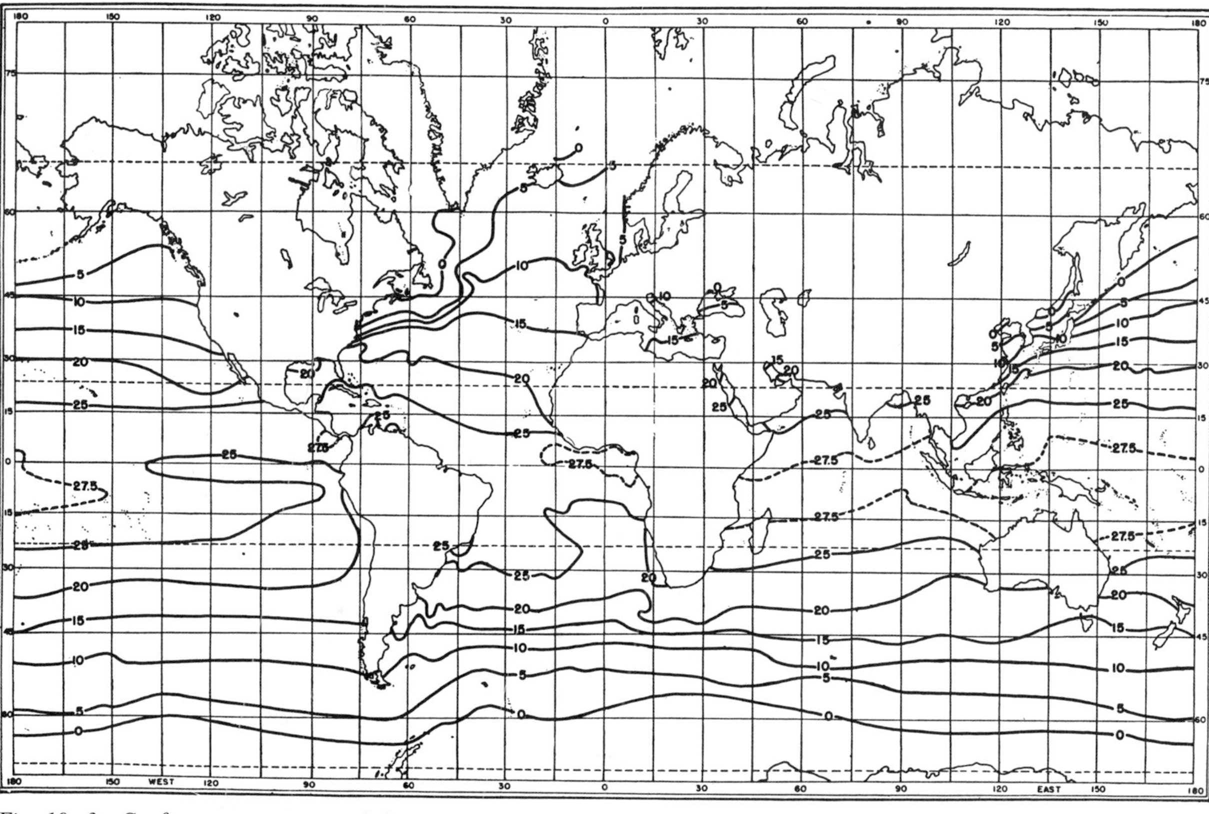

Fig. 19·3 Surface temperatures of the oceans in February (°C). (Haurwitz and Austin, Climatology, modified after Sverdrup, Oceanography for Meteorologists, Prentice-Hall, Inc.)

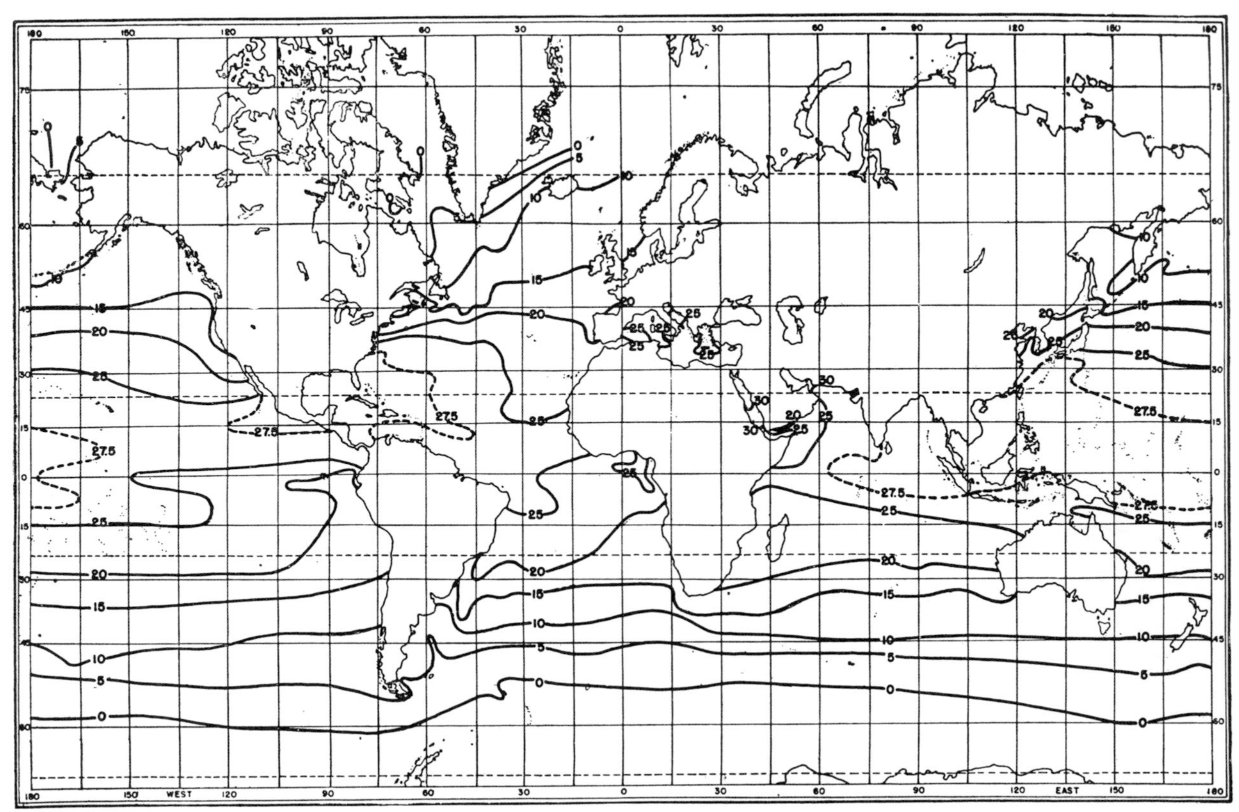

Fig. 19·4 Surface temperatures of the oceans in August (°C). (Haurwitz and Austin, Climatology, modified after Sverdrup, Oceanography for Meteorologists, Prentice-Hall, Inc.)

latitudes this decrease is considerably more rapid than in the higher latitudes. Below 500 fathoms, the temperatures of the oceans regardless of latitude are nearly the same at given levels. In general, the temperature below 400 fathoms is at or below 40°F; any further decrease is very gradual. The waters in the vicinity of the ocean floor are just above freezing and show little variation in temperature. The cold polar-ocean currents creeping equatorward along the ocean floor partly account for the cold subsurface waters of the tropics. Then water, which is a good insulator, prevents the conduction of much heat to any great depth. The average

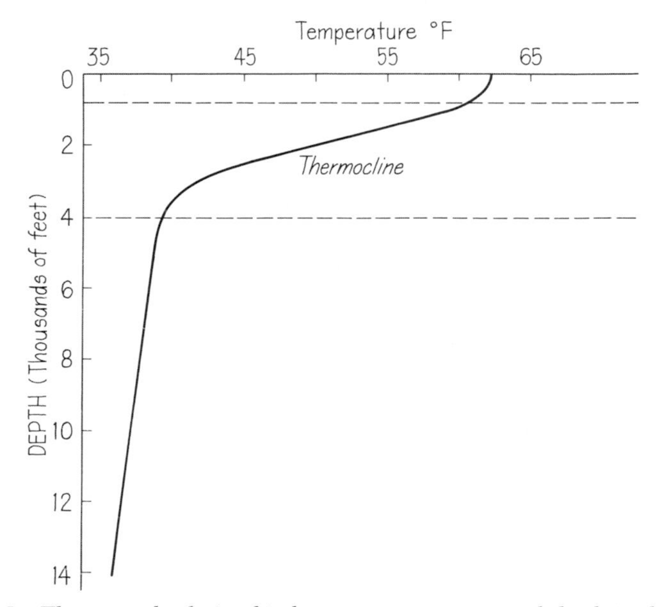

Fig. 19·5 The general relationship between temperature and depth in the oceans.

temperature of the oceans is barely 40° F, so that the oceans are a reservoir of cold water.

Because the distribution of temperature with depth is so important in oceanography, it is important to examine the actual relationship as shown in Fig. 19·5. In an upper zone, which may vary from 500 to nearly 1,000 feet, the temperature decreases very slowly with depth. This *surface* or *mixed layer* develops from the fairly complete mixing of the water from wind stirring, convection, and turbulence. Below the surface layer, temperature falls relatively fast until a level where the temperature of the cold deep water again declines slowly with increasing depth. The *thermocline* or *thermocline layer* refers to the region where the rapid decrease of temperature occurs.

MEASUREMENT OF SEAWATER TEMPERATURES. For purposes of distinction, the upper 3 feet of the ocean waters is to be considered as constituting the surface waters.

There are two methods commonly employed for the determination of surface temperatures: the bucket and the intake method. In the first, a sample of the surface water is hauled aboard in a bucket and the temperature measured by means of an accurate standard thermometer. The sample should be taken from the bow rather than the stern end of the vessel in order to avoid the wake as much as possible. There are many practical difficulties involved in obtaining water samples on the deck of a ship by this technique. In addition, many sources of error are introduced by the use of a small, isolated sample of water. When temperature readings of such a sample are taken, the thermometer should remain immersed in the water. Water samples should obviously be taken as far from discharge outlets as possible.

The condenser-intake method provides a means of obtaining more accurate surface temperature observations. The only serious condition is that the thermometer must be properly located where free movement of the incoming water exists and where it can be read conveniently, without involving any parallax error. An excellent place of exposure is on the intake pipe between the pump and the ship's side. This avoids exposing the thermometer to possible warm-water pockets or currents which may exist farther on in the circulation system.

The method actually utilized depends to a great extent on both the type of vessel and the observer. We emphasize again that the observer must always seek to minimize sources of error.

Because sea-surface temperatures are very important in controlling air temperatures and resulting weather conditions, all reporting vessels usually make these observations. As surface-water temperatures are subject to small variations from a number of factors, they are not read with the precision used in measuring the more constant deeper-water temperatures. These are usually measured by oceanographic or research vessels only.

Deep-water temperatures, which are only taken for scientific purposes, are usually read to 0.01°C by one of three methods. The least reliable consists of lowering a special sampling bottle to the desired depth and reading the water temperature on an expanded scale thermometer on deck after the bottle is recovered.

The reversing thermometer is a more elaborate but more accurate instrument. In principle, the thermometer, which is similar to the maximum thermometer described earlier, is normally attached to a cylindrical water sampler known as a *Nansen bottle.* Several such instruments are attached to a wire cable at fixed intervals and lowered to the desired depths, thus giving temperature and water samples at many levels from a single casting.

After lowering, a weighted messenger is run down the cable. Upon striking the upper part of the first Nansen bottle, a release is triggered, freeing the upper end so that the entire unit flips over, reversing its position as shown in Fig. 19 · 6. When this happens, the mercury thread in the thermometer falls to the bottom (originally top) of the tube, where it remains until restored to its normal position on deck when the temperature is read. The

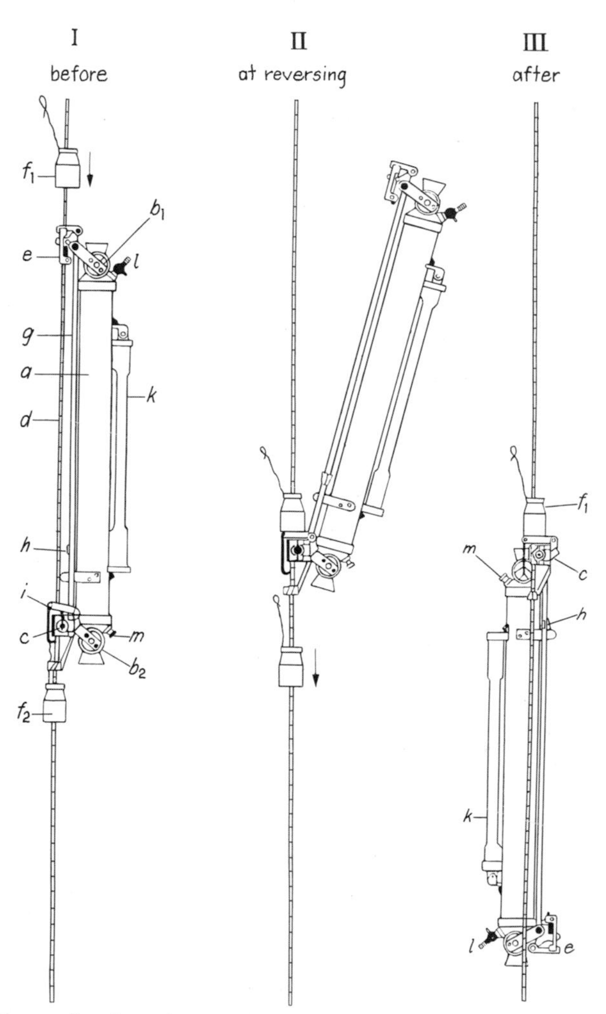

Fig. 19 · 6 Nansen bottle and attached thermometers shown during reversal operation. (G. Dietrich, General Oceanography, John Wiley & Sons, Inc.) (a) *the sampling bottle* (b) *plug valves* (c) *clamping device* (d) *wire cable* (e) *release mechanism* (f) *messengers* (g) *valve guide rod* (h) *locking cone* (i) *lower messenger release* (k) *thermometer housing* (l) *drain valve* (m) *air valve*

reversal process also seals both ends of the sampling tube for water analysis after recovery. As each thermometer reverses, an additional messenger is released to trigger the next underlying unit.

The *bathythermograph* is a device which makes a continuous record of temperature on a smoked slide as the unit is lowered. Although valuable and accurate data are obtained from this temperature instrument, its use is limited to relatively shallow depths by the effect of ocean pressure.

Salinity and Composition

The terms *sea* and *salty* are almost inseparable. Although we now know that the saline content of the oceans has been carried in by the ceaseless flow of different rivers during the billions of years of earth history, numerous legends and stories have arisen in the past to account for the salty seas.

Despite the fact that about 50 elements have been detected in seawater, most of the salt is composed of the relatively few common compounds summarized in Table 19·3. It is interesting to note that while sodium

Table 19·3 Composition of Sea Water (in Solution)

SUBSTANCE	PERCENT
Sodium chloride	77.8
Magnesium chloride	10.9
Magnesium sulphate	4.7
Calcium sulphate	3.6
Potassium sulphate	2.5
Calcium carbonate } Magnesium bromide }	0.5
Others	

chloride (composition of common table salt) is the dominant form of sea salt, it is quite scarce in the world's rivers. But calcium carbonate, the major saline constituent of rivers, is so scarce in seawater as to be unrepresented in the table. The reason for this is that calcium carbonate is relatively insoluble and is readily precipitated from the sea to form beds of limestone, whereas the very soluble sodium chloride is maintained almost indefinitely and keeps increasing in amount.

Seawater salinity is measured and expressed by weight in parts of salt per thousand parts of water (including the salt). The conventional symbol for the measure of parts per thousand, or *parts per mille,* is ‰.

HORIZONTAL DISTRIBUTION OF SALINITY. For rather obvious reasons, the salinity of the surface is known far better than that of subsurface waters.

Although the average surface salinity of the world's oceans is close to 35 ‰, there is actually a considerable horizontal variation. In the open oceans the variation is from 33 to 37.25 ‰, while in restricted seas, bays, and gulfs the range is enormous.

Table 19 · 4 Distribution of Surface Salinity

REGION	SALINITY, ‰
North Atlantic Ocean:	
0–20°	35.57
20–40°	36.72
40–60°	34.29
South Atlantic Ocean:	
0–20°	36.25
20–40°	35.65
40–60°	34.06
North Pacific Ocean:	
0–20°	34.35
20–40°	34.58
40–60°	32.88
South Pacific Ocean:	
0–20°	35.30
20–40°	35.28
40–60°	34.22
North Indian Ocean:	
0–20°	35.15
20–30°	38.24
South Indian Ocean:	
0–20°	34.85
20–40°	35.57
40–60°	34.22
Mediterranean Sea	37–39
Red Sea	37–41
Persian Gulf	37–39
Gulf of Mexico	36
Caribbean Sea	36
Baltic Sea:	
Western part	16–22
Eastern part	6–8
Bering Sea	31–32
Arctic	Less than 30
North Sea	28–35
Japan Sea	33–34.5
Okhotsk Sea	32

Important values of surface salinity are given in Table 19 · 4, whose data emphasize its surface variability. In the open oceans apart from the polar seas the ratio of evaporation to precipitation, which is related to latitude, is the main cause of salinity variation. Excess evaporation concentrates salt content while excess precipitation dilutes it. Hence, belts of average high salinity occur in the horse latitude zones of both hemispheres. A steady decrease in salinity occurs poleward from the horse latitudes as evaporation decreases and precipitation increases.

Other factors, which are particularly important locally or in more restricted sea areas, are melting and formation of sea ice; influx of fresh river water, mixing with either more or less saline deeper water; and melting of high-latitude glaciers. The low salinity of the Baltic Sea is a good example of the effects of high precipitation plus fresh water stream inflow, while the Red Sea and Persian Gulf represent the opposite conditions.

It is worth noting that salinity shows almost no diurnal variation and only very little annual variation compared to that of temperature.

VERTICAL VARIATION OF SALINITY. Although surface salinity can be measured readily and almost continuously along ship routes, observations at depth require that the ship heaves to for lowering of Nansen bottles, the points of observation being known as *hydrographic stations.* Most of the readily available information comes from the historic expeditions of the *Challenger* and *Meteor,* but beginning with the International Geophysical Year (1957–1958), a large number of newer research vessels have added greatly to our store of knowledge. Much of the newer data is not yet readily available.

In general, salinity is highest in the surface or nearsurface waters and decreases slowly with depth. The entire range from the surface downward is within 1 to 2 parts per mille.

Density of Seawater

The most important influence of the temperatures and salinity is the control of the density of seawater. Seawater density, in turn, is important in affecting sea level, stability, and current motion, particularly currents related to the vertical circulation of the oceans.

The density of seawater depends primarily on temperature and salinity and to a lesser extent on pressure. The average density is close to 1.025 grams per cubic centimeter. Variations of density are usually in or beyond the third decimal place, so that density is usually expressed by placing the decimal after the third place. The average value would then be 25.0. As this cannot be in grams per cubic centimeter, the unit *sigma* is used. It is also convenient in oceanography to consider the density of water if brought to the sea surface, so as to remove the pressure factor. The usual notation of density has the expression sigma-t (σt) and refers to the density of seawater as affected only by temperature and salinity. Thus the σt of seawater at the surface, whose density is 1.02531 grams per cubic centimeter, would be 25.31.

In general, the distribution of density varies inversely as the distribution of temperature, which appears to be the most important of the controlling factors. The effect of pressure in influencing density is primarily important below middle depths—say 1,000 fathoms. Variations of temperature and salinity below this level become so small that the pressure effect is the principal cause of increasing density with increasing depth below this zone.

An important method of oceanographic study, also extremely valuable in sea rescue work, has developed from the vertical effects of temperature, pressure, and salinity on the velocity of sound in the oceans. This velocity varies directly with each of these factors, although salinity is of least importance. Recall that the temperature, and hence the sound velocity, decreases rapidly in the thermocline layer. Then, although the temperature changes little below about 1,000 fathoms, the continuous increase in pressure causes a continuous increase in sound velocity. The effect is the existence of a layer of low sound velocity centered a little above 1,000 fathoms. Ordinary wave refraction tends to confine sound waves within this layer, if the sound is generated therein, so that the layer is known as a sound channel. Sound from a small explosion in this channel will thus travel for thousands of miles with relatively little weakening (attenuation). The existence of the sound channel provides a new method known as SOFAR (sound fixing and ranging) for undersea exploration and long-distance signaling in search and rescue operations.

The distribution with latitude of the factors of surface temperature, salinity, density, evaporation, and precipitation is well shown in Fig. 19 · 7, which illustrates mean curves of these values for the Atlantic Ocean. Both the direct and inverse relationships of these properties as referred to above are quite obvious in this diagram, as are the relationships to latitudinal zones.

Ice of the Oceans

The effect on safe navigation of ice in the sea cannot be overstressed. Scores of disasters have marked the end of ill-fated vessels that traversed the zones of floating ice. The classic example of ice catastrophe was the sinking of the "unsinkable" *Titanic* on a clear, calm night in April, 1912, after collision with an iceberg. More than 1,400 persons were lost in this disaster.

SOURCES OF ICE. The perennial cold temperatures of the polar latitudes in both hemispheres are such as to provide for widespread freezing of sea and fresh water in those regions. Nearly all the actual freezing is accom-

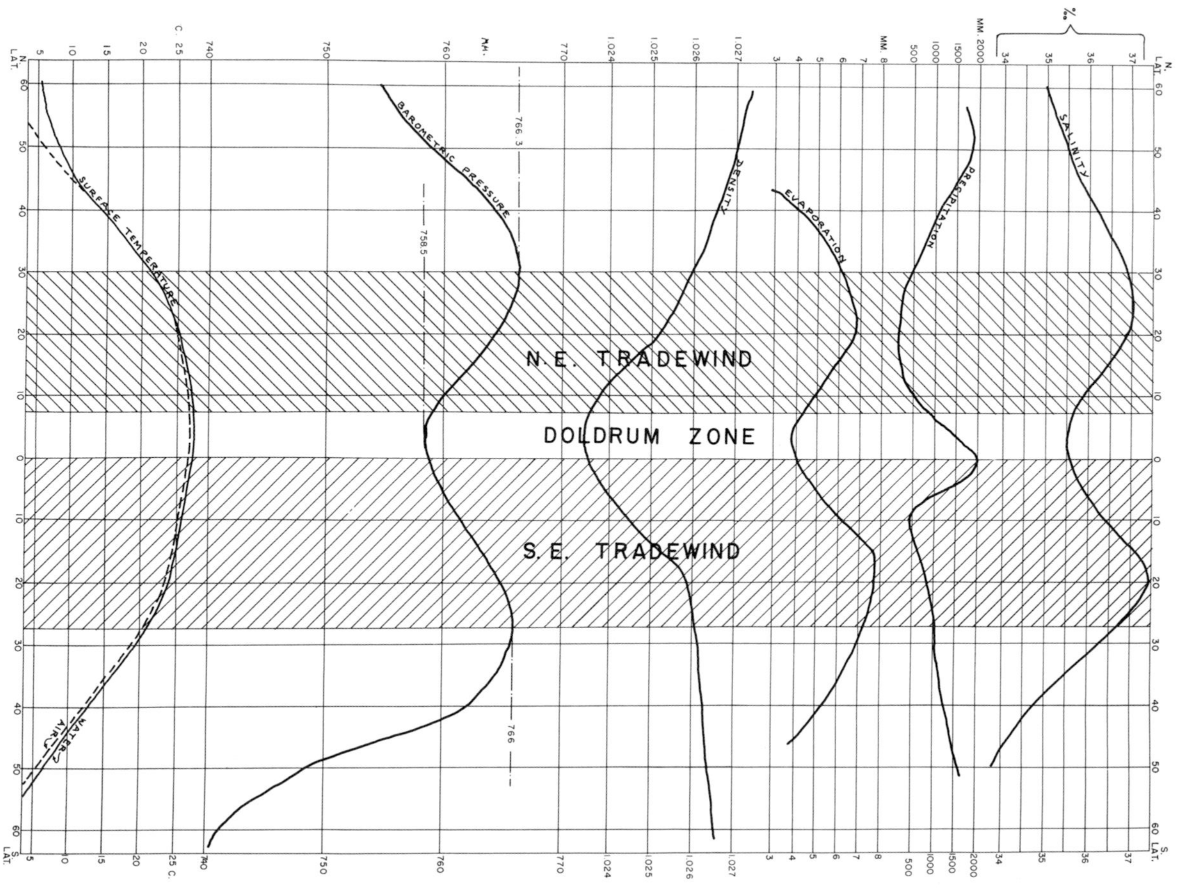

Fig. 19 · 7 Latitudinal relationship of curves of surface temperature, salinity, precipitation, evaporation, and density (plus atmospheric pressure) for the Atlantic Ocean.

plished during the long, dark, polar winters. The ice found in the sea comes from two sources: (1) frozen seawater and (2) frozen freshwater originating as glaciers on neighboring land masses. We shall consider the ice formed from seawater, known as *sea ice,* first, to be followed by a discussion of original land ice reaching the sea, or *icebergs.*

SEA ICE. The factors affecting the formation of sea ice are temperature, salinity, wind, and waves. Clearly the calmer the water, the easier will be the formation of ice. The temperature-salinity relationship in the freezing of seawater is more subtle.

Water bodies are cooled by the loss of heat from the surface—principally by radiation. In the familiar case of freshwater, the maximum density is reached during cooling at 39.2°F (4°C). Below this temperature, the density decreases so that further cooling prevents convection, as the colder water, being lighter, is confined to the surface layer. This tends to prevent a body of fresh water from freezing to the bottom, especially since ice is a very poor heat conductor and blankets the underlying warmer water.

The temperature of maximum density falls rapidly as the salinity increases, and is equal to the freezing point at 29.7°F (−1.3°C) at a salinity of 24.7 ‰. At a greater salinity, the temperature at which maximum density is reached is below the freezing point. Theoretically, then, a water body of uniform salinity greater than 24.7 ‰ will maintain convection, with cold water falling continuously, and can thus freeze to the bottom. Actually, the salinity of the oceans is not uniform, so that convection does become retarded, and also, ice, once formed at the surface, tends to cut off further cooling after reaching a thickness of 5 to 15 feet. The observed thickness may be greater because of snow accumulations on the ice. Seawater with an average salinity of 35 ‰ freezes at 28.5°F (−3.6°C).

It is important to note that when seawater freezes, it is only the pure water that does so, trapping salt water within and among the ice crystals. Thus sea ice is much less saline than the water from which it forms. The lower the temperature of freezing, the higher is the saline property of the ice. A significant amount of salt does not enter the ice until a temperature of about 17.5°F (−8°C), so that sea ice forming at higher temperatures is quite fresh. The salt slowly settles through the ice, which becomes less saline with time.

Because of both the lower salinity and the trapped air bubbles, sea ice is less dense than seawater and remains floating on the surface. The density of pure ice at 32°F (0°C) is 0.917, while the density of seawater, which slightly varies with salinity, is close to 1 for the same temperature.

In the freezing process, elongated ice crystals form first. These freeze together to form roundish patches up to a few feet in diameter known as *pancake ice,* which merge to form *ice floes* of the order of a mile across and several feet thick. With calm waters, the floes join to form a continuous sheet of *field ice* which is given a uniform surface from fresh snow cover. The field ice is known as *pack ice* when formed by the forcible joining together of ice floes which develop strongly upturned or hummocky margins up to 25 feet in height.

ICEBERGS. As troublesome as the floes may be, it is the ice of the second

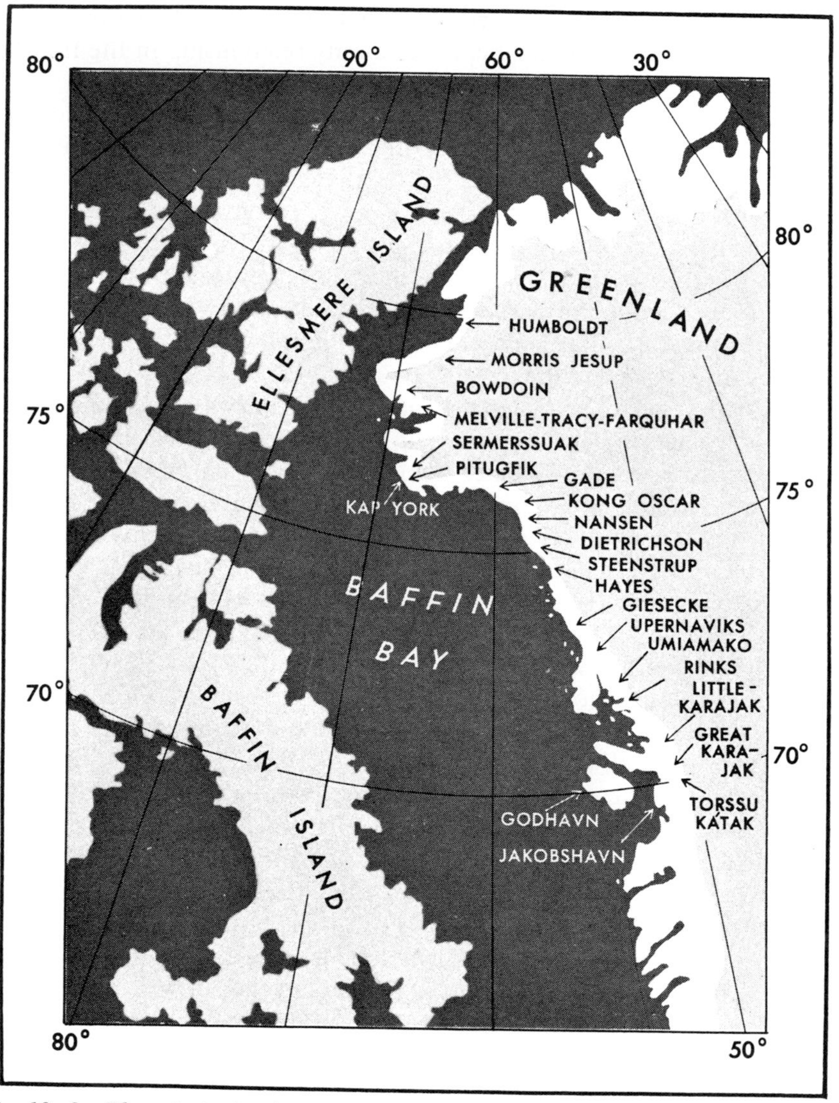

Fig. 19 • 8 The principal iceberg glaciers that discharge into Baffin Bay are the origin of nearly all the icebergs drifting into lower latitudes in the western North Atlantic. (U.S. Oceanographic Office)

source that constitutes the chief menace to navigation, in the form of icebergs. As a result of continued snowfall with little melting, tremendous thicknesses of ice form on the land masses in high latitudes. These accumulations are hundreds of feet thick and are known as *glaciers.* Under the influence of gravity they migrate seaward. Upon reaching the coast, huge sections of ice break into the sea from the parent mass, being then known

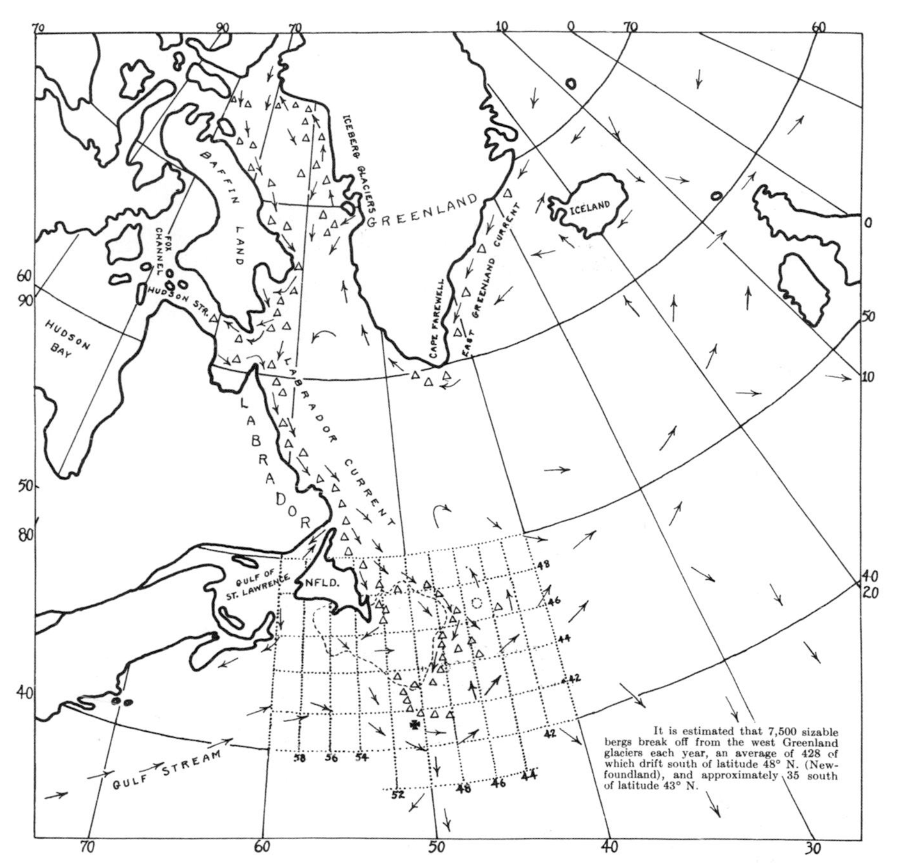

Fig. 19 • 9 Drift of icebergs from their source into the North Atlantic. (U.S. Oceanographic Office) ✠ *Marks place where Titanic sank in 1912.*

as *icebergs.* This process, called "calving," is common after the winter season, when warm weather sets in.

Since ice has nine-tenths the density of water, it follows that nine-tenths of the mass must be below the surface, with one-tenth above. It is erroneous to state that the depth of an iceberg is nine times that of its height above water. This will be true only when it is of uniform cross section. Should the base be more massive than the upper section, this ratio will be much less. Icebergs have been observed extending to 250 feet above the water line. Their length varies from a few tens or hundreds of feet up to a quarter of a mile.

After being severed from the parent glacier, icebergs drift equatorward. They are but little affected by wind motion. Owing to their greater extent below the sea surface, their behavior is influenced solely by existing currents. The shallower ice floes are affected by both winds and currents, so that the motion of icebergs and floes in the same area may at times be at great variance.

Fig. 19 · 10 Picturesque iceberg in the North Atlantic Ocean. (U.S. Coast Guard)

In both hemispheres, ice is rarely encountered in latitudes lower than 40°. The larger floes and icebergs have disappeared through melting and breaking. As they become warmed, cracks develop throughout their masses. An abrupt breakage then occurs along this crack, attended by a sudden explosive report that carries for many miles.

Icebergs are rarely encountered in North Pacific waters, owing to a lack of developing grounds. Most of the North Atlantic icebergs originate from the great continental glacier covering Greenland and drift southward with the Labrador Current.

Ice in the North Atlantic receives much more attention than elsewhere since many common shipping lanes lie in the ice area. Actually, icebergs are far more numerous and common in Antarctic or South Pacific and Atlantic waters. Their danger to shipping, however, is less in these less frequented areas.

During the ice season from January to June or July, in the North Atlantic, the U.S. Coast Guard maintains a constant ice patrol. The latest ice conditions are then issued via radio by the Oceanographic Office. Further information is obtainable in the daily memoranda, weekly bulletins, and monthly pilot charts of this office. Pilot charts contain much valuable information on average ice conditions for a given month.

Water Masses

It is now recognized that the ocean contains large volumes of water having very uniform and distinctive properties of temperature and salinity. These large volumes are known as water masses and are quite equivalent in the oceans to the air masses of the atmosphere described in earlier chapters.

Also, as with air masses, the properties of water masses depend primarily upon source conditions. Examples and brief descriptions of the principal water masses affecting the North and South Atlantic Ocean are given below. The means of identification and description is the temperature-salinity relationship, usually as shown on a *T-S* (temperature-salinity) graph. Since all measurements and references to the temperatures are in degrees Celsius, only these units are given here.

NORTH ATLANTIC CENTRAL WATER. This water mass, which covers a broad area in the mid-latitude zone of the North Atlantic Ocean, is relatively shallow, averaging less than 1,600 feet in thickness. Its temperatures and salinity vary linearly on a *T-S* graph from about 8 to 19°C and from 35.1 to 36.7 ‰, respectively. It forms mainly in the western half of the ocean between about 35 to 45° latitude.

NORTH ATLANTIC INTERMEDIATE WATER. This is a restricted water mass with a temperature of 3.5°C and a salinity of 34.88 ‰. The intermediate water forms east of northern Labrador.

NORTH ATLANTIC DEEP AND BOTTOM WATER. These masses have nearly similar properties of low temperature and high density, so that they occupy the lowest levels of most of the Atlantic basins except for low latitudes, where the colder Antarctic bottom water intrudes across the equator. The deep- and bottom-water temperatures range from 1°C to about 3.5°C, with a salinity spread from 34.8 to 35 ‰. They originate off southern Greenland.

MEDITERRANEAN WATER. This is a dense water body of 6 to 10°C in temperature and 35.3 to 36.4 ‰ in salinity.

SOUTH ATLANTIC CENTRAL WATER. This resembles its northern counterpart in size and region of occurrence, but is somewhat cooler and fresher with a temperature spread of 5 to 16°C and a salinity of 34.3 to 35.6 ‰.

SUBANTARCTIC WATER. This is a surface water layer between about 45 and 52°S which has a temperature range of 3 to 9°C and a salinity spread of 33.8 to 34.5 ‰. It forms between the subtropical and Antarctic convergence zones (to be described later under ocean currents).

ANTARCTIC INTERMEDIATE WATER. This water mass forms near or just north of the Antarctic convergence zone (about 50 to 55°S). It sinks and spreads northward beneath the central water and is distinguished by its relatively low salinity of 34.1 to 34.6 ‰ and a temperature from about 2.5 to 7°C.

SOUTH ATLANTIC DEEP AND BOTTOM WATER. These water masses, which cover some of the deep bottom areas, lie beneath the intermediate water with a temperature range of 0 to 2°C and a salinity of 34.5 to 34.9 ‰.

ANTARCTIC BOTTOM WATER. This is the coldest and densest water mass, with a temperature of about −0.4°C in the lowest parts of the South Atlantic, and a salinity of 34.66 ‰. This water has been observed to spread over the North Atlantic basin area to about 20 to 30°N. It originates as cold surface water in the Weddel Sea of Antarctica.

Sea Level

Sea level is the standard reference level for elevations on the surface of the earth—either above or below the oceans. *Mean sea level* is the level of the seas after averaging out the variations resulting from all stages of the tides. If the earth were a perfect sphere having no topographic irregularities and were composed of a homogeneous or uniformly layered rock, then the mean sea level would also be spherical in form. But forces due to rotation of the earth, large irregularities in topography, differences in rock and water densities, and winds cause mean sea level to be different from place to place. The true surface of the sea is, in detail, a rather irregular and warped surface. This became known during World War II when significant errors in the locations of remote islands or mountainous points became dramatically evident in air navigation. These errors resulted from small but important differences between the real and the supposed shape of sea level.

To the meteorologist and the oceanographer, the important variations of sea level are those that occur periodically from the effects of tides, winds, changes in air pressure, temperature, and salinity. These are unimportant in producing any permanent change in the shape of the sea-level surface because their effects can be averaged out. Since tides and wind waves will be discussed separately in following sections, only the latter three factors will be considered here.

AIR PRESSURE AND SEA LEVEL. The ocean behaves like an inverted barometer with respect to changes in atmospheric pressure. If air pressure decreases, the sea surface rises in the region of lowered pressure, the amount of rise being close to that of the change expected in a water barometer. Hence a decrease of air pressure of 1 millibar causes a sea-level rise of 1 centimeter of water (or in the English system, a 1-inch air-pressure change is equivalent to nearly 12 inches of sea-level change). An increase in atmospheric pressure causes a corresponding decrease in sea level.

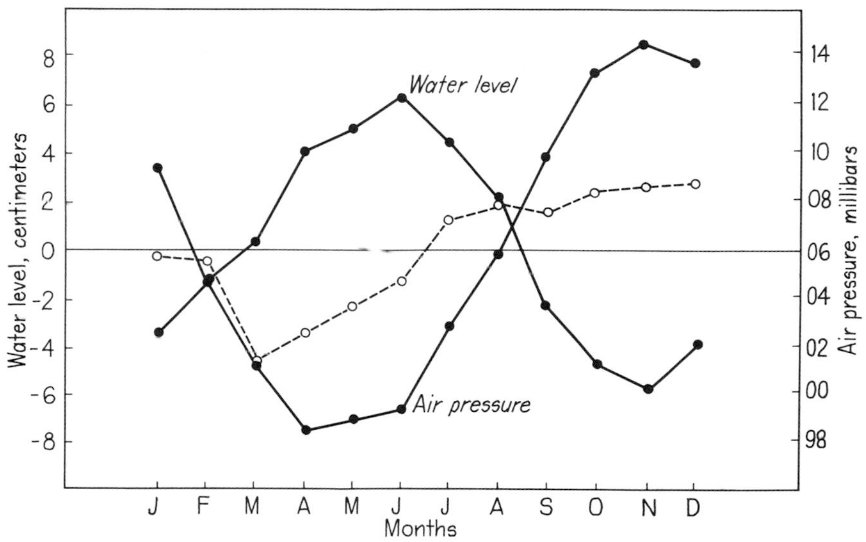

Fig. 19 · 11 Comparison between sea-level and atmospheric pressure at Iceland.

This relationship is shown very well in Fig. 19 · 11, which compares observed monthly average sea level (tidal effect averaged out) with monthly average air pressure at Iceland. The inverse relationship of sea level to air pressure is quite striking. The broken line shows the corrected sea level after the pressure effect is removed. The fact that this line is not at zero indicates that other influences must be considered. The barometric effect is more pronounced at high-latitude stations, where air-pressure variations cover a much greater range, than at low latitudes.

TEMPERATURE-SALINITY AND SEA LEVEL. A volume of water will expand or contract from changes in temperature and changes in salinity. Both of these affect the water density, but temperature plays a much larger part in sea-level variations. Hence, although no actual transfer of water takes place, a seasonal variation in sea level occurs from the simple expansion and contraction of ocean water caused by seasonal variations in temperature. In contrast to the pressure effect, that from temperature is more noticeable in low than in high latitudes.

SECULAR CHANGES IN SEA LEVEL. Changes of sea level that affect the entire world ocean over long periods of time (hundreds to thousands of years) are called secular changes. These may result from relative movements between the sea floor and the continents or from the waxing and waning of glaciers on the lands. Since glaciers store water that is initially evaporated from the seas, and return this water upon melting, sea level varies inversely with glacial conditions. Cyclical variations of hundreds of feet have been involved during the past million years and will continue for some time to come.

Tides

The tides are extremely important in their effect on navigation in coastal waters. For that reason a great wealth of literature dealing with the practical aspects of tides exists. Elaborate tide tables have been developed for the principal seaports and adjacent coastal zones. We shall therefore treat the tides here in brief, simply as a physical feature of the oceans.

The tides are the result of the gravitational forces of the moon and sun on the earth. They are complicated by the relative motions of these three bodies with reference to each other. The motions particularly responsible for the complications of the tides are the rotation of the earth and the revolution of the moon about the earth. Further difficulties are introduced by the fact that the planes of motion of the sun and moon lie at an angle to the equator.

Despite the intricacies of the problem, the tide-producing forces can be worked out separately for the sun and the moon, and the results then superimposed, the net result being the observed tide-producing force. If the earth were covered with a uniform layer of water, the explanation of observed tidal phenomena would be simplified. However, the water areas of the earth are anything but uniform. They vary in size, shape, depth, volume, area, etc. As a consequence, the application of the tide-producing forces to the various observed conditions has not yet completely explained all the observed tidal phenomena.

The influence of the sun and moon in causing tides is identical, with the exception of the magnitude of the results. Although the sun is all-important in controlling the behavior of the earth as a heavenly body and the behavior of the atmosphere, it paradoxically yields to the moon in the production of tides. The tidal force varies directly with the mass of the attracting body, but inversely as the *cube* of the distance of the body. Although the sun is far more massive than the moon, its distance from the earth is so much greater that the tide-producing force of the moon is 2¼ times greater than that of the sun.

TIDE-PRODUCING FORCES. Because the lunar tide is the dominant one, and the mechanics of tide generation are essentially the same for both solar and lunar tides, we will restrict this explanation to the latter. As the earth turns, most places experience two high and two low tides a day despite the fact that the moon is on the meridian only once in about 24 hours. This occurs because two tidal bulges exist, one directed toward the moon and the other away.

The reason for the opposite tides is the distribution of the tide-generating forces within the earth. It is shown in more complete discussions of the

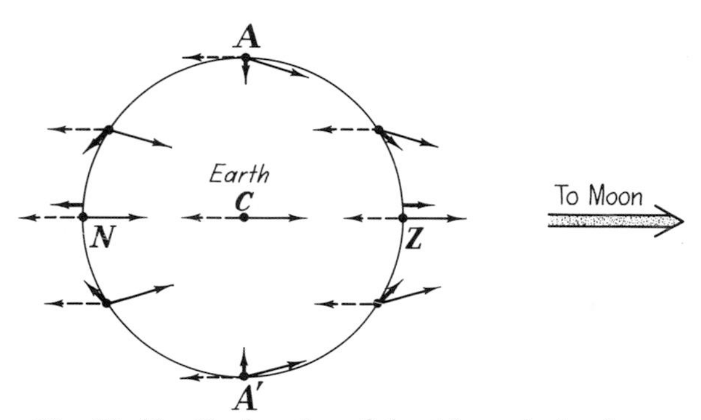

Fig. 19 · 12 Explanation of the tide-producing forces.

subject that these forces are zero at the earth's center, are directed toward the moon on the surface facing the moon, and are directed away from the moon over the earth's surface facing away.

It is also important to note that the upward-directed tidal force on the earth's surface directly beneath the moon is only about one ten-millionth of the earth's gravitation and in itself is thus far too small to raise the sea level significantly. It turns out that the tides are actually produced by the accumulated effect of the horizontal components of the tidal forces shown in Fig. 19 · 12. The fine solid arrows represent forces directed toward the moon. The broken arrows represent forces directed opposite to the moon. At the center C these forces are equal and opposite. Elsewhere, as shown by the heavy arrows, a net force exists known as the tide-producing force.

As we noted earlier, the force at Z (zenith) is too small to be effective. But between points Z to A and A', and N to A and A', a significant horizontal component of the producing force exists. The cumulative effect of these forces on the mobile ocean produces the heaping-up of water at Z and N resulting in the two tidal bulges. Midway between these bulges will be a zone of low water.

THE LUNAR TIDE. Since the main tide depends on the moon, which rises, on the average, 52 minutes later each day, the tidal wave that follows the moon's apparent motion (caused by the earth's rotation) will recur 24 hours and 52 minutes later at a given point. But about 12 hours and 25 minutes later, the high tide opposite the moon will pass, yielding two high-tide levels with two intervening low-water periods. This feature of two high and two low tides a day (lunar) is known as the *semidiurnal tide.*

The tidal bulges are never directly beneath the moon; the complications mentioned above are responsible. There may be a time interval of many hours (in some cases, much longer) between the passage of the moon and the dependent high tide across a given meridian. This interval between meridian passage of the moon and tide is known as the *tidal lag.* The tidal lag is characteristic of a given locality and thus permits the prediction of the tides well in advance.

If a graph is constructed in which time is one coordinate and the height of the tide is the other, the tidal curve results, which shows the variation of tide level with time. Each place has its characteristic tide curve. Even a rapid examination of tidal curves shows that there is a change in tide level from day to day at a given locality and also a pronounced change from place to place. It is noted that some ports experience two equally high tides a day, while others experience a pair of very unequal tides in a day; still others experience but one high and one low tide a day. This latter feature is known as a *diurnal tide.* Further, the tidal range at the same place undergoes great variation from day to day.

Most of these variable features result from the moon's change in declination. Only twice a month is the moon directly over the equator, producing the maximum tidal bulges in the plane of the equator on these occasions. At these times the lunar tide-producing forces are distributed uniformly over the halves of the globe facing toward and away from the moon, respectively. Were the earth uniformly covered with water, all points would have two equally high tides in a day, although the range from high to low water would diminish with latitude. The complications introduced by the great variance of the water bodies of the earth prevent this uniformity from being completely realized even when the moon is over the equator.

The moon's motion places it either above or below the equator and causes the tidal protuberances to have the relationship shown in Fig. 19 · 13. Points on the equator exhibit equally high and low tides. Places above or below the equator, such as P, exhibit unequal tidal ranges during the day. Point P experiences a high tide. But 12 hours later, at P', only a very low high tide is experienced. Points farther north may experience only one high tide a day. Clearly, the same tidal situation will be experienced only when the moon is again in that same declination.

Again, owing to the size, shape, and volume of the different water bodies, further difficulties are introduced by each sea, which complicate the variations due to the moon's declination. Thus, New York City and neighboring coastal waters have two equally high tides a day; the West Coast ports exhibit two high tides which differ greatly in magnitude, while Gulf ports show only the diurnal tide of one high- and one low-water period per day.

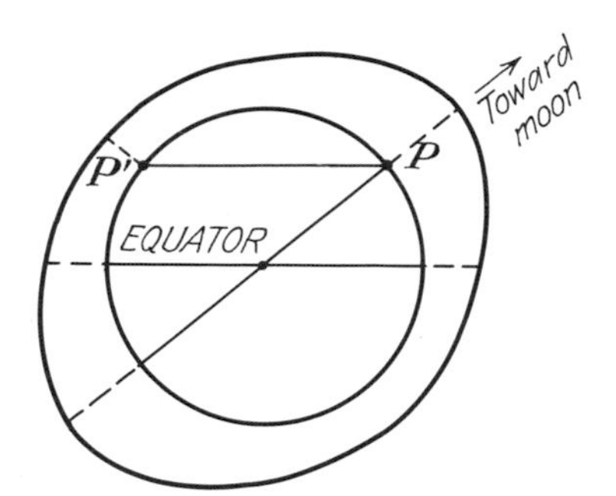

Fig. 19 · 13 The diurnal inequality of the tides owing to the moon's declination.

SOLAR TIDES. In response to solar-tide forces, the seas behave in a similar fashion, but with less response. Solar tides manifest themselves by strengthening or reducing lunar tide effects. Thus, twice a month at full or new moon, when the sun and moon are nearly in a line, the solar and lunar tide-producing forces are added. Hence, twice a month we have the highest of the high tides and the lowest of the low tides, the phenomenon being known as the *spring tide.*

When the moon is at first or last quarter, or at *quadrature,* the maximum solar and lunar tide-producing forces are at right angles. The effect is naturally to lessen the high lunar tide and increase the low-water level, which is in the direction of the solar high tide. Thus we obtain the lowest of the high tides and the highest of the low tides at first and last quarters. This is known as the *neap tide.*

APOGEE AND PERIGEE TIDES. One other cause of variation must be mentioned. The moon's path about the earth is an ellipse, with the earth lying somewhat nearer one end of it than the other, the difference being 26,000 miles. The closest position of the moon is called *perigee* and the farthest position *apogee.* Remember that the tide-producing force varies inversely as the cube of the distance. Since the average lunar distance is only 240,000 miles, the 26,000-mile difference is important. Clearly, high perigee tides will be much higher than high apogee tides. Perigee and apogee tides must occur once a month, respectively. If the perigee tide occurs during spring-tide conditions, the greatest tidal extremes are then realized.

The tidal wave which moves westward through the seas is nearly imperceptible in the open ocean. As the tide reaches continental or island shores, the shape and slope of the sea floor and the shape of the coastline determine the heights to which the tidewaters will rise. In the Bay of Fundy, east of Maine, a range of more than 50 feet is known.

Some of the complexities of the problem of the tides have been indicated in this section. A complete treatment of this difficult problem involves the application of technical mechanics beyond our present scope.

Ocean Waves

Surface waves on the ocean arise from the effect of the wind on the sea. They vary in height from an inch or less to tens of feet, and when very large, they pose a serious threat to vessels of all sizes as well as to coastal installations.

Before entering a discussion of ocean waves in particular, we must review certain definitions appropriate to waves in general. The term *crest* refers to the ridge or top of the wave and *trough* to the lowest level of the wave, as shown in the section through two waves in Fig. 19 · 14. The vertical distance between crest and trough is the *height* and the horizontal distance between two crests (or similar points on adjacent waves) is the *wavelength.* The *wave period* is the time interval between the passage of two full waves and the *frequency* is the number of waves passing a given point in a second. Period and frequency are thus the inverse of each other. Clearly the period and frequency are determined by both the length and the velocity of a wave group. The longer the wave, the greater is the period and the less the frequency for a given velocity. The greater the velocity, the less the period and the higher the frequency for a given length. The relationship between length (L), period (T), and velocity (C) is shown in the following simple formulas:

$$L = CT$$

$$T = \frac{L}{C}$$

and

$$C = \frac{L}{T}$$

Wavelength can be estimated at sea by comparing the ship's length with the distance between successive crests. The ease with which this can be done varies with sea surface conditions. The period can be determined by noting the time interval between successive uplifts of a patch of foam or other floating material. An average over several minutes of observations should be taken. The velocity can be obtained by noting the time needed for the wave to run a known distance, such as along the side of the ship, and then applying a correction for ship's speed. Reliable height measurements are the most difficult at sea and are normally made by comparison with some known elevation on the ship.

NATURE OF OCEAN WAVES. Waves that are still under the influence of the winds that generated them are known as *wind waves.* Waves that have either outrun the storm or traveled in a direction so as to leave the storm area are known as *swell.* Swell are quite regular, even-crested waves in contrast to the more irregular, often choppy waves in the wind-generating area.

It must be remembered that there is no significant forward displacement of water as waves progress through the sea. There is rather the propagation of wave energy which results in the transmission of the wave form through the water, the latter acting essentially as the medium for the wave motion. Each particle of water simply describes an oscillatory, nearly circular motion as a wave passes. The typical picture of the water behavior in wave

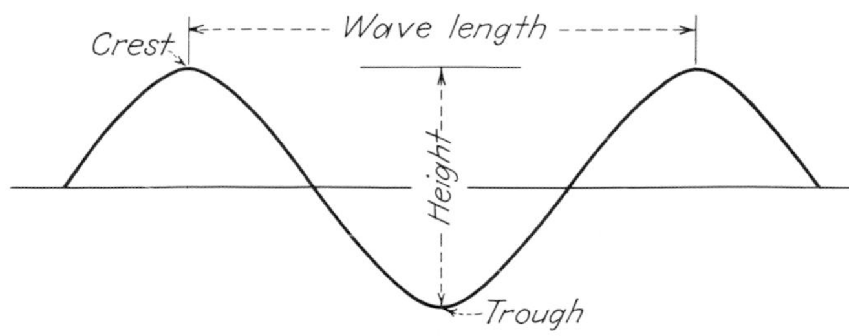

Fig. 19 · 14 Wave nomenclature.

motion shown in Fig. 19 · 15 is fully applicable only to waves of small height in deep water. In shallow water, or for high waves, the water-particle motion becomes distorted from that shown here, which is nevertheless a good working approximation for water waves in general.

In Fig. 19 · 15, the circles show the orbital motion of a water particle as a wave passes. Note that the size of the orbit decreases very rapidly with depth, the decrease being at a rate such that, at a depth of only one-half of the wavelength, the orbital diameter is one-ninth of that for the surface orbit. For this reason, submarines at a depth of 100 to 200 feet feel little disturbance even from strong waves at the surface.

In Fig. 19 · 15, the surface profile of the wave is shown in two positions one-quarter of a wavelength apart, the first by the solid line and the second by the broken line. The movement and positions of water particles are shown for three levels. At the extreme left, corresponding to the crest of the wave in the first position (solid line), the water particles are at their highest positions. To the right, in the direction of wave motion, the particles through which the water surface passes are in positions successively farther from the top. At the wave trough (seventh orbit from the left), the water particles are at the lowest points. As the wave moves to a distance one-quarter of a wavelength to the right, each of the water particles moves in the direction of the arrows to the positions connected by the broken lines. By the continuation and repetition of this particle motion the familiar surface-water wave motion is maintained.

The behavior of particle motion along the vertical is shown by the displacements of the lines which are vertical beneath the wave crest and trough. This effect is nicely shown by the swaying of reeds in shallow water as waves pass.

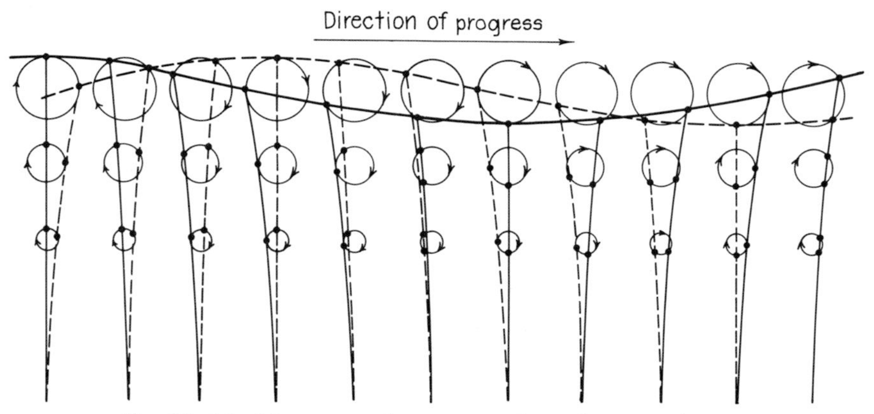

Fig. 19 • 15 Movement of water particles within a single wave.

The velocity with which the wave motion travels through the water depends on the position of the wave with respect to the generating area and the water depth. In the generating area, where the waves are still growing from, or are under the direct influence of, the wind, their velocity is close to that of the wind. Beyond the generating area the waves travel with a speed that depends on the relationship of water depth to wavelength. If the depth is greater than one-half of the wavelength, the wave is called a *deep-water wave* and the velocity C is given by the formula $C = \sqrt{gL/2\pi}$, where g is the acceleration of gravity (32 ft per sec^2 or 980 cm per sec^2). Thus if L, the wavelength, is known, since g and π are constants, the velocity C can be determined. A further simplification can be made by combining this with information from the general wave formula $C = L/T$, to give any two of the terms if the third is known. If C is expressed in knots, L in feet, and T in seconds, then

$$C = 3T$$
$$L = 5.1T$$
$$T = 0.33C$$

If the depth is less than 1/25 of the wavelength, the wave is called a *shallow-water wave,* the type one sees close to and approaching the shore. The velocity is then given simply by the formula $C = \sqrt{gh}$, where h is the water depth. The velocity of deep-water waves thus depends on the wavelength, while that of shallow-water waves depends on depth, regardless of length. Since g, the acceleration of gravity, is a constant term in all water-wave formulas, such waves are known also as *gravity waves.*

FACTORS DETERMINING WAVE GROWTH. The dimensions—height, length, period, etc.—of a wave system depend on several factors:

1. Wind velocity
2. Water depth
3. Duration of the wind velocity
4. Fetch (distance of the wind as it blows with no direction change)
5. Preexisting state of the sea

In the open ocean, the depth is adequate for the maximum development of waves and only becomes important in shallow seas, bays, lakes, etc. The wind velocity in the ocean is the primary factor affecting the growth of waves. A certain minimum value of wind duration or of fetch must exist in order for a wind of given velocity to raise the maximum waves (a fully arisen sea). The longer the wind blows, the higher will be the waves and also the greater distance the wind blows, the higher will be the waves. But for a given velocity, a maximum wave size will develop after a particular duration and distance. If the wind blows for a longer time or over a greater distance, no further wave increase occurs. But should the wind velocity increase, then the duration and fetch required for the generation of waves of greatest magnitude also increase.

Based upon careful observations of waves and the wave theory involved in both the above principles, as well as in others not described here, procedures for forecasting waves in the open ocean and along coasts have been developed by oceanographers. Although the description and application of the methods are too long to be set forth here, they are available in publications of the U.S. Navy Oceanographic Office and of the Beach Erosion Board of the U.S. Army Corps of Engineers.

Some formulas based on observation alone have been developed and may be used for quick, and not unreasonable estimates of waves to be

Fig. 19 · 16 "Smashing Through." A tanker plunges through heavy seas. (Courtesy of The Ship's Bulletin, S.O.N.J.)

Fig. 19 · 17 Rough sea during a North Atlantic gale. (Courtesy of The Ship's Bulletin, S.O.N.J.)

expected under certain conditions. Thus, the height in feet of the maximum waves in the case of very strong winds can be estimated as

$$H = 1.5F$$

where F is the fetch in nautical miles.

Also, the height in feet of the maximum waves for very strong winds has been found to roughly equal the wind velocity in knots. A rough formula for the maximum waves for any wind has also been given as

$$H = 0.025V^2$$

where V is the wind speed.

The relationship between wind speed and wave height is sufficiently close that sea conditions are one of the criteria for estimating wind force at sea, as noted earlier in Chap. 9.

When the height of a wave exceeds one-seventh the length, the wave becomes too steep and breaks. This accounts for the numerous whitecaps on the sea surface, where the height of the waves frequently develops faster than the length.

The presence of foreign matter in the sea has a pronounced effect on the reduction of wave action. The energy expended in overcoming the friction introduced by such matter weakens wave motion very noticeably. But of much greater importance to the mariner is the effect of foreign matter on the surface, rather than within the sea. In particular, heavy animal or vegetable oils are well known for their value in calming stormy waters. Even a relatively thin oil film greatly reduces the danger in rough seas. The calming effect lies in the fact that oil has a lower surface tension than water and a higher viscosity (is thick or syrup-like). The lower surface tension of the oil film causes a steepening of the wave and results in its breaking before it reaches a great height. The effect of the greater viscosity of the oil film is to prevent the development of the waves beyond a relatively small height. Many a vessel and small boat has been saved by the prompt use of oil to calm the surrounding waters.

WAVE REFRACTION. Wave refraction, which is typical of shoal-water areas, especially along shorelines, is quite similar to the more familiar optical refraction of light waves as they travel through media of different density. As waves approach a shore they become converted from deep- to shallow-water waves, so that the velocity depends only on the water depth. If the waves approach with the crests parallel to the shore, they slow uniformly along the entire crest line. But if they travel at an angle to the shore, a very common occurrence, then that part of the wave closest in is slowed the most, while the speed becomes progressively greater, the greater the distance from the shore. Hence the waves tend to "cartwheel" around and become more and more nearly parallel to the shore even if the initial travel is perpendicular to the coast. This is shown quite strikingly in air photographs such as Fig. 19 · 18.

BREAKERS AND SURF. Breakers are the waves that plunge or spill over in the shore zone to produce a sheet of water that rushes up the beach and then returns when the energy is expended. If the breaker zone is broad (from the shore outward) the combined effect is usually known as *surf.* Breakers and surf result from the transformation that takes place when

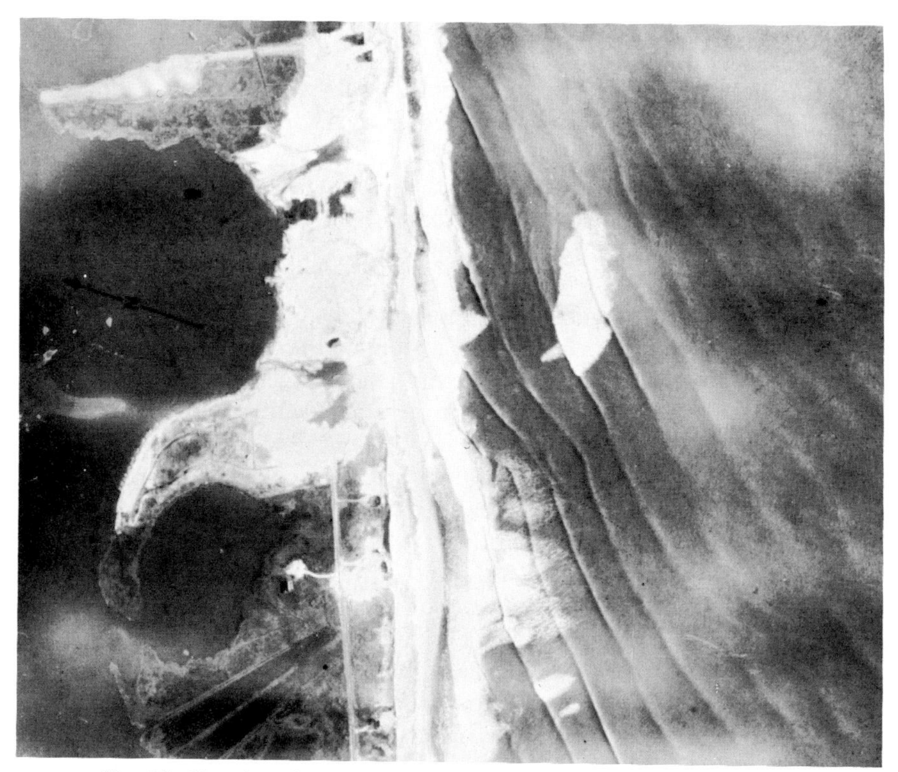

Fig. 19 · 18 Aerial view showing refraction of swell in shallow water.

deep-water waves approach a shoaling bottom and become shallow-water waves. Although in deep water the orbital motion of water particles becomes insignificant before the bottom is reached, in shallow water sufficient motion is still present at the bottom to produce a distortion of the orbital size and velocity. The wave is said to "feel bottom" at this time. With continued approach to the shore, marked changes take place in the wave appearance: (1) the lower orbits become flatter or elliptical, with a marked increase in orbital velocity at crests and troughs; (2) the waves become steeper; and (3) they become higher. All of these effects increase with increasing wavelength so that long swell, possibly insignificant at sea, develop very large breakers in shallow water. Also, the longer the wavelength, the greater is the distance from shore at which long waves break, making them of value in indurating submerged shallow reefs.

Ocean Currents

Ocean currents are among the most significant of the marine factors as regards their effect on atmospheric conditions. The transportation of large quantities of warm and cold water by the movement of the surface layers of the sea determines to a great extent the climatic conditions experienced by many coastal areas.

According to common terminology, a *current* is a progressive movement of a part of the sea. In the case of waves and tides no continuous displacement of the water occurs in a given direction; the motion is cyclical or oscillatory. The direction of motion of a current is known as its *set;* the velocity is known as the *drift.* Mariners often term a slowly moving current a *drift* and reserve the name *current* for movements of seawater having relatively higher velocities.

OBSERVATION OF OCEAN CURRENTS. Owing to the great length and breadth of the seas and the lack of suitable fixed stations, reliable observations on the set and drift of currents are and have been difficult to obtain. Several methods have been employed to this end:

1. Marine charts and logs, after the completion of a run, are often of great value in estimating the features of prevailing currents. The difference between dead-reckoning and astronomical fix positions is for the most part a result of current influence. Thus (Fig. 19 · 19) after a given time, a vessel leaving the coast at *C* might arrive at dead-reckoning position *DR.* However, the fix obtained through celestial observation places the vessel at *F.* Aside from calculation errors, the difference between *F* and *DR* is due mainly to the surface current. If the ship has traveled for 10 hours and the difference in positions is 20 nautical miles, the current drift is 2 knots in the direction shown (the set).
2. Drift bottles provide another source of data of current behavior. These are sealed bottles weighted so as to float just beneath the water surface, thereby avoiding the effects of the winds. A return postal card is enclosed, giving the date and location at the time it is released. The finder is requested to add the new date and location when removing the bottle from the sea and forward the card.
3. Lighthouses, lightships, and vessels temporarily hove to also supply

direct observational indications of the drift and set of the current in their particular vicinity.

4. Neutral buoyancy floats are floats that can be given ballast to remain at a predetermined level and emit signals that enable them to be tracked by research vessels on the surface. This tool has provided the means of discovering the strong subsurface reverse flows beneath the equatorial current and the Gulf Stream.
5. Current meters have been of value in measuring surface currents. They usually consist of a propeller or similar device whose rotations

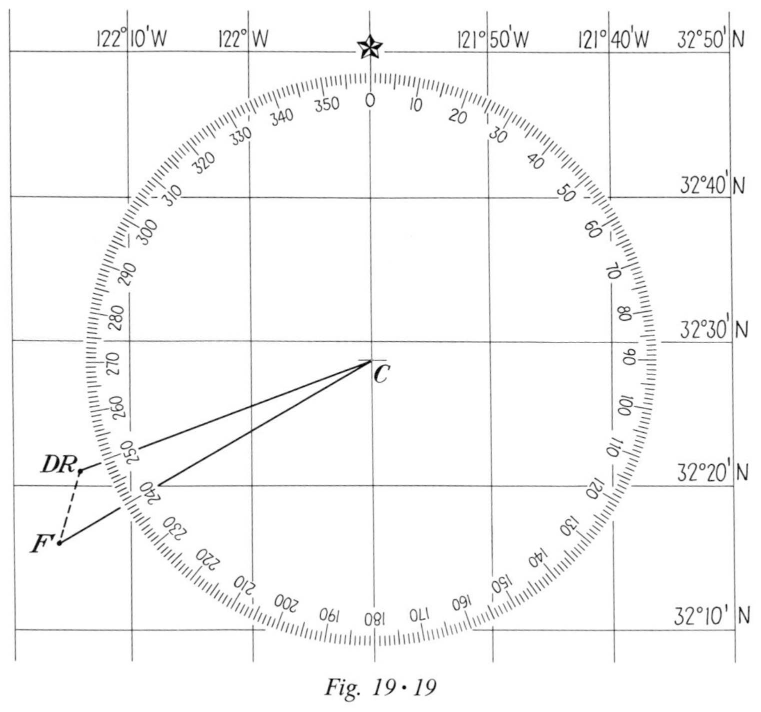

Fig. 19 · 19

in a given time interval indicate the current speed at the instrument.

6. Fluorescent dyes, although mostly in an experimental stage, are being used to trace current motion over a distance long enough to give average velocities and directions.
7. Current motion can often be deduced from theoretical reasoning based on the distribution of pressure and pressure surfaces within the ocean. This is quite analogous to the calculation of the geostrophic wind from a knowledge of atmospheric pressure differences and isobaric surfaces referred to in the chapters on winds. The

pressure distribution with depth can be calculated from a knowledge of the temperature and salinity.

FACTORS CAUSING AND AFFECTING OCEAN CURRENTS. Surface currents are primarily the result of the stress of the wind on the sea surface with modifications produced by the Coriolis deflection (see Chap. 9) and land-mass obstructions. The deeper circulation in the oceans is primarily a result of density differences related to temperature and salinity. Although there is a strong interaction of the results of both of these major causes, that of the wind appears to be more dominant. The strength of the observed deep circulation in the oceans is greater than that predicted from the effects of temperature and salinity (thermohaline effects) alone. The deeper circulation is thus partly maintained by vertical currents related to the convergence and divergence of surface wind-driven currents.

The importance of the wind in driving the surface circulation of the oceans is apparent from the close relationship between this circulation and the planetary wind systems described in Chap. 10. Compare the current flow patterns of Fig. 19 · 20 with the wind patterns of Figs. 10 · 2 and 10 · 3.

Effects of Wind. Currents purely of wind origin are called *drift currents* in distinction to those originating from density differences. Because of observational difficulties, the exact relationship between currents and wind is still not fully understood. For example, it was observed that the German research vessel Fram, during an Arctic Ocean expedition, drifted with the surrounding ice to the right of the prevailing wind. To explain these observations, the German oceanographer Ekman developed a theory according to which surface currents set 45° to the right of the wind. Many observations seemed to support the theoretical conclusion, and in many references the 45° wind-current relationship is reported as factual. However, there are newer observations using objects floating just beneath the surface and fluorescent dyes which diffuse downward that indicate that the immediate surface water may move directly downwind.

Certain coastal effects involve the problem of deviation of current from wind direction. In the Northern Hemisphere, imagine the wind to blow parallel to a coastline with the land to the left, such as the northerly winds of the Pacific high off the coast of California. According to theory, the resulting current should set to the right or west of shore. In the Southern Hemisphere, such as off the coast of Peru, southerly winds should again produce an offshore current. To replace the water leaving the coast, colder and therefore denser water must rise, producing a coastal overturning in which deeper, cooler water rises to the surface. The general term for such an uplift is *upwelling.* The unusually cool waters off southern California, Peru, and southwest Africa suggest that the theoretical effect is actually in operation, but on the whole our knowledge is still very incomplete on this subject.

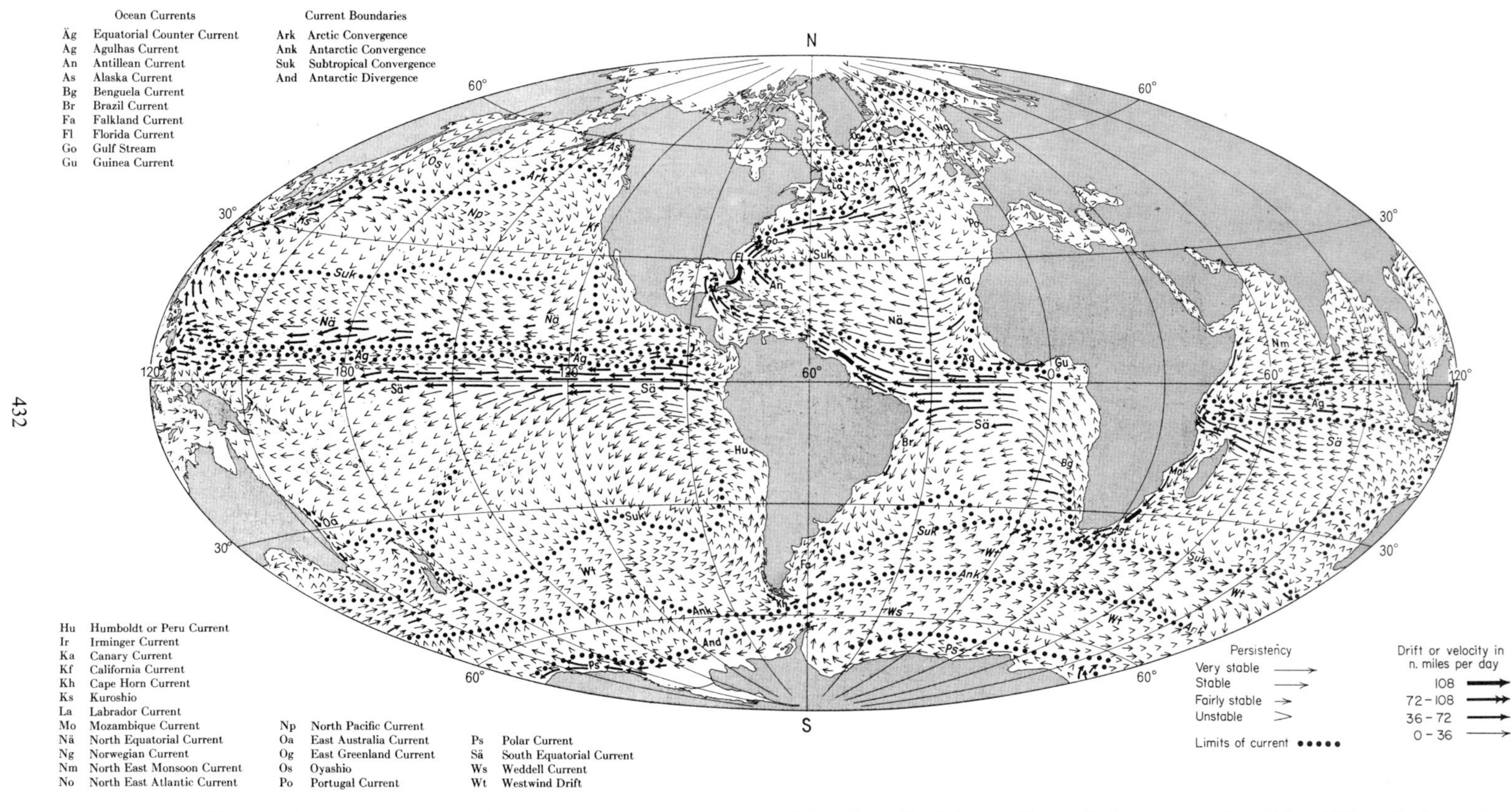

Fig. 19·20 World map showing average surface current patterns. (G. Dietrich, after G. Schott, General Oceanography, John Wiley & Sons, Inc.)

Although the wind affects directly only the actual surface, frictional drag of the surface waters extends downward with decreasing effect for some several hundred meters. According to Ekman, Coriolis deflection should occur, causing a progressive change of direction with depth, as summarized in the classic diagrams known as the *Ekman spiral* (Fig. 19 · 21). The surface waters deviate at 45° to the wind and drag the water just below, which undergoes Coriolis deflection to a direction slightly

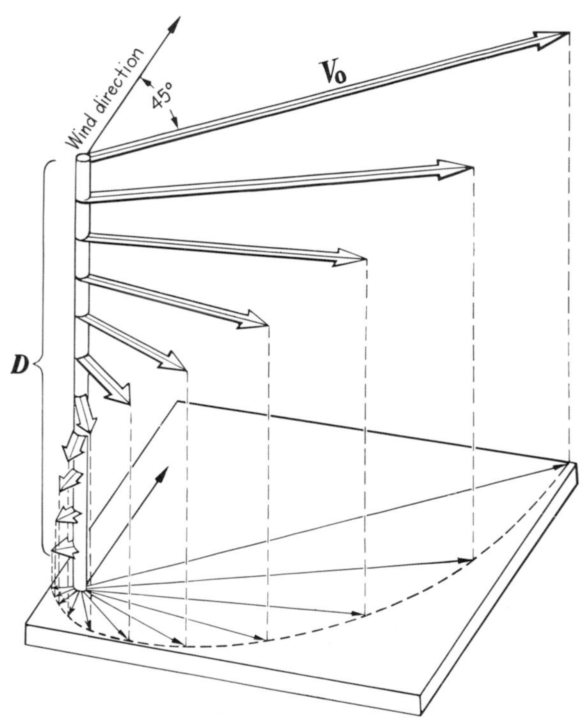

Fig. 19 · 21 The Ekman spiral showing the variation in strength and direction of a wind-induced current with depth for the Northern Hemisphere. Vo *is the direction of the surface current;* D *is the depth of the weak current directly opposite to* Vo.

greater than 45°. Continuation of this drag process produces currents which, although decreasing in velocity, increase in deviation from the original direction. At the bottom of the frictional-drag layer (*D* in Fig. 19 · 21) the weak current can theoretically be deflected until it is opposite to that at the surface. The average direction of flow in the entire frictional layer comes out to be about at right angles to the wind. The degree to which currents in the real ocean conform to the Ekman theory is still far from settled.

Effects of Temperature and Salinity—Convergence and Divergence. Temperature and salinity differences produce density differences in seawater as noted earlier. We also noted that circulation dependent upon such differences is known as *thermohaline* circulation. The movement of water masses described earlier is in large part driven by thermohaline effects, for example, the sinking and low-level spreading of cold, dense waters originating in arctic and antarctic regions.

In some cases contrasts in thermohaline properties between adjacent bodies of seawater are secondary as a consequence of the wind driving water from different source regions. The result is quite analogous to the development of fronts in the process of frontogenesis, in which air masses of different properties are formed on either side of the frontal boundary. Just as surface fronts extend up into the atmosphere, so do oceanic "fronts" extend downward into the sea, acting as water-mass boundaries at depth as well as at the surface. Surface boundaries known as lines or zones of *convergence* develop, along which surface wind-driven currents bring water of different properties together. Water must therefore sink below the convergence zone.

The Antarctic convergence is an excellent example of such an oceanic structure, separating cold and relatively dense polar and subpolar water on the south side from relatively warm and lighter subtropical water on the north side. The sloping surface which separates the two water bodies is known as an oceanic polar front, again analogous to similar structures in the atmosphere. The average position of this convergence (see Fig. 19·20) is roughly about 50°S. The importance of the thermohaline properties here can be seen in the portion of the convergence which lies within the west wind drift. Despite the similarity in current flow, especially in the Atlantic-Indian Ocean regions, striking differences in water exist. These differences originate from the very different sources of the water on opposite sides of the convergence or frontal zone, one from polar and the other from the subtropical latitudes.

It is worth noting that the absence of icebergs in the shipping lanes of the Southern Hemisphere—although they are very numerous in high latitudes—results from the inability of the icebergs to cross the Antarctic convergence. At the convergence the water bodies must descend, thus producing a vertical circulation. As with atmospheric fronts and air masses, the two water bodies cannot penetrate the interface or polar front.

The meridional cross section in Fig. 19·22 shows the thermohaline properties of the water bodies on opposite sides of the Antarctic convergence (broken line), as indicated by lines of constant density (sigma-*t*). The denser water lies to the south (right). Note that the density lines at a relatively great depth to the north of the convergence come close to the surface to the south, indicating the increase in density. Although the surface currents are mostly parallel to the convergence, a meridional vertical circulation must develop at the boundary which is shown schematically by the curving arrows. This vertical circulation is the result of both the surface wind or drift currents and the thermohaline water properties.

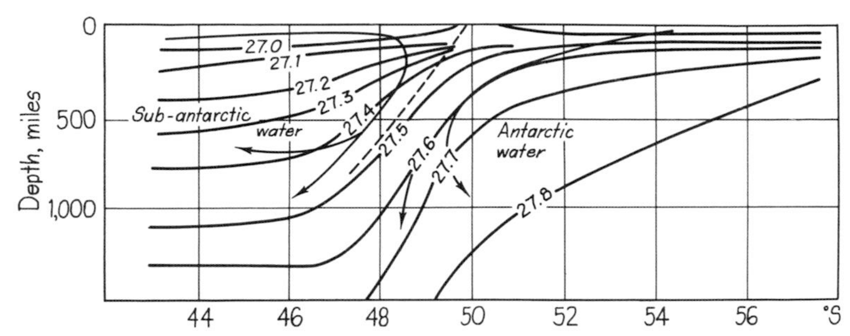

Fig. 19 • 22 Meridional cross section through the Antarctic convergence showing the distribution of density on opposite sides of the polar front—broken line. (Modified after H. U. Sverdrup)

Again as with the atmosphere, there are zones of divergence as well as of convergence. Vertical rising of lower water must occur along zones of divergence in order to compensate for the surface motion. Such uplift, which is again called upwelling, must bring relatively cold, dense water to the surface. In general, zones of convergence and divergence mark the boundaries of the oceanic current systems and form the vertical framework of the circulation.

DENSITY CURRENTS. Differences in temperature and salinity produce differences in density between water masses. Ocean currents known as *density currents* can develop purely from these differences without the existence of a direct wind factor, although the wind may be involved in producing the original contrast in properties. Melting ice, heavy rains, solar warming, etc., are other factors that may produce density differences.

If we imagine two water masses of different density to be in contact along a line of discontinuity (again similar to a front), the lighter water will occupy a greater volume. In Fig. 19 • 23, the volume of the wedge of less dense water increases away from the surface line of discontinuity (broken line), causing a corresponding increase in the height of the water surface. This is the ocean structure that develops when stable current flow has been established. The surfaces labeled P_0, P_1, P_2, etc., represent surfaces of equal pressure or isobaric surfaces in the ocean analogous to those described earlier for the atmosphere. Note that the surfaces are farther apart in the lighter water because a greater thickness is necessary here to register the same pressure as in the dense water.

Again as with the atmosphere, a component of the pressure force is directed down the isobaric surfaces. However, the Coriolis force deflects the resulting current to the right in the Northern Hemisphere (left in the Southern Hemisphere), with the resulting current taking the direction shown by the arrows. Such currents are quite analogous to geostrophic winds and obey the same laws.

A rule that can be generalized from the relationship shown is: Denser water lies to the left of the current motion in the Northern Hemisphere, and to the right in the Southern Hemisphere.

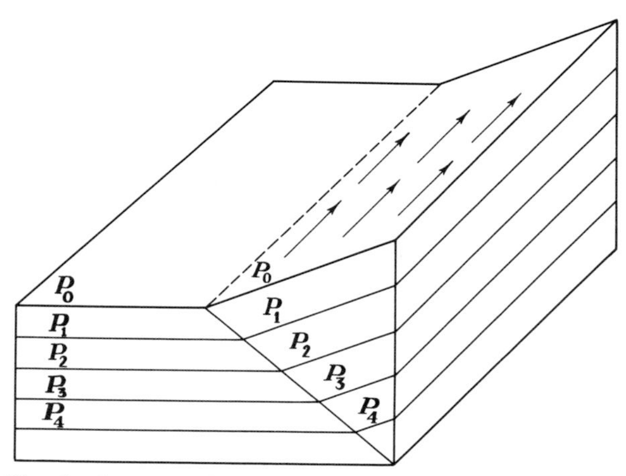

Fig. 19·23 The development of density currents from the presence of adjacent water bodies of differing densities in the Northern Hemisphere. Current flow shown by the arrows would be reversed in the Southern Hemisphere.

Surface Currents of the World

The principal surface circulation of the world ocean is, as noted earlier, a wind-driven effect although only the upper 100 fathoms are affected. However, the direct wind drift is modified by the Coriolis deflection and by land barriers. As a result, the currents take on large circulatory patterns called *gyres* which are narrowest and have the strongest flow on the western sides of the oceans. In the North Atlantic Ocean, for example, the main gyre consists of the Equatorial Current along the southern section, the Gulf Stream along the western, and the North Atlantic and Canaries Currents along the northern and eastern sections, respectively. As the equatorial currents of the oceans are generated by the same wind systems and are so basic to the main gyres of the oceans, we will consider them first and then turn to the additional gyre currents of each of the oceans (Fig. 19·20).

EQUATORIAL CURRENTS. This system of currents comprises a vast, uniform, and persistent flow that would girdle the globe in the equatorial zone were it not for the interruptions of continents. Their great regularity lies in the uniformity and strength of the trade-wind systems of both hemispheres, which generate the equatorial currents. For this reason they are also known as *trade-wind currents.* In the unbounded region of the oceans they set nearly due west with deviations occurring on the western sides of the oceans upon encountering the continents or island archipelagos adjacent to continents. The currents move from about 0.6 knot near their origin to as high as 2.5 knots off the coast of South America.

Two basic divisions of the west-setting currents are recognized: the North Equatorial Current and the South Equatorial Current, separated by the Equatorial Countercurrent.

The Equatorial Current in the Atlantic is split by the eastward promontory of South America, part flowing southwest along the Brazilian and Argentinian coasts and part continuing northwestward, ultimately reaching the Caribbean Sea and the Gulf of Mexico.

The equatorial currents in the Pacific continue westward for great distances before suffering interruptions. The northern branch, after striking the coast of Taiwan, is deflected northward. The southern branch divides near the Fiji Islands, with one part continuing westward and north of west, toward New Guinea, while the other part curves southward along the east coast of Australia.

In the Indian Ocean, the South Equatorial Current is similar to that of the North Pacific and North Atlantic Ocean. It bends southward upon approaching Madagascar and Africa. The North Equatorial Current, being subject to monsoon wind behavior, varies with the seasons in accordance with the monsoon wind variations.

The equatorial countercurrents which lie between the North and South Equatorial Currents are remarkably persistent and strong despite their relatively limited width. They exist as narrow ribbons of water which set to the east at velocities that average close to 1 knot and reach nearly 3 knots at times of maximum development. Their strength, which varies seasonally, is greatest during the northern summer. Although these currents are best developed north of the equator (about 5°N) in the Atlantic and Pacific Oceans, the Indian Ocean counterpart is present south of the equator during the northern winter as a consequence of changes related to the Indian monsoonal circulation. The origin of the countercurrents is still uncompletely resolved but is thought to be a result of the piling up of water due to certain asymmetries of the tropical wind systems.

The dynamics of the wind-induced current systems result in two zones of upwelling and divergence, one at or close to the equator and the second at about the zone of maximum trade-wind strength in the Southern Hemisphere. Regions of upwelling are rich in nutrients for marine life, resulting in a higher fish population plus a higher population of the minute organisms upon which fish feed. Such waters are characteristically green compared to the deep blue of the more biologically barren zones of the adjacent equatorial currents.

CURRENTS OF THE NORTH ATLANTIC. After coursing through the Caribbean and Gulf, and between the islands of the West Indies, the waters of the South Equatorial Current join with the waters of the North Equatorial Current in the Straits of Florida and give rise to the Gulf Stream. So unique is this current that special attention will be given to it.

The Gulf Stream is considered to originate in the Straits of Florida. The magnitude, velocity, and constancy of this current are so pronounced, compared to the other currents of the seas, as to make it truly remarkable. Its effect on weather and climatic conditions of the North Atlantic, as well as on shipping, cannot be emphasized too strongly. The Gulf Stream differs notably from the surrounding ocean waters in its higher temperature, deep blue-indigo color, seaweed content, and direction and velocity of motion.

Although the maximum average velocity of 3.5 knots is reached in the Straits of Florida, some observations of 5- to 6-knot speeds have been made along the axis of the Gulf Stream farther north. A decrease to an average of 1 to 3 knots occurs in the vicinity of Cape Hatteras. In this region the Gulf Stream has its greatest volume transport, consisting of the current originating in the Straits of Florida and water added from that portion of the North Equatorial Current that passes to the North of the West Indies, part of which is referred to as the Antilles Current. Approximately half of the Gulf Stream off Cape Hatteras comes from each of these sources.

The Gulf Stream flows northward to about 31° where it curves, with the coastline, to the northeast, continuing in that general direction until off Newfoundland. The fairly abrupt boundary between warm and cold water along the northwestern border of the Gulf Stream is often referred to as the *cold wall,* and has been related in Chap. 6 to strong advection fog over the Grand Banks.

Recent oceanographic observations show that the Gulf Stream is quite analogous to jet streams in the atmosphere. Its course is not fixed, but shows a meandering motion, with small filaments that turn off the right side. Despite our improved knowledge, many aspects of the motions and behavior of this interesting current are still not fully resolved.

East of Newfoundland the now slowly moving Gulf Stream is usually named the *North Atlantic Drift.* This separates into two branches: one continues northeastward, warming the shores of Iceland, Norway, and northern Great Britain; the second curves to the east and turns southward on approaching the European shore. This equatorward current is frequently termed the *Canaries Current,* or the *Northeast Trade Drift.* It flows along the northwest African coast and unites with the countercurrent off the Guinea coast to form the relatively strong Guinea Current, which sets to the east.

The cold Labrador Current originates in the Davis Strait between Greenland and North America. It sets approximately southward, meeting the Gulf Stream between latitudes 40 and 43°. Upon this meeting, some of the cold current is turned westward and then southward, flowing along the coast, in a parallel but opposite direction to the Gulf Stream. Much of the Labrador Current sinks beneath the warmer Gulf Current and continues southward at lower levels.

The cold wall formed by the meeting of these two contrasting currents is very pronounced. We have already considered the resulting fogs that frequently form from this meeting. The Labrador Current is responsible for most of the dangerous ice being carried southward in the North Atlantic.

THE SARGASSO SEA. Within the central area of the strong North Atlantic whirl, there exists a body of calm water known as the *Sargasso Sea,* the name being derived from the great quantity of sargassum, a type of seaweed contained in these waters. This area corresponds approximately to the position of the Azores high, the North Atlantic portion of the calm horse-latitude belt. The waters of the Sargasso Sea are warm and, if anything, a deeper and clearer blue than the waters of the Gulf Stream. Huge mats of seaweed abound there. It was at one time believed that this material drifted in from the surrounding warm waters, in much the same manner as sediment settles out in the relatively calm center of a stirred liquid. However, study of this marine vegetation showed it to be original to the Sargasso waters, being a deep-sea type of weed which flourishes near the water surface, without anchorage to the floor, as is the case with the common shallow-water seaweed. Despite stories to the contrary, the seaweed here, extensive as it may be, offers little impediment to a ship's speed. Since such tales arose in the sailing ship era, it is very likely that the absence of prevailing winds in this area, rather than the presence of seaweed, accounted for the loss in speed.

CURRENTS OF THE SOUTH ATLANTIC. The southern branch of the South Equatorial Current, flowing southward along Brazil, is known as the Brazil Current, which returns to the east again at about 40° and continues toward Africa, under the name of *Southeast Trade Drift.* The relatively cold current returning northward to the Equatorial Current, along the coast of Africa, is called the *Benguella Current.*

CURRENTS OF THE NORTH PACIFIC. Upon curving northeastward, the North Equatorial Current becomes the warm *Kuroshio,* or Japan Current, which is very similar in many respects to the Gulf Stream. The varying monsoons off Asia have a pronounced influence on the velocity and hence the volume of water transported by the Kuroshio. The Japan Current turns eastward at 40° and is then known as the *North Pacific Current.* Part of this returns southward to the Equatorial Current, just east of the 180th meridian. The other branch, which continues toward North America, splits up on its arrival there. That portion turning southward is relatively cold and is known as the *California Current,* while the northward branch is called the *Alaska Current.*

The cold currents passing southward from the Bering Straits are responsible for the fog so prevalent over the Aleutians and neighboring waters.

CURRENTS OF THE SOUTH PACIFIC. The South Equatorial Current, which turns southward at Australia, known as the *East Australia Current,* is clearly a warm flow. This then joins the prevailing West Wind Drift, returning to South America, where a deflection to the north occurs. The relatively cold current flowing northward along Chile and Peru is known by the names *Humboldt, Chilean,* and *Peruvian* current.

In general there is a pronounced current drift between 40 and 50°S, in the zone of the roaring forties, which seems again to be directly dependent on the strong prevailing westerlies in those latitudes and is known as the *West Wind Drift.*

Further modifications and movements of the surface waters are shown in the generalized current chart of the world (Fig. 19 · 20).

EFFECT OF OCEAN CURRENTS ON AIR MASSES. Clearly, ocean currents must have a marked effect on air-mass conditions, since air masses derive their original and most of their modified properties from the surface beneath. The oceans thus have a direct effect in shaping world weather.

It will be remembered that the cold continents in winter become the centers of strong high-pressure areas, particularly in the region of the horse latitudes. Consequently, a vigorous motion of the prevailing westerlies occurs from the continents to the neighboring oceans to the east or leeward. As outbreaks of cold polar air occur over the continents, this air is carried out over the warm currents, the Gulf Stream, and the Japan Current, in the cases of North America and Asia, respectively. Very marked and vigorous frontal activity develops as this cold continental air moves out across the warm water to the east of the continent. The resulting cyclonic activity has already been examined.

In addition to the effect on frontal activity, conditions within the air masses are greatly modified as a result of passage across a contrasting ocean surface. When a cold air mass traverses a warm surface, warming of the lower layer of the air mass occurs. This produces convection with attendant cumulus and often cumulonimbus clouds, frequently yielding showers.

When a mass of air is warmer than the underlying ocean surface, widespread advection fogs, often accompanied by low stratus clouds, tend to form. As explained previously, the extensive fogs of Newfoundland, the Aleutians, and the California coast have this origin.

Undercurrents

As recently as 1951, a new and strong current, hitherto unsuspected, was discovered by tuna fishermen in the equatorial Pacific waters. Their deep lines drifted to the east instead of to the west with the normal surface drift. Careful oceanographic observations by T. Cromwell led to the establishment of the remarkable current which now bears his name.

The Cromwell Current is centered directly on the equator and extends for about 2° of latitude on each side of the equator. The core, or tubular region of highest velocity (up to 2 knots), lies at a depth of only 55 fathoms or 330 feet, while the total vertical extent ranges from just below the surface to about 100 fathoms or 600 feet. It has now been traced from the mid-Pacific Ocean at about 160°W to the vicinity of the Galapagos Islands, about 90°W.

In the Atlantic Ocean a similar but not nearly as strong undercurrent has been detected, and possibly others will be discovered in time, for the science of oceanography is still in its growth.

EXERCISES

19·1 Describe the principal depth zones from the coast to mid-ocean.

19·2 Construct a temperature profile from the north to the south polar regions along the 135th meridian. How does this profile compare with the one drawn along the 120th meridian in answer to question 3·5*a*? Explain the reason for any striking similarity or difference.

19·3 What is the thermocline? Why does it exist?

19·4 What is the twofold advantage of the reversing process in the reversing thermometer?

19·5 Express the average and the range of surface salinities as a percent of total seawater weight, and list the seven most important constituents in decreasing order of abundance.

19·6 What are the main causes of the horizontal variations in surface salinity? How does the variation with depth compare with the horizontal variation?

19·7 Explain the meaning of sigma-*t*. Express a density value of 1.0261 grams per centimeter cubed in units of sigma-*t*.

19·8 Compare the causes of water density variation at the surface with those at great depth.

19 · 9 Construct a schematic "ray diagram" illustrating the principle of the SOFAR channel.

19 · 10 Compare the freezing process in freshwater with that in salt water. Under what conditions can a body of water freeze throughout?

19 · 11 What is the principal source and track(s) of icebergs in the North Atlantic Ocean?

19 · 12 How do the definitive properties of water masses compare with those of air masses?

19 · 13 Summarize the temperature and salinity properties of surface-, intermediate-, and deep-water masses.

19 · 14 How do temperature, air pressure, and salinity affect sea level? What would the expected equilibrium change of sea level be, in centimeters and inches, for an air-pressure change of 14.5 millibars?

19 · 15 Since the moon and sun, whose gravitational effects produce the tides, are above the earth's surface, why is the actual tide-producing force horizontal rather than vertical?

19 · 16 Distinguish between wind waves and swell. What is the speed in knots of ocean waves having a period of 12 seconds?

19 · 17 Distinguish between deep-water and shallow-water waves, and the basic factor upon which their speeds depend.

19 · 18 What transformations take place when a water wave enters very shallow water?

19 · 19 Distinguish between the primary causes of surface currents and deep currents.

19 · 20 Explain the meaning of convergence and divergence and their importance in the water-mass and circulation system of the oceans.

19 · 21 What is the cause of the current changes described by the Ekman spiral?

19 · 22 What are some causes of upwelling, and where is this effect quite prominent?

19 · 23 How do density currents develop in the absence of any wind?

19 · 24 Compare geostrophic currents with geostrophic winds.

19 · 25 Describe briefly the principal current gyres of the world's oceans.

BIBLIOGRAPHY

Popular Books

Battan, Louis J.: *Cloud Physics and Cloud Seeding,* Anchor Books, Doubleday & Company, Inc., Garden City, N.Y., 1962.

———: *Radar Observes the Weather,* Anchor Books, Doubleday & Company, Inc., Garden City, N.Y., 1962.

Brooks, C. F.: *Why the Weather,* rev. ed., Harcourt, Brace & World, New York, 1935.

Fisher, R. M.: *How About the Weather?* 2d ed., Harper & Row, Publishers, Incorporated, New York, 1958.

Forrester, Frank H.: *1001 Questions Answered about the Weather,* Dodd, Mead & Company, Inc., New York, 1957.

Hare, F. K.: *The Restless Atmosphere,* Harper Torch-books, Harper & Row, Publishers, Incorporated, New York, 1963.

Inwards, R.: *Weather Lore,* E. L. Hawke (ed.), Rider and Company, London, 1950.

Kimble, George H. T.: *Our American Weather,* McGraw-Hill Book Company, New York, 1955.

Lehr, Paul E., R. Will Burnett, and Herbert S. Zim: *Weather: Air Masses, Clouds, Rainfall, Storms, Weather Maps, Climate,* Golden Nature Guide, Simon and Schuster, Inc., New York, 1957.

Spar, Jerome: *The Way of the Weather,* Creative Educational Society, Mankato, Minn., 1957.

———: *Earth, Sea, and Air: A Survey of the Geophysical Sciences,* Addison-Wesley Publishing Company, Inc., Reading, Mass., 1962.

Spilhaus, A. F.: *Weathercraft,* The Viking Press, Inc., New York, 1951.

Stewart, G. R.: *Storm,* Modern Library, Inc., New York, 1947.

Sutton, O. G.: *The Challenge of the Atmosphere,* Harper & Row, Publishers, Incorporated, New York, 1961.

Whelpley, Donald A.: *Weather, Water and Boating,* Cornell Maritime Press, Cambridge, Md., 1961.

Elementary Textbooks

Best, A. C.: *Physics in Meteorology,* Pitman Publishing Corporation, New York, 1957.

Great Britain, Meteorological Office: *Meteorology for Mariners, with a Section on Oceanography,* British Information Services, New York, 1958.

Halpine, C. G., and H. H. Taylor: *Mariner's Meteorology,* D. Van Nostrand Company, Inc., Princeton, N.J., 1956.

Hess, S. L.: *Introduction to Theoretical Meteorology,* Holt, Rinehart and Winston, Inc., New York, 1959.

Neuberger, H.: *Introduction to Physical Meteorology,* The Pennsylvania State University Press, University Park, Pa., 1951.

——— and F. B. Stephens: *Weather and Man,* Prentice-Hall, Inc., Englewood Cliffs, N.J., 1948.

Petterssen, S.: *Introduction to Meteorology,* 2d ed., McGraw-Hill Book Company, New York, 1958.

Taylor, G. F.: *Elementary Meteorology,* Prentice-Hall, Inc., Edgewood Cliffs, N.J., 1954.

Advanced Textbooks

Brooks, C. E. P., and N. Carruthers: *Handbook of Statistical Methods in Meteorology,* British Information Services, New York, 1953.

Byers, H. R.: *General Meteorology,* 3d ed., McGraw-Hill Book Company, New York, 1959.

Godske, C. L., T. Bergeron, J. Bjerknes, and R. C. Bundgaard: *Dynamic Meteorology and Weather Forecasting,* American Meteorological Society, Boston, 1957.

Haltiner, George J., and Frank L. Martin: *Dynamical and Physical Meteorology,* McGraw-Hill Book Company, New York, 1957.

Johnson, John C.: *Physical Meteorology,* John Wiley & Sons, Inc., New York, 1954.

Petterssen, S.: *Weather Analysis and Forecasting,* vol. I, *Motion and Motion Systems,* 2d ed., McGraw-Hill Book Company, New York, 1956.

Riehl, Herbert: *Tropical Meteorology,* McGraw-Hill Book Company, New York, 1954.

Saucier, W. J.: *Principles of Meteorological Analysis,* The University of Chicago Press, Chicago, 1955.

Sutton, O. G.: *Atmospheric Turbulence,* 2d ed., John Wiley & Sons, Inc., New York, 1955.

———: *Micrometeorology: A Study of Physical Processes in the Lowest Layers of the Earth's Atmosphere,* McGraw-Hill Book Company, New York, 1953.

Thompson, Phillip D.: *Numerical Weather Analysis and Prediction,* The Macmillan Company, New York, 1961.

Willett, H. C., and F. Sanders: *Descriptive Meteorology,* 2d ed., Academic Press Inc., New York, 1959.

Special Subjects

Bates, D. R. (ed.): *The Earth and Its Atmosphere,* Science Editions, Inc., 1962.

Battan, Louis J.: *Radar Meteorology,* The University of Chicago Press, Chicago, 1959.

Bell, Corydon: *Wonder of Snow,* Hill and Wang, Inc., New York, 1957.

Bolin, Bert (ed.): *The Atmosphere and the Sea in Motion,* Rockefeller Institute Press and Oxford University Press, New York, 1959.

Byers, H. R. (ed.): *Thunderstorm Electricity,* The University of Chicago Press, Chicago, 1953.

Chalmers, J. A.: *Atmospheric Electricity,* Pergamon Press, New York, 1957.

Dunn, G. E., and B. I. Miller: *Atlantic Hurricanes,* Louisiana State University Press, Baton Rouge, La., 1960.

Fletcher, N. H.: *The Physics of Rain Clouds,* Cambridge University Press, New York, 1962.

Flora, S. D.: *Hailstorms of the United States,* University of Oklahoma Press, Norman, Okla., 1956.

George, J. J.: *Weather Forecasting for Aeronautics,* Academic Press Inc., New York, 1960.

Goody, Richard M.: *The Physics of the Stratosphere,* Cambridge University Press, New York, 1954.

Gregory, P. H.: *The Microbiology of the Atmosphere,* Interscience Publishers, Inc., New York, 1961.

Hoyt, W. G., and W. B. Langbein: *Floods,* Princeton University Press, Princeton, N.J., 1955.

Ludlam, F. H., and R. S. Scorer: *Cloud Study: A Pictorial Guide,* prepared under the auspices of the Royal Meteorological Society, The Macmillan Company, New York, 1957.

Ludlum, David M.: *Early American Hurricanes,* 1492–1870, American Meteorological Society, Boston, 1963.

Mason, Basil John: *The Physics of Clouds,* Oxford University Press, Fair Lawn, N.J., 1957.

Nakaya, Ukichiro: *Snow Crystals—Natural and Artificial,* Harvard University Press, Cambridge, Mass., 1954.

Riehl, Herbert: *Jet Streams of the Atmosphere,* Department of Atmospheric Science Technical Report 32, Colorado State University, Fort Collins, 1962.

Stern, Arthur C. (ed.): *Air Pollution,* vol. I, Academic Press Inc., New York, 1962.

Viemeister, P. E.: *The Lightning Book,* Doubleday & Company, Inc., Garden City, N.Y., 1961.

Weickmann, Helmut (ed.): "Physics of Precipitation," *Proceedings of the Cloud Physics Conference,* Woods Hole, Mass., June, 1959.

World Meteorological Organization: *International Cloud Atlas,* vol. I, *Album* (vol. II, *Abridged Atlas*), Geneva, Switzerland, 1956.

Observing the Weather

Great Britain, Meteorological Office: *Observer's Handbook,* 2d ed., British Information Services, New York, 1956.

———: *Cloud Types for Observers,* British Information Services, New York, 1962.

———: *Handbook of Meteorological Instruments,* part I, *Instruments for Surface Observations,* 1956, part II, *Instruments for Upper Air Observations,* 1961, British Information Services, New York, 1956.

———: *Pictorial Guide for the Maintenance of Meteorological Instruments,* British Information Services, New York, 1963.

Haynes, B. C.: *Techniques of Observing the Weather,* John Wiley & Sons, Inc., New York, 1947.

Middleton, W. E. K.: *Vision through the Atmosphere,* University of Toronto Press, Toronto, Canada, 1952.

——— and A. F. Spilhaus: *Meteorological Instruments,* 3d rev. ed., University of Toronto Press, Toronto, Canada, 1953.

Penman, H. L.: *Humidity,* The Institute of Physics Monographs for Students, London, 1955.

Spencer-Gregory, H., and E. Rourke: *Hygrometry,* Pitman Publishing Corporation, New York, 1957.

Yafte, Charles, Dohrman Byers, and Andrew Hosey (eds.): *Encyclopedia of Instrumentation for Industrial Hygiene,* Institute of Industrial Health, Ann Arbor, Mich., 1956.

Handbooks, Workbooks, and Compendiums

Decker, Fred W.: *The Weather Workbook,* Weather Workbook Co., Corvallis, Ore., 1962.

Great Britain, Meteorological Office: *Handbook of Aviation Meteorology,* British Information Services, New York, 1960.

Handbook of Geophysics, rev. ed., The Macmillan Company, New York, 1960.

Heller, Robert L. (ed.): *Geology and Earth Sciences Sourcebook,* Holt, Rinehart and Winston, Inc., New York, 1962.

Huschke, Ralph E. (ed.): *Glossary of Meteorology,* American Meteorological Society, Boston, 1959.

Kraght, P.: *Meteorology Workbook with Problems,* Cornell Maritime Press, Cambridge, Md., 1943.

Magill, Paul L., Francis R. Holden, and Charles Ackley: *Air Pollution Handbook,* McGraw-Hill Book Company, New York, 1956.

Malone, T. F. (ed.): *Compendium of Meteorology,* American Meteorological Society, Boston, 1951.

Climatology

Ahlmann, H. W.: *Glacier Variations and Climatic Fluctuations,* The American Geographical Society, New York, 1953.

Air Ministry, Meteorological Office: *Tables of Temperature, Relative Humidity, and Precipitation for the World,* 6 parts, British Information Services, New York, 1958.

Aronin, Jeffrey E.: *Climate and Architecture: Progressive Architecture Book,* Reinhold Publishing Corporation, New York, 1953.

Brooks, C. E. P.: *Climate in Everyday Life,* Philosophical Library, Inc., New York, 1951.

Brooks, Frederick A.: *Physical Microclimatology,* rev. ed., University of California, Davis, Calif., 1960.

Clayton, Henry H.: *World Weather Records,* The Smithsonian Institution, 1927. Vol. 1: Earliest Records to 1920. Vol. 2: 1921–1930. Vol. 3: 1931–1940. Vol. 4: 1941–1950, U.S. Government Printing Office, 1959.

Critchfield, H. J.: *General Climatology,* Prentice-Hall, Inc., Englewood Cliffs, N.J., 1960.

Franklin, T. B.: *Climates in Miniature: A Study of Micro-climate and Environment,* Philosophical Library, Inc., New York, 1955.

Geiger, Rudolf: *The Climate Near the Ground,* Harvard University Press, Cambridge, Mass., 1957.

Kendrew, W. G.: *The Climates of the Continents,* 5th ed., Oxford University Press, Fair Lawn, N.J., 1961.

———: *Climatology; Treated Mainly in Relation to Distribution in Time and Place,* 2d ed., Oxford University Press, Fair Lawn, N.J., 1957.

Landsberg, H.: *Physical Climatology,* 2d ed., Gray Printing Co., Dubois, Pa., 1958.

Shapley, Harlow (ed.): *Climatic Change: Evidence, Causes, and Effects,* Harvard University Press, Cambridge, Mass., 1954.

Thomas, Morley K.: *Climatological Atlas*

of Canada, National Research Council, Ottawa, 1953.

Trewartha, Glenn T.: *The Earth's Problem Climates,* The University of Wisconsin Press, Madison, Wis., 1961.

———: *Introduction to Climate,* 3d ed., McGraw-Hill Book Company, New York, 1954.

U.S. Navy: *Marine Climatic Atlas of the World,* vol. 1, *North Atlantic Ocean,* vol. 2, *North Pacific Ocean,* Government Printing Office, 1955–1957.

U.S. Weather Bureau: *The National Atlas of the United States,* Government Printing Office.

Visher, S. S.: *Climatic Atlas of the United States,* Harvard University Press, Cambridge, Mass., 1954.

U.S. Government Publications

Aneroid Barometer, 1957. How to read and set a barometer. Cat. no. 330.2:B26/2.

Average Monthly Weather Résumé and Outlook, semimonthly. Presents in graphic form the observed temperature and precipitation departures from normal for the past month and experimental monthly outlook covering expected temperature and precipitation. Cat. no. C30.46.

Aviation Series, nos. 1–18, 1957. Aimed at helping pilots to apply weather knowledge to practical flight problems. Cat. no. C30.65:1–18.

Climate of the United States. 46 charts of data. Cat. no. A1.10/a:1824.

Climate of the World. Data for 387 stations outside the U.S. Cat. no. 1.10/a: 1822.

Climates of the States. Series of 51 pamphlets containing narrative description of climates and weather, temperature, and precipitation tables.

Climatological Data for the United States by Sections, monthly with annual summaries. Contains weather statistics from 47 separate sections (indicate desired section when ordering). Cat. no. C30.18.

Climatological Data: National Summary, monthly. Contains condensed data for first-order stations. Cat. no. 30.51.

Climatological Observers, rev. ed., 1962. Instructions for climatological observers. Cat. no. C30.4:B/962.

Cloud Code Chart. 36 illustrations of cloud forms, including nomenclature and codes. 31 x 14 in. Cat. no. 30.22:C62/2/958.

Daily Weather Map. Daily weather map, Washington forecasts, and occasional feature articles. Cat. no. C30.12.

The Hurricane. Booklet discusses hurricane in detail. Illustrations. Cat. no. 30.2:H 94/2/956.

Hurricane Tracking Chart. Hurricane area shown with text on warnings and precautions. 19 x 14 in. Cat. no. C30:22:H 94/959.

Hurricane Warnings. Leaflet describing hurricanes, warnings, and precautions. Cat. no. C30.2:H 93/3/958.

Lightning. Causes and effects of lightning, safety precautions, and geographical distribution of thunderstorms. Cat. no. 30:2:L 62.

Local Climatological Data, monthly. Daily and monthly observational data for each of about 270 stations. Specify city desired. Cat. no. C30.56.

Manual of Marine Meteorological Observations, 10th ed., 1960. Instructions for taking observations at sea. Cat. no. C30.4:M/960.

Meteorological Satellites, 1962, Cat. no. Y4.Ae8:M56.

Monthly Climate Data for the World, monthly. Contains surface pressure, temperature, humidity, and precipitation as well as upper-air temperature and dew point for selected world cities. Cat. no. C30.50.

Pilot's Weather Handbook, FAA, 1955. Basic information to enable a pilot to profit by weather forecasts. Cat. no. C31.138:104.

Practical Methods of Weather Analysis and Prognoses. Naval Weather Service, 1955. Discusses the generally accepted methods of forecasting from current map data. Cat. no. D202.6:v37.

Preparation and Use of Weather Maps at Sea, 4th ed., 1959. A guide for mariners. Circular R.

Principal Tracks and Mean Frequencies of Cyclones and Anticyclones in the Northern Hemisphere, by William Klein, 1956. Detailed history of cyclone and anticyclone track charts, etc. Cat. no. C30.29:40.

Storm Data, monthly. Contains storm data

and unusual weather phenomena. Tabulated by states as to time, path, damage, and casualty, with brief descriptions.

Surface Observation, WBAN, 1961. Manual of surface observations. Cat. no. C30.4:N/961.

It Looks Like a Tornado. An aid in distinguishing tornadoes from other cloud forms. Cat. No. 30.6/2:T 63.

Tornadoes—What They Are and What to Do about Them. Cat. no. C30.2:T 63/4/959.

Weather Forecasting, 1952, reprint 1960. Elementary principles of forecasting and some basic facts of meteorology in popular style. Cat. no. 30.2:F76/3.

Weekly Weather and Crop Bulletin: National Summary. Synopsis of weekly climate conditions and their effects on crops and farming. Also contains daily snow, ice, and degree data in winter. Cat. no. C30.11.

Oceanography

Bascom, W.: *Waves and Beaches,* Doubleday & Company, Inc., Garden City, N.Y., 1964.

Bigelow, H. B.: *Oceanography,* Houghton Mifflin Company, Boston, 1931.

Cornish, Vaughn: *Ocean Waves,* Cambridge University Press, London, 1934.

———: *Ocean Waves and Kindred Geophysical Phenomena,* Cambridge University Press, Cambridge, 1934.

Defant, A.: *Physical Oceanography,* 2 vols., Pergamon Press, New York, 1961.

Dietrich, G.: *General Oceanography,* Interscience Publishers, Inc., New York, 1963.

Fomin, L.: *The Dynamic Method in Oceanography,* Elsevier Publishing Company, Amsterdam, 1964.

Hill, M. (ed.): *The Sea,* vol. 1, *Physical Oceanography,* 1962, vol. 2, *Comparative and Descriptive Oceanography,* 1963, vol. 3, *The Earth Beneath the Sea,* 1963, John Wiley & Sons, Inc., New York.

King, C.: *An Introduction to Oceanography,* McGraw-Hill Book Company, New York, 1963.

La Fond, E. C.: *Processing Oceanographic Data,* U.S. Navy Hydrographic Office Publication 614, 1951.

Marmer, H. A.: *The Sea,* Appleton-Century-Crofts, Inc., New York, 1930.

Minikin, R.: *Winds, Waves and Maritime Structures,* Charles Griffin & Company, Ltd., London, 1950.

National Academy of Sciences: *Physical and Chemical Properties of Sea Water,* pub. 600, 1959.

Pierson, W., G. Neumann, and R. James: *Observing and Forecasting Ocean Waves,* U.S. Navy Hydrographic Office Publication 603, 1958.

Russell, R., and D. MacMillan: *Waves and Tides,* Philosophical Library, Inc., New York, 1953.

Sverdrup, H. V.: *Oceanography for Meteorologists,* Prentice-Hall, Inc., Englewood Cliffs, N.J., 1943.

———: M. W. Johnson, and R. H. Fleming: *The Oceans,* Prentice-Hall, Inc., Englewood Cliffs, N.J., New York, 1942.

U.S. Navy Hydrographic Office: *Instruction Manual for Oceanographic Observations,* 2d ed., pub. 607, 2d ed. 1955.

Von Arx, W.: *An Introduction to Physical Oceanography,* Addison-Wesley Publishing Company, Inc., Reading, Mass., 1962.

APPENDIX

AVERAGE MONTHLY WEATHER SUMMARIES FOR PRINCIPAL PORTS AND ISLANDS OF THE WORLD

Line 1.—Mean *daily* maximum temperatures for each month. When preceded by an asterisk (*), the mean *monthly* maximum temperatures are given instead.

Line 2.—Mean *daily* minimum temperatures for each month. When preceded by an asterisk (*), the mean *monthly* minimum temperatures are given instead.

Line 3.—Relative humidity for 0800 or 0900 for each month. When preceded by an asterisk (*), the *mean daily* humidity is given instead.

Line 4.—The number of days with rain in each month (0.01 in. or more).

Line 5.—Average number of days with wind of gale force in each month. When preceded by an asterisk (*), the number of days with strong winds or over are given.

Line 6.—Average number of days with fog for each month.

NOTE: The mean temperature for a period is usually obtained by taking the average for the highest and the lowest temperatures for that period.

	Jan.	Feb.	Mar.	April	May	June	July	Aug.	Sept.	Oct.	Nov.	Dec.
Aberdeen—57°10′N. 2°06′W.												
1	42	40	43	50	55	60	63	63	59	53	47	43
2	33	33	34	37	42	47	50	50	47	42	37	34
3	82	81	81	79	79	77	79	80	81	84	93	83
4	18	17	20	17	17	15	17	18	17	20	19	19
5	0.4	0.3	0.1	0.1	0	0	0	0	0	0.1	0.1	0.6
6	0.5	0.3	1	0.9	3	3	1.5	2	2	2	1	0.7
Acapulco—16°50′N. 99°56′W.												
*1	85	87	87	87	89	89	89	89	88	88	88	87
*2	70	70	70	71	74	76	75	75	75	74	72	70
*3	79	78	77	77	75	79	79	79	82	80	80	80
4	0.5−	0.5−	0.5−	1	4	15	11	14	18	12	4	1
*5	0	0	0	0	0.3	0.5	0	0.2	0	0.2	0.2	0.7
6	0	0	0	0	0	0	0	0	0.1	0	0	0
Accra—5°33′N. 0°12′W.												
1	87	88	89	88	87	84	82	81	82	85	87	88
2	73	74	75	75	74	73	72	70	72	73	73	73
3	79	79	78	78	79	83	83	83	83	80	78	79
4	1	1	3	5	8	9	4	2	4	5	2	1
5	0.7	1	1	2	3	1	0	0	0	0.7	0.7	0.3
6	0	0	0	0	0	0	0	0.3	0.3	0	0	0
Adelaide—34°56′S. 138°35′E.												
1	86	86	81	73	66	60	59	62	68	73	79	83
2	62	62	59	55	50	47	45	46	48	51	45	59
3	38	40	47	56	68	77	76	69	61	51	43	39
4	4	4	6	9	13	16	16	16	14	11	8	6
5	2	1	0.7	1	0.6	1	2	2	0.3	2	2	2
6	0	0	0	0.1	1	3	4	1	2	0	0	0

	Jan.	Feb.	Mar.	April	May	June	July	Aug.	Sept.	Oct.	Nov.	Dec.
					Alexandria—31°12′N. 29°53′E.							
*1	66	67	70	75	80	83	86	87	86	83	77	69
*2	51	51	54	58	64	69	73	74	72	68	62	54
*3	68	66	67	69	72	74	76	74	70	70	67	65
1	11	7	4	1	1	0.1	0	0	0	1	6	10
5	1	1.5	0.6	0.3	0.1	0	0	0	0	0	0.5	1
6†	4	2	2	2	1	0.4	0	0.2	0.8	1	5	6
					Algiers—36°48′N. 3°02′E.							
1	59	60	63	66	71	76	81	82	79	73	66	61
2	47	48	60	53	58	63	67	69	66	59	55	50
3	65	65	55	65	65	65	68	70	69	67	67	68
4	15	13	1	11	9	5	2	2	7	10	13	15
5	3	4	3	4	2	2	1	1	1	2	2	3
6	—	—	—	—	—	—	—	—	—	—	—	—
					Amsterdam—52°23′N. 4°55′E.							
1	41	42	47	54	62	68	70	70	65	57	47	42
2	31	31	35	39	45	51	54	54	49	43	37	33
3	89	87	84	78	75	75	77	80	83	87	89	91
4	10	8	11	8	9	9	11	11	10	13	11	13
5	3	2	2	1	0	0	0	1	1	2	2	2
6	—	—	—	—	—	—	—	—	—	—	—	—
					Antigua—17°05′N. 61°50′W.							
*1	82	83	83	85	86	86	87	87	88	87	85	84
*2	70	70	70	71	73	74	74	75	74	73	72	71
3	73	70	67	69	71	72	73	73	75	74	75	74
4	20	15	14	13	15	16	19	18	18	19	19	20
*5	0	0	0	0.3	0.3	0	0.3	0	0	0.3	0	0.7
6	0	0	0	0	0	0	0	0	0	0	0	0
					Ascension Island—7°55′S. 14°24′W.							
1	85	87	88	87	86	84	83	82	82	82	83	83
2	74	75	76	76	75	74	72	72	71	71	72	73
3	72	69	68	68	67	70	66	68	72	71	71	72
4	3	3	5	7	5	5	5	7	8	9	6	4
5	0	0	0	0	0	0	0	0	0		0.1	0.1
6	0	0	0	0	0	0	0	0	0	0	0	0
					Athens—37°58′N. 23°43′E.							
*1	54	55	60	67	76	84	90	89	83	74	63	57
*2	42	43	46	52	59	67	72	72	66	60	52	46
3	73	71	69	64	60	56	48	48	56	66	72	74
4	13	11	10	9	8	5	3	3	4	8	12	13
5	2	2	1	1	0.7	4	0.7	1	0.7	0.5	1	2
6	4	5	5	3	3	1	0.7	0.6	2	4	4	6

† Number of days with mist or haze.

	Jan.	Feb.	Mar.	April	May	June	July	Aug.	Sept.	Oct.	Nov.	Dec.
						Auckland—36°50′S. 174°50′E.						
*1	73	74	72	68	62	59	57	58	61	63	67	70
*2	59	60	58	55	51	48	46	47	49	51	54	57
3	72	72	74	76	78	79	79	77	76	75	74	73
4	10	10	11	13	19	20	21	19	17	17	15	10
5	3	2	2	2	4	3	3	4	4	4	4	3
6	0.1	0.2	0.2	0.2	0.7	0.8	1.1	0.4	0.4	0.1	0.1	0.1
						Bahia—13°00′S. 38°30′W.						
1	87	87	87	86	83	80	80	80	82	83	85	85
2	73	74	73	73	72	70	69	69	70	71	72	73
3	83	83	83	84	84	84	83	83	83	83	83	84
4	16	18	18	22	24	24	25	21	17	15	15	16
5						Records not given						
6	1	1	2	3	2	2	2	4	4	2	2	2
						Baltimore, Md.—39°18′N. 76°37′W.						
*1	42	43	51	62	73	82	82	83	77	66	54	44
*2	28	28	35	45	55	64	69	67	61	49	39	30
*3	67	65	62	59	50	62	63	66	67	65	66	67
4	11	10	12	11	11	11	11	11	8	8	9	11
5	1	1	1	1	1−	1−	1	1−	1−	1−	1	1
6	3	2	2	1	1−	1−	1−	1−	1	1	2	3
						Bangkok—13°45′N. 100°28′E.						
*1	92	93	95	97	95	93	92	92	91	91	89	89
*2	67	70	73	76	76	76	76	76	75	75	71	67
3	68	60	63	62	65	69	68	66	73	74	68	67
4	1	3	4	6	17	18	19	19	21	17	7	3
5	—	—	—	—	—	—	—	—	—	—	—	—
6	—	—	—	—	—	—	—	—	—	—	—	—
						Barbados—13°06′N. 59°37′W.						
*1	84	86	87	88	89	88	88	88	89	88	87	85
*2	70	70	71	73	74	75	74	74	74	73	73	71
3	69	68	66	65	65	69	72	72	72	73	73	72
4	19	12	11	11	12	18	19	19	16	17	17	17
5	0	0	0	0	0	0	0	0	0	0	0	0
6	0	0	0	0	0	0	0	0	0	0	0	0
						Barcelona—41°23′N. 2°08′E.						
1	55	57	60	64	70	78	82	83	78	71	62	57
2	40	42	44	48	54	60	65	62	62	54	48	65
3	70	69	69	69	68	67	66	68	70	70	71	70
4	5	5	6	8	7	6	3	4	6	8	6	5
5	1	2	2	2	0.8	1	1	0.4	0.2	0.9	2	1.5
6	5	3	2	1	0.9	0.7	0.8	0.8	1	2	3	4

	Jan.	Feb.	Mar.	April	May	June	July	Aug.	Sept.	Oct.	Nov.	Dec.
Bengasi—32°07′N. 20°02′E.												
*1	60	63	68	73	78	81	83	84	84	81	72	64
*2	51	53	56	60	65	69	73	74	72	68	60	55
3	78	74	74	66	71	73	84	80	74	73	76	77
4	11	8	6	2	1	0	0	0	2	4	7	13
5	—	—	—	—	—	—	—	—	—	—	—	—
6	0.7	1	0.7	0.6	0.7	1	2	1.5	0.3	0.5	1	0.9
Bergen—60°24′N. 5°19′E.												
1	39	39	42	48	55	60	64	62	57	50	43	40
2	31	31	32	37	43	48	52	52	47	41	36	33
3	81	79	77	73	75	76	81	82	82	80	80	81
4	21	19	17	16	16	16	16	20	19	20	20	22
5	2	1.5	1.0	0.4	0.2	0.2	0.1	0.2	0.7	1.0	1.0	2.0
6	3	2	3	3	3	3	5	3	4	3	3	3
Beyrouth—33°54′N. 35°28′E.												
*1	62	63	66	72	78	83	87	89	86	81	73	66
*2	51	51	54	58	64	69	73	74	73	69	61	65
3	70	71	71	72	71	69	67	66	66	65	66	59
4	15	14	11	6	3	1	1	1	1	4	9	13
5	1.5	1	0.6	0.4	0.2	0.1	0.1	0	0.2	0.1	0.6	1
6	—	—	—	—	—	—	—	—	—	—	—	—
Bizerte—37°17′N. 9°50′W.												
1	58	59	63	67	73	79	85	85	83	77	68	61
2	46	47	50	53	59	65	71	72	69	63	75	49
3	75	74	72	70	68	65	66	63	67	70	73	75
4	15	12	11	9	6	3	2	2	5	9	12	15
5	3	5	7	5	2	2	2	4	2	5	6	8
6	—	—	—	—	—	—	—	—	—	—	—	—
Block Island, R.I.—41°10′N. 71°33′W.												
*1	37	36	41	49	59	68	74	74	68	60	50	42
*2	25	25	30	39	47	56	63	63	58	50	39	30
*3	76	76	79	80	83	85	85	84	81	76	76	75
4	13	11	12	11	11	10	10	9	8	9	10	11
5	14	9	10	8	3	1	1	1	1	7	10	12
6	3	3	4	4	6	8	7	5	4	2	2	2
Bombay—18°54′N. 72°49′E.												
*1	83	83	86	89	91	88	85	84	85	88	87	85
*2	68	69	73	77	81	80	78	77	77	77	74	70
*3	73	71	75	77	77	82	87	87	86	81	73	72
4	0.2	0.2	0.1	0.1	0.8	14	21	19	13	0.3	0.7	0.1
5	—	—	—	—	—	—	—	—	—	—	—	—
6	—	—	—	—	—	—	—	—	—	—	—	—

	Jan.	Feb.	Mar.	April	May	June	July	Aug.	Sept.	Oct.	Nov.	Dec.
						Bordeaux—44°50′N. 0°36′W.						
*1	50	45	58	62	68	75	79	80	77	67	55	49
*2	35	37	38	42	47	53	56	55	51	46	55	49
*3	87	79	73	70	69	67	67	66	70	80	85	88
4	17	16	11	15	14	10	11	9	8	14	18	17
5	11	8	5	5	5	4	6	5	8	10	8	9
6	6.9	6.3	5.9	6.6	6.7	5.8	5.5	49	4.9	5.6	7.1	7.1
						Boston, Mass.—42°21′N. 71°4′W.						
*1	36	37	43	54	66	75	80	78	71	62	49	40
*2	20	21	28	38	49	58	63	62	55	46	35	24
*3	71	68	64	68	70	72	72	76	76	74	72	72
4	12	10	12	11	11	10	10	10	9	9	10	11
5	1	1	1	1	0	0	0	0	0	0	1	1
6	1	1	1	1	1	1	1	1	2	2	1	1
						Brisbane—27°28′S. 153°02′E.						
1	85	84	82	79	73	69	68	71	76	80	83	85
2	69	68	66	62	55	51	48	50	55	60	64	67
3	67	71	72	74	74	74	75	70	65	61	60	63
4	14	14	15	12	10	9	8	7	9	9	10	12
5	0.7	0.9	0.2	0	0.3	0.4	0.6	0.1	0.3	0.7	1.0	0.2
6	0.4	0.6	1.0	3	4	4	4	2	2	1	0.4	0.3
						Buenos Aires—34°35′S. 58°29′W.						
1	85	83	79	72	64	57	57	60	64	69	76	82
2	63	63	60	53	47	41	42	43	46	50	56	61
3	69	72	76	80	83	84	84	79	78	76	73	69
4	7	6	7	7	6	6	6	6	7	8	8	8
5	1	0.5	0.6	0.8	1.0	0.8	0.6	0.9	1.0	1.5	1.0	0.4
6	1	2	6	8	9	5	7	4	4	3	2	2
						Cadiz—36°3′N. 6°17′W.						
*1	60	62	64	67	72	77	81	83	80	73	67	61
*2	48	50	52	55	59	64	68	69	67	60	55	49
3	75	73	72	71	68	65	66	64	69	71	74	76
4	9	9	10	8	5	2	0.4	0.2	3	7	7	9
5	0.5	0.5	1	0.4	0.6	0.4	0.5	0.6	0.2	0.5	0.5	0.5
6	1	0.8	1	0.3	0.3	0.5	1	0.3	0.5	1	1	2
						Calcutta—22°32′N. 88°24′E.						
*1	77	81	91	95	95	91	89	88	88	87	82	77
*2	56	60	69	76	78	79	79	79	78	75	65	56
3	85	82	80	79	79	85	88	89	87	85	82	81
4	1	2	2	3	7	13	18	18	13	6	1	0.4
5	—	—	—	—	—	—	—	—	—	—	—	—
6	—	—	—	—	—	—	—	—	—	—	—	—

	Jan.	Feb.	Mar.	April	May	June	July	Aug.	Sept.	Oct.	Nov.	Dec.
Cape Town—33°56′S. 18°29′E.												
1	80	80	78	73	67	63	63	63	66	70	73	77
2	60	61	59	55	51	48	47	48	50	53	56	59
3	65	66	69	73	76	79	80	79	76	71	67	64
4	4	4	5	7	11	13	13	12	11	9	7	5
5	4	3	2	1	0.6	2	1	1	0.8	3	0.4	3
6	2	1	4	3	2	1	1	2	1	2	0.6	0.4
Caracas, Venezuela—10°30′N. 66°55′W.												
*1	75	77	78	80	80	78	77	78	79	79	77	75
*2	56	56	57	60	62	62	61	61	61	61	60	58
*3	78	76	76	76	77	80	80	80	80	81	82	81
4	6	2	3	4	9	14	15	15	13	12	13	10
5	—	—	—	—	—	—	—	—	—	—	—	—
6	—	—	—	—	—	—	—	—	—	—	—	—
Carnarvon—24°54′S. 113°39′E.												
1	88	89	88	84	78	73	70	73	75	78	81	85
2	71	72	71	65	58	53	51	53	57	61	65	68
3	59	61	58	58	59	63	62	61	57	57	58	60
4	1	2	1	2	4	7	6	5	3	1	1	0
5	7	1.1	0.7	0.3	0	0	0.3	0.2	0.7	0.2	0.3	0.6
6	0	0	0	0.4	0.3	0	0.4	0.2	0	0.1	0	0
Cebu, Philippines—10°18′N. 123°54′E.												
*1	88	88	90	91	93	92	91	91	91	91	90	89
*2	71	71	72	74	74	73	73	73	73	73	72	72
*3	77	75	73	73	75	77	77	76	77	78	78	78
4	13	11	9	7	11	16	17	16	16	18	15	16
5	—	—	—	—	—	—	—	—	—	—	—	—
6	—	—	—	—	—	—	—	—	—	—	—	—
Charleston, S.C.—32°41′N. 79°53′W.												
*1	57	60	65	71	80	86	88	87	83	75	66	59
*2	43	45	50	57	66	72	75	75	70	61	51	44
*3	74	73	72	70	71	73	75	77	77	73	71	74
4	10	10	9	7	9	11	13	12	10	7	7	9
5	1–	1–	1	1	1–	1–	1–	1–	1–	1–	1–	1–
6	4	3	2	1–	1–	1–	1–	1–	1	1	2	3
Chatham Islands—43°52′S. 176°42′E.												
*1	64	63	62	59	55	51	50	51	53	56	58	61
*2	52	52	51	48	45	43	40	41	43	45	47	40
3	78	79	78	80	82	83	82	81	81	80	78	78
4	12	12	15	15	20	21	23	20	18	15	16	13
5	3	2	2	2	2	2	3	3	3	4	3	2
6	1.7	1.2	1.3	0.3	0.8	0.3	0.3	0.4	1.5	2.4	2.0	2.2

	Jan.	Feb.	Mar.	April	May	June	July	Aug.	Sept.	Oct.	Nov.	Dec.
						Cherbourg—49°39′N. 1°38′W.						
*1	47	47	49	53	58	63	67	68	64	59	52	48
*2	41	40	41	45	49	54	58	59	56	51	45	42
*3	81	79	78	76	77	78	76	77	79	80	81	83
4	16	14	15	13	11	10	8	11	11	17	16	19
5	—	—	—	—	—	—	—	—	—	—	—	—
6	—	—	—	—	—	—	—	—	—	—	—	—
						Constantinople—41°02′N. 28°58′E.						
*1	46	45	52	60	71	78	82	82	75	68	68	51
*2	38	36	40	46	55	62	67	68	62	57	49	43
3	79	77	71	66	64	60	60	61	65	71	76	73
4	10	8	9	7	7	60	3	4	6	7	11	12
5	0.9	0.5	0.4	0.2	0.2	0.2	0.4	0.3	0.4	0.1	0.3	0.6
6	4	4	5	1	1	1	1	1	3	6	5	4
						Copenhagen—55°41′N. 12°39′E.						
1	35	35	39	47	57	65	68	66	60	51	43	32
2	29	29	31	37	45	53	56	56	51	44	37	46
3	88	87	86	77	72	73	75	78	81	84	86	88
4	15	14	15	12	13	12	15	16	14	17	16	17
5	0.1	0.2	0.4	0.1	0.1	0	0	0	0.1	0.3	0.3	0.4
6	9	7	5	3	2	0.9	0.5	0.9	2	4	5	6
						Corfu—39°37′N. 19°57′E.						
1	56	57	61	67	75	82	87	87	82	74	65	59
2	44	45	47	62	58	64	69	69	65	60	53	48
3	76	76	74	74	73	71	67	67	70	76	76	77
4	13	12	9	9	6	5	2	2	5	11	11	13
5	0.3	0.1	0.1	0	0.1	0	0	0.1	0	0.1	0	0.1
6	0.9	0.8	1	1.5	9	0.1	3	1	1	4	3	2
						Cristobal—9°21′N. 79°54′W.						
*1	84	84	85	86	86	86	85	86	86	86	84	85
*2	76	76	77	77	76	76	76	76	75	75	75	76
*3	79	77	77	79	84	85	86	85	85	85	86	83
4	14	11	10	14	20	23	24	24	22	23	25	20
5	—	—	—	—	—	—	—	—	—	—	—	—
6	0	0	0	0.1	0	0.1	0	0.1	0	1	1	1
						Curaçao—12°06′N. 68°56′W.						
*1	83	84	84	86	86	87	87	88	89	88	86	84
*2	75	74	74	76	76	76	77	77	77	77	79	78
3	77	78	76	76	76	76	77	77	77	77	79	78
4	14	8	7	4	4	7	9	8	6	9	15	16
5	—	—	—	—	—	—	—	—	—	—	—	—
6	—	—	—	—	—	—	—	—	—	—	—	—

	Jan.	Feb.	Mar.	April	May	June	July	Aug.	Sept.	Oct.	Nov.	Dec.
					Dakar—14°40′N. 17°26′W.							
*1	81	82	82	82	83	88	88	87	89	91	88	82
*2	64	64	65	66	68	73	76	76	76	77	73	67
3	65	74	81	79	79	78	79	82	82	82	74	67
4	—	—	—	—	—	—	—	—	—	—	—	—
5	—	—	—	—	—	—	—	—	—	—	—	—
6	—	—	—	—	—	—	—	—	—	—	—	—
					Danzig—54°24′N. 18°40′E.							
1	33	35	40	49	59	66	71	69	63	53	32	36
2	25	26	30	36	44	52	56	55	50	42	34	28
3	86	85	82	76	74	73	73	75	78	82	85	87
4	14	12	14	13	13	12	14	15	13	13	14	15
5	3	3	3	0.7	0.8	0.4	1.0	0.6	1.0	2	2	3
6	3	2	3	2	2	1	0.2	0.6	2	3	4	4
					Dover—51°07′N. 1°19′E.							
1	44	45	47	52	59	63	67	67	64	57	51	47
2	35	35	37	41	47	52	56	56	53	47	41	37
3	88	85	80	78	74	74	75	76	76	78	73	85
4	16	14	16	13	10	12	12	13	13	17	16	18
5	0.7	0.6	0	0	0	0.1	0	0	0.1	0.3	1	1
6	4	3	2	0.8	0.2	0.8	0.5	0	0.3	0.8	2	2
					Dublin—53°22′N. 6°21′W.							
1	46	47	49	53	58	64	66	65	62	55	50	47
2	35	34	35	37	42	47	51	50	46	41	38	35
3	87	86	85	82	81	81	82	83	85	86	87	87
4	21	18	19	17	16	15	18	19	16	19	19	21
5	4	2	1	1	0.3	0.4	0.1	0.3	1	1	3	3
6	1.5	3	1.5	2	0	0.6	0.4	1	0.2	2	1	4
					Durban—29°51′S. 31°0 E.							
*1	85	85	84	81	78	76	75	76	76	78	81	83
*2	68	68	66	63	58	54	54	56	59	61	64	63
3	73	73	73	71	69	63	65	63	69	70	73	73
4	14	15	15	13	11	11	8	6	5	5	6	10
5	—	—	—	—	—	—	—	—	—	—	—	—
6	—	—	—	—	—	—	—	—	—	—	—	—
					Eastport—44°54′N. 66°59′W.							
*1	28	28	26	45	55	63	69	62	63	54	43	32
*2	13	14	23	32	40	47	52	53	49	42	31	19
3	77	76	76	76	78	81	84	84	82	80	79	77
4	15	13	14	12	11	12	11	10	10	11	12	14
5	3.5	3.0	2.8	1.9	0.8	0.3	0.2	0.2	0.4	1.3	2.3	3.0
6	2	2	3	3	6	8	12	12	6	4	2	2

	Jan.	Feb.	Mar.	April	May	June	July	Aug.	Sept.	Oct.	Nov.	Dec.
Emden—53°22′N. 7°12′E.												
1	37	39	45	53	61	66	69	68	69	55	46	40
2	29	30	34	38	45	50	55	54	50	43	37	33
3	91	89	85	79	75	76	78	80	83	88	90	92
4	15	14	15	12	13	12	14	16	14	17	16	16
5	3	2	2	1	0.9	0.5	0.6	1	1	2	2	3
6	8	6	4	2	1	1	1	1	3	5	7	7
Falmouth—50°09′N. 5°05′W.												
1	47	47	49	53	58	63	66	65	62	56	51	49
2	40	39	40	43	47	52	55	55	53	48	44	42
*3	84	82	81	80	80	81	82	83	84	85	84	84
4	20	17	18	15	14	13	15	16	15	21	20	23
5	7	6	4	3	0.8	0.3	0.9	0.9	2	5	5	8
6	1	0.6	1	0.3	0.6	0.3	0.3	1	0.6	0.6	0.7	0.1
Galveston, Tex.—29°18′N. 94°47′W.												
*1	63	65	71	76	81	86	87	88	86	80	71	65
*2	48	52	59	66	71	76	77	77	75	67	58	51
*3	79	78	77	78	78	77	76	76	76	75	76	77
4	8	7	7	6	6	6	5	5	9	7	7	8
5	1	1	2	3	2	1−	1−	1−	1	0	1	1
6	3	2	2	1	0	0	0	0	0	1	2	3
Georgetown—6°50′N. 58°12′W.												
1	84	84	84	85	85	85	85	86	87	87	86	84
2	74	74	75	76	75	75	75	75	76	76	76	75
3	80	78	76	77	81	73	82	81	79	78	79	82
4	20	16	17	16	22	25	23	17	8	8	13	22
5	0.1	0.1	0	0	0.1	0	0.1	0.1	0.1	0	0	0
6	0.1	0.1	0.3	0.1	0	0	0.1	0	0.1	0.1	0.1	0.1
Gibraltar—36°06′N. 5°21′W.												
1	61	62	63	67	72	78	82	83	79	72	66	62
2	49	50	51	54	58	63	67	69	66	60	54	50
3	77	77	77	75	73	71	72	72	75	77	79	78
4	10	10	11	9	6	2	0.5	1	4	8	11	11
5	1	1	1	0.3	0.3	0.1	0.1	0	0.1	0.2	0.9	1
6	0.3	0.1	0.3	0.1	0.1	0	0.3	0.3	0.1	0.2	0.1	0.1
Glasgow—55°53′N. 4°18′W.												
1	42	43	45	51	57	63	65	63	59	52	46	43
2	35	35	36	39	43	48	51	51	48	42	38	35
3	85	83	80	76	75	75	78	80	82	84	85	86
4	20	17	18	15	16	15	18	18	17	19	18	21
5	1	1	0.9	0.3	0.1	0.1	0.1	0.1	0.3	0.4	0.7	1
6	3	2	1	0.4	0	0	0	0.1	1	3	4	4

	Jan.	Feb.	Mar.	April	May	June	July	Aug.	Sept.	Oct.	Nov.	Dec.
Grand Banks—47°05′N. 55°46′W.												
*1	32	29	33	40	50	61	67	69	63	53	45	37
*2	21	18	22	29	35	46	53	56	50	41	33	27
3	84	83	87	85	86	88	90	89	88	85	83	86
4	26	20	23	17	13	13	11	13	14	18	25	26
5	4	3	0.5	2	0.5	0	0	0	0.2	0.2	2	1
6	2	1	5	6	6	11	10	8	5	6	6	5
Greenwich—51°29′N. 0°00′.												
1	45	46	50	56	65	69	73	72	67	59	49	45
2	35	35	36	39	45	49	53	53	49	44	37	36
3	85	82	75	71	68	67	67	71	75	80	84	86
4	15	14	15	12	13	11	13	13	11	15	15	16
5	0.7	0.3	0.1	0.2	0.1	0	0	0	0	0.2	0.7	1.0
6	13	10	7	2	0.4	0.1	1	0	2	6	10	12
Halifax—44°39′N. 63°66′W.												
*1	32	31	38	48	59	68	74	74	67	57	46	38
*2	19	15	23	31	40	48	55	56	50	41	32	20
3	89	88	83	77	76	78	84	85	86	85	86	88
4	16	14	15	14	14	14	13	13	12	13	14	15
5	5	4	5	4	2	1	1	1	1	2	4	4
6	3	3	3	4	6	6	7	6	4	4	3	3
Hamburg—53°33′N. 9°59′E.												
1	36	38	43	52	61	67	69	68	63	54	44	39
2	28	30	33	39	46	53	56	56	51	44	36	32
3	90	88	82	73	70	72	76	78	81	85	89	91
4	18	17	19	16	16	15	19	19	15	19	18	19
5	5	3	4	2	1.3	0.8	1.3	2	2	3	3	4
6	14	11	9	6	2	1.1	2	4	8	12	14	14
Hamilton, Bermuda—32°17 N. 64°46 W.												
1	68	68	68	71	76	81	85	86	84	79	74	70
2	58	57	57	59	64	69	73	74	72	69	63	60
3	78	78	77	77	80	80	77	77	76	77	75	78
4	16	15	15	11	11	11	12	14	13	14	15	15
5	0.5	0.6	0.4	0.2	0	0	0.1	0.1	0.5	0.3	0.4	0.3
6	0	0	0	0.1	0	0	0	0	0	0	0	0
Havana, Cuba—23°08′N. 82°21′W.												
1	79	79	81	84	86	88	89	89	88	85	81	80
2	65	65	67	69	72	74	74	75	74	73	69	67
*3	75	73	71	71	74	77	75	76	78	78	75	75
4	8	6	5	5	10	13	12	14	15	15	10	8
5	3	2	3	3	1	2	1	2	2	2	3	1
6	1	2	2	1	1	0.5	0.3	0.4	0.4	0.6	0.5	1

	Jan.	Feb.	Mar.	April	May	June	July	Aug.	Sept.	Oct.	Nov.	Dec.
Helsingfors—60°10′N. 24°57′E.												
1	27	26	30	41	54	64	67	65	56	46	36	65
2	17	14	19	30	40	50	55	53	45	38	29	21
3	88	87	83	78	71	69	74	79	83	86	88	89
4	18	15	14	12	13	12	14	17	17	18	18	18
5	3	5	3	2	2	1	0.7	2	3	7	8	3
6	9	10	13	8	4	3	1	2	11	14	10	11
Hobart, Tasmania—42°53′S. 147°20′E.												
1	71	71	68	63	57	53	52	55	59	54	66	69
2	53	53	51	48	44	41	39	41	43	63	48	51
*3	56	60	61	67	71	76	75	70	64	61	57	55
4	10	8	10	12	13	14	14	14	14	15	14	11
5	9	5	6	8	6	7	6	6	9	10	11	8
6	0	0.3	0.2	1	4	6	5	2	0.2	0	0.2	0.2
Hong Kong—22°18′N. 114°10′E.												
*1	64	63	67	75	82	85	87	87	85	81	74	68
*2	56	55	60	67	74	78	78	78	77	73	65	59
*3	75	79	83	85	84	83	83	83	78	72	68	69
4	6	8	11	12	16	20	19	17	14	8	5	5
5	0	0	0.2	0	0	0.2	0.3	0.5	0.3	0.2	0.2	0
6	4.2	5.0	8.7	7.6	2.4	1	1.1	2.5	2.1	0.6	0.9	2.3
Honolulu—21°19′N. 157°52′W.												
*1	76	76	76	78	80	81	82	83	83	82	79	77
*2	65	66	66	68	70	71	72	73	73	72	70	68
*3	72	72	70	68	68	68	68	68	68	69	70	72
4	14	11	13	13	12	13	13	13	13	14	14	16
5	0.7	0.3	0.2	0.1	0	0	0	0	0	0	0	0.2
6	0	0	0	0	0	0	0	0	0	0	0	0
Inverness—57°28′N. 4°13′W.												
1	43	44	46	51	56	61	64	63	60	53	47	43
2	34	34	35	38	42	48	50	51	47	42	37	34
3	85	85	82	77	76	77	79	79	81	84	86	86
4	14	11	14	11	12	11	14	15	12	13	13	15
5	3	2	1	1	0.2	0.1	0	0.3	0.6	1	3.0	2
6	3	2	2	2	2	3	1	2	4	5	4	6
Jacksonville, Fla.—30°20′N. 81°39′W.												
*1	65	67	72	77	83	88	90	89	85	78	70	65
*2	47	49	54	60	66	72	74	74	72	64	54	48
*3	75	73	70	67	68	73	75	76	78	75	74	76
4	9	8	8	7	9	13	15	15	13	10	7	8
5	1	2	3	1	1	2	3	1	1	1	1	1
6	3	2	1	0	0	0	0	0	0	1	2	3

	Jan.	Feb.	Mar.	April	May	June	July	Aug.	Sept.	Oct.	Nov.	Dec.
Jamaica, Kingston—17°58′N. 76°48′W.												
1	86	86	86	87	87	89	90	90	89	88	87	87
2	67	67	68	70	72	74	73	73	73	73	71	69
*3	78	78	77	78	79	78	76	79	82	84	82	80
4	5	4	4	5	7	6	5	9	10	12	7	5
5	1	0	0	0	1	0	0	3	0	0	0	0
6	—	—	—	—	—	—	—	—	—	—	—	—
Kiel—54°20′N. 10°09′W.												
1	36	37	42	49	59	66	69	67	62	54	43	38
2	29	30	33	37	45	52	55	54	50	44	35	32
3	92	90	87	81	77	79	81	85	87	90	91	92
4	10	9	10	8	8	13	10	9	9	12	11	8
5	4	4	4	2	2	1	1	2	0.8	2	2	4
6	7	5	4	2	1	0.7	0.6	1	2	4	7	8
Le Havre—49°29′N. 0°06′E.												
*1	45	47	52	58	65	70	74	73	70	61	52	47
*2	36	37	38	42	47	53	57	57	52	47	41	38
*3	87	85	81	78	78	79	79	79	82	84	86	88
4	16	16	16	15	13	12	11	14	14	15	15	17
5	—	—	—	—	—	—	—	—	—	—	—	—
6	—	—	—	—	—	—	—	—	—	—	—	—
Leningrad—59°56′N. 30°16′E.												
1	23	24	33	45	58	66	71	66	57	45	34	26
2	12	12	18	31	42	51	57	53	45	37	27	18
3	87	85	89	71	64	63	68	74	68	81	86	87
4	17	15	13	11	12	12	13	15	14	15	17	18
5	0.2	0.2	0.2	0	0	0	0	0.2	0.1	0.2	0.1	0.3
6	3	4	4	4	1.5	0.9	0.9	3	6	6	4	5
Lima—12°04′S. 77°01′W.												
*1	81	83	84	80	75	69	67	67	68	70	74	78
*2	64	66	65	62	59	57	56	56	56	58	60	62
*3	77	78	78	77	83	85	85	84	84	81	80	79
4	2	2	3	5	11	17	22	24	21	13	6	5
5	0.6	0.1	0.2	0.1	0.2	0	0	0.1	0.1	0.1	0.6	0.9
6	6	6	8	9	7	4	6	5	4	4	4	4
Lisbon—38°43′N. 90°09′W.												
*1	56	58	61	64	69	75	79	80	76	69	62	57
*2	46	47	49	52	56	60	63	64	62	57	52	47
3	79	75	72	70	68	65	62	61	66	72	77	79
4	13	12	13	11	9	5	2	2	7	11	13	13
5	0.4	0.5	0.4	0.2	0.0	0.2	0.0	0.1	0.2	0.2	0.4	0.5
6	5	3	2	0.7	0.3	0.3	0.3	0.5	1	2	4	5

	Jan.	Feb.	Mar.	April	May	June	July	Aug.	Sept.	Oct.	Nov.	Dec.
					Liverpool—53°24′N. 3°04′W.							
1	43	44	47	52	58	64	66	65	61	54	49	45
2	35	36	37	40	45	51	54	54	51	45	40	37
3	86	85	83	79	77	76	79	80	80	82	85	87
4	17	16	17	14	15	13	15	17	15	18	18	19
5	1	1	0.6	0.2	0.1	0.2	0	0.2	0.3	0.4	1	1
6	3	7	5	2	2	0	0.8	0.8	3	2	3	5
					London—51°30′N. 0°05′W.							
1	43	45	49	55	62	68	71	70	65	56	49	45
2	35	35	36	40	45	51	54	54	49	44	39	36
*3	85	82	79	75	73	73	73	76	80	85	86	86
4	15	15	14	13	12	12	13	13	12	16	16	16
5	—	—	—	—	—	—	—	—	—	—	—	—
6	—	—	—	—	—	—	—	—	—	—	—	—
					Malta—35°54′N. 14°31′E.							
1	59	60	62	66	71	79	84	85	81	76	68	62
2	51	51	52	56	61	67	72	73	71	66	59	54
3	77	77	75	75	73	70	66	68	71	73	75	78
4	13	10	8	5	3	1	0.2	0.7	3	7	11	14
5	0.1	0	0.1	0	0.1	0	0	0	0	0	0.2	0
6	0.6	0.2	0.4	0.1	0.2	0.2	0.2	0.2	0.1	0.6	0.6	0.6
					Manila, Philippines—14°35′N. 120°59′E.							
*1	86	88	91	93	93	90	88	88	87	88	87	86
*2	63	69	71	73	75	75	75	75	74	74	72	70
*3	78	74	71	70	76	81	85	85	86	84	83	81
4	5	3	3	4	11	16	22	22	21	17	13	9
5	0	0	0	0	1	1	1	7	7	4	1	1
6	—	—	—	—	—	—	—	—	—	—	—	—
					Marseille—43°18′N. 5°23′E.							
1	52	55	59	65	72	78	83	82	77	67	59	53
2	37	38	41	45	51	57	61	60	56	50	43	38
3	68	64	62	60	59	57	54	57	63	69	70	70
4	7	6	8	8	7	3	2	3	6	6	9	7
5	4	4	6	4	3	2	2	2	2	3	3	3
6	2	1	1	1	0	0	0	0	0	1	1	2
					Martinique—14°36′N. 61°05′W.							
*1	83	84	85	86	86	86	86	87	88	87	86	84
*2	69	69	69	71	73	74	74	74	74	73	72	70
3	86	84	84	84	84	85	86	87	87	88	88	87
4	19	15	15	13	18	21	22	22	20	19	20	19
5	—	—	—	—	—	—	—	—	—	—	—	—
6	—	—	—	—	—	—	—	—	—	—	—	—

	Jan.	Feb.	Mar.	April	May	June	July	Aug.	Sept.	Oct.	Nov.	Dec.
					Melbourne—37°49′S. 144°58′E.							
1	78	78	74	68	61	57	55	59	63	67	71	75
2	57	57	55	51	47	44	42	43	46	48	51	54
*3	60	61	65	70	76	77	78	72	68	66	63	60
4	8	7	9	11	13	14	14	14	14	13	11	9
5	4	2	2	2	2	3	4	4	4	4	5	4
6	0.1	0.4	0.6	2	3	4	5	2	0.9	0.4	0.1	0.2
					Memel—55°43′N. 21°07′E.							
1	31	32	37	48	60	67	70	67	61	51	41	34
2	23	23	27	35	44	51	56	55	49	41	33	26
3	90	89	86	79	75	76	78	80	82	85	87	90
4	15	13	14	11	12	10	13	15	14	16	17	18
5	4	3	2	0.7	0.7	0.4	1	3	3	4	5	5
6	5	5	5	3	3	2	1	1	3	3	5	5
					Messina—38°12′N. 15°33′E.							
1	56	56	59	63	68	75	81	82	78	75	65	59
2	47	49	50	55	63	71	76	76	70	65	57	53
3	69	68	66	63	62	62	60	60	63	68	70	70
4	14	13	12	11	7	5	2	4	7	12	14	16
5	—	—	—	—	—	—	—	—	—	—	—	—
6	—	—	—	—	—	—	—	—		—	—	—
					Miami, Fla.—25°47′N. 80°11′W.							
*1	74	75	77	80	83	85	87	87	86	82	78	75
*2	62	61	64	68	71	74	76	76	75	73	67	62
*3	74	72	70	69	72	74	73	73	76	75	71	74
4	9	6	7	7	12	13	15	15	18	16	10	7
5	1–	1–	1–	1–	1–	1–	1–	1–	1–	1–	1–	1–
6	1–	1–	1–	1–	0	0	0	0	0	1–	1–	1
					Montevideo—34°55′S. 56°13′W.							
1	82	82	79	73	65	60	58	59	64	68	75	80
2	61	61	59	54	47	45	43	43	47	50	54	58
3	65	67	68	70	72	74	75	72	71	69	65	64
4	7	6	6	8	8	7	7	8	8	7	8	7
5						Records Not Given						
6	0	0	2	2	6	9	9	6	5	3	1	1
					Montreal—45°30′N. 73°35′W.							
1	26	25	33	42	56	66	73	62	65	54	42	31
2	9	4	19	29	39	49	57	57	50	42	31	18
3	87	86	84	82	76	77	80	81	82	82	84	86
4	12	11	13	12	12	12	11	11	10	12	16	14
5	2	1	2	0	0	0	0	0	0	1	1	1
6	0	1	1	1	0	0	0	0	0	1	1	0

	Jan.	Feb.	Mar.	April	May	June	July	Aug.	Sept.	Oct.	Nov.	Dec.
Nagasaki—32°44′N. 129°52′E.												
1	49	50	57	66	73	78	85	87	82	73	63	53
2	36	36	41	50	57	65	72	74	68	58	48	40
3	70	69	70	73	75	83	82	78	76	71	71	70
4	16	13	15	14	13	17	14	12	14	11	11	15
*5	10	10	12	11	9	9	8	7	5	7	8	5
6	0.2	0.3	0.7	1.0	1.0	2.0	0.4	0.1	0.1	0.2	0.1	0.3
Naha—26°13′N. 127°41′E.												
1	67	70	70	76	80	86	89	89	87	82	76	70
2	55	55	59	64	68	75	77	77	75	70	64	58
3	74	75	77	81	83	85	81	82	81	78	75	73
4	19	18	17	15	18	16	15	18	18	16	15	16
*5	14	13	11	5	4	1.5	3	4	7	9	12	13
6	0.1	0.1	0.1	0.1	0.2	—	0.1	—	0	0	—	—
Nantes—47°15′N. 1°34′W.												
*1	47	50	53	58	64	70	75	74	71	62	52	46
*2	37	38	39	42	47	52	56	56	52	47	41	37
*3	86	83	76	75	74	72	71	75	76	83	88	87
4	19	16	12	15	15	13	13	13	11	18	21	20
5	7	4	2	1	1	0	1	1	3	3	5	8
6	6.6	6.7	6.3	6.5	6.6	6.8	5.8	6.0	5.7	6.3	7.3	7.3
Nantucket, Mass.—41°17′N. 70°06′W.												
*1	39	37	42	50	59	68	74	74	69	60	51	42
*2	26	25	31	38	47	55	62	62	58	49	40	31
*3	78	78	78	77	80	82	83	82	80	76	77	77
4	13	12	13	12	11	10	9	9	9	10	12	13
5	8	4	6	6	3	1	1	1	2	4	6	6
6	3	4	5	5	5	9	10	7	5	2	2	3
Naples—40°52′N. 14°15′E.												
1	51	53	58	63	69	74	83	82	77	69	61	54
2	43	44	47	52	57	67	69	69	65	59	52	46
3	71	71	69	69	67	66	63	64	68	73	73	72
4	12	11	12	11	9	6	3	4	7	11	13	13
5	—	—	—	—	—	—	—	—	—	—	—	—
6	—	—	—	—	—	—	—	—	—	—	—	—
Nassau, Bahamas—25°05′N. 77°21′W.												
1	76	77	78	80	83	86	88	88	87	85	80	78
2	67	67	68	69	72	75	76	76	76	75	71	69
3	76	74	71	71	73	73	72	72	73	74	74	75
4	9	7	6	6	11	13	16	16	17	14	9	8
5	0.5	0.5	0.3	0.3	0.2	0	0	0.2	0.3	0.3	0.5	0.5
6	0.2	0	0	0	0.1	0	0	0	0	0	0	0.2

	Jan.	Feb.	Mar.	April	May	June	July	Aug.	Sept.	Oct.	Nov.	Dec.
Norfolk, Va.—36°50′N. 76°18′W.												
*1	49	50	57	66	75	83	87	85	76	70	59	51
*2	34	34	40	48	58	66	70	70	65	55	44	36
*3	79	69	67	65	67	69	72	73	73	70	69	71
4	11	11	11	10	11	11	12	12	8	8	8	10
5	2	1	2	1	1	1	1	1−	1−	1	1−	1
6	2	2	1	1−	1	1	1−	1−	1	2	2	2
New London, Conn.—41°22′N. 72°06′W.												
*1	37	36	45	55	67	75	80	78	73	63	51	40
*2	22	22	29	38	48	57	63	62	56	45	35	26
*3	75	74	72	70	73	77	78	79	78	76	73	74
4	14	13	14	12	12	10	12	10	11	11	12	12
5	1	1	0	0	0	0	0	0	0	0	1	1
6	2	2	2	2	3	2	1	1	1	1	1	1
New Orleans, La.—29°57′N. 90°04′W.												
*1	62	65	71	77	83	88	89	89	86	78	70	63
*2	47	49	55	61	68	74	75	75	73	64	54	48
*3	76	74	73	71	70	71	73	74	71	73	74	75
4	10	9	9	7	9	13	15	14	10	7	7	10
5	1−	1−	1−	1−	0	1−	1−	1−	1−	0	1−	1−
6	4	2	2	1	1−	1−	1−	1−	1−	0	1−	1−
New York, N.Y.—40°48′N. 73°58′W.												
*1	37	38	45	57	68	77	82	80	74	64	51	41
*2	25	24	30	42	53	61	66	66	60	49	37	29
*3	67	65	63	62	63	66	67	70	70	67	68	67
4	12	10	12	11	11	10	10	10	9	9	9	11
5	12	10	13	11	6	5	4	3	4	7	10	11
6	3	2	2	1	2	1	1	1	1	2	2	3
Odessa—46°29′N. 30°46′E.												
*1	36	37	45	56	70	77	81	80	72	62	50	42
*2	15	19	26	39	51	59	64	63	53	42	31	20
3	88	84	82	73	68	66	62	61	67	77	83	87
4	9	8	9	7	8	9	7	5	5	6	8	10
5	3.2	3.1	2.1	1.5	0.5	0.3	0.3	0.5	0.3	1.8	2	2.6
6	6	6	7	3	2	0	1	1	2	5	5	7
Oran—35°42′N. 0°39′W.												
1	61	63	66	69	73	79	83	84	80	74	69	63
2	45	46	49	53	58	73	68	69	66	59	53	47
3	77	76	75	74	73	72	74	75	77	77	77	77
4	9	7	7	7	5	3	1	1	4	6	8	8
5	1.5	2	2	1	0.3	0.5	0.2	0	0.2	0.3	0.8	1
6	—	—	—	—	—	—	—	—	—	—	—	—

	Jan.	Feb.	Mar.	April	May	June	July	Aug	Sept.	Oct.	Nov.	Dec.
Osaka—34°39′N. 135°26′E.												
1	48	48	54	65	73	79	87	90	82	72	62	52
2	33	32	37	47	55	64	73	74	67	55	44	36
3	72	71	71	72	72	77	77	75	77	76	76	72
4	9	10	13	13	13	15	12	11	14	11	10	9
*5	10	8	8	6	6	5	5	4	3	2	2	2
6	1.0	0.8	0.9	0.5	0.3	0.5	0.2	0.2	0.5	1.0	2.0	2.0
Oslo—59°55′N. 10°43′E.												
1	—	—	—	—	—	—	—	—	—	—	—	—
*2	19	22	26	33	42	51	55	53	46	33	28	23
3	78	74	71	66	62	62	68	74	79	80	81	78
4	7	6	7	6	7	7	10	11	7	10	8	8
5	0.1	0.1	0.2	0.1	0.1	0.1	0.1	0.1	0.1	0.1	0.1	0.2
6	10	8	2	2	0	0	0	0	3	5	8	10
Palermo—38°07′N. 13°21′E.												
1	58	56	60	63	69	75	79	82	77	71	63	58
2	47	47	49	54	61	67	72	74	68	63	55	49
3	76	72	69	68	66	65	62	62	66	70	72	75
4	15	13	13	10	6	4	2	2	6	11	13	16
5	—	—	—	—	—	—	—	—	—	—	—	—
6	—	—	—	—	—	—	—	—	—	—	—	—
Papeete, Tahiti—17°32′S. 149°34′W.												
*1	88	88	87	87	87	84	84	84	86	87	87	87
*2	74	75	75	74	72	70	68	68	71	71	71	72
3	87	75	81	80	80	81	78	78	75	73	78	79
4	16	16	17	10	10	8	5	6	6	9	13	14
5	—	—	—	—	—	—	—	—	—	—	—	—
6	—	—	—	—	—	—	—	—	—	—	—	—
Perth—31°57′S. 115°50′E.												
1	85	85	81	76	69	64	63	64	66	69	75	81
2	63	63	61	57	53	50	48	48	50	53	57	61
3	49	49	51	56	65	71	71	67	63	59	53	50
4	3	3	4	7	14	17	17	18	14	12	6	4
5	1.1	1.1	0.8	1.0	1.6	1.8	2.4	2.7	2.7	2.1	1.5	0.6
6	0.1	0.3	0.5	0.6	0.5	1.1	0.2	0.8	0.2	0.1	0.1	0.1
Philadelphia, Pa.—39°58′N. 75°17′W.												
*1	40	41	49	61	72	80	85	82	76	65	52	43
*2	26	27	34	43	54	63	68	67	61	50	39	30
*3	70	69	67	64	64	66	66	68	69	66	68	69
4	12	11	12	11	11	10	11	11	9	9	9	10
5	2	2	3	3	2	1 –	1	1 –	1	2	2	2
6	1	1	1	1 –	1 –	1 –	1 –	1 –	1	1	1	1

	Jan.	Feb.	Mar.	April	May	June	July	Aug.	Sept.	Oct.	Nov.	Dec.
						Plymouth—50°22′N. 4°08′W.						
1	47	47	50	55	60	65	67	67	64	57	52	49
2	38	38	38	42	47	52	55	55	52	46	42	40
3	87	87	84	80	80	80	81	82	83	86	87	88
4	19	15	16	14	13	12	14	15	14	18	18	22
5	3	1.5	0.6	0.6	0.1	0.1	0.1	0.3	0.4	2	2	3
6	4	3	3	0.5	0.7	0.3	0.5	0.5	0.5	0.8	3	5
						Ponta Delgada, Azores—37°45′N. 25°41′W.						
*1	69	68	73	74	78	81	86	89	85	81	76	71
*2	42	38	40	42	47	49	53	55	52	49	47	43
3	—	—	—	—	—	—	—	—	—	—	—	—
4	19	15	17	13	13	9	9	9	13	14	16	18
5	1.3	1.0	0.8	0.4	0.3	0.1	0	0	0.2	0.4	0.4	1.1
6	—	—	—	—	—	—	—	—	—	—	—	—
						Port-au-Prince—18°34′N. 72°22′W.						
1	87	88	89	89	90	92	94	93	91	90	88	87
2	68	68	69	71	72	73	74	73	73	72	71	69
*3	63	63	63	68	72	67	64	68	72	74	72	67
4	5	7	10	14	16	9	9	13	16	14	9	5
5	—	—	—	—	—	—	—	—	—	—	—	—
6	—	—	—	—	—	—	—	—	—	—	—	—
						Port Darwin—12°28′S. 130°51′E.						
1	90	90	91	93	91	88	87	89	92	94	94	92
2	77	77	77	76	73	69	67	70	74	77	78	78
3	75	77	72	62	55	54	51	53	57	59	62	69
4	19	19	16	8	2	0	0	0	2	6	11	16
*5	5	0	8	2	6	2	0	0	13	0	10	6
6	0	0	0	0	0.2	0.6	1.0	1.6	0.7	0.1	0	0
						Portland, Me.—43°39′N. 70°15′W.						
*1	30	32	39	50	61	71	76	74	67	57	45	34
*2	15	16	24	36	45	54	60	59	52	43	31	21
*3	72	72	70	68	72	73	76	78	79	78	74	72
4	12	11	12	11	12	11	11	11	10	10	11	11
5	1	0	1	0	0	0	0	0	0	1	1	1
6	1	1	2	2	2	3	4	5	4	3	1	1
						Port Moresby—9°29′S. 147°09′E.						
*1	89	88	88	87	85	84	82	82	83	86	87	89
*2	76	76	76	76	76	75	74	73	74	75	76	76
3	71	73	74	75	76	77	77	77	76	75	71	70
4	15	14	15	9	6	5	4	4	6	5	6	9
5	—	—	—	—	—	—	—	—	—	—	—	—
6	—	—	—	—	—	—	—	—	—	—	—	—

	Jan.	Feb.	Mar.	April	May	June	July	Aug.	Sept.	Oct.	Nov.	Dec.
Port Said—31°16′N. 32°19′E.												
*1	66	67	70	74	79	85	88	89	87	83	77	69
*2	51	52	56	60	65	70	74	75	73	70	63	54
*3	76	75	74	72	73	74	76	76	73	73	73	76
4	4	4	2	1	1	0	0	0	0	1	2	4
5	—	—	—	—	—	—	—	—	—	—	—	—
6	—	—	—	—	—	—	—	—	—	—	—	—
Portsmouth—50°48′N. 1°06′W.												
1	45	46	49	55	62	67	70	70	66	58	51	47
2	35	36	37	41	46	52	55	55	52	46	41	38
3	89	87	84	77	74	74	75	77	80	85	87	90
4	15	14	14	12	11	11	12	13	11	16	16	18
5	0.8	0.7	0.7	0.3	0.1	0.2	0.2	0.3	0.6	0.6	0.4	1.0
6	5	2	0.6	0.8	0	0	0	0	0	0.2	2	3
Prince Rupert—54°18′N. 130°18′W.												
*1	39	42	44	50	55	60	62	64	60	53	46	41
*2	31	31	33	37	41	46	50	51	47	42	37	32
3	84	83	83	81	80	83	85	86	84	85	85	86
4	20	17	20	19	17	14	15	15	17	22	22	21
5	3	3	4	1	0.2	0.4	0.1	0.3	0.5	3	4	4
6	0	0.5	0.2	0.2	0.8	3	3	4	3	2	0.5	0.5
Puntarenas—53°10′S. 70°54′W.												
*1	59	58	55	49	43	40	38	40	45	50	54	57
*2	45	44	43	39	35	33	33	33	35	38	40	43
3	64	65	69	72	76	76	72	76	69	64	62	63
4	11	9	12	12	13	8	9	8	8	7	9	10
5	1	1	.09	.06	0	0.6	0.4	0.5	0.9	0.7	1.5	0.6
6	0	0	0	0.7	1.0	0.7	5	2	0.3	1.0	0.7	0.3
Quebec—46°48′N. 71°13′W.												
1	18	20	30	44	61	71	77	73	64	51	35	22
2	2	3	14	29	41	51	57	55	47	37	24	9
3	78	78	82	75	74	76	80	81	83	83	84	82
4	18	10	11	9	9	9	10	8	9	9	12	13
5	9	8	8	7	6	4	2	1	3	4	5	6
6	1	0	1	1	0	0	0	0	1	2	1	1
Rangoon—16°47′N. 96°13′E.												
*1	89	92	96	98	92	86	85	85	86	88	87	87
*2	65	67	71	76	77	76	76	76	76	76	73	67
3	82	84	85	80	86	91	92	93	92	90	86	82
4	0.3	0.3	0.6	2	14	23	25	24	20	10	3	0
5	—	—	—	—	—	—	—	—	—	—	—	—
6	—	—	—	—	—	—	—	—	—	—	—	—

	Jan.	Feb.	Mar.	April	May	June	July	Aug.	Sept.	Oct.	Nov.	Dec.
Richmond, Va.—37°32′N. 77°26′W.												
1	48	48	58	67	77	83	87	85	80	70	59	46
2	30	30	38	46	55	64	69	67	61	49	39	31
3	71	68	65	63	64	67	69	71	72	68	67	71
4	10	10	11	10	12	11	11	11	11	7	7	10
5	1−	1	2	1−	1−	1−	1−	1−	1−	1−	1−	1−
6	2	1	1	0	0	0	0	0	1	3	2	2
Riga—56°57′N. 24°06′E.												
1	39	39	46	62	78	80	82	79	71	61	47	41
2	20	21	25	34	44	52	56	54	47	39	30	22
3	87	85	82	75	68	69	72	77	81	85	88	89
4	14	12	12	11	13	11	14	15	14	14	16	15
5	4	2	3	1.5	2	1	1	2	3	3	3	4
6	4	4	7	5	1	1	0.4	2	5	6	6	6
Rio de Janeiro—22°54′S. 43°10′W.												
1	82	83	81	78	75	74	73	73	74	75	78	81
2	74	76	75	73	69	67	65	66	66	69	71	73
3	78	78	79	79	79	78	78	76	79	79	78	78
4	13	11	12	10	10	7	6	7	11	12	12	14
5	0.1	0	0.3	0	0.1	0	0	0.1	0.2	0.1	0.1	0.2
6	10	11	14	16	19	19	21	21	18	15	11	8
Rotterdam—51°59′N. 4°29′E.												
1	41	43	47	54	63	68	70	70	65	57	48	43
2	31	31	34	39	45	51	54	54	49	42	36	33
3	88	86	82	74	72	73	75	78	81	85	88	89
4	9	8	10	9	9	8	10	12	10	13	11	11
5	2	2	2	1	0	0	0	1	0	1	2	2
6	—	—	—	—	—	—	—	—	—	—	—	—
St. John's—47°34′N. 52°42′W.												
*1	31	29	34	41	51	61	68	68	61	52	43	35
*2	16	15	21	29	35	42	50	52	46	39	31	23
3	79	79	82	81	78	75	76	75	77	79	83	80
4	15	13	14	13	13	12	12	11	12	15	16	15
5	1	1	1	1	0	0	0	0	0	1	1	1
6	2	2	3	4	6	3	4	4	2	3	2	2
Salonika—40°39′N. 22°57′E.												
*1	48	52	58	66	77	84	90	80	81	71	59	52
*2	35	39	44	60	59	66	70	70	64	57	46	41
3	71	69	67	55	63	59	55	57	61	71	72	74
4	7	8	8	8	8	6	4	3	5	8	8	9
5	4	4	3	3	2	3	3	3	2	2	3	3
6	2	1	1.6	1.3	1.1	0	0	0	1.1	1.6	2	3

	Jan.	Feb.	Mar.	April	May	June	July	Aug.	Sept.	Oct.	Nov.	Dec.
San Diego, Calif.—32°43′N. 117°10′W.												
*1	63	63	64	66	67	69	73	74	73	71	69	65
*2	47	48	50	53	56	59	63	64	52	67	52	49
*3	66	71	70	73	75	77	78	78	76	72	64	64
4	7	7	7	4	3	1	1	1	1	3	3	6
5	1–	1–	1–	1–	0	0	0	0	0	0	0	1–
6	2	2	1	1	1	1	1	1	2	4	2	1
San Francisco, Calif.—37°47′N. 122°26′W.												
*1	55	59	61	62	63	66	65	65	69	68	63	51
*2	45	47	48	49	51	52	53	53	55	54	51	46
*3	75	73	70	70	72	74	78	80	73	70	69	74
4	12	11	10	6	4	2	1–	1–	2	4	7	10
5	1	1–	1–	1–	1–	1–	1–	1–	1–	1–	1–	1–
6	2	2	1	1–	1–	1	1	1	1	2	3	2
San Juan, Puerto Rico—18°28′N. 66°07′W.												
1	80	80	81	82	84	85	85	85	86	86	84	81
2	70	69	70	71	73	74	75	75	75	74	73	71
*3	78	77	75	76	77	78	79	78	79	79	79	79
4	21	15	16	14	16	17	19	20	18	17	20	22
5	1.9	0.5	2.0	0.2	0.2	0.1	0.5	0.5	0.8	0.2	1.0	1.2
6	0	0	0	0	0	0	0	0	0	0	0	0
Santos—23°56′S. 46°19′W.												
1	83	85	83	81	77	76	74	74	74	75	78	81
2	71	72	71	68	64	61	60	61	63	65	67	70
*3	81	80	83	82	81	82	81	83	84	82	81	80
4	16	14	16	12	11	10	9	10	12	14	14	15
5	Records Not Given											
6	0.6	1	3	5	5	8	8	8	6	4	2	1
Seattle—47°36′N. 122°20′W.												
*1	44	47	52	58	63	68	73	73	67	59	51	46
*2	36	37	39	42	47	52	55	55	52	47	41	38
3	84	81	81	81	80	81	84	85	86	84	83	84
4	19	16	16	13	12	9	4	5	9	13	17	19
*5	2	1	2	1	0.5	0.5	0	0.5	0.5	1	2	2
6	2	3	2	1	0.5	0.5	0.5	2	4	6	4	3
Sevastopol—44°37′N. 33°31′E.												
*1	36	37	42	50	61	69	74	73	65	57	47	41
2	30	32	36	42	52	60	65	64	57	50	39	35
3	80	80	75	71	71	70	67	66	68	76	78	80
4	9	10	9	7	6	5	4	4	5	6	8	10
5	2	1.4	2	1.4	0.8	0.7	1.1	1.3	1.5	1.3	1.5	2
6	1	1.2	1.3	1.3	0.4	0.1	0	0	0	0.3	1.1	1

	Jan.	Feb.	Mar.	April	May	June	July	Aug.	Sept.	Oct.	Nov.	Dec.
Sidi Barani—31°38′N. 25°58′E.												
*1	64	65	67	72	77	81	82	84	83	80	73	66
*2	44	46	48	52	57	62	67	68	65	61	54	49
3	72	74	72	64	59	68	74	76	68	70	71	77
4	6	4	2	1	0.5	0	0	0	0.5	1	3	5
5	—	—	—	—	—	—	—	—	—	—	—	—
6	—	—	—	—	—	—	—	—	—	—	—	—
Singapore—1°17′N. 103°51′E.												
*1	86	88	88	88	88	88	88	87	88	88	87	86
*2	73	73	74	75	76	76	76	75	75	75	74	74
3	82	80	79	79	80	80	80	80	80	79	80	82
4	16	12	14	15	13	14	12	14	13	16	18	19
5	1.0	0.5	0.5	0.5	2.2	2.7	1.0	0.3	1.0	2.0	0.7	0.0
6	—	—	—	—	—	—	—	—	—	—	—	—
Spitzbergen, Green Harbor—78°02′N. 14°14′E.												
1	5	3	6	18	30	39	44	43	35	25	12	10
2	−15	−16	−16	−3	12	31	36	35	27	14	0	−11
*3	82	82	81	78	76	82	84	86	84	81	82	83
4	13	10	13	10	6	9	8	8	11	16	14	13
5	1.0	6	1.0	0.4	0.3	0.3	0	0.3	0.7	2.0	0.6	1.0
6	0.1	0.6	0.1	0.1	0.4	3.0	4.0	3.0	1.0	0.7	0	0
Stanley, Falkland Islands—51°42′S. 57°51′W.												
*1	56	56	63	49	44	41	40	42	45	48	52	54
*2	42	42	40	37	34	31	31	32	33	35	38	40
3	77	80	81	86	88	84	90	87	85	80	75	76
4	19	16	19	19	21	20	20	19	16	17	17	19
5	2	2	2	3	2	2	2	2	2	2	3	2
6	4	4	4	3	4	4	5	3	4	4	4	4
Stettin—52°26′N. 14°34′E.												
1	34	37	43	53	64	71	73	71	64	53	43	36
2	25	27	31	38	46	52	56	54	49	42	34	29
3	89	85	81	73	70	69	73	75	79	85	88	89
4	15	13	14	13	13	12	15	15	13	13	14	16
5	2	2	2	1	0.8	1	0.6	0.6	1	2	1	2
6	5	4	4	2	0.7	0.7	0.4	2	5	8	8	7
Stockholm—59°21′N. 18°03′E.												
1	31	32	36	46	57	67	71	67	59	48	39	33
2	22	21	24	31	40	49	54	52	46	38	31	24
3	85	83	78	71	63	62	67	73	78	82	85	87
4	15	13	14	11	12	12	15	16	14	16	15	17
5	0.4	0.5	0.4	0.2	0.2	0.2	0.1	0.2	0.2	0.4	0.2	0.3
6	5	5	4	4	1	0.7	0.5	1	4	6	6	6

	Jan.	Feb.	Mar.	April	May	June	July	Aug.	Sept.	Oct.	Nov.	Dec.
					Suva, Fiji Islands—18°08′S. 178°26′E.							
*1	86	86	86	84	82	80	79	79	80	81	83	85
*2	74	74	74	73	71	69	68	68	69	70	71	73
3	78	80	81	81	82	81	80	80	78	76	76	77
4	23	22	24	23	21	18	17	19	19	18	16	21
5	0.1	0.1	0.2	0	0	0	0	0	0	0	0	0.1
6	0	0	0	0	0	0	0	0	0	0	0	0
					Sydney—33°51′S. 151°13′E.							
1	78	78	76	71	65	61	59	63	67	71	74	77
2	65	65	63	58	52	48	46	47	51	56	60	63
*3	71	74	72	72	72	70	69	60	64	64	68	70
4	14	14	15	14	15	13	12	11	12	12	12	13
5	0.7	0.2	0	0	0	0.3	0.7	0.7	1	0.7	0.7	0.7
6	0.3	0.7	2	2	3	3	1.5	3	0.8	0.5	0.8	0.5
					Tamatave—18°09′S. 49°26′E.							
1	88	90	87	85	81	78	77	78	80	83	86	87
2	73	74	77	71	68	65	67	64	66	68	70	72
3	84	84	84	84	84	85	84	83	83	82	82	85
4	17	17	19	17	16	17	20	17	14	12	9	13
5	—	—	—	—	—	—	—	—	—	—	—	—
6	—	—	—	—	—	—	—	—	—	—	—	—
					Tampa, Fla.—27°57′N. 82°27′W.							
*1	70	71	76	80	86	87	89	89	88	82	85	70
*2	52	54	58	62	68	72	74	74	72	66	58	53
*3	73	71	68	66	66	71	73	73	74	72	69	73
4	7	7	6	5	7	14	17	17	15	8	5	7
5	1−	1−	1−	1−	1−	1−	1−	1−	1−	1−	1−	1−
6	4	3	2	1	1−	1−	1−	1−	1−	1	2	3
					Tampico—22°13′N. 97°53′W.							
*1	72	75	78	83	86	88	89	89	87	85	77	73
*2	59	61	64	69	74	75	75	75	75	71	64	60
*3	78	78	77	78	78	78	79	78	79	77	77	78
4	9	7	7	6	8	12	14	13	16	11	10	10
*5	1.7	1.3	2.9	1.3	1.7	1.1	0.7	0.2	0.4	0.3	0.4	1.6
					Tokyo—35°41′N. 139°46′E.							
1	47	47	53	63	70	76	83	85	79	69	60	51
2	30	41	36	47	54	63	69	72	66	54	43	33
3	64	62	67	73	76	82	83	82	83	80	74	66
4	7	8	13	14	14	16	15	13	17	14	10	7
*5	4	5	6	6	5	3	2	2	2	1	2	3
6	0.5	0.2	0.5	1.0	1.5	1.5	2.0	2.0	2.0	1.0	0.4	0.4

	Jan.	Feb.	Mar.	April	May	June	July	Aug.	Sept.	Oct.	Nov.	Dec.
					Trinidad, Port-of-Spain—10°39′N. 63°31′W.							
*1	85	86	87	88	88	87	87	87	88	88	87	86
*2	71	71	71	73	74	74	72	73	73	73	73	72
3	76	71	71	68	71	78	77	82	80	78	79	78
4	14	8	8	7	10	17	20	21	18	16	17	16
*5	0	0	0	0.2	0	0	0	0.3	0	0	0.3	0
6	0	0	0	0	0	0	0	0	0	0	0	0
					Tripoli—32°54′N. 13°11′E.							
*1	60	62	66	71	75	81	85	86	85	80	72	64
*2	47	49	53	57	62	68	72	73	71	67	59	51
3	66	66	64	65	67	67	66	65	64	63	63	65
4	11	6	5	3	2	0	0	0	1	4	6	11
5	—	—	—	—	—	—	—	—	—	—	—	—
6	0.5	0.5	2	2	0.9	3	4	3	2	2	1.5	0.9
					Trondheim—63°26′N. 10°22′E.							
1	44	44	48	58	68	76	76	72	64	59	49	45
2	15	7	11	24	33	39	44	42	36	26	19	10
3	71	69	68	71	71	71	74	77	81	79	75	72
4	21	17	17	17	20	18	22	23	24	22	19	19
5	5.3	4.5	4.9	3.3	2.4	1.2	1.6	1.6	3.4	3.5	5.0	6.1
6	3.9	3.6	3.2	2.6	2.3	2.4	1.5	2.5	5.0	5.0	6.1	6.3
					Tunis—36°48′N. 10°10′E.							
1	58	61	66	70	77	85	92	93	86	78	69	61
2	43	43	46	50	56	63	67	67	64	57	50	44
3	76	75	72	69	65	59	55	59	65	72	74	76
4	11	10	9	8	5	3	1	1	4	8	9	9
5	—	—	—	—	—	—	—	—	—	—	—	—
6	—	—	—	—	—	—	—	—	—	—	—	—
					Valparaiso—33°01′S. 71°38′W.							
*1	73	73	71	68	64	61	60	62	63	66	69	72
*2	56	56	54	52	50	48	47	47	48	50	52	54
3	81	83	84	81	82	81	81	82	83	82	77	77
4	0.6	0.4	0.8	3	6	8	7	6	4	3	1	0.8
5	0.6	0.1	0.2	0.1	0.2	0	0	0.1	0.1	0.1	0.6	0.9
6	6	6	8	9	7	4	6	5	4	4	4	4
					Vancouver, B.C.—49°17′N. 123°05′W.							
*1	40	44	49	56	63	68	73	72	65	59	48	42
*2	32	34	36	40	45	50	53	53	48	44	38	34
3	86	86	80	72	73	72	72	76	81	86	88	87
4	20	16	17	13	13	10	6	7	11	16	21	21
*5	0	0	0	0	0	0	0	0	0	0	0	0
6	3	3	1	0	0	0	0	1	2	5	5	4

	Jan.	Feb.	Mar.	April	May	June	July	Aug.	Sept.	Oct.	Nov.	Dec.
Venice—45°26′N. 12°20′E.												
1	41	46	52	60	70	77	82	81	73	63	51	44
2	32	36	42	49	58	64	68	67	60	52	42	36
3	80	78	75	73	70	68	65	67	72	77	79	79
4	6	7	9	11	9	10	6	7	8	9	8	7
5	1	0.7	0.7	0.8	0.7	0.7	0.3	0.7	0.7	0.9	0.7	0.7
6	9	9	5	0.7	0.5	0	0.1	1	0.7	2	4	8
Veracruz—19°12′N. 96°08′W.												
*1	74	68	78	81	84	85	85	85	85	84	79	76
*2	67	66	69	72	76	76	75	75	75	74	70	67
3	81	83	83	81	80	81	81	79	80	77	78	80
4	6	5	5	3	7	17	20	19	19	13	11	7
*5	12	7	11.5	7	7	9	7	9	10	13	12	13
6	7	6.8	6.8	6.6	5.2	1.7	2.2	2.8	1.7	4.1	5.0	6.5
Viberg—60°43′N. 28°44′E.												
1	25	21	32	43	56	69	72	71	58	46	35	31
2	9	3	16	26	37	49	52	52	44	34	25	20
3	90	87	84	76	67	67	69	75	84	87	90	93
4	17	15	14	11	12	12	13	18	18	18	18	17
5	6	2	1	3	4	2	3	0.4	3	3	2	1
6	0.4	0.6	0.6	0.2	0.2	0	0	0.4	3	3	0.8	0.8
Vladivostok—43°07′N. 131°55′E.												
1	13	23	33	46	54	63	70	76	68	55	36	20
2	0	6	19	34	43	52	60	64	55	42	24	8
3	67	68	70	71	77	86	89	86	79	68	65	66
4	3	3	5	7	10	13	13	13	10	6	4	3
5	2	1	1	1	0.8	0.8	0.5	0.9	1	1	2	2
6	2	2	4	7	12	15	17	12	2	3	2	2
Wellington—41°17′S. 174°46′E.												
*1	70	69	67	63	58	55	53	54	57	60	63	67
*2	56	56	55	51	47	45	42	43	46	48	51	54
3	72	73	73	75	77	79	78	77	75	75	75	73
4	10	9	12	12	16	16	18	17	15	14	12	11
5	5	4	4	5	4	4	3	5	5	7	6	6
6	0	0	0	1	1	2	3	1	0	0	0	0
Yokohama—35°27′N. 139°39′E.												
1	48	48	53	63	70	76	82	85	78	69	60	52
2	32	33	39	48	56	64	71	73	67	57	45	36
3	67	65	69	74	76	81	82	81	81	79	74	68
4	8	9	15	14	15	17	15	13	16	15	11	7
*5	10	10	14	13	13	9	9	9	10	9	9	8
6	2.0	0.9	1.0	0.4	0.7	0.8	0.9	0.6	0.5	1.0	1.0	2.0

	Jan.	Feb.	Mar.	April	May	June	July	Aug.	Sept.	Oct.	Nov.	Dec.
Zanzibar—6°10′S. 39°11′E.												
*1	86	88	88	86	84	83	82	82	83	84	85	86
*2	80	81	80	78	76	75	74	73	74	76	77	79
3	73	72	75	78	78	78	78	77	75	74	75	75
4	7	6	14	15	12	7	4	8	9	10	14	14
5	—	—	—	—	—	—	—	—	—	—	—	—
6	—	—	—	—	—	—	—	—	—	—	—	—

INDEX

D

E

F

G

H

I

J

K

L

M

R

S

T

U

V

W